Rational (Reciprocal) Function

$$f(x) = \frac{1}{x}$$

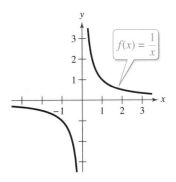

Domain: $(-\infty, 0) \cup (0, \infty)$
Range: $(-\infty, 0) \cup (0, \infty)$
No intercepts
Decreasing on $(-\infty, 0)$ and $(0, \infty)$
Odd function
Origin symmetry
Vertical asymptote: y-axis
Horizontal asymptote: x-axis

Exponential Function

$$f(x) = a^x, \ a > 1$$

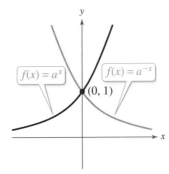

Domain: $(-\infty, \infty)$
Range: $(0, \infty)$
Intercept: $(0, 1)$
Increasing on $(-\infty, \infty)$
   for $f(x) = a^x$
Decreasing on $(-\infty, \infty)$
   for $f(x) = a^{-x}$
Horizontal asymptote: x-axis
Continuous

Log

$$f(x)$$

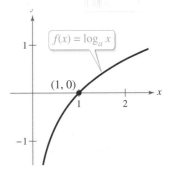

Domain: $(0, \infty)$
Range: $(-\infty, \infty)$
Intercept: $(1, 0)$
Increasing on $(0, \infty)$
Vertical asymptote: y-axis
Continuous
Reflection of graph of $f(x) = a^x$
   in the line $y = x$

Sine Function

$$f(x) = \sin x$$

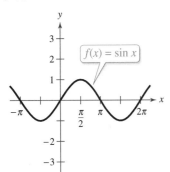

Domain: $(-\infty, \infty)$
Range: $[-1, 1]$
Period: 2π
x-intercepts: $(n\pi, 0)$
y-intercept: $(0, 0)$
Odd function
Origin symmetry

Cosine Function

$$f(x) = \cos x$$

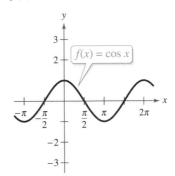

Domain: $(-\infty, \infty)$
Range: $[-1, 1]$
Period: 2π
x-intercepts: $\left(\dfrac{\pi}{2} + n\pi, 0\right)$
y-intercept: $(0, 1)$
Even function
y-axis symmetry

Tangent Function

$$f(x) = \tan x$$

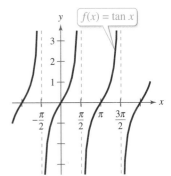

Domain: all $x \neq \dfrac{\pi}{2} + n\pi$

Range: $(-\infty, \infty)$
Period: π
x-intercepts: $(n\pi, 0)$
y-intercept: $(0, 0)$
Vertical asymptotes:

$$x = \frac{\pi}{2} + n\pi$$

Odd function
Origin symmetry

Cosecant Function

$f(x) = \csc x$

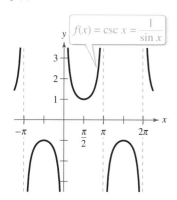

Domain: all $x \neq n\pi$
Range: $(-\infty, -1] \cup [1, \infty)$
Period: 2π
No intercepts
Vertical asymptotes: $x = n\pi$
Odd function
Origin symmetry

Secant Function

$f(x) = \sec x$

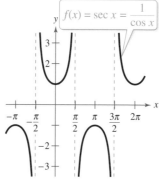

Domain: all $x \neq \dfrac{\pi}{2} + n\pi$

Range: $(-\infty, -1] \cup [1, \infty)$
Period: 2π
y-intercept: $(0, 1)$
Vertical asymptotes:

$$x = \frac{\pi}{2} + n\pi$$

Even function
y-axis symmetry

Cotangent Function

$f(x) = \cot x$

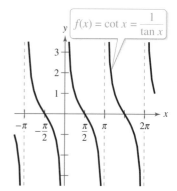

Domain: all $x \neq n\pi$
Range: $(-\infty, \infty)$
Period: π

x-intercepts: $\left(\dfrac{\pi}{2} + n\pi, 0\right)$

Vertical asymptotes: $x = n\pi$
Odd function
Origin symmetry

Inverse Sine Function

$f(x) = \arcsin x$

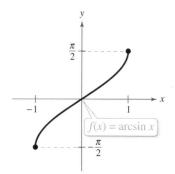

Domain: $[-1, 1]$
Range: $\left[-\dfrac{\pi}{2}, \dfrac{\pi}{2}\right]$
Intercept: $(0, 0)$
Odd function
Origin symmetry

Inverse Cosine Function

$f(x) = \arccos x$

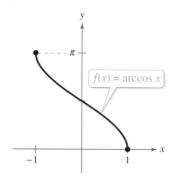

Domain: $[-1, 1]$
Range: $[0, \pi]$

y-intercept: $\left(0, \dfrac{\pi}{2}\right)$

Inverse Tangent Function

$f(x) = \arctan x$

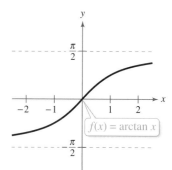

Domain: $(-\infty, \infty)$
Range: $\left(-\dfrac{\pi}{2}, \dfrac{\pi}{2}\right)$
Intercept: $(0, 0)$
Horizontal asymptotes:

$$y = \pm\frac{\pi}{2}$$

Odd function
Origin symmetry

Special Version of Larson Trigonometry for MTH 114 Trigonometry

Contributing Author: Sharon Griffin

Ron Larson

CENGAGE
Learning·

Australia • Brazil • Japan • Korea • Mexico • Singapore • Spain • United Kingdom • United States

Special Version of Larson Trigonometry for MTH 114 Trigonometry: Contributing Author: Sharon Griffin

Source:

Trigonometry, 9th Edition
Ron Larson
© 2014, 2011, 2007 Cengage Learning. All rights reserved.

Senior Project Development Manager:
 Linda deStefano

Market Development Manager:
 Heather Kramer

Senior Production/Manufacturing Manager:
 Donna M. Brown

Production Editorial Manager:
 Kim Fry

Sr. Rights Acquisition Account Manager:
 Todd Osborne

For product information and technology assistance, contact us at
Cengage Learning Customer & Sales Support, 1-800-354-9706
For permission to use material from this text or product,
submit all requests online at **cengage.com/permissions**
Further permissions questions can be emailed to
permissionrequest@cengage.com

This book contains select works from existing Cengage Learning resources and was produced by Cengage Learning Custom Solutions for collegiate use. As such, those adopting and/or contributing to this work are responsible for editorial content accuracy, continuity and completeness.

Compilation © 2013 Cengage Learning
ISBN-13: 978-1-285-90589-1

ISBN-10: 1-285-90589-X

Cengage Learning
5191 Natorp Boulevard
Mason, Ohio 45040
USA
Cengage Learning is a leading provider of customized learning solutions with office locations around the globe, including Singapore, the United Kingdom, Australia, Mexico, Brazil, and Japan. Locate your local office at:
international.cengage.com/region.

Cengage Learning products are represented in Canada by Nelson Education, Ltd.
For your lifelong learning solutions, visit **www.cengage.com/custom.**
Visit our corporate website at **www.cengage.com.**

Printed in the United States of America

Brief Contents

Insert:
MSU Math 114 Coursenote by Griffin, Sikorskii, and MSU Math Department

Preface

Welcome to *Trigonometry*, Ninth Edition. I am proud to present to you this new edition. As with all editions, I have been able to incorporate many useful comments from you, our user. And while much has changed in this revision, you will still find what you expect—a pedagogically sound, mathematically precise, and comprehensive textbook. Additionally, I am pleased and excited to offer you something brand new—a companion website at **LarsonPrecalculus.com.**

My goal for every edition of this textbook is to provide students with the tools that they need to master trigonometry. I hope you find that the changes in this edition, together with **LarsonPrecalculus.com,** will help accomplish just that.

New To This Edition

NEW LarsonPrecalculus.com
This companion website offers multiple tools and resources to supplement your learning. Access to these features is free. View and listen to worked-out solutions of Checkpoint problems in English or Spanish, download data sets, work on chapter projects, watch lesson videos, and much more.

NEW Chapter Opener
Each Chapter Opener highlights real-life applications used in the examples and exercises.

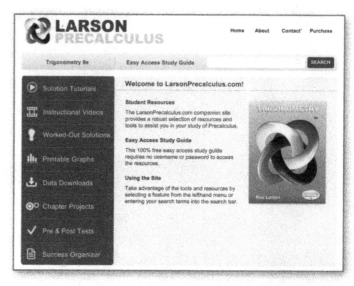

96. **HOW DO YOU SEE IT?** The graph represents the height h of a projectile after t seconds.

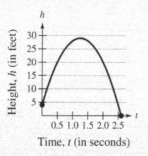

Time, t (in seconds)

(a) Explain why h is a function of t.

(b) Approximate the height of the projectile after 0.5 second and after 1.25 seconds.

(c) Approximate the domain of h.

(d) Is t a function of h? Explain.

NEW Summarize
The Summarize feature at the end of each section helps you organize the lesson's key concepts into a concise summary, providing you with a valuable study tool.

NEW How Do You See It?
The How Do You See It? feature in each section presents a real-life exercise that you will solve by visual inspection using the concepts learned in the lesson. This exercise is excellent for classroom discussion or test preparation.

NEW Checkpoints
Accompanying every example, the Checkpoint problems encourage immediate practice and check your understanding of the concepts presented in the example. View and listen to worked-out solutions of the Checkpoint problems in English or Spanish at LarsonPrecalculus.com.

NEW Data Spreadsheets

Download these editable spreadsheets from LarsonPrecalculus.com, and use the data to solve exercises.

DATA	Year	Number of Tax Returns Made Through E-File
	2003	52.9
	2004	61.5
	2005	68.5
	2006	73.3
	2007	80.0
	2008	89.9
	2009	95.0
	2010	98.7

Spreadsheet at LarsonPrecalculus.com

REVISED Exercise Sets

The exercise sets have been carefully and extensively examined to ensure they are rigorous and relevant and to include all topics our users have suggested. The exercises have been **reorganized and titled** so you can better see the connections between examples and exercises. Multi-step, real-life exercises reinforce problem-solving skills and mastery of concepts by giving you the opportunity to apply the concepts in real-life situations.

REVISED Section Objectives

A bulleted list of learning objectives provides you the opportunity to preview what will be presented in the upcoming section.

REVISED Remark

These hints and tips reinforce or expand upon concepts, help you learn how to study mathematics, caution you about common errors, address special cases, or show alternative or additional steps to a solution of an example.

Calc Chat

For the past several years, an independent website—CalcChat.com—has provided free solutions to all odd-numbered problems in the text. Thousands of students have visited the site for practice and help with their homework. For this edition, I used information from CalcChat.com, including which solutions students accessed most often, to help guide the revision of the exercises.

Trusted Features

Side-By-Side Examples

Throughout the text, we present solutions to many examples from multiple perspectives—algebraically, graphically, and numerically. The side-by-side format of this pedagogical feature helps you to see that a problem can be solved in more than one way and to see that different methods yield the same result. The side-by-side format also addresses many different learning styles.

Algebra Help

Algebra Help directs you to sections of the textbook where you can review algebra skills needed to master the current topic.

Technology

The technology feature gives suggestions for effectively using tools such as calculators, graphing calculators, and spreadsheet programs to help deepen your understanding of concepts, ease lengthy calculations, and provide alternate solution methods for verifying answers obtained by hand.

Historical Notes

These notes provide helpful information regarding famous mathematicians and their work.

Algebra of Calculus

Throughout the text, special emphasis is given to the algebraic techniques used in calculus. Algebra of Calculus examples and exercises are integrated throughout the text and are identified by the symbol $\int$.

Vocabulary Exercises

The vocabulary exercises appear at the beginning of the exercise set for each section. These problems help you review previously learned vocabulary terms that you will use in solving the section exercises.

Project: Department of Defense   The table shows the total numbers of military personnel P (in thousands) on active duty from 1980 through 2010.   *(Source: U.S. Department of Defense)*

Year	Personnel, P	Year	Personnel, P
1980	2051	1995	1518
1981	2083	1996	1472
1982	2109	1997	1439
1983	2123	1998	1407
1984	2138	1999	1386
1985	2151	2000	1384
1986	2169	2001	1385
1987	2174	2002	1414
1988	2138	2003	1434
1989	2130	2004	1427
1990	2044	2005	1389
1991	1986	2006	1385
1992	1807	2007	1380
1993	1705	2008	1402
1994	1610	2009	1419
		2010	1431

(a) Use a graphing utility to plot the data. Let t represent the year, with $t = 0$ corresponding to 1980.

(b) A model that approximates the data is given by

$$P = \frac{9.6518t^2 - 244.743t + 2044.77}{0.0059t^2 - 0.131t + 1}$$

where P is the total number of personnel (in thousands) and t is the year, with $t = 0$ corresponding to 1980. Construct a table showing the actual values of P and the values of P obtained using the model.

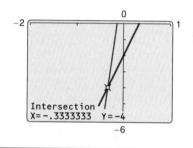

▷ **TECHNOLOGY**   You can use a graphing utility to check that a solution is reasonable. One way to do this is to graph the left side of the equation, then graph the right side of the equation, and determine the point of intersection. For instance, in Example 2, if you graph the equations

$y_1 = 6(x - 1) + 4$ The left side

$y_2 = 3(7x + 1)$ The right side

in the same viewing window, they should intersect at $x = -\frac{1}{3}$, as shown in the graph below.

Intersection
X=-.3333333   Y=-4

Project

The projects at the end of selected sections involve in-depth applied exercises in which you will work with large, real-life data sets, often creating or analyzing models. These projects are offered online at LarsonPrecalculus.com.

Chapter Summaries

The Chapter Summary now includes explanations and examples of the objectives taught in each chapter.

Enhanced WebAssign combines exceptional Precalculus content that you know and love with the most powerful online homework solution, WebAssign. Enhanced WebAssign engages you with immediate feedback, rich tutorial content and interactive, fully customizable eBooks (YouBook) helping you to develop a deeper conceptual understanding of the subject matter.

Instructor Resources

Print

Annotated Instructor's Edition
ISBN-13: 978-1-133-95431-6

This AIE is the complete student text plus point-of-use annotations for you, including extra projects, classroom activities, teaching strategies, and additional examples. Answers to even-numbered text exercises, Vocabulary Checks, and Explorations are also provided.

Complete Solutions Manual
ISBN-13: 978-1-133-95430-9

This manual contains solutions to all exercises from the text, including Chapter Review Exercises, and Chapter Tests.

Media

PowerLecture with ExamView™
ISBN-13: 978-1-133-95428-6

The DVD provides you with dynamic media tools for teaching Trigonometry while using an interactive white board. PowerPoint® lecture slides and art slides of the figures from the text, together with electronic files for the test bank and a link to the Solution Builder, are available. The algorithmic ExamView allows you to create, deliver, and customize tests (both print and online) in minutes with this easy-to-use assessment system. The DVD also provides you with a tutorial on integrating our instructor materials into your interactive whiteboard platform. Enhance how your students interact with you, your lecture, and each other.

Solution Builder
(*www.cengage.com/solutionbuilder*)
This online instructor database offers complete worked-out solutions to all exercises in the text, allowing you to create customized, secure solutions printouts (in PDF format) matched exactly to the problems you assign in class.

www.webassign.net
Printed Access Card: 978-0-538-73810-1
Online Access Code: 978-1-285-18181-3

Exclusively from Cengage Learning, Enhanced WebAssign combines the exceptional mathematics content that you know and love with the most powerful online homework solution, WebAssign. Enhanced WebAssign engages students with immediate feedback, rich tutorial content, and interactive, fully customizable eBooks (YouBook), helping students to develop a deeper conceptual understanding of their subject matter. Online assignments can be built by selecting from thousands of text-specific problems or supplemented with problems from any Cengage Learning textbook.

Student Resources

Print

Student Study and Solutions Manual
ISBN-13: 978-1-133-95429-3

This guide offers step-by-step solutions for all odd-numbered text exercises, Chapter and Cumulative Tests, and Practice Tests with solutions.

Text-Specific DVD
ISBN-13: 978-1-133-95427-9

Keyed to the text by section, these DVDs provide comprehensive coverage of the course—along with additional explanations of concepts, sample problems, and application—to help you review essential topics.

Note Taking Guide
ISBN-13: 978-1-133-95363-0

This innovative study aid, in the form of a notebook organizer, helps you develop a section-by-section summary of key concepts.

Media

www.webassign.net
Printed Access Card: 978-0-538-73810-1
Online Access Code: 978-1-285-18181-3

Enhanced WebAssign (assigned by the instructor) provides you with instant feedback on homework assignments. This online homework system is easy to use and includes helpful links to textbook sections, video examples, and problem-specific tutorials.

CengageBrain.com

Visit *www.cengagebrain.com* to access additional course materials and companion resources. At the CengageBrain.com home page, search for the ISBN of your title (from the back cover of your book) using the search box at the top of the page. This will take you to the product page where free companion resources can be found.

Acknowledgements

I would like to thank the many people who have helped me prepare the text and the supplements package. Their encouragement, criticisms, and suggestions have been invaluable.

Thank you to all of the instructors who took the time to review the changes in this edition and to provide suggestions for improving it. Without your help, this book would not be possible.

Reviewers

Timothy Andrew Brown, *South Georgia College*
Blair E. Caboot, *Keystone College*
Shannon Cornell, *Amarillo College*
Gayla Dance, *Millsaps College*
Paul Finster, *El Paso Community College*
Paul A. Flasch, *Pima Community College West Campus*
Vadas Gintautas, *Chatham University*
Lorraine A. Hughes, *Mississippi State University*
Shu-Jen Huang, *University of Florida*
Renyetta Johnson, *East Mississippi Community College*
George Keihany, *Fort Valley State University*
Mulatu Lemma, *Savannah State University*
William Mays Jr., *Salem Community College*
Marcella Melby, *University of Minnesota*
Jonathan Prewett, *University of Wyoming*
Denise Reid, *Valdosta State University*
David L. Sonnier, *Lyon College*
David H. Tseng, *Miami Dade College – Kendall Campus*
Kimberly Walters, *Mississippi State University*
Richard Weil, *Brown College*
Solomon Willis, *Cleveland Community College*
Bradley R. Young, *Darton College*

My thanks to Robert Hostetler, The Behrend College, The Pennsylvania State University, and David Heyd, The Behrend College, The Pennsylvania State University, for their significant contributions to previous editions of this text.

I would also like to thank the staff at Larson Texts, Inc. who assisted with proofreading the manuscript, preparing and proofreading the art package, and checking and typesetting the supplements.

On a personal level, I am grateful to my spouse, Deanna Gilbert Larson, for her love, patience, and support. Also, a special thanks goes to R. Scott O'Neil. If you have suggestions for improving this text, please feel free to write to me. Over the past two decades I have received many useful comments from both instructors and students, and I value these comments very highly.

Ron Larson, Ph.D.
Professor of Mathematics
Penn State University
www.RonLarson.com

P Prerequisites

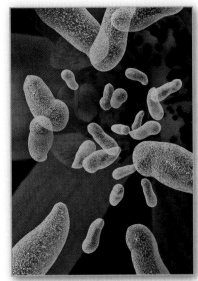

Bacteria *(Example 8, page 98)*

Snowstorm *(Exercise 47, page 84)*

Average Speed *(Example 7, page 72)*

Alternative-Fueled Vehicles
(Example 10, page 60)

Americans with Disabilities Act *(page 46)*

P.1 Review of Real Numbers and Their Properties

Real numbers can represent many real-life quantities. For example, in Exercises 55–58 on page 13, you will use real numbers to represent the federal deficit.

- ■ Represent and classify real numbers.
- ■ Order real numbers and use inequalities.
- ■ Find the absolute values of real numbers and find the distance between two real numbers.
- ■ Evaluate algebraic expressions.
- ■ Use the basic rules and properties of algebra.

Real Numbers

Real numbers can describe quantities in everyday life such as age, miles per gallon, and population. Symbols such as

$$-5, 9, 0, \tfrac{4}{3}, 0.666 \ldots, 28.21, \sqrt{2}, \pi, \text{ and } \sqrt[3]{-32}$$

represent real numbers. Here are some important **subsets** (each member of a subset B is also a member of a set A) of the real numbers. The three dots, called *ellipsis points*, indicate that the pattern continues indefinitely.

$$\{1, 2, 3, 4, \ldots\} \qquad \text{Set of natural numbers}$$

$$\{0, 1, 2, 3, 4, \ldots\} \qquad \text{Set of whole numbers}$$

$$\{\ldots, -3, -2, -1, 0, 1, 2, 3, \ldots\} \qquad \text{Set of integers}$$

A real number is **rational** when it can be written as the ratio p/q of two integers, where $q \neq 0$. For instance, the numbers

$$\tfrac{1}{3} = 0.3333 \ldots = 0.\overline{3}, \tfrac{1}{8} = 0.125, \text{ and } \tfrac{125}{111} = 1.126126 \ldots = 1.\overline{126}$$

are rational. The decimal representation of a rational number either repeats $\left(\text{as in } \tfrac{173}{55} = 3.1\overline{45}\right)$ or terminates $\left(\text{as in } \tfrac{1}{2} = 0.5\right)$. A real number that cannot be written as the ratio of two integers is called **irrational.** Irrational numbers have infinite nonrepeating decimal representations. For instance, the numbers

$$\sqrt{2} = 1.4142135 \ldots \approx 1.41 \quad \text{and} \quad \pi = 3.1415926 \ldots \approx 3.14$$

are irrational. (The symbol $\approx$ means "is approximately equal to.") Figure P.1 shows subsets of real numbers and their relationships to each other.

Figure (flowchart)

Real numbers
→ Irrational numbers
→ Rational numbers
  → Integers
  → Noninteger fractions (positive and negative)
    Integers → Negative integers
    Integers → Whole numbers
      Whole numbers → Natural numbers
      Whole numbers → Zero

Subsets of real numbers
Figure P.1

EXAMPLE 1 Classifying Real Numbers

Determine which numbers in the set $\left\{-13, -\sqrt{5}, -1, -\tfrac{1}{3}, 0, \tfrac{5}{8}, \sqrt{2}, \pi, 7\right\}$ are (a) natural numbers, (b) whole numbers, (c) integers, (d) rational numbers, and (e) irrational numbers.

Solution

a. Natural numbers: $\{7\}$

b. Whole numbers: $\{0, 7\}$

c. Integers: $\{-13, -1, 0, 7\}$

d. Rational numbers: $\left\{-13, -1, -\dfrac{1}{3}, 0, \dfrac{5}{8}, 7\right\}$

e. Irrational numbers: $\left\{-\sqrt{5}, \sqrt{2}, \pi\right\}$

✓ *Checkpoint* 🔊))) Audio-video solution in English & Spanish at LarsonPrecalculus.com.

Repeat Example 1 for the set $\left\{-\pi, -\tfrac{1}{4}, \tfrac{6}{3}, \tfrac{1}{2}\sqrt{2}, -7.5, -1, 8, -22\right\}$.

Real numbers are represented graphically on the **real number line.** When you draw a point on the real number line that corresponds to a real number, you are **plotting** the real number. The point 0 on the real number line is the **origin.** Numbers to the right of 0 are positive, and numbers to the left of 0 are negative, as shown below. The term **nonnegative** describes a number that is either positive or zero.

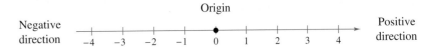

As illustrated below, there is a *one-to-one correspondence* between real numbers and points on the real number line.

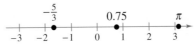

Every real number corresponds to exactly one point on the real number line.

Every point on the real number line corresponds to exactly one real number.

EXAMPLE 2   Plotting Points on the Real Number Line

Plot the real numbers on the real number line.

a. $-\dfrac{7}{4}$

b. 2.3

c. $\dfrac{2}{3}$

d. -1.8

Solution   The following figure shows all four points.

a. The point representing the real number $-\frac{7}{4} = -1.75$ lies between -2 and -1, but closer to -2, on the real number line.

b. The point representing the real number 2.3 lies between 2 and 3, but closer to 2, on the real number line.

c. The point representing the real number $\frac{2}{3} = 0.666\ldots$ lies between 0 and 1, but closer to 1, on the real number line.

d. The point representing the real number -1.8 lies between -2 and -1, but closer to -2, on the real number line. Note that the point representing -1.8 lies slightly to the left of the point representing $-\frac{7}{4}$.

✓ *Checkpoint*  Audio-video solution in English & Spanish at LarsonPrecalculus.com.

Plot the real numbers on the real number line.

a. $\dfrac{5}{2}$ **b.** -1.6

c. $-\dfrac{3}{4}$ **d.** 0.7

Ordering Real Numbers

One important property of real numbers is that they are *ordered*.

> **Definition of Order on the Real Number Line**
>
> If a and b are real numbers, then a is less than b when $b - a$ is positive. The **inequality** $a < b$ denotes the **order** of a and b. This relationship can also be described by saying that b is *greater than* a and writing $b > a$. The inequality $a \leq b$ means that a is *less than or equal to* b, and the inequality $b \geq a$ means that b is *greater than or equal to* a. The symbols $<, >, \leq,$ and $\geq$ are *inequality symbols.*

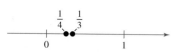

$a < b$ if and only if a lies to the left of b.

Figure P.2

Geometrically, this definition implies that $a < b$ if and only if a lies to the *left* of b on the real number line, as shown in Figure P.2.

EXAMPLE 3 **Ordering Real Numbers**

Place the appropriate inequality symbol ($<$ or $>$) between the pair of real numbers.

a. $-3, 0$ **b.** $-2, -4$ **c.** $\frac{1}{4}, \frac{1}{3}$ **d.** $-\frac{1}{5}, -\frac{1}{2}$

Solution

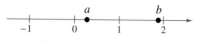

Figure P.3

Figure P.4

Figure P.5

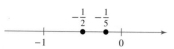

Figure P.6

a. Because -3 lies to the left of 0 on the real number line, as shown in Figure P.3, you can say that -3 is *less than* 0, and write $-3 < 0$.

b. Because -2 lies to the right of -4 on the real number line, as shown in Figure P.4, you can say that -2 is *greater than* -4, and write $-2 > -4$.

c. Because $\frac{1}{4}$ lies to the left of $\frac{1}{3}$ on the real number line, as shown in Figure P.5, you can say that $\frac{1}{4}$ is *less than* $\frac{1}{3}$, and write $\frac{1}{4} < \frac{1}{3}$.

d. Because $-\frac{1}{5}$ lies to the right of $-\frac{1}{2}$ on the real number line, as shown in Figure P.6, you can say that $-\frac{1}{5}$ is *greater than* $-\frac{1}{2}$, and write $-\frac{1}{5} > -\frac{1}{2}$.

✓ *Checkpoint*))) *Audio-video solution in English & Spanish at LarsonPrecalculus.com.*

Place the appropriate inequality symbol ($<$ or $>$) between the pair of real numbers.

a. $1, -5$ **b.** $\frac{3}{2}, 7$ **c.** $-\frac{2}{3}, -\frac{3}{4}$ **d.** $-3.5, 1$

EXAMPLE 4 **Interpreting Inequalities**

Describe the subset of real numbers that the inequality represents.

a. $x \leq 2$ **b.** $-2 \leq x < 3$

Solution

$x \leq 2$

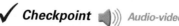

Figure P.7

a. The inequality $x \leq 2$ denotes all real numbers less than or equal to 2, as shown in Figure P.7.

b. The inequality $-2 \leq x < 3$ means that $x \geq -2$ *and* $x < 3$. This "double inequality" denotes all real numbers between -2 and 3, including -2 but not including 3, as shown in Figure P.8.

$-2 \leq x < 3$

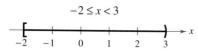

Figure P.8

✓ *Checkpoint*))) *Audio-video solution in English & Spanish at LarsonPrecalculus.com.*

Describe the subset of real numbers that the inequality represents.

a. $x > -3$ **b.** $0 < x \leq 4$

Inequalities can describe subsets of real numbers called **intervals.** In the bounded intervals below, the real numbers a and b are the **endpoints** of each interval. The endpoints of a closed interval are included in the interval, whereas the endpoints of an open interval are not included in the interval.

••**REMARK** The reason that the four types of intervals at the right are called *bounded* is that each has a finite length. An interval that does not have a finite length is *unbounded* (see below).

Bounded Intervals on the Real Number Line

Notation	Interval Type	Inequality	Graph
$[a, b]$	Closed	$a \le x \le b$	
(a, b)	Open	$a < x < b$	
$[a, b)$		$a \le x < b$	
$(a, b]$		$a < x \le b$	

The symbols ∞, **positive infinity,** and $-\infty$, **negative infinity,** do not represent real numbers. They are simply convenient symbols used to describe the unboundedness of an interval such as $(1, \infty)$ or $(-\infty, 3]$.

••**REMARK** Whenever you write an interval containing ∞ or $-\infty$, always use a parenthesis and never a bracket next to these symbols. This is because ∞ and $-\infty$ are never an endpoint of an interval and therefore are not included in the interval.

Unbounded Intervals on the Real Number Line

Notation	Interval Type	Inequality	Graph
$[a, \infty)$		$x \ge a$	
(a, ∞)	Open	$x > a$	
$(-\infty, b]$		$x \le b$	
$(-\infty, b)$	Open	$x < b$	
$(-\infty, \infty)$	Entire real line	$-\infty < x < \infty$	

EXAMPLE 5 Interpreting Intervals

a. The interval $(-1, 0)$ consists of all real numbers greater than -1 and less than 0.

b. The interval $[2, \infty)$ consists of all real numbers greater than or equal to 2.

✓ *Checkpoint* ◀))) *Audio-video solution in English & Spanish at LarsonPrecalculus.com.*

Give a verbal description of the interval $[-2, 5)$.

EXAMPLE 6 Using Inequalities to Represent Intervals

a. The inequality $c \le 2$ can represent the statement "c is at most 2."

b. The inequality $-3 < x \le 5$ can represent "all x in the interval $(-3, 5]$."

✓ *Checkpoint* ◀))) *Audio-video solution in English & Spanish at LarsonPrecalculus.com.*

Use inequality notation to represent the statement "x is greater than -2 and at most 4."

Absolute Value and Distance

The **absolute value** of a real number is its *magnitude,* or the distance between the origin and the point representing the real number on the real number line.

Definition of Absolute Value

If a is a real number, then the absolute value of a is

$$|a| = \begin{cases} a, & \text{if } a \geq 0 \\ -a, & \text{if } a < 0 \end{cases}.$$

Notice in this definition that the absolute value of a real number is never negative. For instance, if $a = -5$, then $|-5| = -(-5) = 5$. The absolute value of a real number is either positive or zero. Moreover, 0 is the only real number whose absolute value is 0. So, $|0| = 0$.

EXAMPLE 7   **Finding Absolute Values**

a. $|-15| = 15$ **b.** $\left|\dfrac{2}{3}\right| = \dfrac{2}{3}$

c. $|-4.3| = 4.3$ **d.** $-|-6| = -(6) = -6$

✓ **Checkpoint**  Audio-video solution in English & Spanish at LarsonPrecalculus.com.

Evaluate each expression.

a. $|1|$ **b.** $-\left|\dfrac{3}{4}\right|$

c. $\dfrac{2}{|-3|}$ **d.** $-|0.7|$

EXAMPLE 8   **Evaluating the Absolute Value of a Number**

Evaluate $\dfrac{|x|}{x}$ for (a) $x > 0$ and (b) $x < 0$.

Solution

a. If $x > 0$, then $|x| = x$ and $\dfrac{|x|}{x} = \dfrac{x}{x} = 1$.

b. If $x < 0$, then $|x| = -x$ and $\dfrac{|x|}{x} = \dfrac{-x}{x} = -1$.

✓ **Checkpoint**  Audio-video solution in English & Spanish at LarsonPrecalculus.com.

Evaluate $\dfrac{|x + 3|}{x + 3}$ for (a) $x > -3$ and (b) $x < -3$. ∎

The **Law of Trichotomy** states that for any two real numbers a and b, *precisely* one of three relationships is possible:

$$a = b, \quad a < b, \quad \text{or} \quad a > b. \qquad \text{Law of Trichotomy}$$

EXAMPLE 9 **Comparing Real Numbers**

Place the appropriate symbol ($<$, $>$, or $=$) between the pair of real numbers.

a. $|-4|$ ▧ $|3|$ **b.** $|-10|$ ▧ $|10|$ **c.** $-|-7|$ ▧ $|-7|$

Solution

a. $|-4| > |3|$ because $|-4| = 4$ and $|3| = 3$, and 4 is greater than 3.
b. $|-10| = |10|$ because $|-10| = 10$ and $|10| = 10$.
c. $-|-7| < |-7|$ because $-|-7| = -7$ and $|-7| = 7$, and -7 is less than 7.

✓ *Checkpoint* ◀))) *Audio-video solution in English & Spanish at LarsonPrecalculus.com.*

Place the appropriate symbol ($<$, $>$, or $=$) between the pair of real numbers.

a. $|-3|$ ▧ $|4|$ **b.** $-|-4|$ ▧ $-|4|$ **c.** $|-3|$ ▧ $-|-3|$ ■

Properties of Absolute Values

1. $|a| \geq 0$ **2.** $|-a| = |a|$

3. $|ab| = |a||b|$ **4.** $\left|\dfrac{a}{b}\right| = \dfrac{|a|}{|b|}, \quad b \neq 0$

Absolute value can be used to define the distance between two points on the real number line. For instance, the distance between -3 and 4 is

$$|-3 - 4| = |-7|$$
$$= 7$$

as shown in Figure P.9.

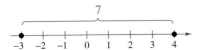

The distance between -3 and 4 is 7.
Figure P.9

Distance Between Two Points on the Real Number Line

Let a and b be real numbers. The **distance between a and b** is

$$d(a, b) = |b - a| = |a - b|.$$

EXAMPLE 10 **Finding a Distance**

Find the distance between -25 and 13.

Solution

The distance between -25 and 13 is

$$|-25 - 13| = |-38| = 38. \qquad \text{Distance between } -25 \text{ and } 13$$

The distance can also be found as follows.

$$|13 - (-25)| = |38| = 38 \qquad \text{Distance between } -25 \text{ and } 13$$

✓ *Checkpoint* ◀))) *Audio-video solution in English & Spanish at LarsonPrecalculus.com.*

a. Find the distance between 35 and -23.
b. Find the distance between -35 and -23.
c. Find the distance between 35 and 23.   ■

One application of finding the distance between two points on the real number line is finding a change in temperature.

Algebraic Expressions

One characteristic of algebra is the use of letters to represent numbers. The letters are **variables,** and combinations of letters and numbers are **algebraic expressions.** Here are a few examples of algebraic expressions.

$$5x, \qquad 2x - 3, \qquad \frac{4}{x^2 + 2}, \qquad 7x + y$$

Definition of an Algebraic Expression

An **algebraic expression** is a collection of letters (**variables**) and real numbers (**constants**) combined using the operations of addition, subtraction, multiplication, division, and exponentiation.

The **terms** of an algebraic expression are those parts that are separated by *addition.* For example, $x^2 - 5x + 8 = x^2 + (-5x) + 8$ has three terms: x^2 and $-5x$ are the **variable terms** and 8 is the **constant term.** The numerical factor of a term is called the **coefficient.** For instance, the coefficient of $-5x$ is -5, and the coefficient of x^2 is 1.

EXAMPLE 11 Identifying Terms and Coefficients

Algebraic Expression	Terms	Coefficients
a. $5x - \dfrac{1}{7}$	$5x, -\dfrac{1}{7}$	$5, -\dfrac{1}{7}$
b. $2x^2 - 6x + 9$	$2x^2, -6x, 9$	$2, -6, 9$
c. $\dfrac{3}{x} + \dfrac{1}{2}x^4 - y$	$\dfrac{3}{x}, \dfrac{1}{2}x^4, -y$	$3, \dfrac{1}{2}, -1$

✓ *Checkpoint* ◀))) *Audio-video solution in English & Spanish at LarsonPrecalculus.com.*

Identify the terms and coefficients of $-2x + 4$.

To **evaluate** an algebraic expression, substitute numerical values for each of the variables in the expression, as shown in the next example.

EXAMPLE 12 Evaluating Algebraic Expressions

Expression	Value of Variable	Substitute.	Value of Expression
a. $-3x + 5$	$x = 3$	$-3(3) + 5$	$-9 + 5 = -4$
b. $3x^2 + 2x - 1$	$x = -1$	$3(-1)^2 + 2(-1) - 1$	$3 - 2 - 1 = 0$
c. $\dfrac{2x}{x + 1}$	$x = -3$	$\dfrac{2(-3)}{-3 + 1}$	$\dfrac{-6}{-2} = 3$

Note that you must substitute the value for *each* occurrence of the variable.

✓ *Checkpoint* ◀))) *Audio-video solution in English & Spanish at LarsonPrecalculus.com.*

Evaluate $4x - 5$ when $x = 0$.

Use the **Substitution Principle** to evaluate algebraic expressions. It states that "If $a = b$, then b can replace a in any expression involving a." In Example 12(a), for instance, 3 is *substituted* for x in the expression $-3x + 5$.

Basic Rules of Algebra

There are four arithmetic operations with real numbers: *addition, multiplication, subtraction,* and *division,* denoted by the symbols $+$, $\times$ or $\cdot$, $-$, and $\div$ or $/$, respectively. Of these, addition and multiplication are the two primary operations. Subtraction and division are the inverse operations of addition and multiplication, respectively.

Definitions of Subtraction and Division

Subtraction: Add the opposite.   **Division:** Multiply by the reciprocal.

$$a - b = a + (-b)$$ If $b \neq 0$, then $a/b = a\left(\dfrac{1}{b}\right) = \dfrac{a}{b}$.

In these definitions, $-b$ is the **additive inverse** (or opposite) of b, and $1/b$ is the **multiplicative inverse** (or reciprocal) of b. In the fractional form a/b, a is the **numerator** of the fraction and b is the **denominator.**

Because the properties of real numbers below are true for variables and algebraic expressions as well as for real numbers, they are often called the **Basic Rules of Algebra.** Try to formulate a verbal description of each property. For instance, the first property states that *the order in which two real numbers are added does not affect their sum.*

Basic Rules of Algebra

Let a, b, and c be real numbers, variables, or algebraic expressions.

Property		**Example**
Commutative Property of Addition:	$a + b = b + a$	$4x + x^2 = x^2 + 4x$
Commutative Property of Multiplication:	$ab = ba$	$(4 - x)x^2 = x^2(4 - x)$
Associative Property of Addition:	$(a + b) + c = a + (b + c)$	$(x + 5) + x^2 = x + (5 + x^2)$
Associative Property of Multiplication:	$(ab)c = a(bc)$	$(2x \cdot 3y)(8) = (2x)(3y \cdot 8)$
Distributive Properties:	$a(b + c) = ab + ac$	$3x(5 + 2x) = 3x \cdot 5 + 3x \cdot 2x$
	$(a + b)c = ac + bc$	$(y + 8)y = y \cdot y + 8 \cdot y$
Additive Identity Property:	$a + 0 = a$	$5y^2 + 0 = 5y^2$
Multiplicative Identity Property:	$a \cdot 1 = a$	$(4x^2)(1) = 4x^2$
Additive Inverse Property:	$a + (-a) = 0$	$5x^3 + (-5x^3) = 0$
Multiplicative Inverse Property:	$a \cdot \dfrac{1}{a} = 1, \quad a \neq 0$	$(x^2 + 4)\left(\dfrac{1}{x^2 + 4}\right) = 1$

Because subtraction is defined as "adding the opposite," the Distributive Properties are also true for subtraction. For instance, the "subtraction form" of $a(b + c) = ab + ac$ is $a(b - c) = ab - ac$. Note that the operations of subtraction and division are neither commutative nor associative. The examples

$$7 - 3 \neq 3 - 7 \quad \text{and} \quad 20 \div 4 \neq 4 \div 20$$

show that subtraction and division are not commutative. Similarly

$$5 - (3 - 2) \neq (5 - 3) - 2 \quad \text{and} \quad 16 \div (4 \div 2) \neq (16 \div 4) \div 2$$

demonstrate that subtraction and division are not associative.

EXAMPLE 13 **Identifying Rules of Algebra**

Identify the rule of algebra illustrated by the statement.

a. $(5x^3)2 = 2(5x^3)$ **b.** $(4x + 3) - (4x + 3) = 0$

c. $7x \cdot \dfrac{1}{7x} = 1, \quad x \neq 0$ **d.** $(2 + 5x^2) + x^2 = 2 + (5x^2 + x^2)$

Solution

a. This statement illustrates the Commutative Property of Multiplication. In other words, you obtain the same result whether you multiply $5x^3$ by 2, or 2 by $5x^3$.

b. This statement illustrates the Additive Inverse Property. In terms of subtraction, this property states that when any expression is subtracted from itself the result is 0.

c. This statement illustrates the Multiplicative Inverse Property. Note that x must be a nonzero number. The reciprocal of x is undefined when x is 0.

d. This statement illustrates the Associative Property of Addition. In other words, to form the sum $2 + 5x^2 + x^2$, it does not matter whether 2 and $5x^2$, or $5x^2$ and x^2 are added first.

✓ *Checkpoint* ◀))) *Audio-video solution in English & Spanish at LarsonPrecalculus.com.*

Identify the rule of algebra illustrated by the statement.

a. $x + 9 = 9 + x$ **b.** $5(x^3 \cdot 2) = (5x^3)2$ **c.** $(2 + 5x^2)y^2 = 2 \cdot y^2 + 5x^2 \cdot y^2$

• • **REMARK** Notice the difference between the *opposite of a number* and a *negative number.* If a is negative, then its opposite, $-a$, is positive. For instance, if $a = -5$, then

$$-a = -(-5) = 5.$$

Properties of Negation and Equality

Let a, b, and c be real numbers, variables, or algebraic expressions.

Property	**Example**
1. $(-1)a = -a$	$(-1)7 = -7$
2. $-(-a) = a$	$-(-6) = 6$
3. $(-a)b = -(ab) = a(-b)$	$(-5)3 = -(5 \cdot 3) = 5(-3)$
4. $(-a)(-b) = ab$	$(-2)(-x) = 2x$
5. $-(a + b) = (-a) + (-b)$	$-(x + 8) = (-x) + (-8)$
	$\quad\quad\quad\quad = -x - 8$
6. If $a = b$, then $a \pm c = b \pm c$.	$\frac{1}{2} + 3 = 0.5 + 3$
7. If $a = b$, then $ac = bc$.	$4^2 \cdot 2 = 16 \cdot 2$
8. If $a \pm c = b \pm c$, then $a = b$.	$1.4 - 1 = \frac{7}{5} - 1 \implies 1.4 = \frac{7}{5}$
9. If $ac = bc$ and $c \neq 0$, then $a = b$.	$3x = 3 \cdot 4 \implies x = 4$

• • **REMARK** The "or" in the Zero-Factor Property includes the possibility that either or both factors may be zero. This is an *inclusive or,* and it is generally the way the word "or" is used in mathematics.

Properties of Zero

Let a and b be real numbers, variables, or algebraic expressions.

1. $a + 0 = a$ and $a - 0 = a$ **2.** $a \cdot 0 = 0$

3. $\dfrac{0}{a} = 0, \quad a \neq 0$ **4.** $\dfrac{a}{0}$ is undefined.

5. Zero-Factor Property: If $ab = 0$, then $a = 0$ or $b = 0$.

Properties and Operations of Fractions

Let a, b, c, and d be real numbers, variables, or algebraic expressions such that $b \neq 0$ and $d \neq 0$.

1. **Equivalent Fractions:** $\dfrac{a}{b} = \dfrac{c}{d}$ if and only if $ad = bc$.

2. **Rules of Signs:** $-\dfrac{a}{b} = \dfrac{-a}{b} = \dfrac{a}{-b}$ and $\dfrac{-a}{-b} = \dfrac{a}{b}$

3. **Generate Equivalent Fractions:** $\dfrac{a}{b} = \dfrac{ac}{bc}$, $c \neq 0$

4. **Add or Subtract with Like Denominators:** $\dfrac{a}{b} \pm \dfrac{c}{b} = \dfrac{a \pm c}{b}$

5. **Add or Subtract with Unlike Denominators:** $\dfrac{a}{b} \pm \dfrac{c}{d} = \dfrac{ad \pm bc}{bd}$

6. **Multiply Fractions:** $\dfrac{a}{b} \cdot \dfrac{c}{d} = \dfrac{ac}{bd}$

7. **Divide Fractions:** $\dfrac{a}{b} \div \dfrac{c}{d} = \dfrac{a}{b} \cdot \dfrac{d}{c} = \dfrac{ad}{bc}$, $c \neq 0$

> **• • REMARK** In Property 1 of fractions, the phrase "if and only if" implies two statements. One statement is: If $a/b = c/d$, then $ad = bc$. The other statement is: If $ad = bc$, where $b \neq 0$ and $d \neq 0$, then $a/b = c/d$.

EXAMPLE 14   **Properties and Operations of Fractions**

a. Equivalent fractions: $\dfrac{x}{5} = \dfrac{3 \cdot x}{3 \cdot 5} = \dfrac{3x}{15}$ **b.** Divide fractions: $\dfrac{7}{x} \div \dfrac{3}{2} = \dfrac{7}{x} \cdot \dfrac{2}{3} = \dfrac{14}{3x}$

✓ *Checkpoint*  *Audio-video solution in English & Spanish at LarsonPrecalculus.com.*

a. Multiply fractions: $\dfrac{3}{5} \cdot \dfrac{x}{6}$ **b.** Add fractions: $\dfrac{x}{10} + \dfrac{2x}{5}$ ∎

> **• • REMARK** The number 1 is neither prime nor composite.

If a, b, and c are integers such that $ab = c$, then a and b are **factors** or **divisors** of c. A **prime number** is an integer that has exactly two positive factors—itself and 1—such as 2, 3, 5, 7, and 11. The numbers 4, 6, 8, 9, and 10 are **composite** because each can be written as the product of two or more prime numbers. The **Fundamental Theorem of Arithmetic** states that every positive integer greater than 1 is a prime number or can be written as the product of prime numbers in precisely one way (disregarding order). For instance, the *prime factorization* of 24 is $24 = 2 \cdot 2 \cdot 2 \cdot 3$.

Summarize   (Section P.1)

1. Describe how to represent and classify real numbers *(pages 2 and 3)*. For examples of representing and classifying real numbers, see Examples 1 and 2.

2. Describe how to order real numbers and use inequalities *(pages 4 and 5)*. For examples of ordering real numbers and using inequalities, see Examples 3–6.

3. State the absolute value of a real number *(page 6)*. For examples of using absolute value, see Examples 7–10.

4. Explain how to evaluate an algebraic expression *(page 8)*. For examples involving algebraic expressions, see Examples 11 and 12.

5. State the basic rules and properties of algebra *(pages 9–11)*. For examples involving the basic rules and properties of algebra, see Examples 13 and 14.

P.1   Exercises

See **CalcChat.com** for tutorial help and worked-out solutions to odd-numbered exercises.

Vocabulary: Fill in the blanks.

1. _____ numbers have infinite nonrepeating decimal representations.
2. The point 0 on the real number line is called the _____.
3. The distance between the origin and a point representing a real number on the real number line is the _____ _____ of the real number.
4. A number that can be written as the product of two or more prime numbers is called a _____ number.
5. The _____ of an algebraic expression are those parts separated by addition.
6. The _____ _____ states that if $ab = 0$, then $a = 0$ or $b = 0$.

Skills and Applications

Classifying Real Numbers In Exercises 7–10, determine which numbers in the set are (a) natural numbers, (b) whole numbers, (c) integers, (d) rational numbers, and (e) irrational numbers.

7. $\left\{ -9, -\frac{7}{2}, 5, \frac{2}{3}, \sqrt{2}, 0, 1, -4, 2, -11 \right\}$
8. $\left\{ \sqrt{5}, -7, -\frac{7}{3}, 0, 3.12, \frac{5}{4}, -3, 12, 5 \right\}$
9. $\{ 2.01, 0.666\ldots, -13, 0.010110111\ldots, 1, -6 \}$
10. $\left\{ 25, -17, -\frac{12}{5}, \sqrt{9}, 3.12, \frac{1}{2}\pi, 7, -11.1, 13 \right\}$

Plotting Points on the Real Number Line In Exercises 11 and 12, plot the real numbers on the real number line.

11. (a) 3   (b) $\frac{7}{2}$ (c) $-\frac{5}{2}$ (d) -5.2
12. (a) 8.5   (b) $\frac{4}{3}$ (c) -4.75 (d) $-\frac{8}{3}$

Plotting and Ordering Real Numbers In Exercises 13–16, plot the two real numbers on the real number line. Then place the appropriate inequality symbol (< or >) between them.

13. $-4, -8$ 14. $1, \frac{16}{3}$ 15. $\frac{5}{6}, \frac{2}{3}$ 16. $-\frac{8}{7}, -\frac{3}{7}$

Interpreting an Inequality or an Interval In Exercises 17–24, (a) give a verbal description of the subset of real numbers represented by the inequality or the interval, (b) sketch the subset on the real number line, and (c) state whether the interval is bounded or unbounded.

17. $x \le 5$ 18. $x < 0$
19. $[4, \infty)$ 20. $(-\infty, 2)$
21. $-2 < x < 2$ 22. $0 < x \le 6$
23. $[-5, 2)$ 24. $(-1, 2]$

Using Inequality and Interval Notation In Exercises 25–30, use inequality notation and interval notation to describe the set.

25. y is nonnegative.   26. y is no more than 25.
27. t is at least 10 and at most 22.
28. k is less than 5 but no less than -3.
29. The dog's weight W is more than 65 pounds.
30. The annual rate of inflation r is expected to be at least 2.5% but no more than 5%.

Evaluating an Absolute Value Expression In Exercises 31–40, evaluate the expression.

31. $|-10|$ 32. $|0|$
33. $|3 - 8|$ 34. $|4 - 1|$
35. $|-1| - |-2|$ 36. $-3 - |-3|$
37. $\dfrac{-5}{|-5|}$ 38. $-3|-3|$
39. $\dfrac{|x + 2|}{x + 2}, \quad x < -2$ 40. $\dfrac{|x - 1|}{x - 1}, \quad x > 1$

Comparing Real Numbers In Exercises 41–44, place the appropriate symbol (<, >, or =) between the two real numbers.

41. $|-4|$ ▢ $|4|$ 42. -5 ▢ $-|5|$
43. $-|-6|$ ▢ $|-6|$ 44. $-|-2|$ ▢ $-|2|$

Finding a Distance In Exercises 45–50, find the distance between a and b.

45. $a = 126, b = 75$ 46. $a = -126, b = -75$
47. $a = -\frac{5}{2}, b = 0$ 48. $a = \frac{1}{4}, b = \frac{11}{4}$
49. $a = \frac{16}{5}, b = \frac{112}{75}$ 50. $a = 9.34, b = -5.65$

Using Absolute Value Notation In Exercises 51–54, use absolute value notation to describe the situation.

51. The distance between x and 5 is no more than 3.
52. The distance between x and -10 is at least 6.
53. y is at most two units from a.
54. The temperature in Bismarck, North Dakota, was 60°F at noon, then 23°F at midnight. What was the change in temperature over the 12-hour period?

• • Federal Deficit • • • • • • • • • • • • • • • •

In Exercises 55–58, use the bar graph, which shows the receipts of the federal government (in billions of dollars) for selected years from 2004 through 2010.

In each exercise you are given the expenditures of the federal government. Find the magnitude of the surplus or deficit for the year. *(Source: U.S. Office of Management and Budget)*

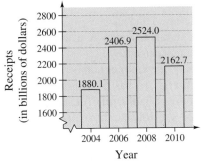

Receipts (in billions of dollars)

Year	Receipts, R	Expenditures, E	$|R - E|$
55. 2004		$2292.8 billion	
56. 2006		$2655.1 billion	
57. 2008		$2982.5 billion	
58. 2010		$3456.2 billion	

Identifying Terms and Coefficients In Exercises 59–62, identify the terms. Then identify the coefficients of the variable terms of the expression.

59. $7x + 4$

60. $6x^3 - 5x$

61. $4x^3 + 0.5x - 5$

62. $3\sqrt{3}x^2 + 1$

Evaluating an Algebraic Expression In Exercises 63–66, evaluate the expression for each value of x. (If not possible, then state the reason.)

Expression	Values
63. $4x - 6$	(a) $x = -1$ (b) $x = 0$
64. $9 - 7x$	(a) $x = -3$ (b) $x = 3$
65. $-x^2 + 5x - 4$	(a) $x = -1$ (b) $x = 1$
66. $(x + 1)/(x - 1)$	(a) $x = 1$ (b) $x = -1$

Identifying Rules of Algebra In Exercises 67–72, identify the rule(s) of algebra illustrated by the statement.

67. $\dfrac{1}{h + 6}(h + 6) = 1, \quad h \neq -6$

68. $(x + 3) - (x + 3) = 0$

69. $2(x + 3) = 2 \cdot x + 2 \cdot 3$

70. $(z - 2) + 0 = z - 2$

71. $x(3y) = (x \cdot 3)y = (3x)y$

72. $\frac{1}{7}(7 \cdot 12) = \left(\frac{1}{7} \cdot 7\right)12 = 1 \cdot 12 = 12$

Operations with Fractions In Exercises 73–76, perform the operation(s). (Write fractional answers in simplest form.)

73. $\frac{5}{8} - \frac{5}{12} + \frac{1}{6}$

74. $-\left(6 \cdot \frac{4}{8}\right)$

75. $\frac{2x}{3} - \frac{x}{4}$

76. $\frac{5x}{6} \cdot \frac{2}{9}$

Exploration

77. Determining the Sign of an Expression Use the real numbers A, B, and C shown on the number line to determine the sign of (a) $-A$, (b) $B - A$, (c) $-C$, and (d) $A - C$.

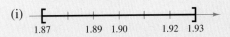

78. **HOW DO YOU SEE IT?** Match each description with its graph. Which types of real numbers shown in Figure P.1 on page 2 may be included in a range of prices? a range of lengths? Explain.

(i)
1.87 1.89 1.90 1.92 1.93

(ii)
1.87 1.88 1.89 1.90 1.91 1.92 1.93

(a) The price of an item is within $0.03 of $1.90.

(b) The distance between the prongs of an electric plug may not differ from 1.9 centimeters by more than 0.03 centimeter.

True or False? In Exercises 79 and 80, determine whether the statement is true or false. Justify your answer.

79. Every nonnegative number is positive.

80. If $a > 0$ and $b < 0$, then $ab > 0$.

81. Conjecture

(a) Use a calculator to complete the table.

n	0.0001	0.01	1	100	10,000
$5/n$					

(b) Use the result from part (a) to make a conjecture about the value of $5/n$ as n (i) approaches 0, and (ii) increases without bound.

P.2 Solving Equations

- Identify different types of equations.
- Solve linear equations in one variable and rational equations.
- Solve quadratic equations by factoring, extracting square roots, completing the square, and using the Quadratic Formula.
- Solve polynomial equations of degree three or greater.
- Solve radical equations.
- Solve absolute value equations.

Linear equations can help you analyze many real-life applications. For example, you can use linear equations in forensics to determine height from femur length. See Exercises 97 and 98 on page 25.

Equations and Solutions of Equations

An **equation** in x is a statement that two algebraic expressions are equal. For example,

$$3x - 5 = 7, \quad x^2 - x - 6 = 0, \quad \text{and} \quad \sqrt{2x} = 4$$

are equations. To **solve** an equation in x means to find all values of x for which the equation is true. Such values are **solutions.** For instance, $x = 4$ is a solution of the equation $3x - 5 = 7$ because $3(4) - 5 = 7$ is a true statement.

The solutions of an equation depend on the kinds of numbers being considered. For instance, in the set of rational numbers, $x^2 = 10$ has no solution because there is no rational number whose square is 10. However, in the set of real numbers, the equation has the two solutions $x = \sqrt{10}$ and $x = -\sqrt{10}$.

An equation that is true for *every* real number in the domain of the variable is called an **identity.** The domain is the set of all real numbers for which the equation is defined. For example,

$$x^2 - 9 = (x + 3)(x - 3) \qquad \text{Identity}$$

is an identity because it is a true statement for any real value of x. The equation

$$\frac{x}{3x^2} = \frac{1}{3x} \qquad \text{Identity}$$

where $x \neq 0$, is an identity because it is true for any nonzero real value of x.

An equation that is true for just *some* (but not all) of the real numbers in the domain of the variable is called a **conditional equation.** For example, the equation

$$x^2 - 9 = 0 \qquad \text{Conditional equation}$$

is conditional because $x = 3$ and $x = -3$ are the only values in the domain that satisfy the equation.

A **contradiction** is an equation that is false for *every* real number in the domain of the variable. For example, the equation

$$2x - 4 = 2x + 1 \qquad \text{Contradiction}$$

is a contradiction because there are no real values of x for which the equation is true.

Linear and Rational Equations

Definition of Linear Equation in One Variable

A **linear equation in one variable** x is an equation that can be written in the standard form

$$ax + b = 0$$

where a and b are real numbers with $a \neq 0$.

A linear equation has exactly one solution. To see this, consider the following steps. (Remember that $a \neq 0$.)

$$ax + b = 0 \qquad \text{Write original equation.}$$

$$ax = -b \qquad \text{Subtract } b \text{ from each side.}$$

$$x = -\frac{b}{a} \qquad \text{Divide each side by } a.$$

To solve a conditional equation in x, isolate x on one side of the equation by a sequence of **equivalent equations,** each having the same solution(s) as the original equation. The operations that yield equivalent equations come from the properties of equality reviewed in Section P.1.

HISTORICAL NOTE

This ancient Egyptian papyrus, discovered in 1858, contains one of the earliest examples of mathematical writing in existence. The papyrus itself dates back to around 1650 B.C., but it is actually a copy of writings from two centuries earlier. The algebraic equations on the papyrus were written in words. Diophantus, a Greek who lived around A.D. 250, is often called the Father of Algebra. He was the first to use abbreviated word forms in equations.

Generating Equivalent Equations

An equation can be transformed into an *equivalent equation* by one or more of the following steps.

	Given Equation	Equivalent Equation
1. Remove symbols of grouping, combine like terms, or simplify fractions on one or both sides of the equation.	$2x - x = 4$	$x = 4$
2. Add (or subtract) the same quantity to (from) *each* side of the equation.	$x + 1 = 6$	$x = 5$
3. Multiply (or divide) *each* side of the equation by the same *nonzero* quantity.	$2x = 6$	$x = 3$
4. Interchange the two sides of the equation.	$2 = x$	$x = 2$

The following example shows the steps for solving a linear equation in one variable x written in standard form.

· · · · · · · · · · · · · · · · · · ▷

· **REMARK** After solving an equation, you should check each solution in the original equation. For instance, you can check the solution of Example 1(a) as follows.

$$3x - 6 = 0 \qquad \text{Write original equation.}$$

$$3(2) - 6 \stackrel{?}{=} 0 \qquad \text{Substitute 2 for } x.$$

$$0 = 0 \qquad \text{Solution checks. } ✓$$

Try checking the solution of Example 1(b).

EXAMPLE 1   Solving a Linear Equation

a. $3x - 6 = 0$ Original equation

     $3x = 6$ Add 6 to each side.

      $x = 2$ Divide each side by 3.

b. $5x + 4 = 3x - 8$ Original equation

   $2x + 4 = -8$ Subtract $3x$ from each side.

     $2x = -12$ Subtract 4 from each side.

      $x = -6$ Divide each side by 2.

✓ *Checkpoint* 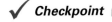 Audio-video solution in English & Spanish at LarsonPrecalculus.com.

Solve each equation.

a. $7 - 2x = 15$ **b.** $7x - 9 = 5x + 7$

▷

REMARK An equation with a *single fraction* on each side can be cleared of denominators by **cross multiplying**. To do this, multiply the left numerator by the right denominator and the right numerator by the left denominator as follows.

$$\frac{a}{b} = \frac{c}{d} \qquad \text{Original equation}$$

$$ad = cb \qquad \text{Cross multiply.}$$

A **rational equation** is an equation that involves one or more fractional expressions. To solve a rational equation, find the least common denominator (LCD) of all terms and multiply every term by the LCD. This process will clear the original equation of fractions and produce a simpler equation.

EXAMPLE 2   **Solving a Rational Equation**

Solve $\dfrac{x}{3} + \dfrac{3x}{4} = 2$.

Solution

$$\frac{x}{3} + \frac{3x}{4} = 2 \qquad\qquad \text{Original equation}$$

$$(12)\frac{x}{3} + (12)\frac{3x}{4} = (12)2 \qquad\qquad \text{Multiply each term by the LCD.}$$

$$4x + 9x = 24 \qquad\qquad \text{Simplify.}$$

$$13x = 24 \qquad\qquad \text{Combine like terms.}$$

$$x = \frac{24}{13} \qquad\qquad \text{Divide each side by 13.}$$

The solution is $x = \frac{24}{13}$. Check this in the original equation.

✓ **Checkpoint**  Audio-video solution in English & Spanish at LarsonPrecalculus.com.

Solve $\dfrac{4x}{9} - \dfrac{1}{3} = x + \dfrac{5}{3}$.

When multiplying or dividing an equation by a *variable expression,* it is possible to introduce an **extraneous solution** that does not satisfy the original equation.

EXAMPLE 3   **An Equation with an Extraneous Solution**

Solve $\dfrac{1}{x - 2} = \dfrac{3}{x + 2} - \dfrac{6x}{x^2 - 4}$.

▷

REMARK Recall that the least common denominator of two or more fractions consists of the product of all prime factors in the denominators, with each factor given the highest power of its occurrence in any denominator. For instance, in Example 3, by factoring each denominator you can determine that the LCD is $(x + 2)(x - 2)$.

Solution   The LCD is $x^2 - 4 = (x + 2)(x - 2)$. Multiply each term by this LCD.

$$\frac{1}{x - 2}(x + 2)(x - 2) = \frac{3}{x + 2}(x + 2)(x - 2) - \frac{6x}{x^2 - 4}(x + 2)(x - 2)$$

$$x + 2 = 3(x - 2) - 6x, \quad x \neq \pm 2$$

$$x + 2 = 3x - 6 - 6x$$

$$x + 2 = -3x - 6$$

$$4x = -8$$

$$x = -2 \qquad\qquad \text{Extraneous solution}$$

In the original equation, $x = -2$ yields a denominator of zero. So, $x = -2$ is an extraneous solution, and the original equation has *no solution.*

✓ **Checkpoint**  Audio-video solution in English & Spanish at LarsonPrecalculus.com.

Solve $\dfrac{3x}{x - 4} = 5 + \dfrac{12}{x - 4}$.

Quadratic Equations

A **quadratic equation** in x is an equation that can be written in the general form

$$ax^2 + bx + c = 0$$

where a, b, and c are real numbers with $a \neq 0$. A quadratic equation in x is also called a **second-degree polynomial equation** in x.

You should be familiar with the following four methods of solving quadratic equations.

Solving a Quadratic Equation

Factoring

If $ab = 0$, then $a = 0$ or $b = 0$.

Example: $\qquad\qquad\qquad x^2 - x - 6 = 0$

$$(x - 3)(x + 2) = 0$$

$$x - 3 = 0 \implies x = 3$$

$$x + 2 = 0 \implies x = -2$$

Square Root Principle

If $u^2 = c$, where $c > 0$, then $u = \pm\sqrt{c}$.

Example: $\qquad\qquad (x + 3)^2 = 16$

$$x + 3 = \pm 4$$

$$x = -3 \pm 4$$

$$x = 1 \quad \text{or} \quad x = -7$$

Completing the Square

If $x^2 + bx = c$, then

$$x^2 + bx + \left(\frac{b}{2}\right)^2 = c + \left(\frac{b}{2}\right)^2 \qquad \text{Add } \left(\frac{b}{2}\right)^2 \text{ to each side.}$$

$$\left(x + \frac{b}{2}\right)^2 = c + \frac{b^2}{4}.$$

Example: $\qquad\qquad\qquad x^2 + 6x = 5$

$$x^2 + 6x + 3^2 = 5 + 3^2 \qquad \text{Add } \left(\frac{6}{2}\right)^2 \text{ to each side.}$$

$$(x + 3)^2 = 14$$

$$x + 3 = \pm\sqrt{14}$$

$$x = -3 \pm \sqrt{14}$$

Quadratic Formula

If $ax^2 + bx + c = 0$, then $x = \dfrac{-b \pm \sqrt{b^2 - 4ac}}{2a}$.

Example: $\qquad 2x^2 + 3x - 1 = 0$

$$x = \frac{-3 \pm \sqrt{3^2 - 4(2)(-1)}}{2(2)}$$

$$= \frac{-3 \pm \sqrt{17}}{4}$$

•• **REMARK** The Square Root Principle is also referred to as *extracting square roots.*

•• **REMARK** You can solve every quadratic equation by completing the square or using the Quadratic Formula.

| EXAMPLE 4 | **Solving a Quadratic Equation by Factoring** |

a. $2x^2 + 9x + 7 = 3$ Original equation

$2x^2 + 9x + 4 = 0$ Write in general form.

$(2x + 1)(x + 4) = 0$ Factor.

$2x + 1 = 0$ ⟹ $x = -\frac{1}{2}$ Set 1st factor equal to 0.

$x + 4 = 0$ ⟹ $x = -4$ Set 2nd factor equal to 0.

The solutions are $x = -\frac{1}{2}$ and $x = -4$. Check these in the original equation.

b. $6x^2 - 3x = 0$ Original equation

$3x(2x - 1) = 0$ Factor.

$3x = 0$ ⟹ $x = 0$ Set 1st factor equal to 0.

$2x - 1 = 0$ ⟹ $x = \frac{1}{2}$ Set 2nd factor equal to 0.

The solutions are $x = 0$ and $x = \frac{1}{2}$. Check these in the original equation.

✓ **Checkpoint** *Audio-video solution in English & Spanish at LarsonPrecalculus.com.*

Solve $2x^2 - 3x + 1 = 6$ by factoring.

Note that the method of solution in Example 4 is based on the Zero-Factor Property from Section P.1. This property applies *only* to equations written in general form (in which the right side of the equation is zero). So, all terms must be collected on one side *before* factoring. For instance, in the equation $(x - 5)(x + 2) = 8$, it is *incorrect* to set each factor equal to 8. Try to solve this equation correctly.

| EXAMPLE 5 | **Extracting Square Roots** |

Solve each equation by extracting square roots.

a. $4x^2 = 12$

b. $(x - 3)^2 = 7$

Solution

a. $4x^2 = 12$ Write original equation.

$x^2 = 3$ Divide each side by 4.

$x = \pm\sqrt{3}$ Extract square roots.

The solutions are $x = \sqrt{3}$ and $x = -\sqrt{3}$. Check these in the original equation.

b. $(x - 3)^2 = 7$ Write original equation.

$x - 3 = \pm\sqrt{7}$ Extract square roots.

$x = 3 \pm \sqrt{7}$ Add 3 to each side.

The solutions are $x = 3 \pm \sqrt{7}$. Check these in the original equation.

✓ **Checkpoint** *Audio-video solution in English & Spanish at LarsonPrecalculus.com.*

Solve each equation by extracting square roots.

a. $3x^2 = 36$

b. $(x - 1)^2 = 10$

When solving quadratic equations by completing the square, you must add

$$\left(\frac{b}{2}\right)^2$$

to *each side* in order to maintain equality. When the leading coefficient is *not* 1, you must divide each side of the equation by the leading coefficient *before* completing the square, as shown in Example 7.

EXAMPLE 6   **Completing the Square: Leading Coefficient Is 1**

Solve $x^2 + 2x - 6 = 0$ by completing the square.

Solution

$x^2 + 2x - 6 = 0$	Write original equation.
$x^2 + 2x = 6$	Add 6 to each side.
$x^2 + 2x + 1^2 = 6 + 1^2$	Add 1^2 to each side.

(half of 2)^2

$(x + 1)^2 = 7$	Simplify.
$x + 1 = \pm\sqrt{7}$	Extract square roots.
$x = -1 \pm \sqrt{7}$	Subtract 1 from each side.

The solutions are $x = -1 \pm \sqrt{7}$. Check these in the original equation.

✓ *Checkpoint*  *Audio-video solution in English & Spanish at LarsonPrecalculus.com.*

Solve $x^2 - 4x - 1 = 0$ by completing the square.

EXAMPLE 7   **Completing the Square: Leading Coefficient Is Not 1**

Solve $3x^2 - 4x - 5 = 0$ by completing the square.

Solution

$3x^2 - 4x - 5 = 0$	Write original equation.
$3x^2 - 4x = 5$	Add 5 to each side.
$x^2 - \dfrac{4}{3}x = \dfrac{5}{3}$	Divide each side by 3.
$x^2 - \dfrac{4}{3}x + \left(-\dfrac{2}{3}\right)^2 = \dfrac{5}{3} + \left(-\dfrac{2}{3}\right)^2$	Add $\left(-\frac{2}{3}\right)^2$ to each side.

$\left(\text{half of } -\frac{4}{3}\right)^2$

$\left(x - \dfrac{2}{3}\right)^2 = \dfrac{19}{9}$	Simplify.
$x - \dfrac{2}{3} = \pm\dfrac{\sqrt{19}}{3}$	Extract square roots.
$x = \dfrac{2}{3} \pm \dfrac{\sqrt{19}}{3}$	Add $\frac{2}{3}$ to each side.

✓ *Checkpoint*  *Audio-video solution in English & Spanish at LarsonPrecalculus.com.*

Solve $3x^2 - 10x - 2 = 0$ by completing the square.

EXAMPLE 8   The Quadratic Formula: Two Distinct Solutions

Use the Quadratic Formula to solve

$$x^2 + 3x = 9.$$

Solution

$x^2 + 3x = 9$	Write original equation.
$x^2 + 3x - 9 = 0$	Write in general form.
$x = \dfrac{-b \pm \sqrt{b^2 - 4ac}}{2a}$	Quadratic Formula
$x = \dfrac{-3 \pm \sqrt{(3)^2 - 4(1)(-9)}}{2(1)}$	Substitute $a = 1$, $b = 3$, and $c = -9$.
$x = \dfrac{-3 \pm \sqrt{45}}{2}$	Simplify.
$x = \dfrac{-3 \pm 3\sqrt{5}}{2}$	Simplify.

•••••••••••••••••▷

REMARK When using the Quadratic Formula, remember that *before* applying the formula, you must first write the quadratic equation in general form.

The two solutions are

$$x = \frac{-3 + 3\sqrt{5}}{2} \quad \text{and} \quad x = \frac{-3 - 3\sqrt{5}}{2}.$$

Check these in the original equation.

✓ **Checkpoint** 🔊)) *Audio-video solution in English & Spanish at LarsonPrecalculus.com.*

Use the Quadratic Formula to solve $3x^2 + 2x - 10 = 0.$

EXAMPLE 9   The Quadratic Formula: One Solution

Use the Quadratic Formula to solve $8x^2 - 24x + 18 = 0.$

Solution

$8x^2 - 24x + 18 = 0$	Write original equation
$4x^2 - 12x + 9 = 0$	Divide out common factor of 2.
$x = \dfrac{-b \pm \sqrt{b^2 - 4ac}}{2a}$	Quadratic Formula
$x = \dfrac{-(-12) \pm \sqrt{(-12)^2 - 4(4)(9)}}{2(4)}$	Substitute $a = 4$, $b = -12$, and $c = 9$.
$x = \dfrac{12 \pm \sqrt{0}}{8}$	Simplify.
$x = \dfrac{3}{2}$	Simplify.

This quadratic equation has only one solution: $x = \frac{3}{2}$. Check this in the original equation.

✓ **Checkpoint** 🔊)) *Audio-video solution in English & Spanish at LarsonPrecalculus.com.*

Use the Quadratic Formula to solve $18x^2 - 48x + 32 = 0.$ ∎

Note that you could have solved Example 9 without first dividing out a common factor of 2. Substituting $a = 8$, $b = -24$, and $c = 18$ into the Quadratic Formula produces the same result.

Polynomial Equations of Higher Degree

The methods used to solve quadratic equations can sometimes be extended to solve polynomial equations of higher degree.

EXAMPLE 10 **Solving a Polynomial Equation by Factoring**

Solve $3x^4 = 48x^2$.

Solution First write the polynomial equation in general form with zero on one side. Then factor the other side, set each factor equal to zero, and solve.

$$3x^4 = 48x^2 \qquad \text{Write original equation.}$$

$$3x^4 - 48x^2 = 0 \qquad \text{Write in general form.}$$

$$3x^2(x^2 - 16) = 0 \qquad \text{Factor out common factor.}$$

$$3x^2(x + 4)(x - 4) = 0 \qquad \text{Write in factored form.}$$

$$3x^2 = 0 \implies x = 0 \qquad \text{Set 1st factor equal to 0.}$$

$$x + 4 = 0 \implies x = -4 \qquad \text{Set 2nd factor equal to 0.}$$

$$x - 4 = 0 \implies x = 4 \qquad \text{Set 3rd factor equal to 0.}$$

You can check these solutions by substituting in the original equation, as follows.

Check

$$3(0)^4 = 48(0)^2 \qquad \text{0 checks. } \checkmark$$

$$3(-4)^4 = 48(-4)^2 \qquad -4 \text{ checks. } \checkmark$$

$$3(4)^4 = 48(4)^2 \qquad 4 \text{ checks. } \checkmark$$

So, the solutions are

$$x = 0, \quad x = -4, \quad \text{and} \quad x = 4.$$

✓ *Checkpoint*  *Audio-video solution in English & Spanish at LarsonPrecalculus.com.*

Solve $9x^4 - 12x^2 = 0$.

EXAMPLE 11 **Solving a Polynomial Equation by Factoring**

Solve $x^3 - 3x^2 - 3x + 9 = 0$.

Solution

$$x^3 - 3x^2 - 3x + 9 = 0 \qquad \text{Write original equation.}$$

$$x^2(x - 3) - 3(x - 3) = 0 \qquad \text{Factor by grouping.}$$

$$(x - 3)(x^2 - 3) = 0 \qquad \text{Distributive Property}$$

$$x - 3 = 0 \implies x = 3 \qquad \text{Set 1st factor equal to 0.}$$

$$x^2 - 3 = 0 \implies x = \pm\sqrt{3} \qquad \text{Set 2nd factor equal to 0.}$$

The solutions are $x = 3$, $x = \sqrt{3}$, and $x = -\sqrt{3}$. Check these in the original equation.

✓ *Checkpoint*  *Audio-video solution in English & Spanish at LarsonPrecalculus.com.*

Solve each equation.

a. $x^3 - 5x^2 - 2x + 10 = 0$

b. $6x^3 - 27x^2 - 54x = 0$

Radical Equations

A **radical equation** is an equation that involves one or more radical expressions.

EXAMPLE 12   **Solving Radical Equations**

a. $\sqrt{2x + 7} - x = 2$ Original equation

$\qquad \sqrt{2x + 7} = x + 2$ Isolate radical.

$\qquad 2x + 7 = x^2 + 4x + 4$ Square each side.

$\qquad 0 = x^2 + 2x - 3$ Write in general form.

$\qquad 0 = (x + 3)(x - 1)$ Factor.

$x + 3 = 0 \implies x = -3$ Set 1st factor equal to 0.

$x - 1 = 0 \implies x = 1$ Set 2nd factor equal to 0.

By checking these values, you can determine that the only solution is $x = 1$.

b. $\sqrt{2x - 5} - \sqrt{x - 3} = 1$ Original equation

$\qquad \sqrt{2x - 5} = \sqrt{x - 3} + 1$ Isolate $\sqrt{2x - 5}$.

$\qquad 2x - 5 = x - 3 + 2\sqrt{x - 3} + 1$ Square each side.

$\qquad 2x - 5 = x - 2 + 2\sqrt{x - 3}$ Combine like terms.

$\qquad x - 3 = 2\sqrt{x - 3}$ Isolate $2\sqrt{x - 3}$.

$\qquad x^2 - 6x + 9 = 4(x - 3)$ Square each side.

$\qquad x^2 - 10x + 21 = 0$ Write in general form.

$\qquad (x - 3)(x - 7) = 0$ Factor.

$x - 3 = 0 \implies x = 3$ Set 1st factor equal to 0.

$x - 7 = 0 \implies x = 7$ Set 2nd factor equal to 0.

The solutions are $x = 3$ and $x = 7$. Check these in the original equation.

✓ *Checkpoint*  *Audio-video solution in English & Spanish at LarsonPrecalculus.com.*

Solve $-\sqrt{40 - 9x} + 2 = x$.

EXAMPLE 13   **Solving an Equation Involving a Rational Exponent**

Solve $(x - 4)^{2/3} = 25$.

Solution

$\qquad (x - 4)^{2/3} = 25$ Write original equation.

$\qquad \sqrt[3]{(x - 4)^2} = 25$ Rewrite in radical form.

$\qquad (x - 4)^2 = 15{,}625$ Cube each side.

$\qquad x - 4 = \pm 125$ Extract square roots.

$\qquad x = 129, \ x = -121$ Add 4 to each side.

✓ *Checkpoint*  *Audio-video solution in English & Spanish at LarsonPrecalculus.com.*

Solve $(x - 5)^{2/3} = 16$.

Absolute Value Equations

An **absolute value equation** is an equation that involves one or more absolute value expressions. To solve an absolute value equation, remember that the expression inside the absolute value bars can be positive or negative. This results in *two* separate equations, each of which must be solved. For instance, the equation

$$|x - 2| = 3$$

results in the two equations $x - 2 = 3$ and $-(x - 2) = 3$, which implies that the equation has two solutions: $x = 5$ and $x = -1$.

EXAMPLE 14   **Solving an Absolute Value Equation**

Solve $|x^2 - 3x| = -4x + 6$.

Solution   Because the variable expression inside the absolute value bars can be positive or negative, you must solve the following two equations.

First Equation

$x^2 - 3x = -4x + 6$	Use positive expression.
$x^2 + x - 6 = 0$	Write in general form.
$(x + 3)(x - 2) = 0$	Factor.
$x + 3 = 0 \implies x = -3$	Set 1st factor equal to 0.
$x - 2 = 0 \implies x = 2$	Set 2nd factor equal to 0.

Second Equation

$-(x^2 - 3x) = -4x + 6$	Use negative expression.
$x^2 - 7x + 6 = 0$	Write in general form.
$(x - 1)(x - 6) = 0$	Factor.
$x - 1 = 0 \implies x = 1$	Set 1st factor equal to 0.
$x - 6 = 0 \implies x = 6$	Set 2nd factor equal to 0.

Check the values in the original equation to determine that the only solutions are $x = -3$ and $x = 1$.

✓ Checkpoint  Audio-video solution in English & Spanish at LarsonPrecalculus.com.

Solve $|x^2 + 4x| = 5x + 12$.

Summarize   (Section P.2)

1. State the definition of an identity, a conditional equation, and a contradiction *(page 14)*.

2. State the definition of a linear equation in one variable *(page 14)*. For examples of solving linear equations in one variable and rational equations that lead to linear equations, see Examples 1–3.

3. List the four methods of solving quadratic equations discussed in this section *(page 17)*. For examples of solving quadratic equations, see Examples 4–9.

4. Explain how to solve polynomial equations of degree three or greater *(page 21)*, radical equations *(page 22)*, and absolute value equations *(page 23)*. For examples of solving these types of equations, see Examples 10–14.

P.2 Exercises

See **CalcChat.com** for tutorial help and worked-out solutions to odd-numbered exercises.

Vocabulary: Fill in the blanks.

1. An _____ is a statement that equates two algebraic expressions.
2. A linear equation in one variable x is an equation that can be written in the standard form _____.
3. An _____ solution is a solution that does not satisfy the original equation.
4. Four methods that can be used to solve a quadratic equation are _____, extracting _____ _____, _____ the _____, and the _____ _____.

Skills and Applications

Solving a Linear Equation In Exercises 5–12, solve the equation and check your solution. (If not possible, explain why.)

5. $x + 11 = 15$
6. $7 - x = 19$
7. $7 - 2x = 25$
8. $3x - 5 = 2x + 7$
9. $4y + 2 - 5y = 7 - 6y$
10. $0.25x + 0.75(10 - x) = 3$
11. $x - 3(2x + 3) = 8 - 5x$
12. $9x - 10 = 5x + 2(2x - 5)$

Solving a Rational Equation In Exercises 13–24, solve the equation and check your solution. (If not possible, explain why.)

13. $\dfrac{3x}{8} - \dfrac{4x}{3} = 4$

14. $\dfrac{5x}{4} + \dfrac{1}{2} = x - \dfrac{1}{2}$

15. $\dfrac{5x - 4}{5x + 4} = \dfrac{2}{3}$

16. $\dfrac{10x + 3}{5x + 6} = \dfrac{1}{2}$

17. $10 - \dfrac{13}{x} = 4 + \dfrac{5}{x}$

18. $\dfrac{1}{x} + \dfrac{2}{x - 5} = 0$

19. $\dfrac{x}{x + 4} + \dfrac{4}{x + 4} + 2 = 0$

20. $\dfrac{7}{2x + 1} - \dfrac{8x}{2x - 1} = -4$

21. $\dfrac{2}{(x - 4)(x - 2)} = \dfrac{1}{x - 4} + \dfrac{2}{x - 2}$

22. $\dfrac{4}{x - 1} + \dfrac{6}{3x + 1} = \dfrac{15}{3x + 1}$

23. $\dfrac{1}{x - 3} + \dfrac{1}{x + 3} = \dfrac{10}{x^2 - 9}$

24. $\dfrac{1}{x - 2} + \dfrac{3}{x + 3} = \dfrac{4}{x^2 + x - 6}$

Solving a Quadratic Equation by Factoring In Exercises 25–34, solve the quadratic equation by factoring.

25. $6x^2 + 3x = 0$
26. $9x^2 - 1 = 0$
27. $x^2 - 2x - 8 = 0$
28. $x^2 - 10x + 9 = 0$
29. $x^2 + 10x + 25 = 0$
30. $4x^2 + 12x + 9 = 0$
31. $x^2 + 4x = 12$
32. $-x^2 + 8x = 12$
33. $\frac{3}{4}x^2 + 8x + 20 = 0$
34. $\frac{1}{8}x^2 - x - 16 = 0$

Extracting Square Roots In Exercises 35–42, solve the equation by extracting square roots. When a solution is irrational, list both the exact solution *and* its approximation rounded to two decimal places.

35. $x^2 = 49$
36. $x^2 = 32$
37. $3x^2 = 81$
38. $9x^2 = 36$
39. $(x - 12)^2 = 16$
40. $(x + 9)^2 = 24$
41. $(2x - 1)^2 = 18$
42. $(x - 7)^2 = (x + 3)^2$

Completing the Square In Exercises 43–50, solve the quadratic equation by completing the square.

43. $x^2 + 4x - 32 = 0$
44. $x^2 - 2x - 3 = 0$
45. $x^2 + 6x + 2 = 0$
46. $x^2 + 8x + 14 = 0$
47. $9x^2 - 18x = -3$
48. $7 + 2x - x^2 = 0$
49. $2x^2 + 5x - 8 = 0$
50. $3x^2 - 4x - 7 = 0$

Using the Quadratic Formula In Exercises 51–64, use the Quadratic Formula to solve the equation.

51. $2x^2 + x - 1 = 0$
52. $2x^2 - x - 1 = 0$
53. $2 + 2x - x^2 = 0$
54. $x^2 - 10x + 22 = 0$
55. $2x^2 - 3x - 4 = 0$
56. $3x + x^2 - 1 = 0$
57. $12x - 9x^2 = -3$
58. $9x^2 - 37 = 6x$
59. $9x^2 + 30x + 25 = 0$
60. $28x - 49x^2 = 4$
61. $8t = 5 + 2t^2$
62. $25h^2 + 80h + 61 = 0$
63. $(y - 5)^2 = 2y$
64. $(z + 6)^2 = -2z$

Using the Quadratic Formula   In Exercises 65–68, use the Quadratic Formula to solve the equation. (Round your answer to three decimal places.)

65. $5.1x^2 - 1.7x - 3.2 = 0$

66. $-0.005x^2 + 0.101x - 0.193 = 0$

67. $422x^2 - 506x - 347 = 0$

68. $-3.22x^2 - 0.08x + 28.651 = 0$

Choosing a Method   In Exercises 69–76, solve the equation using any convenient method.

69. $x^2 - 2x - 1 = 0$ **70.** $11x^2 + 33x = 0$

71. $(x + 3)^2 = 81$ **72.** $x^2 - 14x + 49 = 0$

73. $x^2 - x - \frac{11}{4} = 0$

74. $x^2 + 3x - \frac{3}{4} = 0$

75. $(x + 1)^2 = x^2$

76. $3x + 4 = 2x^2 - 7$

Solving a Polynomial Equation   In Exercises 77–80, solve the equation. Check your solutions.

77. $6x^4 - 14x^2 = 0$

78. $36x^3 - 100x = 0$

79. $5x^3 + 30x^2 + 45x = 0$

80. $x^3 - 3x^2 - x = -3$

Solving a Radical Equation   In Exercises 81–88, solve the equation. Check your solutions.

81. $\sqrt{3x} - 12 = 0$

82. $\sqrt{x - 10} - 4 = 0$

83. $\sqrt[3]{2x + 5} + 3 = 0$

84. $\sqrt[3]{3x + 1} - 5 = 0$

85. $-\sqrt{26 - 11x} + 4 = x$

86. $x + \sqrt{31 - 9x} = 5$

87. $\sqrt{x} - \sqrt{x - 5} = 1$

88. $2\sqrt{x + 1} - \sqrt{2x + 3} = 1$

Solving an Equation Involving a Rational Exponent   In Exercises 89–92, solve the equation. Check your solutions.

89. $(x - 5)^{3/2} = 8$ **90.** $(x + 2)^{2/3} = 9$

91. $(x^2 - 5)^{3/2} = 27$

92. $(x^2 - x - 22)^{3/2} = 27$

Solving an Absolute Value Equation   In Exercises 93–96, solve the equation. Check your solutions.

93. $|2x - 5| = 11$

94. $|3x + 2| = 7$

95. $|x^2 + 6x| = 3x + 18$

96. $|x - 15| = x^2 - 15x$

• • **Forensics** • • • • • × • • • • • ○ • • • • • • •

In Exercises 97 and 98, use the following information. The relationship between the length of an adult's femur (thigh bone) and the height of the adult can be approximated by the linear equations

$y = 0.432x - 10.44$ Female

$y = 0.449x - 12.15$ Male

where y is the length of the femur in inches and x is the height of the adult in inches (see figure).

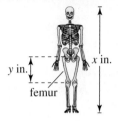

97. A crime scene investigator discovers a femur belonging to an adult human female. The bone is 18 inches long. Estimate the height of the female.

98. Officials search a forest for a missing man who is 6 feet 2 inches tall. They find an adult male femur that is 21 inches long. Is it possible that the femur belongs to the missing man?

Exploration

True or False?   In Exercises 99–101, determine whether the statement is true or false. Justify your answer.

99. An equation can never have more than one extraneous solution.

100. The equation $2(x - 3) + 1 = 2x - 5$ has no solution.

101. The equation $\sqrt{x + 10} - \sqrt{x - 10} = 0$ has no solution.

102. **HOW DO YOU SEE IT?**   The figure shows a glass cube partially filled with water.

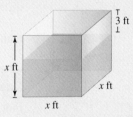

(a) What does the expression $x^2(x - 3)$ represent?

(b) Given $x^2(x - 3) = 320$, explain how you can find the capacity of the cube.

103. Think About It   What is meant by *equivalent* equations? Give an example of two equivalent equations.

P.3   The Cartesian Plane and Graphs of Equations

The Cartesian plane can help you visualize relationships between two variables. For instance, in Exercise 35 on page 37, given how far north and west one city is from another, plotting points to represent the cities can help you visualize these distances and determine the flying distance between the cities.

- Plot points in the Cartesian plane.
- Use the Distance Formula to find the distance between two points.
- Use the Midpoint Formula to find the midpoint of a line segment.
- Use a coordinate plane to model and solve real-life problems.
- Sketch graphs of equations.
- Find x- and y-intercepts of graphs of equations.
- Use symmetry to sketch graphs of equations.
- Write equations of and sketch graphs of circles.

The Cartesian Plane

Just as you can represent real numbers by points on a real number line, you can represent ordered pairs of real numbers by points in a plane called the **rectangular coordinate system,** or the **Cartesian plane,** named after the French mathematician René Descartes (1596–1650).

Two real number lines intersecting at right angles form the Cartesian plane, as shown in Figure P.10. The horizontal real number line is usually called the **x-axis,** and the vertical real number line is usually called the **y-axis.** The point of intersection of these two axes is the **origin,** and the two axes divide the plane into four parts called **quadrants.**

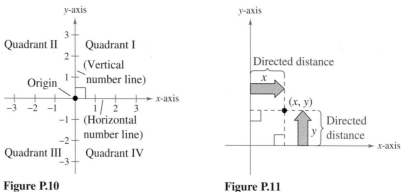

Figure P.10   Figure P.11

Each point in the plane corresponds to an **ordered pair** (x, y) of real numbers x and y, called **coordinates** of the point. The **x-coordinate** represents the directed distance from the y-axis to the point, and the **y-coordinate** represents the directed distance from the x-axis to the point, as shown in Figure P.11.

The notation (x, y) denotes both a point in the plane and an open interval on the real number line. The context will tell you which meaning is intended.

EXAMPLE 1   Plotting Points in the Cartesian Plane

Plot the points $(-1, 2)$, $(3, 4)$, $(0, 0)$, $(3, 0)$, and $(-2, -3)$.

Solution   To plot the point $(-1, 2)$, imagine a vertical line through -1 on the x-axis and a horizontal line through 2 on the y-axis. The intersection of these two lines is the point $(-1, 2)$. Plot the other four points in a similar way, as shown in Figure P.12.

✓ *Checkpoint*))) *Audio-video solution in English & Spanish at LarsonPrecalculus.com.*

Plot the points $(-3, 2)$, $(4, -2)$, $(3, 1)$, $(0, -2)$, and $(-1, -2)$.

Figure P.12

The beauty of a rectangular coordinate system is that it allows you to *see* relationships between two variables. It would be difficult to overestimate the importance of Descartes's introduction of coordinates in the plane. Today, his ideas are in common use in virtually every scientific and business-related field.

EXAMPLE 2   **Sketching a Scatter Plot**

The table shows the numbers N (in millions) of subscribers to a cellular telecommunication service in the United States from 2001 through 2010, where t represents the year. Sketch a scatter plot of the data.   *(Source: CTIA-The Wireless Association)*

Solution   To sketch a *scatter plot* of the data shown in the table, represent each pair of values by an ordered pair (t, N) and plot the resulting points, as shown below. For instance, the ordered pair (2001, 128.4) represents the first pair of values. Note that the break in the t-axis indicates omission of the years before 2001.

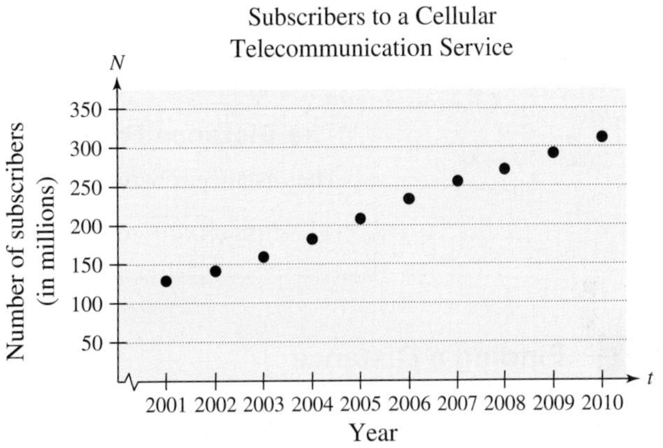

Year, t	Subscribers, N
2001	128.4
2002	140.8
2003	158.7
2004	182.1
2005	207.9
2006	233.0
2007	255.4
2008	270.3
2009	290.9
2010	311.0

Spreadsheet at LarsonPrecalculus.com

Subscribers to a Cellular Telecommunication Service

✓ *Checkpoint* ◀))) Audio-video solution in English & Spanish at LarsonPrecalculus.com.

The table shows the numbers N (in thousands) of cellular telecommunication service employees in the United States from 2001 through 2010, where t represents the year. Sketch a scatter plot of the data.   *(Source: CTIA-The Wireless Association)*

t	N
2001	203.6
2002	192.4
2003	205.6
2004	226.0
2005	233.1
2006	253.8
2007	266.8
2008	268.5
2009	249.2
2010	250.4

Spreadsheet at LarsonPrecalculus.com

▷ **TECHNOLOGY** The scatter plot in Example 2 is only one way to represent the data graphically. You could also represent the data using a bar graph or a line graph. Try using a graphing utility to represent the data given in Example 2 graphically.

In Example 2, you could have let $t = 1$ represent the year 2001. In that case, there would not have been a break in the horizontal axis, and the labels 1 through 10 (instead of 2001 through 2010) would have been on the tick marks.

The Distance Formula

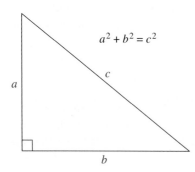

Figure P.13

Recall from the Pythagorean Theorem that, for a right triangle with hypotenuse of length c and sides of lengths a and b, you have

$$a^2 + b^2 = c^2 \qquad \text{Pythagorean Theorem}$$

as shown in Figure P.13. (The converse is also true. That is, if $a^2 + b^2 = c^2$, then the triangle is a right triangle.)

Suppose you want to determine the distance d between two points (x_1, y_1) and (x_2, y_2) in the plane. These two points can form a right triangle, as shown in Figure P.14. The length of the vertical side of the triangle is $|y_2 - y_1|$ and the length of the horizontal side is $|x_2 - x_1|$.

By the Pythagorean Theorem,

$$d^2 = |x_2 - x_1|^2 + |y_2 - y_1|^2$$
$$d = \sqrt{|x_2 - x_1|^2 + |y_2 - y_1|^2}$$
$$= \sqrt{(x_2 - x_1)^2 + (y_2 - y_1)^2}.$$

This result is the **Distance Formula.**

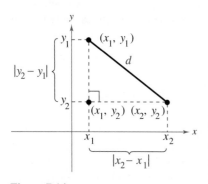

Figure P.14

The Distance Formula

The distance d between the points (x_1, y_1) and (x_2, y_2) in the plane is

$$d = \sqrt{(x_2 - x_1)^2 + (y_2 - y_1)^2}.$$

EXAMPLE 3   **Finding a Distance**

Find the distance between the points $(-2, 1)$ and $(3, 4)$.

Algebraic Solution

Let

$$(x_1, y_1) = (-2, 1) \quad \text{and} \quad (x_2, y_2) = (3, 4).$$

Then apply the Distance Formula.

$$d = \sqrt{(x_2 - x_1)^2 + (y_2 - y_1)^2} \qquad \text{Distance Formula}$$
$$= \sqrt{[3 - (-2)]^2 + (4 - 1)^2} \qquad \text{Substitute for } x_1, y_1, x_2, \text{ and } y_2.$$
$$= \sqrt{(5)^2 + (3)^2} \qquad \text{Simplify.}$$
$$= \sqrt{34} \qquad \text{Simplify.}$$
$$\approx 5.83 \qquad \text{Use a calculator.}$$

So, the distance between the points is about 5.83 units. Use the Pythagorean Theorem to check that the distance is correct.

$$d^2 \overset{?}{=} 5^2 + 3^2 \qquad \text{Pythagorean Theorem}$$
$$\left(\sqrt{34}\right)^2 \overset{?}{=} 5^2 + 3^2 \qquad \text{Substitute for } d.$$
$$34 = 34 \qquad \text{Distance checks. } \checkmark$$

Graphical Solution

Use centimeter graph paper to plot the points $A(-2, 1)$ and $B(3, 4)$. Carefully sketch the line segment from A to B. Then use a centimeter ruler to measure the length of the segment.

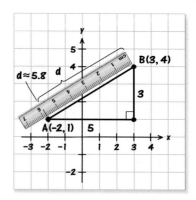

The line segment measures about 5.8 centimeters. So, the distance between the points is about 5.8 units.

✓ **Checkpoint** 🔊))) *Audio-video solution in English & Spanish at LarsonPrecalculus.com.*

Find the distance between the points $(3, 1)$ and $(-3, 0)$.

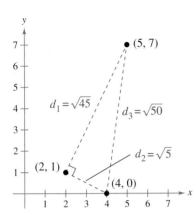

Figure P.15

EXAMPLE 4 Verifying a Right Triangle

Show that the points

$$(2, 1), \quad (4, 0), \quad \text{and} \quad (5, 7)$$

are vertices of a right triangle.

Solution The three points are plotted in Figure P.15. Using the Distance Formula, the lengths of the three sides are as follows.

$$d_1 = \sqrt{(5 - 2)^2 + (7 - 1)^2} = \sqrt{9 + 36} = \sqrt{45}$$

$$d_2 = \sqrt{(4 - 2)^2 + (0 - 1)^2} = \sqrt{4 + 1} = \sqrt{5}$$

$$d_3 = \sqrt{(5 - 4)^2 + (7 - 0)^2} = \sqrt{1 + 49} = \sqrt{50}$$

Because $(d_1)^2 + (d_2)^2 = 45 + 5 = 50 = (d_3)^2$, you can conclude by the Pythagorean Theorem that the triangle must be a right triangle.

✓ **Checkpoint** Audio-video solution in English & Spanish at LarsonPrecalculus.com.

Show that the points $(2, -1)$, $(5, 5)$, and $(6, -3)$ are vertices of a right triangle. ▪

The Midpoint Formula

To find the **midpoint** of the line segment that joins two points in a coordinate plane, you can find the average values of the respective coordinates of the two endpoints using the **Midpoint Formula.**

The Midpoint Formula

The midpoint of the line segment joining the points (x_1, y_1) and (x_2, y_2) is given by the Midpoint Formula

$$\text{Midpoint} = \left(\frac{x_1 + x_2}{2}, \frac{y_1 + y_2}{2} \right).$$

For a proof of the Midpoint Formula, see Proofs in Mathematics on page 118.

EXAMPLE 5 Finding a Line Segment's Midpoint

Find the midpoint of the line segment joining the points

$$(-5, -3) \quad \text{and} \quad (9, 3).$$

Solution Let $(x_1, y_1) = (-5, -3)$ and $(x_2, y_2) = (9, 3)$.

$$\text{Midpoint} = \left(\frac{x_1 + x_2}{2}, \frac{y_1 + y_2}{2} \right) \quad \text{Midpoint Formula}$$

$$= \left(\frac{-5 + 9}{2}, \frac{-3 + 3}{2} \right) \quad \text{Substitute for } x_1, y_1, x_2, \text{ and } y_2.$$

$$= (2, 0) \quad \text{Simplify.}$$

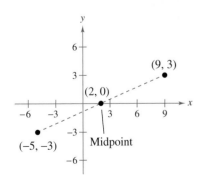

Figure P.16

The midpoint of the line segment is $(2, 0)$, as shown in Figure P.16.

✓ **Checkpoint** Audio-video solution in English & Spanish at LarsonPrecalculus.com.

Find the midpoint of the line segment joining the points $(-2, 8)$ and $(4, -10)$. ▪

Applications

EXAMPLE 6 **Finding the Length of a Pass**

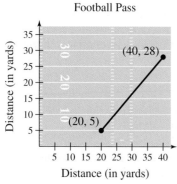

Football Pass

Distance (in yards)

Figure P.17

A football quarterback throws a pass from the 28-yard line, 40 yards from the sideline. A wide receiver catches the pass on the 5-yard line, 20 yards from the same sideline, as shown in Figure P.17. How long is the pass?

Solution You can find the length of the pass by finding the distance between the points (40, 28) and (20, 5).

$$d = \sqrt{(x_2 - x_1)^2 + (y_2 - y_1)^2}$$ Distance Formula

$$= \sqrt{(40 - 20)^2 + (28 - 5)^2}$$ Substitute for x_1, y_1, x_2, and y_2.

$$= \sqrt{20^2 + 23^2}$$ Simplify.

$$= \sqrt{400 + 529}$$ Simplify.

$$= \sqrt{929}$$ Simplify.

$$\approx 30$$ Use a calculator.

So, the pass is about 30 yards long.

✓ **Checkpoint**))) *Audio-video solution in English & Spanish at LarsonPrecalculus.com.*

A football quarterback throws a pass from the 10-yard line, 10 yards from the sideline. A wide receiver catches the pass on the 32-yard line, 25 yards from the same sideline. How long is the pass?

In Example 6, the scale along the goal line does not normally appear on a football field. However, when you use coordinate geometry to solve real-life problems, you are free to place the coordinate system in any way that is convenient for the solution of the problem.

EXAMPLE 7 **Estimating Annual Sales**

Starbucks Corporation had annual sales of approximately $9.8 billion in 2009 and $11.7 billion in 2011. Without knowing any additional information, what would you estimate the 2010 sales to have been? *(Source: Starbucks Corporation)*

Solution One solution to the problem is to assume that sales followed a linear pattern. With this assumption, you can estimate the 2010 sales by finding the midpoint of the line segment connecting the points (2009, 9.8) and (2011, 11.7).

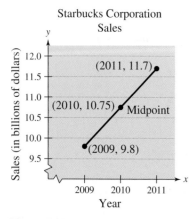

Starbucks Corporation Sales

Figure P.18

$$\text{Midpoint} = \left(\frac{x_1 + x_2}{2}, \frac{y_1 + y_2}{2}\right)$$ Midpoint Formula

$$= \left(\frac{2009 + 2011}{2}, \frac{9.8 + 11.7}{2}\right)$$ Substitute for x_1, x_2, y_1, and y_2.

$$= (2010, 10.75)$$ Simplify.

So, you would estimate the 2010 sales to have been about $10.75 billion, as shown in Figure P.18. (The actual 2010 sales were about $10.71 billion.)

✓ **Checkpoint**))) *Audio-video solution in English & Spanish at LarsonPrecalculus.com.*

Yahoo! Inc. had annual revenues of approximately $7.2 billion in 2008 and $6.3 billion in 2010. Without knowing any additional information, what would you estimate the 2009 revenue to have been? *(Source: Yahoo! Inc.)*

The Graph of an Equation

Earlier in this section, you used a coordinate system to represent graphically the relationship between two quantities. There, the graphical picture consisted of a collection of points in a coordinate plane (see Example 2).

Frequently, a relationship between two quantities is expressed as an **equation in two variables.** For instance, $y = 7 - 3x$ is an equation in x and y. An ordered pair (a, b) is a **solution** or **solution point** of an equation in x and y when the substitutions $x = a$ and $y = b$ result in a true statement. For instance, $(1, 4)$ is a solution of $y = 7 - 3x$ because $4 = 7 - 3(1)$ is a true statement.

In the remainder of this section, you will review some basic procedures for sketching the graph of an equation in two variables. The **graph of an equation** is the set of all points that are solutions of the equation. The basic technique used for sketching the graph of an equation is the **point-plotting method.** To sketch a graph using the point-plotting method, first, when possible, isolate one of the variables. Next, construct a table of values showing several solution points. Then, plot the points from your table in a rectangular coordinate system. Finally, connect the points with a smooth curve or line.

EXAMPLE 8   **Sketching the Graph of an Equation**

Sketch the graph of

$$y = x^2 - 2.$$

Solution

Because the equation is already solved for y, begin by constructing a table of values.

x	-2	-1	0	1	2	3
$y = x^2 - 2$	2	-1	-2	-1	2	7
(x, y)	$(-2, 2)$	$(-1, -1)$	$(0, -2)$	$(1, -1)$	$(2, 2)$	$(3, 7)$

Next, plot the points given in the table, as shown in Figure P.19. Finally, connect the points with a smooth curve, as shown in Figure P.20.

> • • REMARK   One of your goals in this course is to learn to classify the basic shape of a graph from its equation. For instance, you will learn that a *linear equation* has the form
>
> $$y = mx + b$$
>
> and its graph is a line. Similarly, the *quadratic equation* in Example 8 has the form
>
> $$y = ax^2 + bx + c$$
>
> and its graph is a parabola.

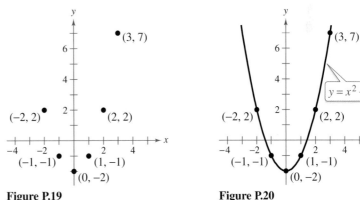

Figure P.19   Figure P.20

✓ **Checkpoint** 🔊))) *Audio-video solution in English & Spanish at LarsonPrecalculus.com.*

Sketch the graph of each equation.

a. $y = x^2 + 3$

b. $y = 1 - x^2$

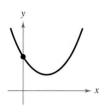

No *x*-intercepts; one *y*-intercept

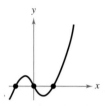

Three *x*-intercepts; one *y*-intercept

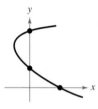

One *x*-intercept; two *y*-intercepts

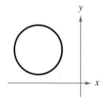

No intercepts
Figure P.21

▷ **TECHNOLOGY** To graph an equation involving *x* and *y* on a graphing utility, use the following procedure.

1. Rewrite the equation so that *y* is isolated on the left side.
2. Enter the equation into the graphing utility.
3. Determine a *viewing window* that shows all important features of the graph.
4. Graph the equation.

Intercepts of a Graph

It is often easy to determine the solution points that have zero as either the *x*-coordinate or the *y*-coordinate. These points are called **intercepts** because they are the points at which the graph intersects or touches the *x*- or *y*-axis. It is possible for a graph to have no intercepts, one intercept, or several intercepts, as shown in Figure P.21.

Note that an *x*-intercept can be written as the ordered pair $(a, 0)$ and a *y*-intercept can be written as the ordered pair $(0, b)$. Some texts denote the *x*-intercept as the *x*-coordinate of the point $(a, 0)$ [and the *y*-intercept as the *y*-coordinate of the point $(0, b)$] rather than the point itself. Unless it is necessary to make a distinction, the term *intercept* will refer to either the point or the coordinate.

> **Finding Intercepts**
>
> 1. To find *x*-intercepts, let *y* be zero and solve the equation for *x*.
> 2. To find *y*-intercepts, let *x* be zero and solve the equation for *y*.

EXAMPLE 9   **Finding *x*- and *y*-Intercepts**

To find the *x*-intercepts of the graph of $y = x^3 - 4x$, let $y = 0$. Then $0 = x^3 - 4x = x(x^2 - 4)$ has solutions $x = 0$ and $x = \pm2$.

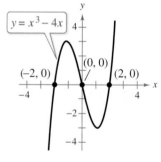

   x-intercepts: $(0, 0), (2, 0), (-2, 0)$ See figure.

To find the *y*-intercept of the graph of $y = x^3 - 4x$, let $x = 0$. Then $y = (0)^3 - 4(0)$ has one solution, $y = 0$.

   y-intercept: $(0, 0)$ See figure.

✓ **Checkpoint** ◄))) *Audio-video solution in English & Spanish at LarsonPrecalculus.com.*

Find the *x*- and *y*-intercepts of the graph of $y = -x^2 - 5x$ shown in the figure below.

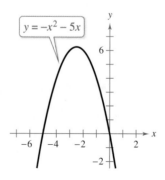

Symmetry

Graphs of equations can have **symmetry** with respect to one of the coordinate axes or with respect to the origin. Symmetry with respect to the x-axis means that when the Cartesian plane is folded along the x-axis, the portion of the graph above the x-axis coincides with the portion below the x-axis. Symmetry with respect to the y-axis or the origin can be described in a similar manner, as shown below.

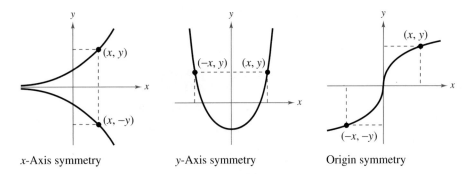

x-Axis symmetry y-Axis symmetry   Origin symmetry

Knowing the symmetry of a graph *before* attempting to sketch it is helpful, because then you need only half as many solution points to sketch the graph.

Graphical Tests for Symmetry

1. A graph is **symmetric with respect to the x-axis** if, whenever (x, y) is on the graph, $(x, -y)$ is also on the graph.

2. A graph is **symmetric with respect to the y-axis** if, whenever (x, y) is on the graph, $(-x, y)$ is also on the graph.

3. A graph is **symmetric with respect to the origin** if, whenever (x, y) is on the graph, $(-x, -y)$ is also on the graph.

EXAMPLE 10   **Testing for Symmetry**

The graph of $y = x^2 - 2$ is symmetric with respect to the y-axis because the point $(-x, y)$ is also on the graph of $y = x^2 - 2$. (See figure.) The table below confirms that the graph is symmetric with respect to the y-axis.

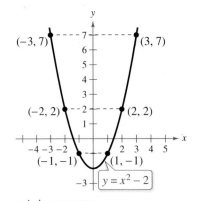

y-Axis symmetry

x	-3	-2	-1
y	7	2	-1
(x, y)	$(-3, 7)$	$(-2, 2)$	$(-1, -1)$

x	1	2	3
y	-1	2	7
(x, y)	$(1, -1)$	$(2, 2)$	$(3, 7)$

✓ **Checkpoint**   *Audio-video solution in English & Spanish at LarsonPrecalculus.com.*

Determine the symmetry of the graph of $y^2 = 6 - x$.

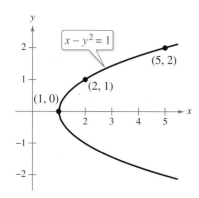

Figure P.22

> **Algebraic Tests for Symmetry**
>
> 1. The graph of an equation is symmetric with respect to the *x*-axis when replacing *y* with −*y* yields an equivalent equation.
>
> 2. The graph of an equation is symmetric with respect to the *y*-axis when replacing *x* with −*x* yields an equivalent equation.
>
> 3. The graph of an equation is symmetric with respect to the origin when replacing *x* with −*x* and *y* with −*y* yields an equivalent equation.

EXAMPLE 11   **Using Symmetry as a Sketching Aid**

Use symmetry to sketch the graph of $x - y^2 = 1$.

Solution   Of the three tests for symmetry, the only one that is satisfied is the test for *x*-axis symmetry because $x - (-y)^2 = 1$ is equivalent to $x - y^2 = 1$. So, the graph is symmetric with respect to the *x*-axis. Using symmetry, you only need to find the solution points above the *x*-axis and then reflect them to obtain the graph, as shown in Figure P.22.

✓ **Checkpoint** ◀))) *Audio-video solution in English & Spanish at LarsonPrecalculus.com.*

Use symmetry to sketch the graph of $y = x^2 - 4$.

EXAMPLE 12   **Sketching the Graph of an Equation**

Sketch the graph of $y = |x - 1|$.

Solution   This equation fails all three tests for symmetry, and consequently its graph is not symmetric with respect to either axis or to the origin. The absolute value bars indicate that *y* is always nonnegative. Construct a table of values. Then plot and connect the points, as shown in Figure P.23. From the table, you can see that $x = 0$ when $y = 1$. So, the *y*-intercept is $(0, 1)$. Similarly, $y = 0$ when $x = 1$. So, the *x*-intercept is $(1, 0)$.

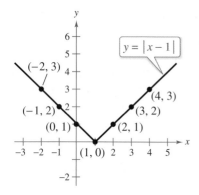

Figure P.23

x	-2	-1	0	1	2	3	4
$y = |x - 1|$	3	2	1	0	1	2	3
(x, y)	$(-2, 3)$	$(-1, 2)$	$(0, 1)$	$(1, 0)$	$(2, 1)$	$(3, 2)$	$(4, 3)$

✓ **Checkpoint** ◀))) *Audio-video solution in English & Spanish at LarsonPrecalculus.com.*

Sketch the graph of $y = |x - 2|$.

Circles

Throughout this course, you will learn to recognize several types of graphs from their equations. For instance, you will learn to recognize that the graph of a second-degree equation of the form $y = ax^2 + bx + c$ is a parabola (see Example 8). The graph of a **circle** is also easy to recognize.

Consider the circle shown in Figure P.24. A point (x, y) lies on the circle if and only if its distance from the center (h, k) is *r*. By the Distance Formula,

$$\sqrt{(x - h)^2 + (y - k)^2} = r.$$

By squaring each side of this equation, you obtain the **standard form of the equation of a circle.**

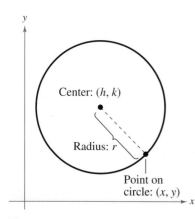

Figure P.24

Standard Form of the Equation of a Circle

A point (x, y) lies on the circle of **radius** r and **center** (h, k) if and only if

$$(x - h)^2 + (y - k)^2 = r^2.$$

REMARK Be careful when you are finding h and k from the standard form of the equation of a circle. For instance, to find h and k from the equation of the circle in Example 13, rewrite the quantities $(x + 1)^2$ and $(y - 2)^2$ using subtraction.

$$(x + 1)^2 = [x - (-1)]^2,$$

$$(y - 2)^2 = [y - (2)]^2$$

So, $h = -1$ and $k = 2$.

From this result, you can see that the standard form of the equation of a circle *with its center at the origin*, $(h, k) = (0, 0)$, is simply $x^2 + y^2 = r^2$.

EXAMPLE 13 **Writing the Equation of a Circle**

The point $(3, 4)$ lies on a circle whose center is at $(-1, 2)$, as shown in Figure P.25. Write the standard form of the equation of this circle.

Solution

The radius of the circle is the distance between $(-1, 2)$ and $(3, 4)$.

$$r = \sqrt{(x - h)^2 + (y - k)^2} \qquad \text{Distance Formula}$$

$$= \sqrt{[3 - (-1)]^2 + (4 - 2)^2} \qquad \text{Substitute for } x, y, h, \text{ and } k.$$

$$= \sqrt{20} \qquad \text{Radius}$$

Using $(h, k) = (-1, 2)$ and $r = \sqrt{20}$, the equation of the circle is

$$(x - h)^2 + (y - k)^2 = r^2 \qquad \text{Equation of circle}$$

$$[x - (-1)]^2 + (y - 2)^2 = \left(\sqrt{20}\right)^2 \qquad \text{Substitute for } h, k, \text{ and } r.$$

$$(x + 1)^2 + (y - 2)^2 = 20. \qquad \text{Standard form}$$

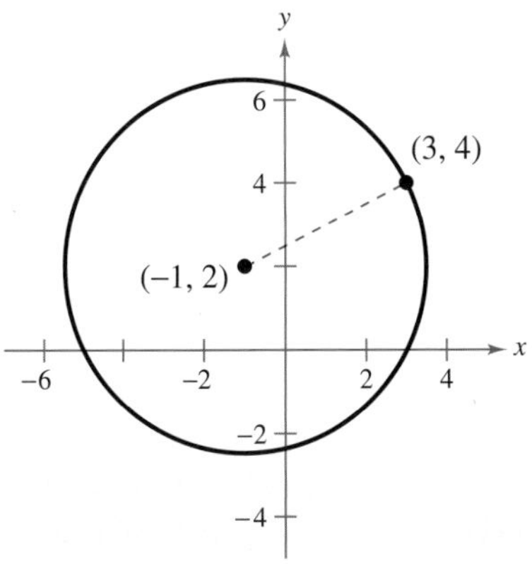

 ✓ ***Checkpoint*** ◆))) *Audio-video solution in English & Spanish at LarsonPrecalculus.com.*

The point $(1, -2)$ lies on a circle whose center is at $(-3, -5)$. Write the standard form of the equation of this circle.

Figure P.25

Summarize (Section P.3)

1. Describe the Cartesian plane *(page 26)*. For an example of plotting points in the Cartesian plane, see Example 1.

2. State the Distance Formula *(page 28)*. For examples of using the Distance Formula to find the distance between two points, see Examples 3 and 4.

3. State the Midpoint Formula *(page 29)*. For an example of using the Midpoint Formula to find the midpoint of a line segment, see Example 5.

4. Describe examples of how to use a coordinate plane to model and solve real-life problems *(page 30, Examples 6 and 7)*.

5. Describe how to sketch the graph of an equation *(page 31)*. For an example of sketching the graph of an equation, see Example 8.

6. Describe how to find the *x*- and *y*-intercepts of a graph *(page 32)*. For an example of finding *x*- and *y*-intercepts, see Example 9.

7. Describe how to use symmetry to graph an equation *(pages 33 and 34)*. For an example of using symmetry to graph an equation, see Example 11.

8. State the standard form of the equation of a circle *(page 35)*. For an example of writing the standard form of the equation of a circle, see Example 13.

P.3   Exercises

See CalcChat.com for tutorial help and worked-out solutions to odd-numbered exercises.

Vocabulary: **Fill in the blanks.**

1. An ordered pair of real numbers can be represented in a plane called the rectangular coordinate system or the _____ plane.

2. The _____ _____ is a result derived from the Pythagorean Theorem.

3. Finding the average values of the representative coordinates of the two endpoints of a line segment in a coordinate plane is also known as using the _____ _____.

4. An ordered pair (a, b) is a _____ of an equation in x and y when the substitutions $x = a$ and $y = b$ result in a true statement.

5. The set of all solution points of an equation is the _____ of the equation.

6. The points at which a graph intersects or touches an axis are called the _____ of the graph.

7. A graph is symmetric with respect to the _____ if, whenever (x, y) is on the graph, $(-x, y)$ is also on the graph.

8. The equation $(x - h)^2 + (y - k)^2 = r^2$ is the standard form of the equation of a _____ with center _____ and radius _____.

Skills and Applications

Approximating Coordinates of Points In Exercises 9 and 10, approximate the coordinates of the points.

9.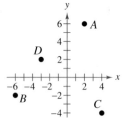

10.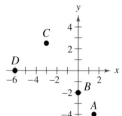

Plotting Points in the Cartesian Plane In Exercises 11 and 12, plot the points in the Cartesian plane.

11. $(-4, 2), (-3, -6), (0, 5), (1, -4), (0, 0), (3, 1)$

12. $\left(1, -\frac{1}{3}\right), (0.5, -1), \left(\frac{3}{7}, 3\right), \left(-\frac{4}{3}, -\frac{3}{7}\right), (-2, 2.5)$

Finding the Coordinates of a Point In Exercises 13 and 14, find the coordinates of the point.

13. The point is located three units to the left of the y-axis and four units above the x-axis.

14. The point is on the x-axis and 12 units to the left of the y-axis.

Determining Quadrant(s) for a Point In Exercises 15–18, determine the quadrant(s) in which (x, y) is located so that the condition(s) is (are) satisfied.

15. $x > 0$ and $y < 0$

16. $x = -4$ and $y > 0$

17. $x < 0$ and $-y > 0$

18. $xy > 0$

Sketching a Scatter Plot In Exercises 19 and 20, sketch a scatter plot of the data.

19. The table shows the number y of Wal-Mart stores for each year x from 2003 through 2010. *(Source: Wal-Mart Stores, Inc.)*

DATA	Year, x	Number of Stores, y
	2003	4906
	2004	5289
	2005	6141
	2006	6779
	2007	7262
	2008	7720
	2009	8416
	2010	8970

Spreadsheet at LarsonPrecalculus.com

20. **Meteorology** The following data points (x, y) represent the lowest temperatures on record y (in degrees Fahrenheit) in Duluth, Minnesota, for each month x, where $x = 1$ represents January. *(Source: NOAA).*

$(1, -39), (2, -39), (3, -29), (4, -5), (5, 17), (6, 27),$
$(7, 35), (8, 32), (9, 22), (10, 8), (11, -23), (12, -34)$

Finding a Distance In Exercises 21–24, find the distance between the points.

21. $(-2, 6), (3, -6)$

22. $(8, 5), (0, 20)$

23. $(1, 4), (-5, -1)$

24. $(9.5, -2.6), (-3.9, 8.2)$

Verifying a Right Triangle   In Exercises 25 and 26, (a) find the length of each side of the right triangle, and (b) show that these lengths satisfy the Pythagorean Theorem.

25.
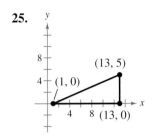

26.
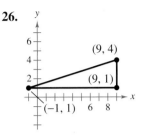

Verifying a Polygon   In Exercises 27–30, show that the points form the vertices of the indicated polygon.

27. Right triangle: $(4, 0), (2, 1), (-1, -5)$

28. Right triangle: $(-1, 3), (3, 5), (5, 1)$

29. Isosceles triangle: $(1, -3), (3, 2), (-2, 4)$

30. Isosceles triangle: $(2, 3), (4, 9), (-2, 7)$

Plotting, Distance, and Midpoint   In Exercises 31–34, (a) plot the points, (b) find the distance between the points, and (c) find the midpoint of the line segment joining the points.

31. $(6, -3), (6, 5)$ 32. $(1, 1), (9, 7)$

33. $(-1, 2), (5, 4)$ 34. $\left(\frac{1}{2}, 1\right), \left(-\frac{5}{2}, \frac{4}{3}\right)$

• • **35. Flying Distance** • • • • • • • • • • • • • •

An airplane flies from Naples, Italy, in a straight line to Rome, Italy, which is 120 kilometers north and 150 kilometers west of Naples. How far does the plane fly?

36. **Sports**   A soccer player passes the ball from a point that is 18 yards from the endline and 12 yards from the sideline. A teammate who is 42 yards from the same endline and 50 yards from the same sideline receives the pass. (See figure.) How long is the pass?

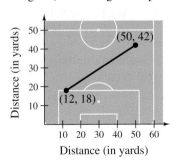

Distance (in yards)

37. **Sales**   The Coca-Cola Company had sales of $19,564 million in 2002 and $35,123 million in 2010. Use the Midpoint Formula to estimate the sales in 2006. Assume that the sales followed a linear pattern. *(Source: The Coca-Cola Company)*

38. **Earnings per Share**   The earnings per share for Big Lots, Inc. were $1.89 in 2008 and $2.83 in 2010. Use the Midpoint Formula to estimate the earnings per share in 2009. Assume that the earnings per share followed a linear pattern.   *(Source: Big Lots, Inc.)*

Determining Solution Points   In Exercises 39–44, determine whether each point lies on the graph of the equation.

Equation	Points	
39. $y = \sqrt{x + 4}$	(a) $(0, 2)$	(b) $(5, 3)$
40. $y = 4 - |x - 2|$	(a) $(1, 5)$	(b) $(6, 0)$
41. $y = x^2 - 3x + 2$	(a) $(2, 0)$	(b) $(-2, 8)$
42. $2x - y - 3 = 0$	(a) $(1, 2)$	(b) $(1, -1)$
43. $x^2 + y^2 = 20$	(a) $(3, -2)$	(b) $(-4, 2)$
44. $y = \frac{1}{3}x^3 - 2x^2$	(a) $\left(2, -\frac{16}{3}\right)$	(b) $(-3, 9)$

Sketching the Graph of an Equation   In Exercises 45–48, complete the table. Use the resulting solution points to sketch the graph of the equation.

45. $y = -2x + 5$

x	-1	0	1	2	$\frac{5}{2}$
y					
(x, y)					

46. $y = \frac{3}{4}x - 1$

x	-2	0	1	$\frac{4}{3}$	2
y					
(x, y)					

47. $y = x^2 - 3x$

x	-1	0	1	2	3
y					
(x, y)					

48. $y = 5 - x^2$

x	-2	-1	0	1	2
y					
(x, y)					

Finding x- and y-Intercepts In Exercises 49–60, find the x- and y-intercepts of the graph of the equation.

49. $y = 16 - 4x^2$

50. $y = (x + 3)^2$

51. $y = 5x - 6$

52. $y = 8 - 3x$

53. $y = \sqrt{x + 4}$

54. $y = \sqrt{2x - 1}$

55. $y = |3x - 7|$

56. $y = -|x + 10|$

57. $y = 2x^3 - 4x^2$

58. $y = x^4 - 25$

59. $y^2 = 6 - x$

60. $y^2 = x + 1$

Using Symmetry as a Sketching Aid In Exercises 61–64, assume that the graph has the indicated type of symmetry. Sketch the complete graph of the equation. To print an enlarged copy of the graph, go to *MathGraphs.com.*

61.
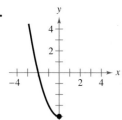
y-Axis symmetry

62.
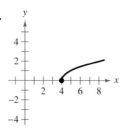
x-Axis symmetry

63.
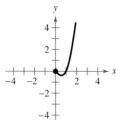
Origin symmetry

64.
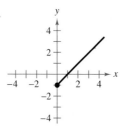
y-Axis symmetry

Testing for Symmetry In Exercises 65–72, use the algebraic tests to check for symmetry with respect to both axes and the origin.

65. $x^2 - y = 0$

66. $x - y^2 = 0$

67. $y = x^3$

68. $y = x^4 - x^2 + 3$

69. $y = \dfrac{x}{x^2 + 1}$

70. $y = \dfrac{1}{x^2 + 1}$

71. $xy^2 + 10 = 0$

72. $xy = 4$

Sketching the Graph of an Equation In Exercises 73–82, identify any intercepts and test for symmetry. Then sketch the graph of the equation.

73. $y = -3x + 1$

74. $y = 2x - 3$

75. $y = x^2 - 2x$

76. $x = y^2 - 1$

77. $y = x^3 + 3$

78. $y = x^3 - 1$

79. $y = \sqrt{x - 3}$

80. $y = \sqrt{1 - x}$

81. $y = |x - 6|$

82. $y = 1 - |x|$

Writing the Equation of a Circle In Exercises 83–88, write the standard form of the equation of the circle with the given characteristics.

83. Center: $(0, 0)$; Radius: 4

84. Center: $(-7, -4)$; Radius: 7

85. Center: $(-1, 2)$; Solution point: $(0, 0)$

86. Center: $(3, -2)$; Solution point: $(-1, 1)$

87. Endpoints of a diameter: $(0, 0), (6, 8)$

88. Endpoints of a diameter: $(-4, -1), (4, 1)$

Sketching the Graph of a Circle In Exercises 89–92, find the center and radius of the circle. Then sketch the graph of the circle.

89. $x^2 + y^2 = 25$

90. $x^2 + (y - 1)^2 = 1$

91. $\left(x - \frac{1}{2}\right)^2 + \left(y - \frac{1}{2}\right)^2 = \frac{9}{4}$

92. $(x - 2)^2 + (y + 3)^2 = \frac{16}{9}$

93. **Depreciation** A hospital purchases a new magnetic resonance imaging (MRI) machine for $500,000. The depreciated value y (reduced value) after t years is given by $y = 500,000 - 40,000t$, $0 \le t \le 8$. Sketch the graph of the equation.

94. **Consumerism** You purchase an all-terrain vehicle (ATV) for $8000. The depreciated value y after t years is given by $y = 8000 - 900t$, $0 \le t \le 6$. Sketch the graph of the equation.

95. **Electronics** The resistance y (in ohms) of 1000 feet of solid copper wire at 68 degrees Fahrenheit is

$$y = \frac{10,370}{x^2}$$

where x is the diameter of the wire in mils (0.001 inch).

(a) Complete the table.

x	5	10	20	30	40	50
y						

x	60	70	80	90	100
y					

(b) Use the table of values in part (a) to sketch a graph of the model. Then use your graph to estimate the resistance when $x = 85.5$.

(c) Use the model to confirm algebraically the estimate you found in part (b).

(d) What can you conclude in general about the relationship between the diameter of the copper wire and the resistance?

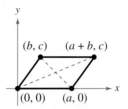

96. Population Statistics   The table shows the life expectancies of a child (at birth) in the United States for selected years from 1930 through 2000.   *(Source: U.S. National Center for Health Statistics)*

Year	Life Expectancy, y
1930	59.7
1940	62.9
1950	68.2
1960	69.7
1970	70.8
1980	73.7
1990	75.4
2000	76.8

DATA

Spreadsheet at LarsonPrecalculus.com

A model for the life expectancy during this period is

$$y = -0.002t^2 + 0.50t + 46.6, \quad 30 \le t \le 100$$

where y represents the life expectancy and t is the time in years, with $t = 30$ corresponding to 1930.

(a) Use a graphing utility to graph the data from the table and the model in the same viewing window. How well does the model fit the data? Explain.

(b) Determine the life expectancy in 1990 both graphically and algebraically.

(c) Use the graph to determine the year when life expectancy was approximately 76.0. Verify your answer algebraically.

(d) One projection for the life expectancy of a child born in 2015 is 78.9. How does this compare with the projection given by the model?

(e) Do you think this model can be used to predict the life expectancy of a child 50 years from now? Explain.

Exploration

True or False?   **In Exercises 97–100, determine whether the statement is true or false. Justify your answer.**

97. In order to divide a line segment into 16 equal parts, you would have to use the Midpoint Formula 16 times.

98. The points $(-8, 4)$, $(2, 11)$, and $(-5, 1)$ represent the vertices of an isosceles triangle.

99. A graph is symmetric with respect to the x-axis if, whenever (x, y) is on the graph, $(-x, y)$ is also on the graph.

100. A graph of an equation can have more than one y-intercept.

101. Think About It   What is the y-coordinate of any point on the x-axis? What is the x-coordinate of any point on the y-axis?

102. Think About It   When plotting points on the rectangular coordinate system, is it true that the scales on the x- and y-axes must be the same? Explain.

103. Proof   Prove that the diagonals of the parallelogram in the figure intersect at their midpoint.

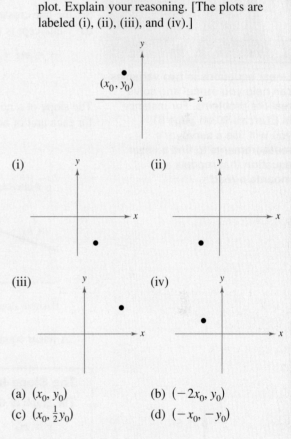

104. HOW DO YOU SEE IT?   Use the plot of the point (x_0, y_0) in the figure. Match the transformation of the point with the correct plot. Explain your reasoning. [The plots are labeled (i), (ii), (iii), and (iv).]

(a) (x_0, y_0)
(b) $(-2x_0, y_0)$
(c) $(x_0, \frac{1}{2}y_0)$
(d) $(-x_0, -y_0)$

105. Using the Midpoint Formula   A line segment has (x_1, y_1) as one endpoint and (x_m, y_m) as its midpoint. Find the other endpoint (x_2, y_2) of the line segment in terms of x_1, y_1, and y_m. Then use the result to find the coordinates of the endpoint of a line segment when the coordinates of the other endpoint and midpoint are, respectively,

(a) $(1, -2)$, $(4, -1)$ and (b) $(-5, 11)$, $(2, 4)$.

The symbol 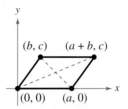 indicates an exercise or a part of an exercise in which you are instructed to use a graphing utility.

P.4 Linear Equations in Two Variables

- Use slope to graph linear equations in two variables.
- Find the slope of a line given two points on the line.
- Write linear equations in two variables.
- Use slope to identify parallel and perpendicular lines.
- Use slope and linear equations in two variables to model and solve real-life problems.

Using Slope

The simplest mathematical model for relating two variables is the **linear equation in two variables** $y = mx + b$. The equation is called *linear* because its graph is a line. (In mathematics, the term *line* means *straight line*.) By letting $x = 0$, you obtain

$$y = m(0) + b = b.$$

So, the line crosses the y-axis at $y = b$, as shown in the figures below. In other words, the y-intercept is $(0, b)$. The steepness or slope of the line is m.

$$y = mx + b$$

Slope ⎵ ⎵ y-Intercept

The **slope** of a nonvertical line is the number of units the line rises (or falls) vertically for each unit of horizontal change from left to right, as shown below.

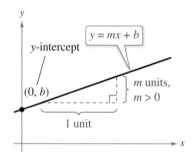

Positive slope, line rises.

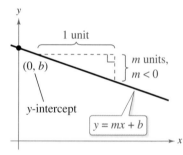

Negative slope, line falls.

A linear equation written in **slope-intercept form** has the form $y = mx + b$.

Linear equations in two variables can help you model and solve real-life problems. For instance, in Exercise 90 on page 51, you will use a surveyor's measurements to find a linear equation that models a mountain road.

The Slope-Intercept Form of the Equation of a Line

The graph of the equation

$$y = mx + b$$

is a line whose slope is m and whose y-intercept is $(0, b)$.

Once you have determined the slope and the y-intercept of a line, it is a relatively simple matter to sketch its graph. In the next example, note that none of the lines is vertical. A vertical line has an equation of the form

$$x = a. \qquad \text{Vertical line}$$

The equation of a vertical line cannot be written in the form $y = mx + b$ because the slope of a vertical line is undefined, as indicated in Figure P.26.

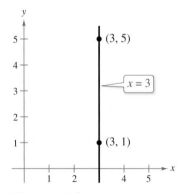

Slope is undefined.

Figure P.26

EXAMPLE 1   **Graphing a Linear Equation**

Sketch the graph of each linear equation.

a. $y = 2x + 1$

b. $y = 2$

c. $x + y = 2$

Solution

a. Because $b = 1$, the y-intercept is $(0, 1)$.
Moreover, because the slope is $m = 2$,
the line *rises* two units for each unit the
line moves to the right.

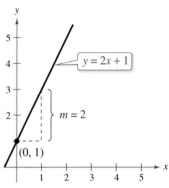

$y = 2x + 1$

$m = 2$

$(0, 1)$

When m is positive, the line rises.

b. By writing this equation in the form
$y = (0)x + 2$, you can see that the
y-intercept is $(0, 2)$ and the slope is
zero. A zero slope implies that the
line is horizontal—that is, it does
not rise *or* fall.

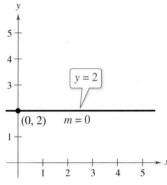

$(0, 2)$ $m = 0$

$y = 2$

When m is 0, the line is horizontal.

c. By writing this equation in
slope-intercept form

$x + y = 2$ Write original equation.

$y = -x + 2$ Subtract x from each side.

$y = (-1)x + 2$ Write in slope-intercept form.

you can see that the y-intercept is $(0, 2)$.
Moreover, because the slope is $m = -1$,
the line *falls* one unit for each unit the
line moves to the right.

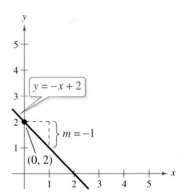

$y = -x + 2$

$m = -1$

$(0, 2)$

When m is negative, the line falls.

✓ *Checkpoint* *Audio-video solution in English & Spanish at LarsonPrecalculus.com.*

Sketch the graph of each linear equation.

a. $y = -3x + 2$

b. $y = -3$

c. $4x + y = 5$

Finding the Slope of a Line

Given an equation of a line, you can find its slope by writing the equation in slope-intercept form. If you are not given an equation, then you can still find the slope of a line. For instance, suppose you want to find the slope of the line passing through the points (x_1, y_1) and (x_2, y_2), as shown below.

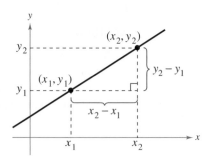

As you move from left to right along this line, a change of $(y_2 - y_1)$ units in the vertical direction corresponds to a change of $(x_2 - x_1)$ units in the horizontal direction.

$$y_2 - y_1 = \text{the change in } y = \text{rise}$$

and

$$x_2 - x_1 = \text{the change in } x = \text{run}$$

The ratio of $(y_2 - y_1)$ to $(x_2 - x_1)$ represents the slope of the line that passes through the points (x_1, y_1) and (x_2, y_2).

$$\text{Slope} = \frac{\text{change in } y}{\text{change in } x} = \frac{\text{rise}}{\text{run}} = \frac{y_2 - y_1}{x_2 - x_1}$$

The Slope of a Line Passing Through Two Points

The **slope** m of the nonvertical line through (x_1, y_1) and (x_2, y_2) is

$$m = \frac{y_2 - y_1}{x_2 - x_1}$$

where $x_1 \neq x_2$.

When using the formula for slope, the *order of subtraction* is important. Given two points on a line, you are free to label either one of them as (x_1, y_1) and the other as (x_2, y_2). However, once you have done this, you must form the numerator and denominator using the same order of subtraction.

$$m = \frac{y_2 - y_1}{x_2 - x_1} \qquad m = \frac{y_1 - y_2}{x_1 - x_2} \qquad m = \frac{y_2 - y_1}{x_1 - x_2}$$

$\quad\quad\quad$ Correct $\quad\quad\quad\quad\quad$ Correct $\quad\quad\quad\quad\quad$ Incorrect

For instance, the slope of the line passing through the points $(3, 4)$ and $(5, 7)$ can be calculated as

$$m = \frac{7 - 4}{5 - 3} = \frac{3}{2}$$

or, reversing the subtraction order in both the numerator and denominator, as

$$m = \frac{4 - 7}{3 - 5} = \frac{-3}{-2} = \frac{3}{2}.$$

EXAMPLE 2   **Finding the Slope of a Line Through Two Points**

Find the slope of the line passing through each pair of points.

a. $(-2, 0)$ and $(3, 1)$ **b.** $(-1, 2)$ and $(2, 2)$

c. $(0, 4)$ and $(1, -1)$ **d.** $(3, 4)$ and $(3, 1)$

Solution

a. Letting $(x_1, y_1) = (-2, 0)$ and $(x_2, y_2) = (3, 1)$, you obtain a slope of

$$m = \frac{y_2 - y_1}{x_2 - x_1} = \frac{1 - 0}{3 - (-2)} = \frac{1}{5}.$$ See Figure P.27.

b. The slope of the line passing through $(-1, 2)$ and $(2, 2)$ is

$$m = \frac{2 - 2}{2 - (-1)} = \frac{0}{3} = 0.$$ See Figure P.28.

c. The slope of the line passing through $(0, 4)$ and $(1, -1)$ is

$$m = \frac{-1 - 4}{1 - 0} = \frac{-5}{1} = -5.$$ See Figure P.29.

d. The slope of the line passing through $(3, 4)$ and $(3, 1)$ is

$$m = \frac{1 - 4}{3 - 3} = \frac{-3}{0}.$$ See Figure P.30.

Because division by 0 is undefined, the slope is undefined and the line is vertical.

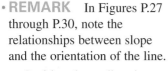

• • **REMARK**   In Figures P.27
through P.30, note the
relationships between slope
and the orientation of the line.

a. Positive slope: line rises
    from left to right

b. Zero slope: line is horizontal

c. Negative slope: line falls
    from left to right

d. Undefined slope: line is
    vertical

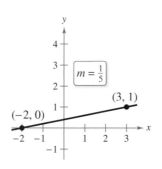

Figure P.27

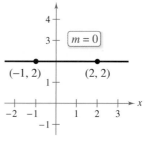

Figure P.28

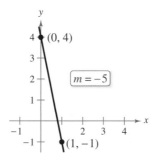

Figure P.29

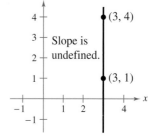

Figure P.30

✓ **Checkpoint** 🔊)))   *Audio-video solution in English & Spanish at LarsonPrecalculus.com.*

Find the slope of the line passing through each pair of points.

a. $(-5, -6)$ and $(2, 8)$ **b.** $(4, 2)$ and $(2, 5)$

c. $(0, 0)$ and $(0, -6)$ **d.** $(0, -1)$ and $(3, -1)$

Writing Linear Equations in Two Variables

If (x_1, y_1) is a point on a line of slope m and (x, y) is *any other* point on the line, then

$$\frac{y - y_1}{x - x_1} = m.$$

This equation involving the variables x and y, rewritten in the form

$$y - y_1 = m(x - x_1)$$

is the **point-slope form** of the equation of a line.

> ### Point-Slope Form of the Equation of a Line
> The equation of the line with slope m passing through the point (x_1, y_1) is
> $$y - y_1 = m(x - x_1).$$

The point-slope form is most useful for *finding* the equation of a line. You should remember this form.

EXAMPLE 3   Using the Point-Slope Form

Find the slope-intercept form of the equation of the line that has a slope of 3 and passes through the point $(1, -2)$.

Solution   Use the point-slope form with $m = 3$ and $(x_1, y_1) = (1, -2)$.

$y - y_1 = m(x - x_1)$	Point-slope form
$y - (-2) = 3(x - 1)$	Substitute for m, x_1, and y_1.
$y + 2 = 3x - 3$	Simplify.
$y = 3x - 5$	Write in slope-intercept form.

The slope-intercept form of the equation of the line is $y = 3x - 5$. Figure P.31 shows the graph of this equation.

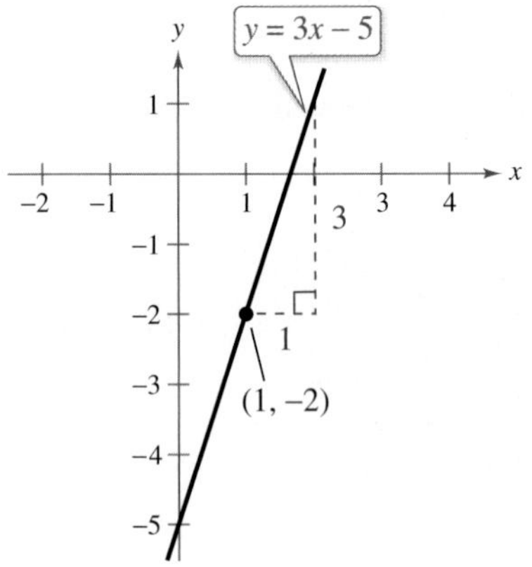

✓ *Checkpoint* *Audio-video solution in English & Spanish at LarsonPrecalculus.com.*

Find the slope-intercept form of the equation of the line that has the given slope and passes through the given point.

a. $m = 2$, $(3, -7)$

b. $m = -\frac{2}{3}$, $(1, 1)$

c. $m = 0$, $(1, 1)$

The point-slope form can be used to find an equation of the line passing through two points (x_1, y_1) and (x_2, y_2). To do this, first find the slope of the line

$$m = \frac{y_2 - y_1}{x_2 - x_1}, \quad x_1 \neq x_2$$

and then use the point-slope form to obtain the equation

$$y - y_1 = \frac{y_2 - y_1}{x_2 - x_1}(x - x_1). \qquad \text{Two-point form}$$

This is sometimes called the **two-point form** of the equation of a line.

Figure P.31

[Graph showing line $y = 3x - 5$ passing through point $(1, -2)$, with a slope triangle showing rise over run of 3 over 1]

•• **REMARK**   When you find an equation of the line that passes through two given points, you only need to substitute the coordinates of one of the points in the point-slope form. It does not matter which point you choose because both points will yield the same result.

Parallel and Perpendicular Lines

Slope can tell you whether two nonvertical lines in a plane are parallel, perpendicular, or neither.

> **Parallel and Perpendicular Lines**
>
> **1.** Two distinct nonvertical lines are **parallel** if and only if their slopes are equal. That is,
> $$m_1 = m_2.$$
>
> **2.** Two nonvertical lines are **perpendicular** if and only if their slopes are negative reciprocals of each other. That is,
> $$m_1 = \frac{-1}{m_2}.$$

EXAMPLE 4   **Finding Parallel and Perpendicular Lines**

Find the slope-intercept form of the equations of the lines that pass through the point $(2, -1)$ and are (a) parallel to and (b) perpendicular to the line $2x - 3y = 5$.

Solution   By writing the equation of the given line in slope-intercept form

$2x - 3y = 5$	Write original equation.
$-3y = -2x + 5$	Subtract $2x$ from each side.
$y = \frac{2}{3}x - \frac{5}{3}$	Write in slope-intercept form.

you can see that it has a slope of $m = \frac{2}{3}$, as shown in Figure P.32.

a. Any line parallel to the given line must also have a slope of $\frac{2}{3}$. So, the line through $(2, -1)$ that is parallel to the given line has the following equation.

$y - (-1) = \frac{2}{3}(x - 2)$	Write in point-slope form.
$3(y + 1) = 2(x - 2)$	Multiply each side by 3.
$3y + 3 = 2x - 4$	Distributive Property
$y = \frac{2}{3}x - \frac{7}{3}$	Write in slope-intercept form.

b. Any line perpendicular to the given line must have a slope of $-\frac{3}{2}$ (because $-\frac{3}{2}$ is the negative reciprocal of $\frac{2}{3}$). So, the line through $(2, -1)$ that is perpendicular to the given line has the following equation.

$y - (-1) = -\frac{3}{2}(x - 2)$	Write in point-slope form.
$2(y + 1) = -3(x - 2)$	Multiply each side by 2.
$2y + 2 = -3x + 6$	Distributive Property
$y = -\frac{3}{2}x + 2$	Write in slope-intercept form.

✓ *Checkpoint* �))) *Audio-video solution in English & Spanish at LarsonPrecalculus.com.*

Find the slope-intercept form of the equations of the lines that pass through the point $(-4, 1)$ and are (a) parallel to and (b) perpendicular to the line $5x - 3y = 8$. ■

Notice in Example 4 how the slope-intercept form is used to obtain information about the graph of a line, whereas the point-slope form is used to write the equation of a line.

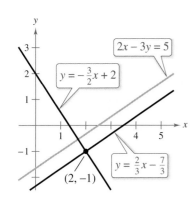

Figure P.32

▷ TECHNOLOGY On a graphing utility, lines will not appear to have the correct slope unless you use a viewing window that has a square setting. For instance, try graphing the lines in Example 4 using the standard setting $-10 \le x \le 10$ and $-10 \le y \le 10$. Then reset the viewing window with the square setting $-9 \le x \le 9$ and $-6 \le y \le 6$. On which setting do the lines $y = \frac{2}{3}x - \frac{5}{3}$ and $y = -\frac{3}{2}x + 2$ appear to be perpendicular?

Applications

In real-life problems, the slope of a line can be interpreted as either a *ratio* or a *rate*. If the *x*-axis and *y*-axis have the same unit of measure, then the slope has no units and is a **ratio.** If the *x*-axis and *y*-axis have different units of measure, then the slope is a **rate** or **rate of change.**

The Americans with Disabilities Act (ADA) became law on July 26, 1990. It is the most comprehensive formulation of rights for persons with disabilities in U.S. (and world) history.

EXAMPLE 5   Using Slope as a Ratio

The maximum recommended slope of a wheelchair ramp is $\frac{1}{12}$. A business is installing a wheelchair ramp that rises 22 inches over a horizontal length of 24 feet. Is the ramp steeper than recommended? *(Source: ADA Standards for Accessible Design)*

Solution   The horizontal length of the ramp is 24 feet or $12(24) = 288$ inches, as shown below. So, the slope of the ramp is

$$\text{Slope} = \frac{\text{vertical change}}{\text{horizontal change}} = \frac{22 \text{ in.}}{288 \text{ in.}} \approx 0.076.$$

Because $\frac{1}{12} \approx 0.083$, the slope of the ramp is not steeper than recommended.

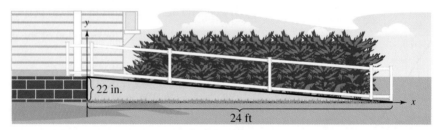

22 in.

24 ft

✓ *Checkpoint* 🔊)) *Audio-video solution in English & Spanish at LarsonPrecalculus.com.*

The business in Example 5 installs a second ramp that rises 36 inches over a horizontal length of 32 feet. Is the ramp steeper than recommended?

EXAMPLE 6   Using Slope as a Rate of Change

A kitchen appliance manufacturing company determines that the total cost C (in dollars) of producing x units of a blender is

$$C = 25x + 3500. \qquad \text{Cost equation}$$

Describe the practical significance of the *y*-intercept and slope of this line.

Solution   The *y*-intercept $(0, 3500)$ tells you that the cost of producing zero units is $3500. This is the *fixed cost* of production—it includes costs that must be paid regardless of the number of units produced. The slope of $m = 25$ tells you that the cost of producing each unit is $25, as shown in Figure P.33. Economists call the cost per unit the *marginal cost*. If the production increases by one unit, then the "margin," or extra amount of cost, is $25. So, the cost increases at a rate of $25 per unit.

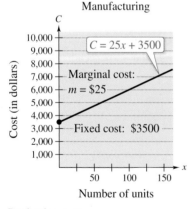

Manufacturing

$C = 25x + 3500$

Marginal cost: $m = \$25$

Fixed cost: $3500

Number of units

Production cost
Figure P.33

✓ *Checkpoint* 🔊)) *Audio-video solution in English & Spanish at LarsonPrecalculus.com.*

An accounting firm determines that the value V (in dollars) of a copier t years after its purchase is

$$V = -300t + 1500.$$

Describe the practical significance of the *y*-intercept and slope of this line.

Businesses can deduct most of their expenses in the same year they occur. One exception is the cost of property that has a useful life of more than 1 year. Such costs must be *depreciated* (decreased in value) over the useful life of the property. Depreciating the *same amount* each year is called *linear* or *straight-line depreciation*. The *book value* is the difference between the original value and the total amount of depreciation accumulated to date.

EXAMPLE 7   Straight-Line Depreciation

A college purchased exercise equipment worth $12,000 for the new campus fitness center. The equipment has a useful life of 8 years. The salvage value at the end of 8 years is $2000. Write a linear equation that describes the book value of the equipment each year.

Solution   Let V represent the value of the equipment at the end of year t. Represent the initial value of the equipment by the data point $(0, 12{,}000)$ and the salvage value of the equipment by the data point $(8, 2000)$. The slope of the line is

$$m = \frac{2000 - 12{,}000}{8 - 0}$$

$$= -\$1250$$

which represents the annual depreciation in *dollars per year*. Using the point-slope form, you can write the equation of the line as follows.

$$V - 12{,}000 = -1250(t - 0) \qquad \text{Write in point-slope form.}$$

$$V = -1250t + 12{,}000 \qquad \text{Write in slope-intercept form.}$$

The table shows the book value at the end of each year, and Figure P.34 shows the graph of the equation.

Useful Life of Equipment

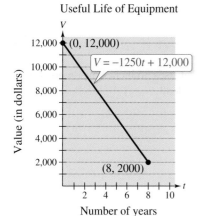

Straight-line depreciation
Figure P.34

Year, t	Value, V
0	12,000
1	10,750
2	9500
3	8250
4	7000
5	5750
6	4500
7	3250
8	2000

✓ *Checkpoint*  Audio-video solution in English & Spanish at LarsonPrecalculus.com.

A manufacturing firm purchased a machine worth $24,750. The machine has a useful life of 6 years. After 6 years, the machine will have to be discarded and replaced. That is, it will have no salvage value. Write a linear equation that describes the book value of the machine each year.

In many real-life applications, the two data points that determine the line are often given in a disguised form. Note how the data points are described in Example 7.

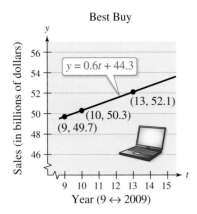

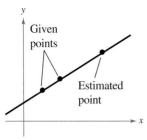

Figure P.35

EXAMPLE 8 **Predicting Sales**

The sales for Best Buy were approximately $49.7 billion in 2009 and $50.3 billion in 2010. Using only this information, write a linear equation that gives the sales in terms of the year. Then predict the sales in 2013. *(Source: Best Buy Company, Inc.)*

Solution Let $t = 9$ represent 2009. Then the two given values are represented by the data points $(9, 49.7)$ and $(10, 50.3)$. The slope of the line through these points is

$$m = \frac{50.3 - 49.7}{10 - 9} = 0.6.$$

You can find the equation that relates the sales y and the year t to be

$$y - 49.7 = 0.6(t - 9) \qquad \text{Write in point-slope form.}$$
$$y = 0.6t + 44.3. \qquad \text{Write in slope-intercept form.}$$

According to this equation, the sales in 2013 will be

$$y = 0.6(13) + 44.3 = 7.8 + 44.3 = \$52.1 \text{ billion. (See Figure P.35.)}$$

✓ *Checkpoint* ◀))) *Audio-video solution in English & Spanish at LarsonPrecalculus.com.*

The sales for Nokia were approximately $58.6 billion in 2009 and $56.6 billion in 2010. Repeat Example 8 using this information. *(Source: Nokia Corporation)*

The prediction method illustrated in Example 8 is called **linear extrapolation.** Note in Figure P.36 that an extrapolated point does not lie between the given points. When the estimated point lies between two given points, as shown in Figure P.37, the procedure is called **linear interpolation.**

Because the slope of a vertical line is not defined, its equation cannot be written in slope-intercept form. However, every line has an equation that can be written in the **general form** $Ax + By + C = 0$, where A and B are not both zero.

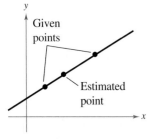

Linear extrapolation
Figure P.36

Linear interpolation
Figure P.37

Summary of Equations of Lines

1. General form: $Ax + By + C = 0$
2. Vertical line: $x = a$
3. Horizontal line: $y = b$
4. Slope-intercept form: $y = mx + b$
5. Point-slope form: $y - y_1 = m(x - x_1)$
6. Two-point form: $y - y_1 = \dfrac{y_2 - y_1}{x_2 - x_1}(x - x_1)$

Summarize (Section P.4)

1. Explain how to use slope to graph a linear equation in two variables *(page 40)* and how to find the slope of a line passing through two points *(page 42)*. For examples of using and finding slopes, see Examples 1 and 2.

2. State the point-slope form of the equation of a line *(page 44)*. For an example of using point-slope form, see Example 3.

3. Explain how to use slope to identify parallel and perpendicular lines *(page 45)*. For an example of finding parallel and perpendicular lines, see Example 4.

4. Describe examples of how to use slope and linear equations in two variables to model and solve real-life problems *(pages 46–48, Examples 5–8)*.

P.4   Exercises

See **CalcChat.com** for tutorial help and worked-out solutions to odd-numbered exercises.

Vocabulary: Fill in the blanks.

1. The simplest mathematical model for relating two variables is the _____ equation in two variables $y = mx + b$.

2. For a line, the ratio of the change in y to the change in x is called the _____ of the line.

3. The _____-_____ form of the equation of a line with slope m passing through (x_1, y_1) is $y - y_1 = m(x - x_1)$.

4. Two lines are _____ if and only if their slopes are equal.

5. Two lines are _____ if and only if their slopes are negative reciprocals of each other.

6. When the x-axis and y-axis have different units of measure, the slope can be interpreted as a _____.

7. The prediction method _____ _____ is the method used to estimate a point on a line when the point does not lie between the given points.

8. Every line has an equation that can be written in _____ form.

Skills and Applications

Identifying Lines In Exercises 9 and 10, identify the line that has each slope.

9. (a) $m = \frac{2}{3}$
   (b) m is undefined.
   (c) $m = -2$

10. (a) $m = 0$
    (b) $m = -\frac{3}{4}$
    (c) $m = 1$

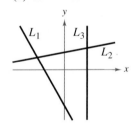

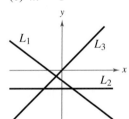

Sketching Lines In Exercises 11 and 12, sketch the lines through the point with the indicated slopes on the same set of coordinate axes.

Point	Slopes
11. $(2, 3)$	(a) 0 (b) 1
	(c) 2 (d) -3
12. $(-4, 1)$	(a) 3 (b) -3
	(c) $\frac{1}{2}$ (d) Undefined

Estimating the Slope of a Line In Exercises 13 and 14, estimate the slope of the line.

13.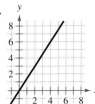

14.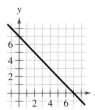

Graphing a Linear Equation In Exercises 15–24, find the slope and y-intercept (if possible) of the equation of the line. Sketch the line.

15. $y = 5x + 3$
16. $y = -x - 10$
17. $y = -\frac{1}{2}x + 4$
18. $y = \frac{3}{2}x + 6$
19. $y - 3 = 0$
20. $x + 5 = 0$
21. $5x - 2 = 0$
22. $3y + 5 = 0$
23. $7x - 6y = 30$
24. $2x + 3y = 9$

Finding the Slope of a Line Through Two Points In Exercises 25–34, plot the points and find the slope of the line passing through the pair of points.

25. $(0, 9), (6, 0)$
26. $(12, 0), (0, -8)$
27. $(-3, -2), (1, 6)$
28. $(2, 4), (4, -4)$
29. $(5, -7), (8, -7)$
30. $(-2, 1), (-4, -5)$
31. $(-6, -1), (-6, 4)$
32. $(0, -10), (-4, 0)$
33. $(4.8, 3.1), (-5.2, 1.6)$
34. $\left(\frac{11}{2}, -\frac{4}{3}\right), \left(-\frac{3}{2}, -\frac{1}{3}\right)$

Using a Point and Slope In Exercises 35–42, use the point on the line and the slope m of the line to find three additional points through which the line passes. (There are many correct answers.)

35. $(2, 1),   m = 0$
36. $(3, -2),   m = 0$
37. $(-8, 1),   m$ is undefined.
38. $(1, 5),   m$ is undefined.
39. $(-5, 4),   m = 2$
40. $(0, -9),   m = -2$
41. $(-1, -6),   m = -\frac{1}{2}$
42. $(7, -2),   m = \frac{1}{2}$

Finding an Equation of a Line In Exercises 43–54, find an equation of the line that passes through the given point and has the indicated slope m. Sketch the line.

43. $(0, -2)$, $m = 3$

44. $(0, 10)$, $m = -1$

45. $(-3, 6)$, $m = -2$

46. $(0, 0)$, $m = 4$

47. $(4, 0)$, $m = -\frac{1}{3}$

48. $(8, 2)$, $m = \frac{1}{4}$

49. $(2, -3)$, $m = -\frac{1}{2}$

50. $(-2, -5)$, $m = \frac{3}{4}$

51. $(6, -1)$, m is undefined.

52. $(-10, 4)$, m is undefined.

53. $\left(4, \frac{5}{2}\right)$, $m = 0$

54. $(-5.1, 1.8)$, $m = 5$

Finding an Equation of a Line In Exercises 55–64, find an equation of the line passing through the points. Sketch the line.

55. $(5, -1), (-5, 5)$

56. $(4, 3), (-4, -4)$

57. $(-8, 1), (-8, 7)$

58. $(-1, 4), (6, 4)$

59. $\left(2, \frac{1}{2}\right), \left(\frac{1}{2}, \frac{5}{4}\right)$

60. $(1, 1), \left(6, -\frac{2}{3}\right)$

61. $(1, 0.6), (-2, -0.6)$

62. $(-8, 0.6), (2, -2.4)$

63. $(2, -1), \left(\frac{1}{3}, -1\right)$

64. $\left(\frac{7}{3}, -8\right), \left(\frac{7}{3}, 1\right)$

Parallel and Perpendicular Lines In Exercises 65–68, determine whether the lines are parallel, perpendicular, or neither.

65. $L_1\colon y = \frac{1}{3}x - 2$
    $L_2\colon y = \frac{1}{3}x + 3$

66. $L_1\colon y = 4x - 1$
    $L_2\colon y = 4x + 7$

67. $L_1\colon y = \frac{1}{2}x - 3$
    $L_2\colon y = -\frac{1}{2}x + 1$

68. $L_1\colon y = -\frac{4}{5}x - 5$
    $L_2\colon y = \frac{5}{4}x + 1$

Parallel and Perpendicular Lines In Exercises 69–72, determine whether the lines L_1 and L_2 passing through the pairs of points are parallel, perpendicular, or neither.

69. $L_1\colon (0, -1), (5, 9)$
    $L_2\colon (0, 3), (4, 1)$

70. $L_1\colon (-2, -1), (1, 5)$
    $L_2\colon (1, 3), (5, -5)$

71. $L_1\colon (3, 6), (-6, 0)$
    $L_2\colon (0, -1), \left(5, \frac{7}{3}\right)$

72. $L_1\colon (4, 8), (-4, 2)$
    $L_2\colon (3, -5), \left(-1, \frac{1}{3}\right)$

Finding Parallel and Perpendicular Lines In Exercises 73–80, write equations of the lines through the given point (a) parallel to and (b) perpendicular to the given line.

73. $4x - 2y = 3$, $(2, 1)$

74. $x + y = 7$, $(-3, 2)$

75. $3x + 4y = 7$, $\left(-\frac{2}{3}, \frac{7}{8}\right)$

76. $5x + 3y = 0$, $\left(\frac{7}{8}, \frac{3}{4}\right)$

77. $y + 3 = 0$, $(-1, 0)$

78. $x - 4 = 0$, $(3, -2)$

79. $x - y = 4$, $(2.5, 6.8)$

80. $6x + 2y = 9$, $(-3.9, -1.4)$

Intercept Form of the Equation of a Line In Exercises 81–86, use the *intercept form* to find the equation of the line with the given intercepts. The intercept form of the equation of a line with intercepts $(a, 0)$ and $(0, b)$ is

$$\frac{x}{a} + \frac{y}{b} = 1, \ a \neq 0, \ b \neq 0.$$

81. x-intercept: $(2, 0)$
    y-intercept: $(0, 3)$

82. x-intercept: $(-3, 0)$
    y-intercept: $(0, 4)$

83. x-intercept: $\left(-\frac{1}{6}, 0\right)$
    y-intercept: $\left(0, -\frac{2}{3}\right)$

84. x-intercept: $\left(\frac{2}{3}, 0\right)$
    y-intercept: $(0, -2)$

85. Point on line: $(1, 2)$
    x-intercept: $(c, 0)$
    y-intercept: $(0, c)$, $c \neq 0$

86. Point on line: $(-3, 4)$
    x-intercept: $(d, 0)$
    y-intercept: $(0, d)$, $d \neq 0$

87. **Sales** The following are the slopes of lines representing annual sales y in terms of time x in years. Use the slopes to interpret any change in annual sales for a one-year increase in time.

    (a) The line has a slope of $m = 135$.

    (b) The line has a slope of $m = 0$.

    (c) The line has a slope of $m = -40$.

88. **Sales** The graph shows the sales (in billions of dollars) for Apple Inc. in the years 2004 through 2010. (*Source: Apple Inc.*)

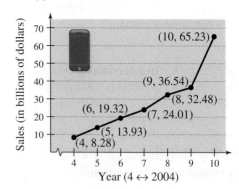

    (a) Use the slopes of the line segments to determine the years in which the sales showed the greatest increase and the least increase.

    (b) Find the slope of the line segment connecting the points for the years 2004 and 2010.

    (c) Interpret the meaning of the slope in part (b) in the context of the problem.

89. Road Grade You are driving on a road that has a 6% uphill grade. This means that the slope of the road is $\frac{6}{100}$. Approximate the amount of vertical change in your position when you drive 200 feet.

90. Road Grade

From the top of a mountain road, a surveyor takes several horizontal measurements x and several vertical measurements y, as shown in the table (x and y are measured in feet).

x	300	600	900	1200
y	-25	-50	-75	-100

x	1500	1800	2100
y	-125	-150	-175

(a) Sketch a scatter plot of the data.

(b) Use a straightedge to sketch the line that you think best fits the data.

(c) Find an equation for the line you sketched in part (b).

(d) Interpret the meaning of the slope of the line in part (c) in the context of the problem.

(e) The surveyor needs to put up a road sign that indicates the steepness of the road. For instance, a surveyor would put up a sign that states "8% grade" on a road with a downhill grade that has a slope of $-\frac{8}{100}$. What should the sign state for the road in this problem?

Rate of Change In Exercises 91 and 92, you are given the dollar value of a product in 2013 and the rate at which the value of the product is expected to change during the next 5 years. Use this information to write a linear equation that gives the dollar value V of the product in terms of the year t. (Let $t = 13$ represent 2013.)

2013 Value	**Rate**
91. $2540	$125 decrease per year
92. $156	$4.50 increase per year

93. Cost The cost C of producing n computer laptop bags is given by

$$C = 1.25n + 15{,}750, \quad 0 < n.$$

Explain what the C-intercept and the slope measure.

94. Monthly Salary A pharmaceutical salesperson receives a monthly salary of $2500 plus a commission of 7% of sales. Write a linear equation for the salesperson's monthly wage W in terms of monthly sales S.

95. Depreciation A sub shop purchases a used pizza oven for $875. After 5 years, the oven will have to be discarded and replaced. Write a linear equation giving the value V of the equipment during the 5 years it will be in use.

96. Depreciation A school district purchases a high-volume printer, copier, and scanner for $24,000. After 10 years, the equipment will have to be replaced. Its value at that time is expected to be $2000. Write a linear equation giving the value V of the equipment during the 10 years it will be in use.

97. Temperature Conversion Write a linear equation that expresses the relationship between the temperature in degrees Celsius C and degrees Fahrenheit F. Use the fact that water freezes at 0°C (32°F) and boils at 100°C (212°F).

98. Brain Weight The average weight of a male child's brain is 970 grams at age 1 and 1270 grams at age 3. (*Source: American Neurological Association*)

(a) Assuming that the relationship between brain weight y and age t is linear, write a linear model for the data.

(b) What is the slope and what does it tell you about brain weight?

(c) Use your model to estimate the average brain weight at age 2.

(d) Use your school's library, the Internet, or some other reference source to find the actual average brain weight at age 2. How close was your estimate?

(e) Do you think your model could be used to determine the average brain weight of an adult? Explain.

99. Cost, Revenue, and Profit A roofing contractor purchases a shingle delivery truck with a shingle elevator for $42,000. The vehicle requires an average expenditure of $9.50 per hour for fuel and maintenance, and the operator is paid $11.50 per hour.

(a) Write a linear equation giving the total cost C of operating this equipment for t hours. (Include the purchase cost of the equipment.)

(b) Assuming that customers are charged $45 per hour of machine use, write an equation for the revenue R derived from t hours of use.

(c) Use the formula for profit $P = R - C$ to write an equation for the profit derived from t hours of use.

(d) Use the result of part (c) to find the break-even point—that is, the number of hours this equipment must be used to yield a profit of 0 dollars.

100. Geometry The length and width of a rectangular garden are 15 meters and 10 meters, respectively. A walkway of width x surrounds the garden.

    (a) Draw a diagram that gives a visual representation of the problem.

    (b) Write the equation for the perimeter y of the walkway in terms of x.

    (c) Use a graphing utility to graph the equation for the perimeter.

    (d) Determine the slope of the graph in part (c). For each additional one-meter increase in the width of the walkway, determine the increase in its perimeter.

Exploration

True or False? **In Exercises 101 and 102, determine whether the statement is true or false. Justify your answer.**

101. A line with a slope of $-\frac{5}{7}$ is steeper than a line with a slope of $-\frac{6}{7}$.

102. The line through $(-8, 2)$ and $(-1, 4)$ and the line through $(0, -4)$ and $(-7, 7)$ are parallel.

103. Right Triangle Explain how you could use slope to show that the points $A(-1, 5)$, $B(3, 7)$, and $C(5, 3)$ are the vertices of a right triangle.

104. Vertical Line Explain why the slope of a vertical line is said to be undefined.

105. Think About It With the information shown in the graphs, is it possible to determine the slope of each line? Is it possible that the lines could have the same slope? Explain.

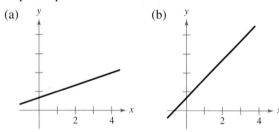

106. Perpendicular Segments Find d_1 and d_2 in terms of m_1 and m_2, respectively (see figure). Then use the Pythagorean Theorem to find a relationship between m_1 and m_2.

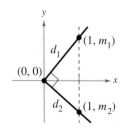

107. Think About It Is it possible for two lines with positive slopes to be perpendicular? Explain.

108. Slope and Steepness The slopes of two lines are -4 and $\frac{5}{2}$. Which is steeper? Explain.

109. Comparing Slopes Use a graphing utility to compare the slopes of the lines $y = mx$, where $m = 0.5$, 1, 2, and 4. Which line rises most quickly? Now, let $m = -0.5$, -1, -2, and -4. Which line falls most quickly? Use a square setting to obtain a true geometric perspective. What can you conclude about the slope and the "rate" at which the line rises or falls?

110. **HOW DO YOU SEE IT?** Match the description of the situation with its graph. Also determine the slope and y-intercept of each graph and interpret the slope and y-intercept in the context of the situation. [The graphs are labeled (i), (ii), (iii), and (iv).]

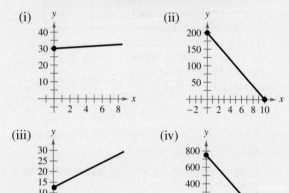

    (a) A person is paying $20 per week to a friend to repay a $200 loan.

    (b) An employee receives $12.50 per hour plus $2 for each unit produced per hour.

    (c) A sales representative receives $30 per day for food plus $0.32 for each mile traveled.

    (d) A computer that was purchased for $750 depreciates $100 per year.

Finding a Relationship for Equidistance **In Exercises 111–114, find a relationship between x and y such that (x, y) is equidistant (the same distance) from the two points.**

111. $(4, -1), (-2, 3)$ **112.** $(6, 5), (1, -8)$

113. $\left(3, \frac{5}{2}\right), (-7, 1)$ **114.** $\left(-\frac{1}{2}, -4\right), \left(\frac{7}{2}, \frac{5}{4}\right)$

Project: Bachelor's Degrees To work an extended application analyzing the numbers of bachelor's degrees earned by women in the United States from 1998 through 2009, visit this text's website at *LarsonPrecalculus.com*. (*Source: National Center for Education Statistics*)

P.5   Functions

Functions can help you model and solve real-life problems. For instance, in Exercise 74 on page 65, you will use a function to model the force of water against the face of a dam.

■ Determine whether relations between two variables are functions, and use function notation.
■ Find the domains of functions.
■ Use functions to model and solve real-life problems.
■ Evaluate difference quotients.

Introduction to Functions and Function Notation

Many everyday phenomena involve two quantities that are related to each other by some rule of correspondence. The mathematical term for such a rule of correspondence is a **relation.** In mathematics, equations and formulas often represent relations. For instance, the simple interest I earned on \$1000 for 1 year is related to the annual interest rate r by the formula $I = 1000r$.

The formula $I = 1000r$ represents a special kind of relation that matches each item from one set with *exactly one* item from a different set. Such a relation is called a **function.**

Definition of Function

A **function** f from a set A to a set B is a relation that assigns to each element x in the set A exactly one element y in the set B. The set A is the **domain** (or set of inputs) of the function f, and the set B contains the **range** (or set of outputs).

To help understand this definition, look at the function below, which relates the time of day to the temperature.

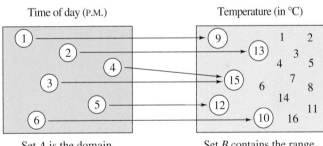

Time of day (P.M.)

Temperature (in °C)

Set A is the domain.
Inputs: 1, 2, 3, 4, 5, 6

Set B contains the range.
Outputs: 9, 10, 12, 13, 15

The following ordered pairs can represent this function. The first coordinate (x-value) is the input and the second coordinate (y-value) is the output.

$$\{(1, 9), (2, 13), (3, 15), (4, 15), (5, 12), (6, 10)\}$$

Characteristics of a Function from Set *A* to Set *B*

1. Each element in A must be matched with an element in B.

2. Some elements in B may not be matched with any element in A.

3. Two or more elements in A may be matched with the same element in B.

4. An element in A (the domain) cannot be matched with two different elements in B.

Four common ways to represent functions are as follows.

Four Ways to Represent a Function

1. *Verbally* by a sentence that describes how the input variable is related to the output variable

2. *Numerically* by a table or a list of ordered pairs that matches input values with output values

3. *Graphically* by points on a graph in a coordinate plane in which the horizontal axis represents the input values and the vertical axis represents the output values

4. *Algebraically* by an equation in two variables

To determine whether a relation is a function, you must decide whether each input value is matched with exactly one output value. When any input value is matched with two or more output values, the relation is not a function.

EXAMPLE 1   **Testing for Functions**

Determine whether the relation represents *y* as a function of *x*.

a. The input value *x* is the number of representatives from a state, and the output value *y* is the number of senators.

b.

Input, x	Output, y
2	11
2	10
3	8
4	5
5	1

c.

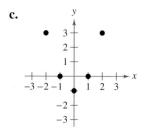

Solution

a. This verbal description *does* describe *y* as a function of *x*. Regardless of the value of *x*, the value of *y* is always 2. Such functions are called *constant functions*.

b. This table *does not* describe *y* as a function of *x*. The input value 2 is matched with two different *y*-values.

c. The graph *does* describe *y* as a function of *x*. Each input value is matched with exactly one output value.

✓ *Checkpoint* 🔊)) *Audio-video solution in English & Spanish at LarsonPrecalculus.com.*

Determine whether the relation represents *y* as a function of *x*.

a. *Domain, x* *Range, y*

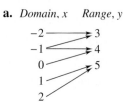

b.

Input, x	0	1	2	3	4
Output, y	−4	−2	0	2	4

Representing functions by sets of ordered pairs is common in *discrete mathematics*. In algebra, however, it is more common to represent functions by equations or formulas involving two variables. For instance, the equation

$$y = x^2 \qquad \text{\small{y is a function of x.}}$$

represents the variable y as a function of the variable x. In this equation, x is the **independent variable** and y is the **dependent variable.** The domain of the function is the set of all values taken on by the independent variable x, and the range of the function is the set of all values taken on by the dependent variable y.

EXAMPLE 2　**Testing for Functions Represented Algebraically**

Which of the equations represent(s) y as a function of x?

a. $x^2 + y = 1$

b. $-x + y^2 = 1$

Solution　To determine whether y is a function of x, try to solve for y in terms of x.

a. Solving for y yields

$$x^2 + y = 1 \qquad \text{\small{Write original equation.}}$$
$$y = 1 - x^2. \qquad \text{\small{Solve for y.}}$$

To each value of x there corresponds exactly one value of y. So, y is a function of x.

b. Solving for y yields

$$-x + y^2 = 1 \qquad \text{\small{Write original equation.}}$$
$$y^2 = 1 + x \qquad \text{\small{Add x to each side.}}$$
$$y = \pm\sqrt{1 + x}. \qquad \text{\small{Solve for y.}}$$

The $\pm$ indicates that to a given value of x there correspond two values of y. So, y is not a function of x.

✓ **Checkpoint** 🔊))) Audio-video solution in English & Spanish at LarsonPrecalculus.com.

Which of the equations represent(s) y as a function of x?

a. $x^2 + y^2 = 8$　　**b.** $y - 4x^2 = 36$

When using an equation to represent a function, it is convenient to name the function for easy reference. For example, you know that the equation $y = 1 - x^2$ describes y as a function of x. Suppose you give this function the name "f." Then you can use the following **function notation.**

Input	Output	Equation
x	$f(x)$	$f(x) = 1 - x^2$

The symbol $f(x)$ is read as *the value of f at x* or simply *f of x.* The symbol $f(x)$ corresponds to the y-value for a given x. So, you can write $y = f(x)$. Keep in mind that f is the *name* of the function, whereas $f(x)$ is the *value* of the function at x. For instance, the function $f(x) = 3 - 2x$ has *function values* denoted by $f(-1), f(0), f(2)$, and so on. To find these values, substitute the specified input values into the given equation.

For $x = -1$, 　　$f(-1) = 3 - 2(-1) = 3 + 2 = 5.$

For $x = 0$, 　　　$f(0) = 3 - 2(0) = 3 - 0 = 3.$

For $x = 2$, 　　　$f(2) = 3 - 2(2) = 3 - 4 = -1.$

Although f is often used as a convenient function name and x is often used as the independent variable, you can use other letters. For instance,

$$f(x) = x^2 - 4x + 7, \quad f(t) = t^2 - 4t + 7, \quad \text{and} \quad g(s) = s^2 - 4s + 7$$

all define the same function. In fact, the role of the independent variable is that of a "placeholder." Consequently, the function could be described by

$$f(\blacksquare) = (\blacksquare)^2 - 4(\blacksquare) + 7.$$

EXAMPLE 3   **Evaluating a Function**

Let $g(x) = -x^2 + 4x + 1$. Find each function value.

a. $g(2)$ **b.** $g(t)$ **c.** $g(x + 2)$

Solution

a. Replacing x with 2 in $g(x) = -x^2 + 4x + 1$ yields the following.

$$g(2) = -(2)^2 + 4(2) + 1$$

$$= -4 + 8 + 1$$

$$= 5$$

b. Replacing x with t yields the following.

$$g(t) = -(t)^2 + 4(t) + 1$$

$$= -t^2 + 4t + 1$$

c. Replacing x with $x + 2$ yields the following.

$$g(x + 2) = -(x + 2)^2 + 4(x + 2) + 1$$

$$= -(x^2 + 4x + 4) + 4x + 8 + 1$$

$$= -x^2 - 4x - 4 + 4x + 8 + 1$$

$$= -x^2 + 5$$

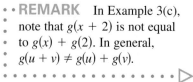 **··REMARK** In Example 3(c), note that $g(x + 2)$ is not equal to $g(x) + g(2)$. In general, $g(u + v) \neq g(u) + g(v)$.

✓ *Checkpoint* 🔊))) *Audio-video solution in English & Spanish at LarsonPrecalculus.com.*

Let $f(x) = 10 - 3x^2$. Find each function value.

a. $f(2)$ **b.** $f(-4)$ **c.** $f(x - 1)$ ◼

A function defined by two or more equations over a specified domain is called a **piecewise-defined function.**

EXAMPLE 4   **A Piecewise-Defined Function**

Evaluate the function when $x = -1, 0,$ and 1.

$$f(x) = \begin{cases} x^2 + 1, & x < 0 \\ x - 1, & x \geq 0 \end{cases}$$

Solution Because $x = -1$ is less than 0, use $f(x) = x^2 + 1$ to obtain $f(-1) = (-1)^2 + 1 = 2$. For $x = 0$, use $f(x) = x - 1$ to obtain $f(0) = (0) - 1 = -1$. For $x = 1$, use $f(x) = x - 1$ to obtain $f(1) = (1) - 1 = 0$.

✓ *Checkpoint* 🔊))) *Audio-video solution in English & Spanish at LarsonPrecalculus.com.*

Evaluate the function given in Example 4 when $x = -2, 2,$ and 3.   ◼

EXAMPLE 5   **Finding Values for Which $f(x) = 0$**

Find all real values of x such that $f(x) = 0$.

a. $f(x) = -2x + 10$ **b.** $f(x) = x^2 - 5x + 6$

Solution   For each function, set $f(x) = 0$ and solve for x.

a. $-2x + 10 = 0$ Set $f(x)$ equal to 0.

            $-2x = -10$ Subtract 10 from each side.

                $x = 5$ Divide each side by -2.

So, $f(x) = 0$ when $x = 5$.

b. $x^2 - 5x + 6 = 0$ Set $f(x)$ equal to 0.

     $(x - 2)(x - 3) = 0$ Factor.

          $x - 2 = 0$ $x = 2$ Set 1st factor equal to 0.

          $x - 3 = 0$ $x = 3$ Set 2nd factor equal to 0.

So, $f(x) = 0$ when $x = 2$ or $x = 3$.

✓ *Checkpoint* *Audio-video solution in English & Spanish at LarsonPrecalculus.com.*

Find all real values of x such that $f(x) = 0$, where $f(x) = x^2 - 16$.

EXAMPLE 6   **Finding Values for Which $f(x) = g(x)$**

Find the values of x for which $f(x) = g(x)$.

a. $f(x) = x^2 + 1$ and $g(x) = 3x - x^2$

b. $f(x) = x^2 - 1$ and $g(x) = -x^2 + x + 2$

Solution

a. $x^2 + 1 = 3x - x^2$ Set $f(x)$ equal to $g(x)$.

      $2x^2 - 3x + 1 = 0$ Write in general form.

   $(2x - 1)(x - 1) = 0$ Factor.

          $2x - 1 = 0$ $x = \frac{1}{2}$ Set 1st factor equal to 0.

            $x - 1 = 0$ $x = 1$ Set 2nd factor equal to 0.

So, $f(x) = g(x)$ when $x = \dfrac{1}{2}$ or $x = 1$.

b. $x^2 - 1 = -x^2 + x + 2$ Set $f(x)$ equal to $g(x)$.

      $2x^2 - x - 3 = 0$ Write in general form.

   $(2x - 3)(x + 1) = 0$ Factor.

          $2x - 3 = 0$ $x = \frac{3}{2}$ Set 1st factor equal to 0.

            $x + 1 = 0$ $x = -1$ Set 2nd factor equal to 0.

So, $f(x) = g(x)$ when $x = \dfrac{3}{2}$ or $x = -1$.

✓ *Checkpoint* *Audio-video solution in English & Spanish at LarsonPrecalculus.com.*

Find the values of x for which $f(x) = g(x)$, where $f(x) = x^2 + 6x - 24$ and $g(x) = 4x - x^2$. ∎

The Domain of a Function

The domain of a function can be described explicitly or it can be *implied* by the expression used to define the function. The **implied domain** is the set of all real numbers for which the expression is defined. For instance, the function

$$f(x) = \frac{1}{x^2 - 4} \qquad \text{Domain excludes } x\text{-values that result in division by zero.}$$

has an implied domain consisting of all real x other than $x = \pm 2$. These two values are excluded from the domain because division by zero is undefined. Another common type of implied domain is that used to avoid even roots of negative numbers. For example, the function

$$f(x) = \sqrt{x} \qquad \text{Domain excludes } x\text{-values that result in even roots of negative numbers.}$$

is defined only for $x \geq 0$. So, its implied domain is the interval $[0, \infty)$. In general, the domain of a function *excludes* values that would cause division by zero *or* that would result in the even root of a negative number.

EXAMPLE 7   Finding the Domain of a Function

Find the domain of each function.

a. f: $\{(-3, 0), (-1, 4), (0, 2), (2, 2), (4, -1)\}$ **b.** $g(x) = \dfrac{1}{x + 5}$

c. Volume of a sphere: $V = \frac{4}{3}\pi r^3$ **d.** $h(x) = \sqrt{4 - 3x}$

Solution

a. The domain of f consists of all first coordinates in the set of ordered pairs.

$$\text{Domain} = \{-3, -1, 0, 2, 4\}$$

b. Excluding x-values that yield zero in the denominator, the domain of g is the set of all real numbers x except $x = -5$.

c. Because this function represents the volume of a sphere, the values of the radius r must be positive. So, the domain is the set of all real numbers r such that $r > 0$.

d. This function is defined only for x-values for which

$$4 - 3x \geq 0.$$

By solving this inequality, you can conclude that $x \leq \frac{4}{3}$. So, the domain is the interval $\left(-\infty, \frac{4}{3}\right]$.

✓ *Checkpoint*  Audio-video solution in English & Spanish at LarsonPrecalculus.com.

Find the domain of each function.

a. f: $\{(-2, 2), (-1, 1), (0, 3), (1, 1), (2, 2)\}$ **b.** $g(x) = \dfrac{1}{3 - x}$

c. Circumference of a circle: $C = 2\pi r$ **d.** $h(x) = \sqrt{x - 16}$

    In Example 7(c), note that the domain of a function may be implied by the physical context. For instance, from the equation

$$V = \frac{4}{3}\pi r^3$$

you would have no reason to restrict r to positive values, but the physical context implies that a sphere cannot have a negative or zero radius.

Applications

EXAMPLE 8   **The Dimensions of a Container**

You work in the marketing department of a soft-drink company and are experimenting with a new can for iced tea that is slightly narrower and taller than a standard can. For your experimental can, the ratio of the height to the radius is 4.

a. Write the volume of the can as a function of the radius r.

b. Write the volume of the can as a function of the height h.

Solution

a. $V(r) = \pi r^2 h = \pi r^2 (4r) = 4\pi r^3$ — Write V as a function of r.

b. $V(h) = \pi r^2 h = \pi \left(\dfrac{h}{4}\right)^2 h = \dfrac{\pi h^3}{16}$ — Write V as a function of h.

✓ **Checkpoint**  *Audio-video solution in English & Spanish at LarsonPrecalculus.com.*

For the experimental can described in Example 8, write the *surface area* as a function of (a) the radius r and (b) the height h.

EXAMPLE 9   **The Path of a Baseball**

A batter hits a baseball at a point 3 feet above ground at a velocity of 100 feet per second and an angle of 45°. The path of the baseball is given by the function

$$f(x) = -0.0032x^2 + x + 3$$

where $f(x)$ is the height of the baseball (in feet) and x is the horizontal distance from home plate (in feet). Will the baseball clear a 10-foot fence located 300 feet from home plate?

Algebraic Solution

When $x = 300$, you can find the height of the baseball as follows.

$f(x) = -0.0032x^2 + x + 3$ — Write original function.

$f(300) = -0.0032(300)^2 + 300 + 3$ — Substitute 300 for x.

$= 15$ — Simplify.

When $x = 300$, the height of the baseball is 15 feet. So, the baseball will clear a 10-foot fence.

Graphical Solution

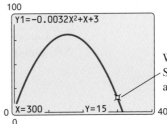

When $x = 300$, $y = 15$. So, the ball will clear a 10-foot fence.

✓ **Checkpoint**  *Audio-video solution in English & Spanish at LarsonPrecalculus.com.*

A second baseman throws a baseball toward the first baseman 60 feet away. The path of the baseball is given by

$$f(x) = -0.004x^2 + 0.3x + 6$$

where $f(x)$ is the height of the baseball (in feet) and x is the horizontal distance from the second baseman (in feet). The first baseman can reach 8 feet high. Can the first baseman catch the baseball without jumping?

Alternative fuels for vehicles include electricity, ethanol, hydrogen, compressed natural gas, liquefied natural gas, and liquefied petroleum gas.

EXAMPLE 10   **Alternative-Fueled Vehicles**

The number V (in thousands) of alternative-fueled vehicles in the United States increased in a linear pattern from 2003 through 2005, and then increased in a different linear pattern from 2006 through 2009, as shown in the bar graph. These two patterns can be approximated by the function

$$V(t) = \begin{cases} 29.05t + 447.7, & 3 \le t \le 5 \\ 65.50t + 241.9, & 6 \le t \le 9 \end{cases}$$

where t represents the year, with $t = 3$ corresponding to 2003. Use this function to approximate the number of alternative-fueled vehicles for each year from 2003 through 2009.   *(Source: U.S. Energy Information Administration)*

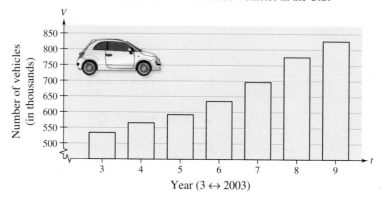

Number of Alternative-Fueled Vehicles in the U.S.

Year (3 ↔ 2003)

Solution   From 2003 through 2005, use $V(t) = 29.05t + 447.7$.

534.9   563.9   593.0

2003   2004   2005

From 2006 to 2009, use $V(t) = 65.50t + 241.9$.

634.9   700.4   765.9   831.4

2006   2007   2008   2009

✓ *Checkpoint* ◀))))   *Audio-video solution in English & Spanish at LarsonPrecalculus.com.*

The number V (in thousands) of 85%-ethanol-fueled vehicles in the United States from 2003 through 2009 can be approximated by the function

$$V(t) = \begin{cases} 33.65t + 77.8, & 3 \le t \le 5 \\ 70.75t - 126.6, & 6 \le t \le 9 \end{cases}$$

where t represents the year, with $t = 3$ corresponding to 2003. Use this function to approximate the number of 85%-ethanol-fueled vehicles for each year from 2003 through 2009.   *(Source: U.S. Energy Information Administration)* ■

Difference Quotients

One of the basic definitions in calculus employs the ratio

$$\frac{f(x + h) - f(x)}{h}, \quad h \ne 0.$$

This ratio is called a **difference quotient,** as illustrated in Example 11.

EXAMPLE 11   **Evaluating a Difference Quotient**

•• REMARK You may find it
easier to calculate the difference
quotient in Example 11 by
first finding $f(x + h)$, and
then substituting the resulting
expression into the difference
quotient

$$\frac{f(x + h) - f(x)}{h}.$$

For $f(x) = x^2 - 4x + 7$, find $\dfrac{f(x + h) - f(x)}{h}$.

Solution

$$\frac{f(x + h) - f(x)}{h} = \frac{[(x + h)^2 - 4(x + h) + 7] - (x^2 - 4x + 7)}{h}$$

$$= \frac{x^2 + 2xh + h^2 - 4x - 4h + 7 - x^2 + 4x - 7}{h}$$

$$= \frac{2xh + h^2 - 4h}{h} = \frac{h(2x + h - 4)}{h} = 2x + h - 4, \quad h \neq 0$$

✓ *Checkpoint*   *Audio-video solution in English & Spanish at LarsonPrecalculus.com.*

For $f(x) = x^2 + 2x - 3$, find $\dfrac{f(x + h) - f(x)}{h}$.

Summary of Function Terminology

Function: A **function** is a relationship between two variables such that to each value of the independent variable there corresponds exactly one value of the dependent variable.

Function Notation: $y = f(x)$

  f is the *name* of the function.

  y is the **dependent variable.**

  x is the **independent variable.**

  $f(x)$ is the *value of the function at x.*

Domain: The **domain** of a function is the set of all values (inputs) of the independent variable for which the function is defined. If x is in the domain of f, then f is said to be *defined* at x. If x is not in the domain of f, then f is said to be *undefined* at x.

Range: The **range** of a function is the set of all values (outputs) assumed by the dependent variable (that is, the set of all function values).

Implied Domain: If f is defined by an algebraic expression and the domain is not specified, then the **implied domain** consists of all real numbers for which the expression is defined.

Summarize   **(Section P.5)**

1. State the definition of a function and describe function notation *(pages 53–56)*. For examples of determining functions and using function notation, see Examples 1–6.

2. State the definition of the implied domain of a function *(page 58)*. For an example of finding the domains of functions, see Example 7.

3. Describe examples of how functions can model real-life problems *(pages 59 and 60, Examples 8–10)*.

4. State the definition of a difference quotient *(page 60)*. For an example of evaluating a difference quotient, see Example 11.

P.5   Exercises

See **CalcChat.com** for tutorial help and worked-out solutions to odd-numbered exercises.

Vocabulary: Fill in the blanks.

1. A relation that assigns to each element x from a set of inputs, or _____, exactly one element y in a set of outputs, or _____, is called a _____.

2. For an equation that represents y as a function of x, the set of all values taken on by the _____ variable x is the domain, and the set of all values taken on by the _____ variable y is the range.

3. If the domain of the function f is not given, then the set of values of the independent variable for which the expression is defined is called the _____ _____.

4. In calculus, one of the basic definitions is that of a _____ _____, given by $\dfrac{f(x + h) - f(x)}{h}$, $h \neq 0$.

Skills and Applications

Testing for Functions   In Exercises 5–8, determine whether the relation represents y as a function of x.

5. Domain, x Range, y

6. Domain, x Range, y

7.

Input, x	10	7	4	7	10
Output, y	3	6	9	12	15

8.

Input, x	0	3	9	12	15
Output, y	3	3	3	3	3

Testing for Functions   In Exercises 9 and 10, which sets of ordered pairs represent functions from A to B? Explain.

9. $A = \{0, 1, 2, 3\}$ and $B = \{-2, -1, 0, 1, 2\}$
   (a) $\{(0, 1), (1, -2), (2, 0), (3, 2)\}$
   (b) $\{(0, -1), (2, 2), (1, -2), (3, 0), (1, 1)\}$
   (c) $\{(0, 0), (1, 0), (2, 0), (3, 0)\}$
   (d) $\{(0, 2), (3, 0), (1, 1)\}$

10. $A = \{a, b, c\}$ and $B = \{0, 1, 2, 3\}$
    (a) $\{(a, 1), (c, 2), (c, 3), (b, 3)\}$
    (b) $\{(a, 1), (b, 2), (c, 3)\}$
    (c) $\{(1, a), (0, a), (2, c), (3, b)\}$
    (d) $\{(c, 0), (b, 0), (a, 3)\}$

Testing for Functions Represented Algebraically
In Exercises 11–20, determine whether the equation represents y as a function of x.

11. $x^2 + y^2 = 4$

12. $x^2 + y = 4$

13. $2x + 3y = 4$

14. $(x - 2)^2 + y^2 = 4$

15. $y = \sqrt{16 - x^2}$

16. $y = \sqrt{x + 5}$

17. $y = |4 - x|$

18. $|y| = 4 - x$

19. $y = -75$

20. $x - 1 = 0$

Evaluating a Function   In Exercises 21–32, evaluate (if possible) the function at each specified value of the independent variable and simplify.

21. $f(x) = 2x - 3$
    (a) $f(1)$ (b) $f(-3)$ (c) $f(x - 1)$

22. $V(r) = \frac{4}{3}\pi r^3$
    (a) $V(3)$ (b) $V\left(\frac{3}{2}\right)$ (c) $V(2r)$

23. $g(t) = 4t^2 - 3t + 5$
    (a) $g(2)$ (b) $g(t - 2)$ (c) $g(t) - g(2)$

24. $h(t) = t^2 - 2t$
    (a) $h(2)$ (b) $h(1.5)$ (c) $h(x + 2)$

25. $f(y) = 3 - \sqrt{y}$
    (a) $f(4)$ (b) $f(0.25)$ (c) $f(4x^2)$

26. $f(x) = \sqrt{x + 8} + 2$
    (a) $f(-8)$ (b) $f(1)$ (c) $f(x - 8)$

27. $q(x) = 1/(x^2 - 9)$
    (a) $q(0)$ (b) $q(3)$ (c) $q(y + 3)$

28. $q(t) = (2t^2 + 3)/t^2$
    (a) $q(2)$ (b) $q(0)$ (c) $q(-x)$

29. $f(x) = |x|/x$
    (a) $f(2)$ (b) $f(-2)$ (c) $f(x - 1)$

30. $f(x) = |x| + 4$
    (a) $f(2)$ (b) $f(-2)$ (c) $f(x^2)$

31. $f(x) = \begin{cases} 2x + 1, & x < 0 \\ 2x + 2, & x \geq 0 \end{cases}$
    (a) $f(-1)$ (b) $f(0)$ (c) $f(2)$

32. $f(x) = \begin{cases} 4 - 5x, & x \leq -2 \\ 0, & -2 < x < 2 \\ x^2 + 1, & x \geq 2 \end{cases}$
    (a) $f(-3)$ (b) $f(4)$ (c) $f(-1)$

Evaluating a Function   **In Exercises 33–36, complete the table.**

33. $f(x) = x^2 - 3$

x	-2	-1	0	1	2
$f(x)$					

34. $h(t) = \frac{1}{2}|t + 3|$

t	-5	-4	-3	-2	-1
$h(t)$					

35. $f(x) = \begin{cases} -\frac{1}{2}x + 4, & x \le 0 \\ (x - 2)^2, & x > 0 \end{cases}$

x	-2	-1	0	1	2
$f(x)$					

36. $f(x) = \begin{cases} 9 - x^2, & x < 3 \\ x - 3, & x \ge 3 \end{cases}$

x	1	2	3	4	5
$f(x)$					

Finding Values for Which $f(x) = 0$ **In Exercises 37–44, find all real values of x such that $f(x) = 0$.**

37. $f(x) = 15 - 3x$

38. $f(x) = 5x + 1$

39. $f(x) = \dfrac{3x - 4}{5}$

40. $f(x) = \dfrac{12 - x^2}{5}$

41. $f(x) = x^2 - 9$

42. $f(x) = x^2 - 8x + 15$

43. $f(x) = x^3 - x$

44. $f(x) = x^3 - x^2 - 4x + 4$

Finding Values for Which $f(x) = g(x)$ **In Exercises 45–48, find the value(s) of x for which $f(x) = g(x)$.**

45. $f(x) = x^2$, $g(x) = x + 2$

46. $f(x) = x^2 + 2x + 1$, $g(x) = 7x - 5$

47. $f(x) = x^4 - 2x^2$, $g(x) = 2x^2$

48. $f(x) = \sqrt{x} - 4$, $g(x) = 2 - x$

Finding the Domain of a Function   **In Exercises 49–60, find the domain of the function.**

49. $f(x) = 5x^2 + 2x - 1$

50. $g(x) = 1 - 2x^2$

51. $h(t) = \dfrac{4}{t}$

52. $s(y) = \dfrac{3y}{y + 5}$

53. $g(y) = \sqrt{y - 10}$

54. $f(t) = \sqrt[3]{t + 4}$

55. $g(x) = \dfrac{1}{x} - \dfrac{3}{x + 2}$

56. $h(x) = \dfrac{10}{x^2 - 2x}$

57. $f(s) = \dfrac{\sqrt{s - 1}}{s - 4}$

58. $f(x) = \dfrac{\sqrt{x + 6}}{6 + x}$

59. $f(x) = \dfrac{x - 4}{\sqrt{x}}$

60. $f(x) = \dfrac{x + 2}{\sqrt{x - 10}}$

61. Maximum Volume   An open box of maximum volume is to be made from a square piece of material 24 centimeters on a side by cutting equal squares from the corners and turning up the sides (see figure).

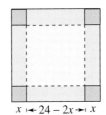
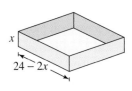

(a) The table shows the volumes V (in cubic centimeters) of the box for various heights x (in centimeters). Use the table to estimate the maximum volume.

Height, x	1	2	3	4	5	6
Volume, V	484	800	972	1024	980	864

(b) Plot the points (x, V) from the table in part (a). Does the relation defined by the ordered pairs represent V as a function of x?

(c) Given that V is a function of x, write the function and determine its domain.

62. Maximum Profit   The cost per unit in the production of an MP3 player is $60. The manufacturer charges $90 per unit for orders of 100 or less. To encourage large orders, the manufacturer reduces the charge by $0.15 per MP3 player for each unit ordered in excess of 100 (for example, there would be a charge of $87 per MP3 player for an order size of 120).

(a) The table shows the profits P (in dollars) for various numbers of units ordered, x. Use the table to estimate the maximum profit.

Units, x	130	140	150	160	170
Profit, P	3315	3360	3375	3360	3315

(b) Plot the points (x, P) from the table in part (a). Does the relation defined by the ordered pairs represent P as a function of x?

(c) Given that P is a function of x, write the function and determine its domain. (*Note: $P = R - C$,* where R is revenue and C is cost.)

63. Geometry Write the area A of a square as a function of its perimeter P.

64. Geometry Write the area A of a circle as a function of its circumference C.

65. Path of a Ball The height y (in feet) of a baseball thrown by a child is

$$y = -\frac{1}{10}x^2 + 3x + 6$$

where x is the horizontal distance (in feet) from where the ball was thrown. Will the ball fly over the head of another child 30 feet away trying to catch the ball? (Assume that the child who is trying to catch the ball holds a baseball glove at a height of 5 feet.)

66. Postal Regulations A rectangular package to be sent by the U.S. Postal Service can have a maximum combined length and girth (perimeter of a cross section) of 108 inches (see figure).

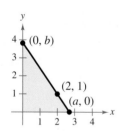

(a) Write the volume V of the package as a function of x. What is the domain of the function?

(b) Use a graphing utility to graph the function. Be sure to use an appropriate window setting.

(c) What dimensions will maximize the volume of the package? Explain your answer.

67. Geometry A right triangle is formed in the first quadrant by the x- and y-axes and a line through the point $(2, 1)$ (see figure). Write the area A of the triangle as a function of x, and determine the domain of the function.

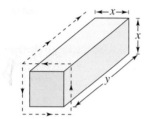

Figure for 67

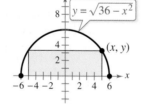

Figure for 68

68. Geometry A rectangle is bounded by the x-axis and the semicircle $y = \sqrt{36 - x^2}$ (see figure). Write the area A of the rectangle as a function of x, and graphically determine the domain of the function.

69. Prescription Drugs The percents p of prescriptions filled with generic drugs in the United States from 2004 through 2010 (see figure) can be approximated by the model

$$p(t) = \begin{cases} 4.57t + 27.3, & 4 \le t \le 7 \\ 3.35t + 37.6, & 8 \le t \le 10 \end{cases}$$

where t represents the year, with $t = 4$ corresponding to 2004. Use this model to find the percent of prescriptions filled with generic drugs in each year from 2004 through 2010. (*Source: National Association of Chain Drug Stores*)

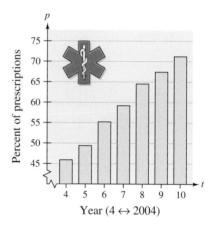

70. Median Sale Price The median sale prices p (in thousands of dollars) of an existing one-family home in the United States from 2000 through 2010 (see figure) can be approximated by the model

$$p(t) = \begin{cases} 0.438t^2 + 10.81t + 145.9, & 0 \le t \le 6 \\ 5.575t^2 - 110.67t + 720.8, & 7 \le t \le 10 \end{cases}$$

where t represents the year, with $t = 0$ corresponding to 2000. Use this model to find the median sale price of an existing one-family home in each year from 2000 through 2010. (*Source: National Association of Realtors*)

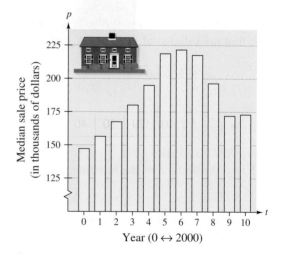

71. Cost, Revenue, and Profit A company produces a product for which the variable cost is $12.30 per unit and the fixed costs are $98,000. The product sells for $17.98. Let x be the number of units produced and sold.

(a) The total cost for a business is the sum of the variable cost and the fixed costs. Write the total cost C as a function of the number of units produced.

(b) Write the revenue R as a function of the number of units sold.

(c) Write the profit P as a function of the number of units sold. (*Note: $P = R - C$*)

72. Average Cost The inventor of a new game believes that the variable cost for producing the game is $0.95 per unit and the fixed costs are $6000. The inventor sells each game for $1.69. Let x be the number of games sold.

(a) The total cost for a business is the sum of the variable cost and the fixed costs. Write the total cost C as a function of the number of games sold.

(b) Write the average cost per unit $\overline{C} = C/x$ as a function of x.

73. Height of a Balloon A balloon carrying a transmitter ascends vertically from a point 3000 feet from the receiving station.

(a) Draw a diagram that gives a visual representation of the problem. Let h represent the height of the balloon and let d represent the distance between the balloon and the receiving station.

(b) Write the height of the balloon as a function of d. What is the domain of the function?

74. Physics

The function $F(y) = 149.76\sqrt{10}\,y^{5/2}$ estimates the force F (in tons) of water against the face of a dam, where y is the depth of the water (in feet).

(a) Complete the table. What can you conclude from the table?

y	5	10	20	30	40
$F(y)$					

(b) Use the table to approximate the depth at which the force against the dam is 1,000,000 tons.

(c) Find the depth at which the force against the dam is 1,000,000 tons algebraically.

75. Transportation For groups of 80 or more people, a charter bus company determines the rate per person according to the formula

$$\text{Rate} = 8 - 0.05(n - 80), \quad n \geq 80$$

where the rate is given in dollars and n is the number of people.

(a) Write the revenue R for the bus company as a function of n.

(b) Use the function in part (a) to complete the table. What can you conclude?

n	90	100	110	120	130	140	150
$R(n)$							

76. E-Filing The table shows the numbers of tax returns (in millions) made through e-file from 2003 through 2010. Let $f(t)$ represent the number of tax returns made through e-file in the year t.   (*Source: Internal Revenue Service*)

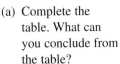

DATA	Year	Number of Tax Returns Made Through E-File
	2003	52.9
	2004	61.5
	2005	68.5
	2006	73.3
	2007	80.0
	2008	89.9
	2009	95.0
	2010	98.7

(a) Find $\dfrac{f(2010) - f(2003)}{2010 - 2003}$ and interpret the result in the context of the problem.

(b) Make a scatter plot of the data.

(c) Find a linear model for the data algebraically. Let N represent the number of tax returns made through e-file and let $t = 3$ correspond to 2003.

(d) Use the model found in part (c) to complete the table.

t	3	4	5	6	7	8	9	10
N								

(e) Compare your results from part (d) with the actual data.

(f) Use a graphing utility to find a linear model for the data. Let $x = 3$ correspond to 2003. How does the model you found in part (c) compare with the model given by the graphing utility?

Lester Lefkowitz/CORBIS

ƒ Evaluating a Difference Quotient In Exercises 77–84, find the difference quotient and simplify your answer.

77. $f(x) = x^2 - x + 1$, $\dfrac{f(2 + h) - f(2)}{h}$, $h \neq 0$

78. $f(x) = 5x - x^2$, $\dfrac{f(5 + h) - f(5)}{h}$, $h \neq 0$

79. $f(x) = x^3 + 3x$, $\dfrac{f(x + h) - f(x)}{h}$, $h \neq 0$

80. $f(x) = 4x^2 - 2x$, $\dfrac{f(x + h) - f(x)}{h}$, $h \neq 0$

81. $g(x) = \dfrac{1}{x^2}$, $\dfrac{g(x) - g(3)}{x - 3}$, $x \neq 3$

82. $f(t) = \dfrac{1}{t - 2}$, $\dfrac{f(t) - f(1)}{t - 1}$, $t \neq 1$

83. $f(x) = \sqrt{5x}$, $\dfrac{f(x) - f(5)}{x - 5}$, $x \neq 5$

84. $f(x) = x^{2/3} + 1$, $\dfrac{f(x) - f(8)}{x - 8}$, $x \neq 8$

Matching and Determining Constants In Exercises 85–88, match the data with one of the following functions

$$f(x) = cx, \quad g(x) = cx^2, \quad h(x) = c\sqrt{|x|}, \quad \text{and} \quad r(x) = \dfrac{c}{x}$$

and determine the value of the constant c that will make the function fit the data in the table.

85.

x	-4	-1	0	1	4
y	-32	-2	0	-2	-32

86.

x	-4	-1	0	1	4
y	-1	$-\frac{1}{4}$	0	$\frac{1}{4}$	1

87.

x	-4	-1	0	1	4
y	-8	-32	Undefined	32	8

88.

x	-4	-1	0	1	4
y	6	3	0	3	6

Exploration

True or False? In Exercises 89–92, determine whether the statement is true or false. Justify your answer.

89. Every relation is a function.

90. Every function is a relation.

91. For the function

$$f(x) = x^4 - 1$$

the domain is $(-\infty, \infty)$ and the range is $(0, \infty)$.

92. The set of ordered pairs $\{(-8, -2), (-6, 0), (-4, 0), (-2, 2), (0, 4), (2, -2)\}$ represents a function.

93. **Think About It** Consider

$$f(x) = \sqrt{x - 1} \quad \text{and} \quad g(x) = \dfrac{1}{\sqrt{x - 1}}.$$

Why are the domains of f and g different?

94. **Think About It** Consider

$$f(x) = \sqrt{x - 2} \quad \text{and} \quad g(x) = \sqrt[3]{x - 2}.$$

Why are the domains of f and g different?

95. **Think About It** Given

$$f(x) = x^2$$

is f the independent variable? Why or why not?

96. **HOW DO YOU SEE IT?** The graph represents the height h of a projectile after t seconds.

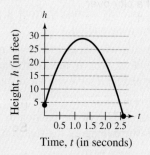

Time, t (in seconds)

(a) Explain why h is a function of t.

(b) Approximate the height of the projectile after 0.5 second and after 1.25 seconds.

(c) Approximate the domain of h.

(d) Is t a function of h? Explain.

Think About It In Exercises 97 and 98, determine whether the statements use the word *function* in ways that are mathematically correct. Explain your reasoning.

97. (a) The sales tax on a purchased item is a function of the selling price.

(b) Your score on the next algebra exam is a function of the number of hours you study the night before the exam.

98. (a) The amount in your savings account is a function of your salary.

(b) The speed at which a free-falling baseball strikes the ground is a function of the height from which it was dropped.

P.6 Analyzing Graphs of Functions

■ Use the Vertical Line Test for functions.
■ Find the zeros of functions.
■ Determine intervals on which functions are increasing or decreasing and determine relative maximum and relative minimum values of functions.
■ Determine the average rate of change of a function.
■ Identify even and odd functions.

The Graph of a Function

In Section P.5, you studied functions from an algebraic point of view. In this section, you will study functions from a graphical perspective.

The **graph of a function** f is the collection of ordered pairs $(x, f(x))$ such that x is in the domain of f. As you study this section, remember that

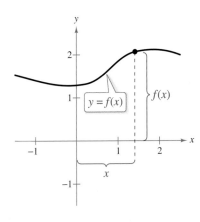

x = the directed distance from the y-axis

$y = f(x)$ = the directed distance from the x-axis

as shown in the figure at the right.

Graphs of functions can help you visualize relationships between variables in real life. For instance, in Exercise 90 on page 77, you will use the graph of a function to visually represent the temperature of a city over a 24-hour period.

Figure P.38

EXAMPLE 1   **Finding the Domain and Range of a Function**

Use the graph of the function f, shown in Figure P.38, to find (a) the domain of f, (b) the function values $f(-1)$ and $f(2)$, and (c) the range of f.

Solution

a. The closed dot at $(-1, 1)$ indicates that $x = -1$ is in the domain of f, whereas the open dot at $(5, 2)$ indicates that $x = 5$ is not in the domain. So, the domain of f is all x in the interval $[-1, 5)$.

b. Because $(-1, 1)$ is a point on the graph of f, it follows that $f(-1) = 1$. Similarly, because $(2, -3)$ is a point on the graph of f, it follows that $f(2) = -3$.

c. Because the graph does not extend below $f(2) = -3$ or above $f(0) = 3$, the range of f is the interval $[-3, 3]$.

✓ **Checkpoint** 🔊)) *Audio-video solution in English & Spanish at LarsonPrecalculus.com.*

Use the graph of the function f to find (a) the domain of f, (b) the function values $f(0)$ and $f(3)$, and (c) the range of f.

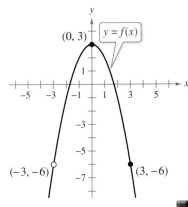

• • • • • • • • • • • • • ▷

•• **REMARK** The use of dots (open or closed) at the extreme left and right points of a graph indicates that the graph does not extend beyond these points. If such dots are not on the graph, then assume that the graph extends beyond these points.

By the definition of a function, at most one *y*-value corresponds to a given *x*-value. This means that the graph of a function cannot have two or more different points with the same *x*-coordinate, and no two points on the graph of a function can be vertically above or below each other. It follows, then, that a vertical line can intersect the graph of a function at most once. This observation provides a convenient visual test called the **Vertical Line Test** for functions.

Vertical Line Test for Functions

A set of points in a coordinate plane is the graph of *y* as a function of *x* if and only if no *vertical* line intersects the graph at more than one point.

EXAMPLE 2 **Vertical Line Test for Functions**

Use the Vertical Line Test to decide whether each of the following graphs represents *y* as a function of *x*.

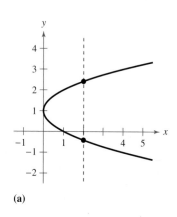

(a)

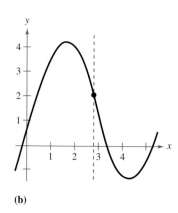

(b)

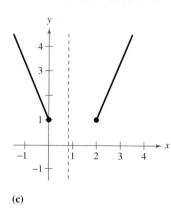

(c)

Solution

a. This *is not* a graph of *y* as a function of *x*, because there are vertical lines that intersect the graph twice. That is, for a particular input *x*, there is more than one output *y*.

b. This *is* a graph of *y* as a function of *x*, because every vertical line intersects the graph at most once. That is, for a particular input *x*, there is at most one output *y*.

c. This *is* a graph of *y* as a function of *x*. (Note that when a vertical line does not intersect the graph, it simply means that the function is undefined for that particular value of *x*.) That is, for a particular input *x*, there is at most one output *y*.

✓ **Checkpoint** ◀))) Audio-video solution in English & Spanish at LarsonPrecalculus.com.

Use the Vertical Line Test to decide whether the graph represents *y* as a function of *x*.

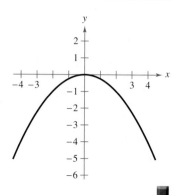

▷ **TECHNOLOGY** Most graphing utilities graph functions of *x* more easily than other types of equations. For instance, the graph shown in (a) above represents the equation $x - (y - 1)^2 = 0$. To use a graphing utility to duplicate this graph, you must first solve the equation for *y* to obtain $y = 1 \pm \sqrt{x}$, and then graph the two equations $y_1 = 1 + \sqrt{x}$ and $y_2 = 1 - \sqrt{x}$ in the same viewing window.

Zeros of a Function

If the graph of a function of x has an x-intercept at $(a, 0)$, then a is a **zero** of the function.

> **Zeros of a Function**
>
> The **zeros of a function** f of x are the x-values for which $f(x) = 0$.

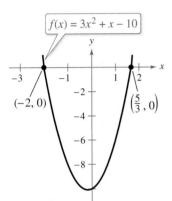

Zeros of f: $x = -2, x = \frac{5}{3}$

Figure P.39

EXAMPLE 3   Finding the Zeros of a Function

Find the zeros of each function.

a. $f(x) = 3x^2 + x - 10$

b. $g(x) = \sqrt{10 - x^2}$

c. $h(t) = \dfrac{2t - 3}{t + 5}$

Solution To find the zeros of a function, set the function equal to zero and solve for the independent variable.

a.
$$3x^2 + x - 10 = 0 \qquad \text{Set } f(x) \text{ equal to 0.}$$
$$(3x - 5)(x + 2) = 0 \qquad \text{Factor.}$$
$$3x - 5 = 0 \implies x = \tfrac{5}{3} \qquad \text{Set 1st factor equal to 0.}$$
$$x + 2 = 0 \implies x = -2 \qquad \text{Set 2nd factor equal to 0.}$$

The zeros of f are $x = \frac{5}{3}$ and $x = -2$. In Figure P.39, note that the graph of f has $\left(\frac{5}{3}, 0\right)$ and $(-2, 0)$ as its x-intercepts.

b.
$$\sqrt{10 - x^2} = 0 \qquad \text{Set } g(x) \text{ equal to 0.}$$
$$10 - x^2 = 0 \qquad \text{Square each side.}$$
$$10 = x^2 \qquad \text{Add } x^2 \text{ to each side.}$$
$$\pm\sqrt{10} = x \qquad \text{Extract square roots.}$$

The zeros of g are $x = -\sqrt{10}$ and $x = \sqrt{10}$. In Figure P.40, note that the graph of g has $\left(-\sqrt{10}, 0\right)$ and $\left(\sqrt{10}, 0\right)$ as its x-intercepts.

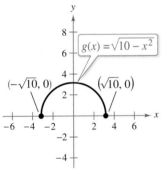

Zeros of g: $x = \pm\sqrt{10}$

Figure P.40

c.
$$\dfrac{2t - 3}{t + 5} = 0 \qquad \text{Set } h(t) \text{ equal to 0.}$$
$$2t - 3 = 0 \qquad \text{Multiply each side by } t + 5.$$
$$2t = 3 \qquad \text{Add 3 to each side.}$$
$$t = \dfrac{3}{2} \qquad \text{Divide each side by 2.}$$

The zero of h is $t = \frac{3}{2}$. In Figure P.41, note that the graph of h has $\left(\frac{3}{2}, 0\right)$ as its t-intercept.

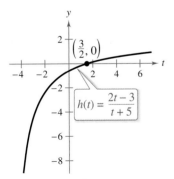

Zero of h: $t = \frac{3}{2}$

Figure P.41

✓ *Checkpoint*  *Audio-video solution in English & Spanish at LarsonPrecalculus.com.*

Find the zeros of each function.

a. $f(x) = 2x^2 + 13x - 24$ **b.** $g(t) = \sqrt{t - 25}$ **c.** $h(x) = \dfrac{x^2 - 2}{x - 1}$

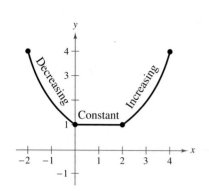

Figure P.42

Increasing and Decreasing Functions

The more you know about the graph of a function, the more you know about the function itself. Consider the graph shown in Figure P.42. As you move from *left to right*, this graph falls from $x = -2$ to $x = 0$, is constant from $x = 0$ to $x = 2$, and rises from $x = 2$ to $x = 4$.

> **Increasing, Decreasing, and Constant Functions**
>
> A function f is **increasing** on an interval when, for any x_1 and x_2 in the interval,
>
> $$x_1 < x_2 \quad \text{implies} \quad f(x_1) < f(x_2).$$
>
> A function f is **decreasing** on an interval when, for any x_1 and x_2 in the interval,
>
> $$x_1 < x_2 \quad \text{implies} \quad f(x_1) > f(x_2).$$
>
> A function f is **constant** on an interval when, for any x_1 and x_2 in the interval,
>
> $$f(x_1) = f(x_2).$$

EXAMPLE 4   **Describing Function Behavior**

Use the graphs to describe the increasing, decreasing, or constant behavior of each function.

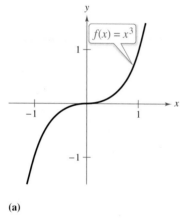

(a)

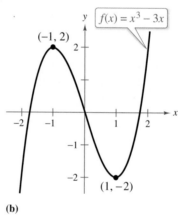

(b)

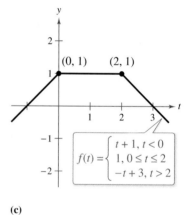

(c)

Solution

a. This function is increasing over the entire real line.

b. This function is increasing on the interval $(-\infty, -1)$, decreasing on the interval $(-1, 1)$, and increasing on the interval $(1, \infty)$.

c. This function is increasing on the interval $(-\infty, 0)$, constant on the interval $(0, 2)$, and decreasing on the interval $(2, \infty)$.

✓ *Checkpoint*    *Audio-video solution in English & Spanish at LarsonPrecalculus.com.*

Graph the function

$$f(x) = x^3 + 3x^2 - 1.$$

Then use the graph to describe the increasing and decreasing behavior of the function.

To help you decide whether a function is increasing, decreasing, or constant on an interval, you can evaluate the function for several values of x. However, you need calculus to determine, for certain, all intervals on which a function is increasing, decreasing, or constant.

The points at which a function changes its increasing, decreasing, or constant behavior are helpful in determining the **relative minimum** or **relative maximum** values of the function.

· · REMARK A relative minimum or relative maximum is also referred to as a local minimum or local maximum.

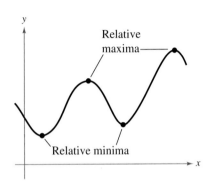

Figure P.43

> **Definitions of Relative Minimum and Relative Maximum**
>
> A function value $f(a)$ is called a **relative minimum** of f when there exists an interval (x_1, x_2) that contains a such that
>
> $$x_1 < x < x_2 \quad \text{implies} \quad f(a) \leq f(x).$$
>
> A function value $f(a)$ is called a **relative maximum** of f when there exists an interval (x_1, x_2) that contains a such that
>
> $$x_1 < x < x_2 \quad \text{implies} \quad f(a) \geq f(x).$$

Figure P.43 shows several different examples of relative minima and relative maxima. By writing a second-degree equation in standard form, $y = a(x - h)^2 + k$, you can find the *exact point* (h, k) at which it has a relative minimum or relative maximum. For the time being, however, you can use a graphing utility to find reasonable approximations of these points.

EXAMPLE 5   **Approximating a Relative Minimum**

Use a graphing utility to approximate the relative minimum of the function

$$f(x) = 3x^2 - 4x - 2.$$

Solution   The graph of f is shown in Figure P.44. By using the *zoom* and *trace* features or the *minimum* feature of a graphing utility, you can estimate that the function has a relative minimum at the point

$$(0.67, -3.33). \qquad \text{Relative minimum}$$

By writing this equation in standard form, $f(x) = 3\left(x - \frac{2}{3}\right)^2 - \frac{10}{3}$, you can determine that the exact point at which the relative minimum occurs is $\left(\frac{2}{3}, -\frac{10}{3}\right)$.

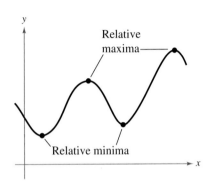

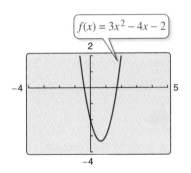

Figure P.44

✓ ***Checkpoint*** 🔊))   *Audio-video solution in English & Spanish at LarsonPrecalculus.com.*

Use a graphing utility to approximate the relative maximum of the function

$$f(x) = -4x^2 - 7x + 3. \qquad ■$$

You can also use the *table* feature of a graphing utility to numerically approximate the relative minimum of the function in Example 5. Using a table that begins at 0.6 and increments the value of x by 0.01, you can approximate that the minimum of $f(x) = 3x^2 - 4x - 2$ occurs at the point $(0.67, -3.33)$.

▷ **TECHNOLOGY**   When you use a graphing utility to estimate the x- and y-values of a relative minimum or relative maximum, the *zoom* feature will often produce graphs that are nearly flat. To overcome this problem, you can manually change the vertical setting of the viewing window. The graph will stretch vertically when the values of Ymin and Ymax are closer together.

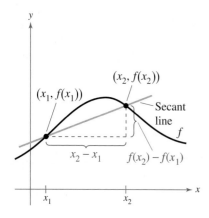

Figure P.45

Average Rate of Change

In Section P.4, you learned that the slope of a line can be interpreted as a *rate of change.* For a nonlinear graph whose slope changes at each point, the **average rate of change** between any two points $(x_1, f(x_1))$ and $(x_2, f(x_2))$ is the slope of the line through the two points (see Figure P.45). The line through the two points is called the **secant line,** and the slope of this line is denoted as $m_{\sec}$.

$$\text{Average rate of change of } f \text{ from } x_1 \text{ to } x_2 = \frac{f(x_2) - f(x_1)}{x_2 - x_1}$$

$$= \frac{\text{change in } y}{\text{change in } x}$$

$$= m_{\sec}$$

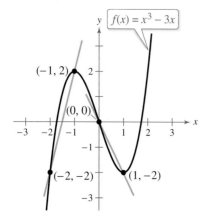

Figure P.46

EXAMPLE 6   Average Rate of Change of a Function

Find the average rates of change of $f(x) = x^3 - 3x$ (a) from $x_1 = -2$ to $x_2 = -1$ and (b) from $x_1 = 0$ to $x_2 = 1$ (see Figure P.46).

Solution

a. The average rate of change of f from $x_1 = -2$ to $x_2 = -1$ is

$$\frac{f(x_2) - f(x_1)}{x_2 - x_1} = \frac{f(-1) - f(-2)}{-1 - (-2)} = \frac{2 - (-2)}{1} = 4. \qquad \text{Secant line has positive slope.}$$

b. The average rate of change of f from $x_1 = 0$ to $x_2 = 1$ is

$$\frac{f(x_2) - f(x_1)}{x_2 - x_1} = \frac{f(1) - f(0)}{1 - 0} = \frac{-2 - 0}{1} = -2. \qquad \text{Secant line has negative slope.}$$

✔ **Checkpoint**))) *Audio-video solution in English & Spanish at LarsonPrecalculus.com.*

Find the average rates of change of $f(x) = x^2 + 2x$ (a) from $x_1 = -3$ to $x_2 = -2$ and (b) from $x_1 = -2$ to $x_2 = 0$.

EXAMPLE 7   Finding Average Speed

The distance s (in feet) a moving car is from a stoplight is given by the function

$$s(t) = 20t^{3/2}$$

where t is the time (in seconds). Find the average speed of the car (a) from $t_1 = 0$ to $t_2 = 4$ seconds and (b) from $t_1 = 4$ to $t_2 = 9$ seconds.

Solution

a. The average speed of the car from $t_1 = 0$ to $t_2 = 4$ seconds is

$$\frac{s(t_2) - s(t_1)}{t_2 - t_1} = \frac{s(4) - s(0)}{4 - 0} = \frac{160 - 0}{4} = 40 \text{ feet per second.}$$

b. The average speed of the car from $t_1 = 4$ to $t_2 = 9$ seconds is

$$\frac{s(t_2) - s(t_1)}{t_2 - t_1} = \frac{s(9) - s(4)}{9 - 4} = \frac{540 - 160}{5} = 76 \text{ feet per second.}$$

✔ **Checkpoint**))) *Audio-video solution in English & Spanish at LarsonPrecalculus.com.*

In Example 7, find the average speed of the car (a) from $t_1 = 0$ to $t_2 = 1$ second and (b) from $t_1 = 1$ second to $t_2 = 4$ seconds.

Average speed is an average rate of change.

Even and Odd Functions

In Section P.3, you studied different types of symmetry of a graph. In the terminology of functions, a function is said to be **even** when its graph is symmetric with respect to the *y*-axis and **odd** when its graph is symmetric with respect to the origin. The symmetry tests in Section P.3 yield the following tests for even and odd functions.

> **Tests for Even and Odd Functions**
>
> A function $y = f(x)$ is **even** when, for each x in the domain of f, $f(-x) = f(x)$.
>
> A function $y = f(x)$ is **odd** when, for each x in the domain of f, $f(-x) = -f(x)$.

EXAMPLE 8 Even and Odd Functions

a. The function $g(x) = x^3 - x$ is odd because $g(-x) = -g(x)$, as follows.

$$g(-x) = (-x)^3 - (-x) \qquad \text{Substitute } -x \text{ for } x.$$

$$= -x^3 + x \qquad \text{Simplify.}$$

$$= -(x^3 - x) \qquad \text{Distributive Property}$$

$$= -g(x) \qquad \text{Test for odd function}$$

b. The function $h(x) = x^2 + 1$ is even because $h(-x) = h(x)$, as follows.

$$h(-x) = (-x)^2 + 1 \qquad \text{Substitute } -x \text{ for } x.$$

$$= x^2 + 1 \qquad \text{Simplify.}$$

$$= h(x) \qquad \text{Test for even function}$$

Figure P.47 shows the graphs and symmetry of these two functions.

 ✓ **Checkpoint** ◀))) *Audio-video solution in English & Spanish at LarsonPrecalculus.com.*

Determine whether the function is even, odd, or neither. Then describe the symmetry.

a. $f(x) = 5 - 3x$ **b.** $g(x) = x^4 - x^2 - 1$ **c.** $h(x) = 2x^3 + 3x$ ■

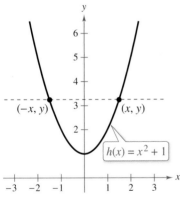

(a) Symmetric to origin: Odd Function

(b) Symmetric to *y*-axis: Even Function

Figure P.47

> ## Summarize   (Section P.6)
>
> 1. State the Vertical Line Test for functions *(page 68)*. For an example of using the Vertical Line Test, see Example 2.
>
> 2. Explain how to find the zeros of a function *(page 69)*. For an example of finding the zeros of functions, see Example 3.
>
> 3. Explain how to determine intervals on which functions are increasing or decreasing *(page 70)* and how to determine relative maximum and relative minimum values of functions *(page 71)*. For an example of describing function behavior, see Example 4. For an example of approximating a relative minimum, see Example 5.
>
> 4. Explain how to determine the average rate of change of a function *(page 72)*. For examples of determining average rates of change, see Examples 6 and 7.
>
> 5. State the definitions of an even function and an odd function *(page 73)*. For an example of identifying even and odd functions, see Example 8.

P.6   Exercises

See **CalcChat.com** for tutorial help and worked-out solutions to odd-numbered exercises.

Vocabulary: Fill in the blanks.

1. The ____ ____ ____ is used to determine whether the graph of an equation is a function of y in terms of x.

2. The ____ of a function f are the values of x for which $f(x) = 0$.

3. A function f is ____ on an interval when, for any x_1 and x_2 in the interval, $x_1 < x_2$ implies $f(x_1) > f(x_2)$.

4. A function value $f(a)$ is a relative ____ of f when there exists an interval (x_1, x_2) containing a such that $x_1 < x < x_2$ implies $f(a) \geq f(x)$.

5. The ____ ____ ____ between any two points $(x_1, f(x_1))$ and $(x_2, f(x_2))$ is the slope of the line through the two points, and this line is called the ____ line.

6. A function f is ____ when, for each x in the domain of f, $f(-x) = -f(x)$.

Skills and Applications

Domain, Range, and Values of a Function In Exercises 7–10, use the graph of the function to find the domain and range of f and the indicated function values.

7. (a) $f(-2)$ (b) $f(-1)$
   (c) $f\left(\frac{1}{2}\right)$ (d) $f(1)$

8. (a) $f(-1)$ (b) $f(2)$
   (c) $f(0)$ (d) $f(1)$

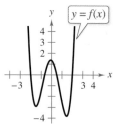
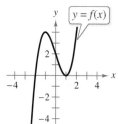

9. (a) $f(2)$ (b) $f(1)$
   (c) $f(3)$ (d) $f(-1)$

10. (a) $f(-2)$ (b) $f(1)$
    (c) $f(0)$ (d) $f(2)$

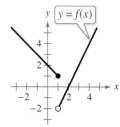
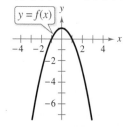

Vertical Line Test for Functions In Exercises 11–14, use the Vertical Line Test to determine whether y is a function of x. To print an enlarged copy of the graph, go to *MathGraphs.com*.

11. $y = \frac{1}{4}x^3$

12. $x - y^2 = 1$

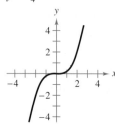
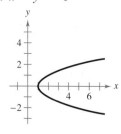

13. $x^2 + y^2 = 25$

14. $x^2 = 2xy - 1$

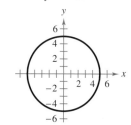
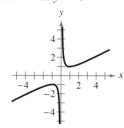

Finding the Zeros of a Function In Exercises 15–24, find the zeros of the function algebraically.

15. $f(x) = 2x^2 - 7x - 30$

16. $f(x) = 3x^2 + 22x - 16$

17. $f(x) = \dfrac{x}{9x^2 - 4}$

18. $f(x) = \dfrac{x^2 - 9x + 14}{4x}$

19. $f(x) = \frac{1}{2}x^3 - x$

20. $f(x) = 9x^4 - 25x^2$

21. $f(x) = x^3 - 4x^2 - 9x + 36$

22. $f(x) = 4x^3 - 24x^2 - x + 6$

23. $f(x) = \sqrt{2x} - 1$

24. $f(x) = \sqrt{3x + 2}$

Graphing and Finding Zeros In Exercises 25–30, (a) use a graphing utility to graph the function and find the zeros of the function and (b) verify your results from part (a) algebraically.

25. $f(x) = 3 + \dfrac{5}{x}$

26. $f(x) = x(x - 7)$

27. $f(x) = \sqrt{2x + 11}$

28. $f(x) = \sqrt{3x - 14} - 8$

29. $f(x) = \dfrac{3x - 1}{x - 6}$

30. $f(x) = \dfrac{2x^2 - 9}{3 - x}$

Describing Function Behavior   In Exercises 31–38, determine the intervals on which the function is increasing, decreasing, or constant.

31. $f(x) = \frac{3}{2}x$

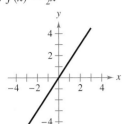

32. $f(x) = x^2 - 4x$

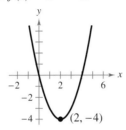

33. $f(x) = x^3 - 3x^2 + 2$

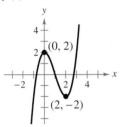

34. $f(x) = \sqrt{x^2 - 1}$

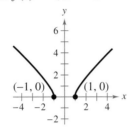

35. $f(x) = |x + 1| + |x - 1|$

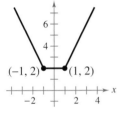

36. $f(x) = \dfrac{x^2 + x + 1}{x + 1}$

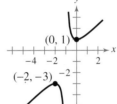

37. $f(x) = \begin{cases} x + 3, & x \le 0 \\ 3, & 0 < x \le 2 \\ 2x + 1, & x > 2 \end{cases}$

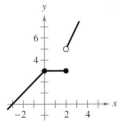

38. $f(x) = \begin{cases} 2x + 1, & x \le -1 \\ x^2 - 2, & x > -1 \end{cases}$

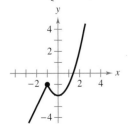

Describing Function Behavior   In Exercises 39–46, (a) use a graphing utility to graph the function and visually determine the intervals on which the function is increasing, decreasing, or constant, and (b) make a table of values to verify whether the function is increasing, decreasing, or constant on the intervals you identified in part (a).

39. $f(x) = 3$ **40.** $g(x) = x$

41. $g(s) = \dfrac{s^2}{4}$ **42.** $f(x) = 3x^4 - 6x^2$

43. $f(x) = \sqrt{1 - x}$ **44.** $f(x) = x\sqrt{x + 3}$

45. $f(x) = x^{3/2}$ **46.** $f(x) = x^{2/3}$

Approximating Relative Minima or Maxima   In Exercises 47–54, use a graphing utility to graph the function and approximate (to two decimal places) any relative minima or maxima.

47. $f(x) = 3x^2 - 2x - 5$ **48.** $f(x) = -x^2 + 3x - 2$

49. $f(x) = -2x^2 + 9x$ **50.** $f(x) = x(x - 2)(x + 3)$

51. $f(x) = x^3 - 3x^2 - x + 1$

52. $h(x) = x^3 - 6x^2 + 15$

53. $h(x) = (x - 1)\sqrt{x}$ **54.** $g(x) = x\sqrt{4 - x}$

Graphical Analysis   In Exercises 55–60, graph the function and determine the interval(s) for which $f(x) \ge 0$.

55. $f(x) = 4 - x$ **56.** $f(x) = 4x + 2$

57. $f(x) = 9 - x^2$ **58.** $f(x) = x^2 - 4x$

59. $f(x) = \sqrt{x - 1}$ **60.** $f(x) = -\left(1 + |x|\right)$

Average Rate of Change of a Function   In Exercises 61–64, find the average rate of change of the function from x_1 to x_2.

Function	x-Values
61. $f(x) = -2x + 15$	$x_1 = 0, x_2 = 3$
62. $f(x) = x^2 - 2x + 8$	$x_1 = 1, x_2 = 5$
63. $f(x) = x^3 - 3x^2 - x$	$x_1 = 1, x_2 = 3$
64. $f(x) = -x^3 + 6x^2 + x$	$x_1 = 1, x_2 = 6$

65. Research and Development   The amounts y (in millions of dollars) the U.S. Department of Energy spent for research and development from 2005 through 2010 can be approximated by the model

$$y = 56.77t^2 - 366.8t + 8916, \quad 5 \le t \le 10$$

where t represents the year, with $t = 5$ corresponding to 2005. (*Source: American Association for the Advancement of Science*)

(a) Use a graphing utility to graph the model.

(b) Find the average rate of change of the model from 2005 to 2010. Interpret your answer in the context of the problem.

66. Finding Average Speed Use the information in Example 7 to find the average speed of the car from $t_1 = 0$ to $t_2 = 9$ seconds. Explain why the result is less than the value obtained in part (b) of Example 7.

Physics In Exercises 67–70, (a) use the position equation $s = -16t^2 + v_0t + s_0$ to write a function that represents the situation, (b) use a graphing utility to graph the function, (c) find the average rate of change of the function from t_1 to t_2, (d) describe the slope of the secant line through t_1 and t_2, (e) find the equation of the secant line through t_1 and t_2, and (f) graph the secant line in the same viewing window as your position function.

67. An object is thrown upward from a height of 6 feet at a velocity of 64 feet per second.
$$t_1 = 0, t_2 = 3$$

68. An object is thrown upward from a height of 6.5 feet at a velocity of 72 feet per second.
$$t_1 = 0, t_2 = 4$$

69. An object is thrown upward from ground level at a velocity of 120 feet per second.
$$t_1 = 3, t_2 = 5$$

70. An object is dropped from a height of 80 feet.
$$t_1 = 1, t_2 = 2$$

Even, Odd, or Neither? In Exercises 71–76, determine whether the function is even, odd, or neither. Then describe the symmetry.

71. $f(x) = x^6 - 2x^2 + 3$ **72.** $g(x) = x^3 - 5x$
73. $f(x) = x\sqrt{1 - x^2}$ **74.** $h(x) = x\sqrt{x + 5}$
75. $f(s) = 4s^{3/2}$ **76.** $g(s) = 4s^{2/3}$

Even, Odd, or Neither? In Exercises 77–82, sketch a graph of the function and determine whether it is even, odd, or neither. Verify your answer algebraically.

77. $f(x) = -9$ **78.** $f(x) = 5 - 3x$
79. $f(x) = -|x - 5|$ **80.** $h(x) = x^2 - 4$
81. $f(x) = \sqrt{1 - x}$ **82.** $g(t) = \sqrt[3]{t - 1}$

Height of a Rectangle In Exercises 83 and 84, write the height h of the rectangle as a function of x.

83.

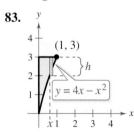

84.

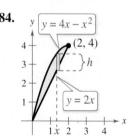

Length of a Rectangle In Exercises 85 and 86, write the length L of the rectangle as a function of y.

85.

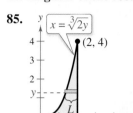

86.

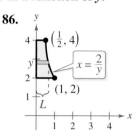

87. Lumens The number of lumens (time rate of flow of light) L from a fluorescent lamp can be approximated by the model

$$L = -0.294x^2 + 97.744x - 664.875, \quad 20 \le x \le 90$$

where x is the wattage of the lamp.

(a) Use a graphing utility to graph the function.

(b) Use the graph from part (a) to estimate the wattage necessary to obtain 2000 lumens.

88. Geometry Corners of equal size are cut from a square with sides of length 8 meters (see figure).

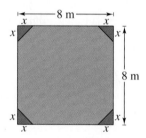

(a) Write the area A of the resulting figure as a function of x. Determine the domain of the function.

(b) Use a graphing utility to graph the area function over its domain. Use the graph to find the range of the function.

(c) Identify the figure that results when x is the maximum value in the domain of the function. What would be the length of each side of the figure?

89. Coordinate Axis Scale Each function described below models the specified data for the years 2003 through 2013, with $t = 3$ corresponding to 2003. Estimate a reasonable scale for the vertical axis (e.g., hundreds, thousands, millions, etc.) of the graph and justify your answer. (There are many correct answers.)

(a) $f(t)$ represents the average salary of college professors.

(b) $f(t)$ represents the U.S. population.

(c) $f(t)$ represents the percent of the civilian work force that is unemployed.

• •90. Data Analysis: Temperature • • • • • • • •

The table shows the temperatures y (in degrees Fahrenheit) in a city over a 24-hour period. Let x represent the time of day, where $x = 0$ corresponds to 6 A.M.

DATA	Time, x	Temperature, y
	0	34
	2	50
	4	60
	6	64
	8	63
	10	59
	12	53
	14	46
	16	40
	18	36
	20	34
	22	37
	24	45

Spreadsheet at LarsonPrecalculus.com

A model that represents these data is given by

$$y = 0.026x^3 - 1.03x^2 + 10.2x + 34, \quad 0 \le x \le 24.$$

(a) Use a graphing utility to create a scatter plot of the data. Then graph the model in the same viewing window.

(b) How well does the model fit the data?

(c) Use the graph to approximate the times when the temperature was increasing and decreasing.

(d) Use the graph to approximate the maximum and minimum temperatures during this 24-hour period.

(e) Could this model predict the temperatures in the city during the next 24-hour period? Why or why not?

91. Writing Use a graphing utility to graph each function. Write a paragraph describing any similarities and differences you observe among the graphs.

(a) $y = x$

(b) $y = x^2$

(c) $y = x^3$

(d) $y = x^4$

(e) $y = x^5$

(f) $y = x^6$

92. HOW DO YOU SEE IT?   Use the graph of the function to answer (a)–(e).

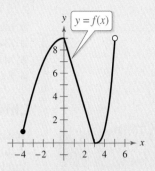

(a) Find the domain and range of f.

(b) Find the zero(s) of f.

(c) Determine the intervals over which f is increasing, decreasing, or constant.

(d) Approximate any relative minimum or relative maximum values of f.

(e) Is f even, odd, or neither?

Exploration

True or False? **In Exercises 93 and 94, determine whether the statement is true or false. Justify your answer.**

93. A function with a square root cannot have a domain that is the set of real numbers.

94. It is possible for an odd function to have the interval $[0, \infty)$ as its domain.

Think About It **In Exercises 95 and 96, find the coordinates of a second point on the graph of a function f when the given point is on the graph and the function is (a) even and (b) odd.**

95. $\left(-\frac{5}{3}, -7\right)$ **96.** $(2a, 2c)$

97. Graphical Reasoning Graph each of the functions with a graphing utility. Determine whether the function is even, odd, or neither.

$f(x) = x^2 - x^4$ $g(x) = 2x^3 + 1$

$h(x) = x^5 - 2x^3 + x$ $j(x) = 2 - x^6 - x^8$

$k(x) = x^5 - 2x^4 + x - 2$ $p(x) = x^9 + 3x^5 - x^3 + x$

What do you notice about the equations of functions that are odd? What do you notice about the equations of functions that are even? Can you describe a way to identify a function as odd or even by inspecting the equation? Can you describe a way to identify a function as neither odd nor even by inspecting the equation?

98. Even, Odd, or Neither?   If f is an even function, determine whether g is even, odd, or neither. Explain.

(a) $g(x) = -f(x)$ (b) $g(x) = f(-x)$

(c) $g(x) = f(x) - 2$ (d) $g(x) = f(x - 2)$

P.7   A Library of Parent Functions

Piecewise-defined functions can help you model real-life situations. For instance, in Exercise 47 on page 84, you will write a piecewise-defined function to model the depth of snow during a snowstorm.

■ Identify and graph linear and squaring functions.
■ Identify and graph cubic, square root, and reciprocal functions.
■ Identify and graph step and other piecewise-defined functions.
■ Recognize graphs of parent functions.

Linear and Squaring Functions

One of the goals of this text is to enable you to recognize the basic shapes of the graphs of different types of functions. For instance, you know that the graph of the **linear function** $f(x) = ax + b$ is a line with slope $m = a$ and y-intercept at $(0, b)$. The graph of the linear function has the following characteristics.

- The domain of the function is the set of all real numbers.
- When $m \neq 0$, the range of the function is the set of all real numbers.
- The graph has an x-intercept at $(-b/m, 0)$ and a y-intercept at $(0, b)$.
- The graph is increasing when $m > 0$, decreasing when $m < 0$, and constant when $m = 0$.

EXAMPLE 1   **Writing a Linear Function**

Write the linear function f for which $f(1) = 3$ and $f(4) = 0$.

Solution   To find the equation of the line that passes through $(x_1, y_1) = (1, 3)$ and $(x_2, y_2) = (4, 0)$, first find the slope of the line.

$$m = \frac{y_2 - y_1}{x_2 - x_1} = \frac{0 - 3}{4 - 1} = \frac{-3}{3} = -1$$

Next, use the point-slope form of the equation of a line.

$$y - y_1 = m(x - x_1) \qquad \text{Point-slope form}$$
$$y - 3 = -1(x - 1) \qquad \text{Substitute for } x_1, y_1, \text{ and } m.$$
$$y = -x + 4 \qquad \text{Simplify.}$$
$$f(x) = -x + 4 \qquad \text{Function notation}$$

The figure below shows the graph of this function.

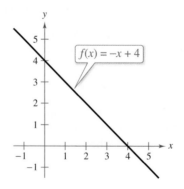

$$f(x) = -x + 4$$

✔ *Checkpoint* ◀))) *Audio-video solution in English & Spanish at LarsonPrecalculus.com.*

Write the linear function f for which $f(-2) = 6$ and $f(4) = -9$.

There are two special types of linear functions, the **constant function** and the **identity function.** A constant function has the form

$$f(x) = c$$

and has the domain of all real numbers with a range consisting of a single real number c. The graph of a constant function is a horizontal line, as shown in Figure P.48. The identity function has the form

$$f(x) = x.$$

Its domain and range are the set of all real numbers. The identity function has a slope of $m = 1$ and a y-intercept at $(0, 0)$. The graph of the identity function is a line for which each x-coordinate equals the corresponding y-coordinate. The graph is always increasing, as shown in Figure P.49.

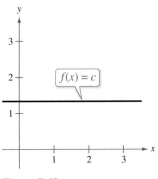

Figure P.48

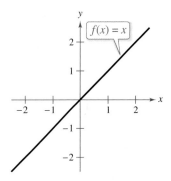

Figure P.49

The graph of the **squaring function**

$$f(x) = x^2$$

is a U-shaped curve with the following characteristics.

- The domain of the function is the set of all real numbers.
- The range of the function is the set of all nonnegative real numbers.
- The function is even.
- The graph has an intercept at $(0, 0)$.
- The graph is decreasing on the interval $(-\infty, 0)$ and increasing on the interval $(0, \infty)$.
- The graph is symmetric with respect to the y-axis.
- The graph has a relative minimum at $(0, 0)$.

The figure below shows the graph of the squaring function.

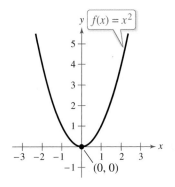

Cubic, Square Root, and Reciprocal Functions

The following summarizes the basic characteristics of the graphs of the **cubic, square root,** and **reciprocal functions.**

1. The graph of the *cubic* function

$$f(x) = x^3$$

has the following characteristics.

- The domain of the function is the set of all real numbers.
- The range of the function is the set of all real numbers.
- The function is odd.
- The graph has an intercept at $(0, 0)$.
- The graph is increasing on the interval $(-\infty, \infty)$.
- The graph is symmetric with respect to the origin.

The figure shows the graph of the cubic function.

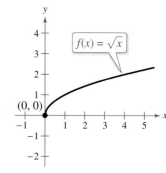

Cubic function

2. The graph of the *square root* function

$$f(x) = \sqrt{x}$$

has the following characteristics.

- The domain of the function is the set of all nonnegative real numbers.
- The range of the function is the set of all nonnegative real numbers.
- The graph has an intercept at $(0, 0)$.
- The graph is increasing on the interval $(0, \infty)$.

The figure shows the graph of the square root function.

Square root function

3. The graph of the *reciprocal* function

$$f(x) = \frac{1}{x}$$

has the following characteristics.

- The domain of the function is $(-\infty, 0) \cup (0, \infty)$.
- The range of the function is $(-\infty, 0) \cup (0, \infty)$.
- The function is odd.
- The graph does not have any intercepts.
- The graph is decreasing on the intervals $(-\infty, 0)$ and $(0, \infty)$.
- The graph is symmetric with respect to the origin.

The figure shows the graph of the reciprocal function.

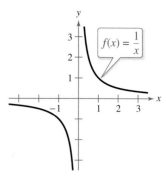

Reciprocal function

Step and Piecewise-Defined Functions

Functions whose graphs resemble sets of stairsteps are known as **step functions.** The most famous of the step functions is the **greatest integer function,** denoted by $[\![x]\!]$ and defined as

$$f(x) = [\![x]\!] = \textit{the greatest integer less than or equal to } x.$$

Some values of the greatest integer function are as follows.

$$[\![-1]\!] = (\text{greatest integer} \le -1) = -1$$
$$\left[\!\!\left[-\tfrac{1}{2}\right]\!\!\right] = \left(\text{greatest integer} \le -\tfrac{1}{2}\right) = -1$$
$$\left[\!\!\left[\tfrac{1}{10}\right]\!\!\right] = \left(\text{greatest integer} \le \tfrac{1}{10}\right) = 0$$
$$[\![1.5]\!] = (\text{greatest integer} \le 1.5) = 1$$
$$[\![1.9]\!] = (\text{greatest integer} \le 1.9) = 1$$

The graph of the greatest integer function

$$f(x) = [\![x]\!]$$

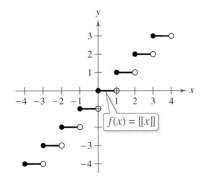

has the following characteristics, as shown in Figure P.50.

- The domain of the function is the set of all real numbers.
- The range of the function is the set of all integers.
- The graph has a y-intercept at $(0, 0)$ and x-intercepts in the interval $[0, 1)$.
- The graph is constant between each pair of consecutive integer values of x.
- The graph jumps vertically one unit at each integer value of x.

Figure P.50

▷ TECHNOLOGY   When using your graphing utility to graph a step function, you should set your graphing utility to *dot* mode.

EXAMPLE 2   **Evaluating a Step Function**

Evaluate the function when $x = -1, 2,$ and $\tfrac{3}{2}$.

$$f(x) = [\![x]\!] + 1$$

Solution   For $x = -1$, the greatest integer ≤ -1 is -1, so

$$f(-1) = [\![-1]\!] + 1 = -1 + 1 = 0.$$

For $x = 2$, the greatest integer ≤ 2 is 2, so

$$f(2) = [\![2]\!] + 1 = 2 + 1 = 3.$$

For $x = \tfrac{3}{2}$, the greatest integer $\le \tfrac{3}{2}$ is 1, so

$$f\!\left(\tfrac{3}{2}\right) = \left[\!\!\left[\tfrac{3}{2}\right]\!\!\right] + 1 = 1 + 1 = 2.$$

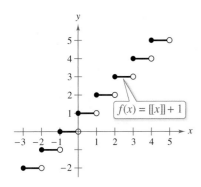

Figure P.51

Verify your answers by examining the graph of $f(x) = [\![x]\!] + 1$ shown in Figure P.51.

✓ **Checkpoint** ◀))) *Audio-video solution in English & Spanish at LarsonPrecalculus.com.*

Evaluate the function when $x = -\tfrac{3}{2}, 1,$ and $-\tfrac{5}{2}$.

$$f(x) = [\![x + 2]\!]$$

Recall from Section P.5 that a piecewise-defined function is defined by two or more equations over a specified domain. To graph a piecewise-defined function, graph each equation separately over the specified domain, as shown in Example 3.

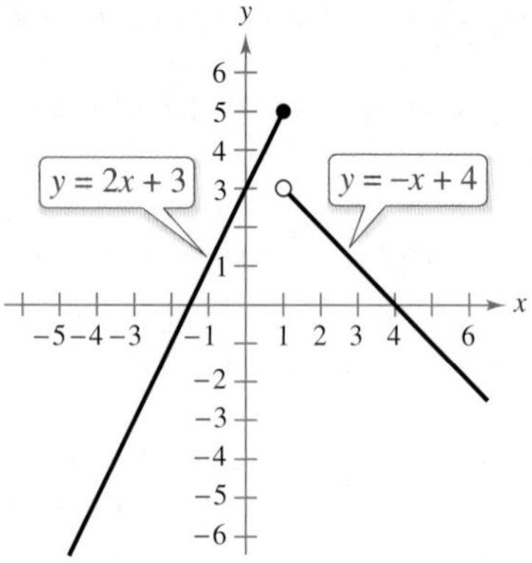

Figure P.52

EXAMPLE 3 Graphing a Piecewise-Defined Function

Sketch the graph of $f(x) = \begin{cases} 2x + 3, & x \le 1 \\ -x + 4, & x > 1 \end{cases}$.

Solution This piecewise-defined function consists of two linear functions. At $x = 1$ and to the left of $x = 1$, the graph is the line $y = 2x + 3$, and to the right of $x = 1$ the graph is the line $y = -x + 4$, as shown in Figure P.52. Notice that the point $(1, 5)$ is a solid dot and the point $(1, 3)$ is an open dot. This is because $f(1) = 2(1) + 3 = 5$.

✓ ***Checkpoint*** *Audio-video solution in English & Spanish at LarsonPrecalculus.com.*

Sketch the graph of $f(x) = \begin{cases} -\frac{1}{2}x - 6, & x \le -4 \\ x + 5, & x > -4 \end{cases}$.

Parent Functions

The eight graphs shown below represent the most commonly used functions in algebra. Familiarity with the basic characteristics of these simple graphs will help you analyze the shapes of more complicated graphs—in particular, graphs obtained from these graphs by the rigid and nonrigid transformations studied in the next section.

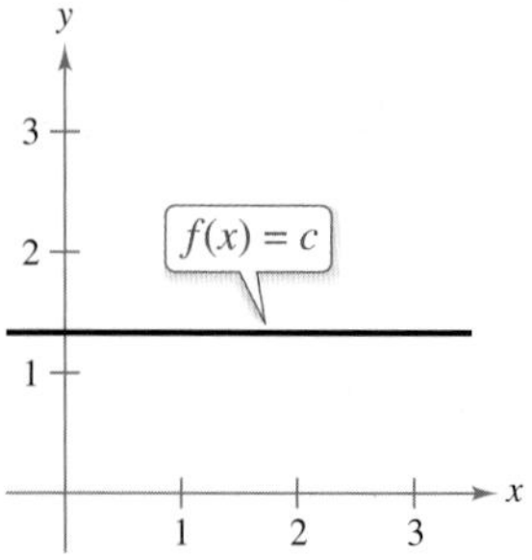

(a) Constant Function

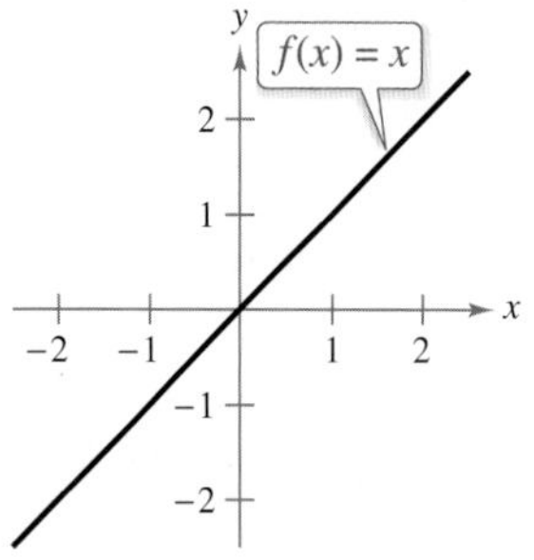

(b) Identity Function

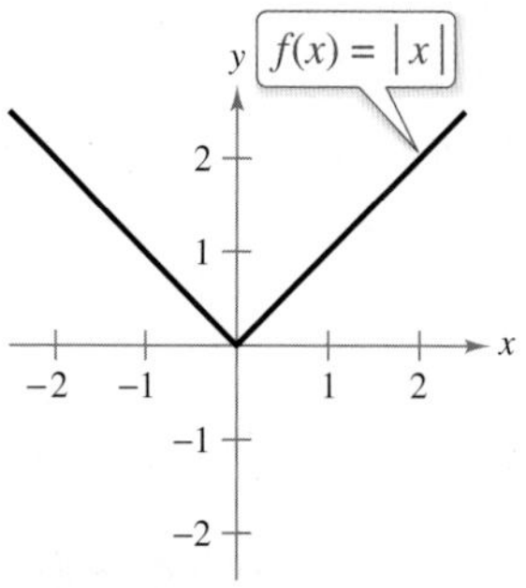

(c) Absolute Value Function

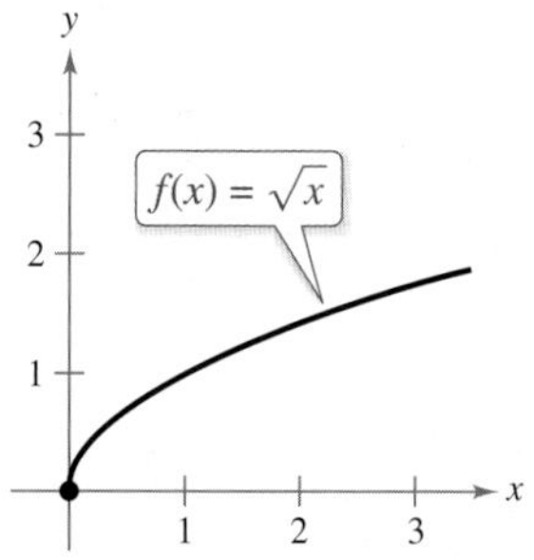

(d) Square Root Function

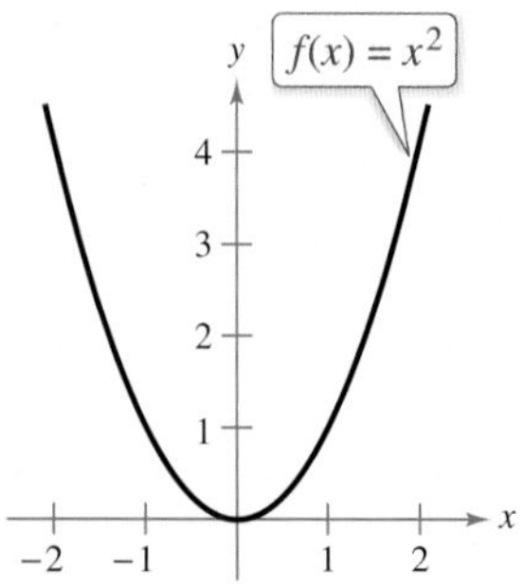

(e) Quadratic Function

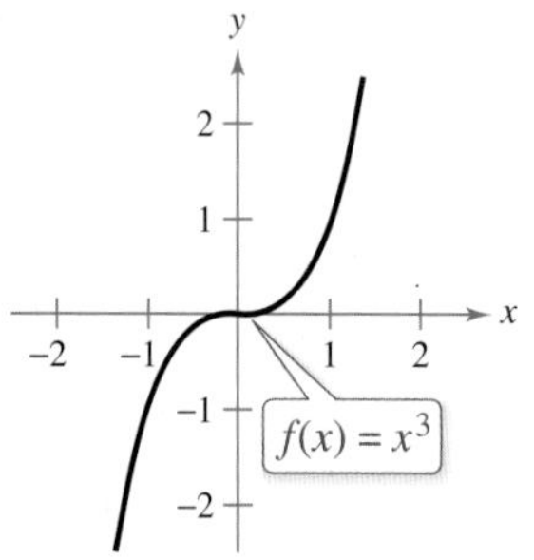

(f) Cubic Function

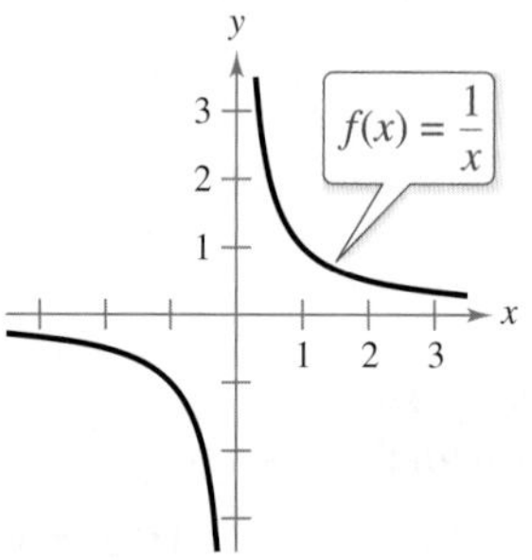

(g) Reciprocal Function

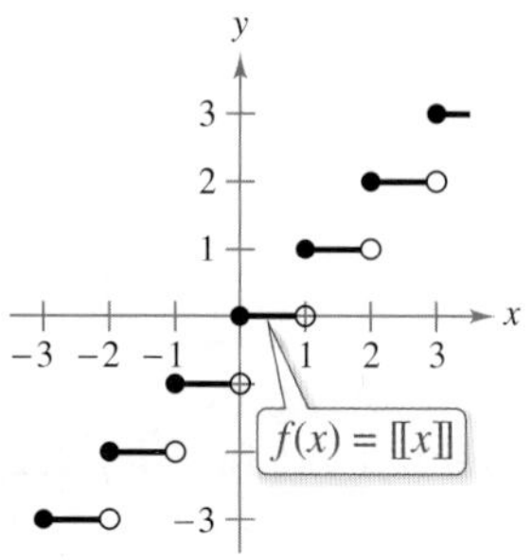

(h) Greatest Integer Function

Summarize (Section P.7)

1. Explain how to identify and graph linear and squaring functions *(pages 78 and 79)*. For an example involving a linear function, see Example 1.
2. Explain how to identify and graph cubic, square root, and reciprocal functions *(page 80)*.
3. Explain how to identify and graph step and other piecewise-defined functions *(page 81)*. For an example involving a step function, see Example 2. For an example of graphing a piecewise-defined function, see Example 3.
4. State and sketch the graphs of parent functions *(page 82)*.

P.7 Exercises

See CalcChat.com for tutorial help and worked-out solutions to odd-numbered exercises.

Vocabulary

In Exercises 1–9, match each function with its name.

1. $f(x) = [\![x]\!]$
2. $f(x) = x$
3. $f(x) = 1/x$
4. $f(x) = x^2$
5. $f(x) = \sqrt{x}$
6. $f(x) = c$
7. $f(x) = |x|$
8. $f(x) = x^3$
9. $f(x) = ax + b$

  (a) squaring function
  (b) square root function
  (c) cubic function

  (d) linear function
  (e) constant function
  (f) absolute value function

  (g) greatest integer function
  (h) reciprocal function
  (i) identity function

10. Fill in the blank: The constant function and the identity function are two special types of _____ functions.

Skills and Applications

Writing a Linear Function In Exercises 11–14, (a) write the linear function f such that it has the indicated function values and (b) sketch the graph of the function.

11. $f(1) = 4, \quad f(0) = 6$
12. $f(-3) = -8, \quad f(1) = 2$
13. $f(-5) = -1, \quad f(5) = -1$
14. $f\left(\frac{2}{3}\right) = -\frac{15}{2}, \quad f(-4) = -11$

Graphing a Function In Exercises 15–26, use a graphing utility to graph the function. Be sure to choose an appropriate viewing window.

15. $f(x) = 2.5x - 4.25$
16. $f(x) = \frac{5}{6} - \frac{2}{3}x$
17. $g(x) = -2x^2$
18. $f(x) = 3x^2 - 1.75$
19. $f(x) = x^3 - 1$
20. $f(x) = (x - 1)^3 + 2$
21. $f(x) = 4 - 2\sqrt{x}$
22. $h(x) = \sqrt{x + 2} + 3$
23. $f(x) = 4 + (1/x)$
24. $k(x) = 1/(x - 3)$
25. $g(x) = |x| - 5$
26. $f(x) = |x - 1|$

Evaluating a Step Function In Exercises 27–30, evaluate the function for the indicated values.

27. $f(x) = [\![x]\!]$
  (a) $f(2.1)$ (b) $f(2.9)$ (c) $f(-3.1)$ (d) $f\left(\frac{7}{2}\right)$

28. $h(x) = [\![x + 3]\!]$
  (a) $h(-2)$ (b) $h\left(\frac{1}{2}\right)$ (c) $h(4.2)$ (d) $h(-21.6)$

29. $k(x) = \left[\!\left[\frac{1}{2}x + 6\right]\!\right]$
  (a) $k(5)$ (b) $k(-6.1)$ (c) $k(0.1)$ (d) $k(15)$

30. $g(x) = -7[\![x + 4]\!] + 6$
  (a) $g\left(\frac{1}{8}\right)$ (b) $g(9)$ (c) $g(-4)$ (d) $g\left(\frac{3}{2}\right)$

Graphing a Step Function In Exercises 31–34, sketch the graph of the function.

31. $g(x) = -[\![x]\!]$
32. $g(x) = 4[\![x]\!]$
33. $g(x) = [\![x]\!] - 1$
34. $g(x) = [\![x - 3]\!]$

Graphing a Piecewise-Defined Function In Exercises 35–40, sketch the graph of the function.

35. $g(x) = \begin{cases} x + 6, & x \le -4 \\ \frac{1}{2}x - 4, & x > -4 \end{cases}$

36. $f(x) = \begin{cases} \sqrt{4 + x}, & x < 0 \\ \sqrt{4 - x}, & x \ge 0 \end{cases}$

37. $f(x) = \begin{cases} 1 - (x - 1)^2, & x \le 2 \\ \sqrt{x - 2}, & x > 2 \end{cases}$

38. $f(x) = \begin{cases} x^2 + 5, & x \le 1 \\ -x^2 + 4x + 3, & x > 1 \end{cases}$

39. $h(x) = \begin{cases} 4 - x^2, & x < -2 \\ 3 + x, & -2 \le x < 0 \\ x^2 + 1, & x \ge 0 \end{cases}$

40. $k(x) = \begin{cases} 2x + 1, & x \le -1 \\ 2x^2 - 1, & -1 < x \le 1 \\ 1 - x^2, & x > 1 \end{cases}$

Graphing a Function In Exercises 41 and 42, (a) use a graphing utility to graph the function and (b) state the domain and range of the function.

41. $s(x) = 2\left(\frac{1}{4}x - \left[\!\left[\frac{1}{4}x\right]\!\right]\right)$
42. $k(x) = 4\left(\frac{1}{2}x - \left[\!\left[\frac{1}{2}x\right]\!\right]\right)^2$

43. **Wages** A mechanic's pay is $14.00 per hour for regular time and time-and-a-half for overtime. The weekly wage function is

$$W(h) = \begin{cases} 14h, & 0 < h \le 40 \\ 21(h - 40) + 560, & h > 40 \end{cases}$$

where h is the number of hours worked in a week.

  (a) Evaluate $W(30)$, $W(40)$, $W(45)$, and $W(50)$.

  (b) The company increased the regular work week to 45 hours. What is the new weekly wage function?

44. Revenue The table shows the monthly revenue y (in thousands of dollars) of a landscaping business for each month of the year 2013, with $x = 1$ representing January.

DATA	Month, x	Revenue, y
	1	5.2
	2	5.6
	3	6.6
	4	8.3
	5	11.5
	6	15.8
	7	12.8
	8	10.1
	9	8.6
	10	6.9
	11	4.5
	12	2.7

Spreadsheet at LarsonPrecalculus.com

A mathematical model that represents these data is

$$f(x) = \begin{cases} -1.97x + 26.3 \\ 0.505x^2 - 1.47x + 6.3. \end{cases}$$

(a) Use a graphing utility to graph the model. What is the domain of each part of the piecewise-defined function? How can you tell? Explain your reasoning.

(b) Find $f(5)$ and $f(11)$, and interpret your results in the context of the problem.

(c) How do the values obtained from the model in part (a) compare with the actual data values?

45. Fluid Flow The intake pipe of a 100-gallon tank has a flow rate of 10 gallons per minute, and two drainpipes have flow rates of 5 gallons per minute each. The figure shows the volume V of fluid in the tank as a function of time t. Determine the combination of the input pipe and drain pipes in which the fluid is flowing in specific subintervals of the 1 hour of time shown on the graph. (There are many correct answers.)

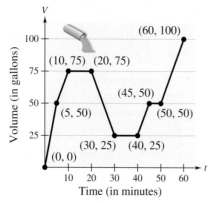

nulinukas/Shutterstock.com

46. Delivery Charges The cost of sending an overnight package from New York to Atlanta is $26.10 for a package weighing up to, but not including, 1 pound and $4.35 for each additional pound or portion of a pound.

(a) Use the greatest integer function to create a model for the cost C of overnight delivery of a package weighing x pounds, $x > 0$.

(b) Sketch the graph of the function.

47. Snowstorm

During a nine-hour snowstorm, it snows at a rate of 1 inch per hour for the first 2 hours, at a rate of 2 inches per hour for the next 6 hours, and at a rate of 0.5 inch per hour for the final hour. Write and graph a piecewise-defined function that gives the depth of the snow during the snowstorm. How many inches of snow accumulated from the storm?

Exploration

48. HOW DO YOU SEE IT? For each graph of f shown below, answer (a)–(d).

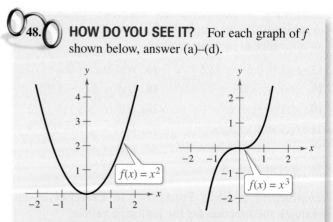

(a) Find the domain and range of f.

(b) Find the x- and y-intercepts of the graph of f.

(c) Determine the intervals on which f is increasing, decreasing, or constant.

(d) Determine whether f is even, odd, or neither. Then describe the symmetry.

True or False? In Exercises 49 and 50, determine whether the statement is true or false. Justify your answer.

49. A piecewise-defined function will always have at least one x-intercept or at least one y-intercept.

50. A linear equation will always have an x-intercept and a y-intercept.

P.8   Transformations of Functions

Transformations of functions can help you model real-life applications. For instance, Exercise 69 on page 92 shows how a transformation of a function can model the number of horsepower required to overcome wind drag on an automobile.

■ Use vertical and horizontal shifts to sketch graphs of functions.
■ Use reflections to sketch graphs of functions.
■ Use nonrigid transformations to sketch graphs of functions.

Shifting Graphs

Many functions have graphs that are transformations of the parent graphs summarized in Section P.7. For example, you can obtain the graph of

$$h(x) = x^2 + 2$$

by shifting the graph of $f(x) = x^2$ *up* two units, as shown in Figure P.53. In function notation, h and f are related as follows.

$$h(x) = x^2 + 2 = f(x) + 2 \qquad \text{Upward shift of two units}$$

Similarly, you can obtain the graph of

$$g(x) = (x - 2)^2$$

by shifting the graph of $f(x) = x^2$ to the *right* two units, as shown in Figure P.54. In this case, the functions g and f have the following relationship.

$$g(x) = (x - 2)^2 = f(x - 2) \qquad \text{Right shift of two units}$$

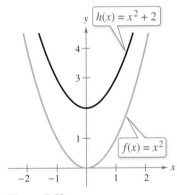

Figure P.53

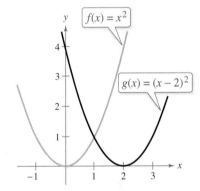

Figure P.54

The following list summarizes this discussion about horizontal and vertical shifts.

• • REMARK   In items 3 and 4, be sure you see that $h(x) = f(x - c)$ corresponds to a *right* shift and $h(x) = f(x + c)$ corresponds to a *left* shift for $c > 0$.

> ### Vertical and Horizontal Shifts
>
> Let c be a positive real number. **Vertical and horizontal shifts** in the graph of $y = f(x)$ are represented as follows.
>
> **1.** Vertical shift c units *up:* $h(x) = f(x) + c$
>
> **2.** Vertical shift c units *down:* $h(x) = f(x) - c$
>
> **3.** Horizontal shift c units to the *right:* $h(x) = f(x - c)$
>
> **4.** Horizontal shift c units to the *left:* $h(x) = f(x + c)$

Some graphs can be obtained from combinations of vertical and horizontal shifts, as demonstrated in Example 1(b). Vertical and horizontal shifts generate a *family of functions,* each with the same shape but at a different location in the plane.

EXAMPLE 1   **Shifts in the Graph of a Function**

Use the graph of $f(x) = x^3$ to sketch the graph of each function.

a. $g(x) = x^3 - 1$

b. $h(x) = (x + 2)^3 + 1$

Solution

a. Relative to the graph of $f(x) = x^3$, the graph of

$$g(x) = x^3 - 1$$

is a downward shift of one unit, as shown below.

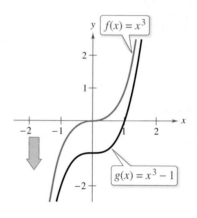

• • • • • • • • • • • • • • • ▷
• •**REMARK** In Example 1(a),
note that $g(x) = f(x) - 1$ and
in Example 1(b),
$h(x) = f(x + 2) + 1.$

b. Relative to the graph of $f(x) = x^3$, the graph of

$$h(x) = (x + 2)^3 + 1$$

involves a left shift of two units and an upward shift of one unit, as shown below.

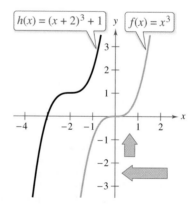

✓ *Checkpoint* ◀))) *Audio-video solution in English & Spanish at LarsonPrecalculus.com.*

Use the graph of $f(x) = x^3$ to sketch the graph of each function.

a. $h(x) = x^3 + 5$

b. $g(x) = (x - 3)^3 + 2$

In Example 1(b), you obtain the same result when the vertical shift precedes the horizontal shift *or* when the horizontal shift precedes the vertical shift.

Reflecting Graphs

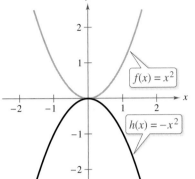

Another common type of transformation is a **reflection.** For instance, if you consider the x-axis to be a mirror, then the graph of $h(x) = -x^2$ is the mirror image (or reflection) of the graph of $f(x) = x^2$, as shown in Figure P.55.

Figure P.55

Reflections in the Coordinate Axes

Reflections in the coordinate axes of the graph of $y = f(x)$ are represented as follows.

1. Reflection in the x-axis: $h(x) = -f(x)$

2. Reflection in the y-axis: $h(x) = f(-x)$

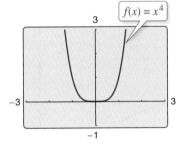

Figure P.56

EXAMPLE 2   Writing Equations from Graphs

The graph of the function

$$f(x) = x^4$$

is shown in Figure P.56. Each of the graphs below is a transformation of the graph of f. Write an equation for each of these functions.

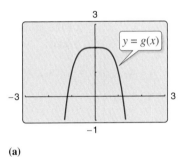

(a)

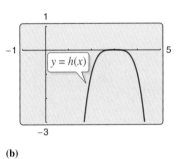

(b)

Solution

a. The graph of g is a reflection in the x-axis *followed by* an upward shift of two units of the graph of $f(x) = x^4$. So, the equation for g is

$$g(x) = -x^4 + 2.$$

b. The graph of h is a horizontal shift of three units to the right *followed by* a reflection in the x-axis of the graph of $f(x) = x^4$. So, the equation for h is

$$h(x) = -(x - 3)^4.$$

✓ ***Checkpoint*** ◀))) Audio-video solution in English & Spanish at *LarsonPrecalculus.com.*

The graph is a transformation of the graph of $f(x) = x^4$. Write an equation for the function.

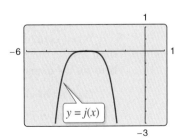

| EXAMPLE 3 | **Reflections and Shifts** |

Compare the graph of each function with the graph of $f(x) = \sqrt{x}$.

a. $g(x) = -\sqrt{x}$ **b.** $h(x) = \sqrt{-x}$ **c.** $k(x) = -\sqrt{x+2}$

Algebraic Solution

a. The graph of g is a reflection of the graph of f in the x-axis because

$$g(x) = -\sqrt{x}$$

$$= -f(x).$$

b. The graph of h is a reflection of the graph of f in the y-axis because

$$h(x) = \sqrt{-x}$$

$$= f(-x).$$

c. The graph of k is a left shift of two units followed by a reflection in the x-axis because

$$k(x) = -\sqrt{x+2}$$

$$= -f(x+2).$$

Graphical Solution

a. Graph f and g on the same set of coordinate axes. From the graph, you can see that the graph of g is a reflection of the graph of f in the x-axis.

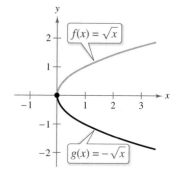

b. Graph f and h on the same set of coordinate axes. From the graph, you can see that the graph of h is a reflection of the graph of f in the y-axis.

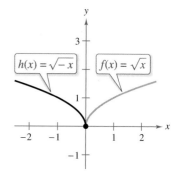

c. Graph f and k on the same set of coordinate axes. From the graph, you can see that the graph of k is a left shift of two units of the graph of f, followed by a reflection in the x-axis.

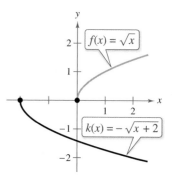

✓ *Checkpoint* *Audio-video solution in English & Spanish at LarsonPrecalculus.com.*

Compare the graph of each function with the graph of

$$f(x) = \sqrt{x-1}.$$

a. $g(x) = -\sqrt{x-1}$ **b.** $h(x) = \sqrt{-x-1}$

When sketching the graphs of functions involving square roots, remember that you must restrict the domain to exclude negative numbers inside the radical. For instance, here are the domains of the functions in Example 3.

Domain of $g(x) = -\sqrt{x}$: $x \geq 0$

Domain of $h(x) = \sqrt{-x}$: $x \leq 0$

Domain of $k(x) = -\sqrt{x+2}$: $x \geq -2$

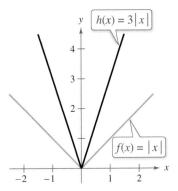

Figure P.57

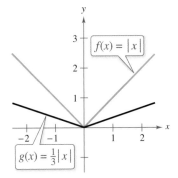

Figure P.58

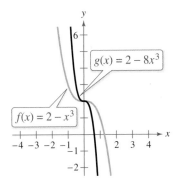

Figure P.59

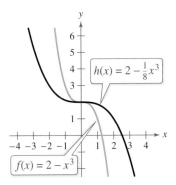

Figure P.60

Nonrigid Transformations

Horizontal shifts, vertical shifts, and reflections are **rigid transformations** because the basic shape of the graph is unchanged. These transformations change only the *position* of the graph in the coordinate plane. **Nonrigid transformations** are those that cause a *distortion*—a change in the shape of the original graph. For instance, a nonrigid transformation of the graph of $y = f(x)$ is represented by $g(x) = cf(x)$, where the transformation is a **vertical stretch** when $c > 1$ and a **vertical shrink** when $0 < c < 1$. Another nonrigid transformation of the graph of $y = f(x)$ is represented by $h(x) = f(cx)$, where the transformation is a **horizontal shrink** when $c > 1$ and a **horizontal stretch** when $0 < c < 1$.

EXAMPLE 4 Nonrigid Transformations

Compare the graph of each function with the graph of $f(x) = |x|$.

a. $h(x) = 3|x|$ **b.** $g(x) = \frac{1}{3}|x|$

Solution

a. Relative to the graph of $f(x) = |x|$, the graph of $h(x) = 3|x| = 3f(x)$ is a vertical stretch (each y-value is multiplied by 3) of the graph of f. (See Figure P.57.)

b. Similarly, the graph of $g(x) = \frac{1}{3}|x| = \frac{1}{3}f(x)$ is a vertical shrink $\left(\text{each } y\text{-value is multiplied by } \frac{1}{3}\right)$ of the graph of f. (See Figure P.58.)

✓ *Checkpoint* Audio-video solution in English & Spanish at *LarsonPrecalculus.com*.

Compare the graph of each function with the graph of $f(x) = x^2$.

a. $g(x) = 4x^2$ **b.** $h(x) = \frac{1}{4}x^2$

EXAMPLE 5 Nonrigid Transformations

Compare the graph of each function with the graph of $f(x) = 2 - x^3$.

a. $g(x) = f(2x)$ **b.** $h(x) = f\left(\frac{1}{2}x\right)$

Solution

a. Relative to the graph of $f(x) = 2 - x^3$, the graph of $g(x) = f(2x) = 2 - (2x)^3 = 2 - 8x^3$ is a horizontal shrink $(c > 1)$ of the graph of f. (See Figure P.59.)

b. Similarly, the graph of $h(x) = f\left(\frac{1}{2}x\right) = 2 - \left(\frac{1}{2}x\right)^3 = 2 - \frac{1}{8}x^3$ is a horizontal stretch $(0 < c < 1)$ of the graph of f. (See Figure P.60.)

✓ *Checkpoint* Audio-video solution in English & Spanish at *LarsonPrecalculus.com*.

Compare the graph of each function with the graph of $f(x) = x^2 + 3$.

a. $g(x) = f(2x)$ **b.** $h(x) = f\left(\frac{1}{2}x\right)$ ■

Summarize (Section P.8)

1. Describe how to shift the graph of a function vertically and horizontally *(page 85)*. For an example of shifting the graph of a function, see Example 1.

2. Describe how to reflect the graph of a function in the x-axis and in the y-axis *(page 87)*. For examples of reflecting graphs of functions, see Examples 2 and 3.

3. Describe nonrigid transformations of the graph of a function *(page 89)*. For examples of nonrigid transformations, see Examples 4 and 5.

P.8   Exercises

See **CalcChat.com** for tutorial help and worked-out solutions to odd-numbered exercises.

Vocabulary

In Exercises 1–3, fill in the blanks.

1. Horizontal shifts, vertical shifts, and reflections are called _____ transformations.

2. A reflection in the x-axis of $y = f(x)$ is represented by $h(x) = $ _____, while a reflection in the y-axis of $y = f(x)$ is represented by $h(x) = $ _____.

3. A nonrigid transformation of $y = f(x)$ represented by $g(x) = cf(x)$ is a _____ _____ when $c > 1$ and a _____ _____ when $0 < c < 1$.

4. Match the rigid transformation of $y = f(x)$ with the correct representation of the graph of h, where $c > 0$.
   (a) $h(x) = f(x) + c$ (i) A horizontal shift of f, c units to the right
   (b) $h(x) = f(x) - c$ (ii) A vertical shift of f, c units down
   (c) $h(x) = f(x + c)$ (iii) A horizontal shift of f, c units to the left
   (d) $h(x) = f(x - c)$ (iv) A vertical shift of f, c units up

Skills and Applications

5. **Shifts in the Graph of a Function** For each function, sketch (on the same set of coordinate axes) a graph of each function for $c = -1$, 1, and 3.
   (a) $f(x) = |x| + c$ (b) $f(x) = |x - c|$

6. **Shifts in the Graph of a Function** For each function, sketch (on the same set of coordinate axes) a graph of each function for $c = -3$, -1, 1, and 3.
   (a) $f(x) = \sqrt{x} + c$ (b) $f(x) = \sqrt{x - c}$

7. **Shifts in the Graph of a Function** For each function, sketch (on the same set of coordinate axes) a graph of each function for $c = -2$, 0, and 2.
   (a) $f(x) = [\![x]\!] + c$ (b) $f(x) = [\![x + c]\!]$

8. **Shifts in the Graph of a Function** For each function, sketch (on the same set of coordinate axes) a graph of each function for $c = -3$, -1, 1, and 3.

   (a) $f(x) = \begin{cases} x^2 + c, & x < 0 \\ -x^2 + c, & x \ge 0 \end{cases}$

   (b) $f(x) = \begin{cases} (x + c)^2, & x < 0 \\ -(x + c)^2, & x \ge 0 \end{cases}$

Sketching Transformations **In Exercises 9 and 10, use the graph of f to sketch each graph. To print an enlarged copy of the graph, go to *MathGraphs.com*.**

9. (a) $y = f(-x)$
   (b) $y = f(x) + 4$
   (c) $y = 2f(x)$
   (d) $y = -f(x - 4)$
   (e) $y = f(x) - 3$
   (f) $y = -f(x) - 1$
   (g) $y = f(2x)$

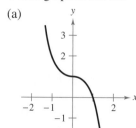

10. (a) $y = f(x - 5)$
    (b) $y = -f(x) + 3$
    (c) $y = \frac{1}{3}f(x)$
    (d) $y = -f(x + 1)$
    (e) $y = f(-x)$
    (f) $y = f(x) - 10$
    (g) $y = f\left(\frac{1}{3}x\right)$

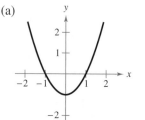

11. **Writing Equations from Graphs** Use the graph of $f(x) = x^2$ to write an equation for each function whose graph is shown.
    (a)   (b)

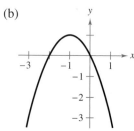

12. **Writing Equations from Graphs** Use the graph of $f(x) = x^3$ to write an equation for each function whose graph is shown.
    (a)   (b)

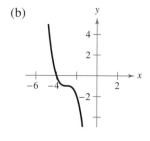

13. Writing Equations from Graphs   Use the graph of $f(x) = |x|$ to write an equation for each function whose graph is shown.

(a)

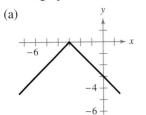

(b)

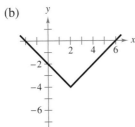

14. Writing Equations from Graphs   Use the graph of $f(x) = \sqrt{x}$ to write an equation for each function whose graph is shown.

(a)

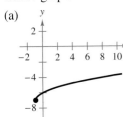

(b)

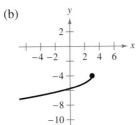

Identifying a Parent Function   In Exercises 15–20, identify the parent function and the transformation shown in the graph. Write an equation for the function shown in the graph.

15.

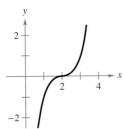

16.

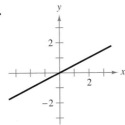

17.

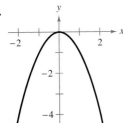

18.

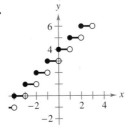

19.

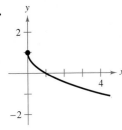

20.

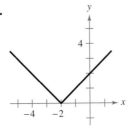

Identifying a Parent Function   In Exercises 21–46, g is related to one of the parent functions described in Section P.7. (a) Identify the parent function f. (b) Describe the sequence of transformations from f to g. (c) Sketch the graph of g. (d) Use function notation to write g in terms of f.

21. $g(x) = 12 - x^2$
22. $g(x) = (x - 8)^2$
23. $g(x) = x^3 + 7$
24. $g(x) = -x^3 - 1$
25. $g(x) = \frac{2}{3}x^2 + 4$
26. $g(x) = 2(x - 7)^2$
27. $g(x) = 2 - (x + 5)^2$
28. $g(x) = -\frac{1}{4}(x + 2)^2 - 2$
29. $g(x) = \sqrt{3x}$
30. $g(x) = \sqrt{\frac{1}{4}x}$
31. $g(x) = (x - 1)^3 + 2$
32. $g(x) = (x + 3)^3 - 10$
33. $g(x) = 3(x - 2)^3$
34. $g(x) = -\frac{1}{2}(x + 1)^3$
35. $g(x) = -|x| - 2$
36. $g(x) = 6 - |x + 5|$
37. $g(x) = -|x + 4| + 8$
38. $g(x) = |-x + 3| + 9$
39. $g(x) = -2|x - 1| - 4$
40. $g(x) = \frac{1}{2}|x - 2| - 3$
41. $g(x) = 3 - [\![x]\!]$
42. $g(x) = 2[\![x + 5]\!]$
43. $g(x) = \sqrt{x - 9}$
44. $g(x) = \sqrt{x + 4} + 8$
45. $g(x) = \sqrt{7 - x} - 2$
46. $g(x) = \sqrt{3x} + 1$

Writing an Equation from a Description   In Exercises 47–54, write an equation for the function described by the given characteristics.

47. The shape of $f(x) = x^2$, but shifted three units to the right and seven units down

48. The shape of $f(x) = x^2$, but shifted two units to the left, nine units up, and then reflected in the x-axis

49. The shape of $f(x) = x^3$, but shifted 13 units to the right

50. The shape of $f(x) = x^3$, but shifted six units to the left, six units down, and then reflected in the y-axis

51. The shape of $f(x) = |x|$, but shifted 12 units up and then reflected in the x-axis

52. The shape of $f(x) = |x|$, but shifted four units to the left and eight units down

53. The shape of $f(x) = \sqrt{x}$, but shifted six units to the left and then reflected in both the x-axis and the y-axis

54. The shape of $f(x) = \sqrt{x}$, but shifted nine units down and then reflected in both the x-axis and the y-axis

55. Writing Equations from Graphs   Use the graph of $f(x) = x^2$ to write an equation for each function whose graph is shown.

(a)

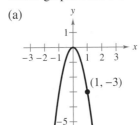

(1, -3)

(b)

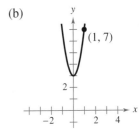

(1, 7)

56. Writing Equations from Graphs   Use the graph of

$$f(x) = x^3$$

to write an equation for each function whose graph is shown.

(a)

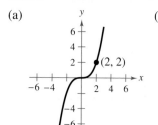

(2, 2)

(b)

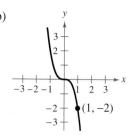

(1, −2)

57. Writing Equations from Graphs   Use the graph of

$$f(x) = |x|$$

to write an equation for each function whose graph is shown.

(a)

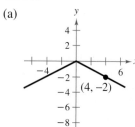

(4, −2)

(b)

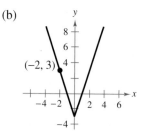

(−2, 3)

58. Writing Equations from Graphs   Use the graph of

$$f(x) = \sqrt{x}$$

to write an equation for each function whose graph is shown.

(a)

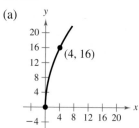

(4, 16)

(b)

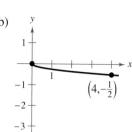

$\left(4, -\frac{1}{2}\right)$

Identifying a Parent Function   In Exercises 59–64, identify the parent function and the transformation shown in the graph. Write an equation for the function shown in the graph. Then use a graphing utility to verify your answer.

59.

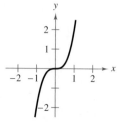

60.

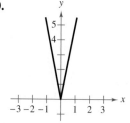

61.

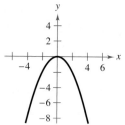

62.

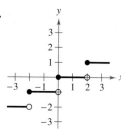

63.

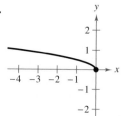

64.

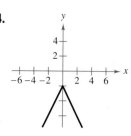

Graphical Analysis   In Exercises 65–68, use the viewing window shown to write a possible equation for the transformation of the parent function.

65.

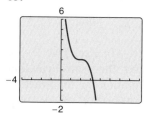

66.

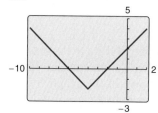

67.

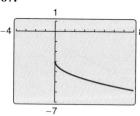

68.

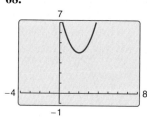

69. Automobile Aerodynamics

The number of horsepower H required to overcome wind drag on an automobile is approximated by

$$H(x) = 0.002x^2 + 0.005x - 0.029, \quad 10 \le x \le 100$$

where x is the speed of the car (in miles per hour).

(a) Use a graphing utility to graph the function.

(b) Rewrite the horsepower function so that x represents the speed in kilometers per hour. [Find $H(x/1.6)$.] Identify the type of transformation applied to the graph of the horsepower function.

70. Households The numbers N (in millions) of households in the United States from 2003 through 2010 can be approximated by

$$N = -0.068(x - 13.68)^2 + 119, \quad 3 \le t \le 10$$

where t represents the year, with $t = 3$ corresponding to 2003. *(Source: U.S. Census Bureau)*

(a) Describe the transformation of the parent function $f(x) = x^2$. Then use a graphing utility to graph the function over the specified domain.

(b) Find the average rate of change of the function from 2003 to 2010. Interpret your answer in the context of the problem.

(c) Use the model to predict the number of households in the United States in 2018. Does your answer seem reasonable? Explain.

Exploration

True or False? **In Exercises 71–74, determine whether the statement is true or false. Justify your answer.**

71. The graph of $y = f(-x)$ is a reflection of the graph of $y = f(x)$ in the x-axis.

72. The graph of $y = -f(x)$ is a reflection of the graph of $y = f(x)$ in the y-axis.

73. The graphs of $f(x) = |x| + 6$ and $f(x) = |-x| + 6$ are identical.

74. If the graph of the parent function $f(x) = x^2$ is shifted six units to the right, three units up, and reflected in the x-axis, then the point $(-2, 19)$ will lie on the graph of the transformation.

75. Finding Points on a Graph The graph of $y = f(x)$ passes through the points $(0, 1)$, $(1, 2)$, and $(2, 3)$. Find the corresponding points on the graph of $y = f(x + 2) - 1$.

76. Think About It You can use either of two methods to graph a function: plotting points or translating a parent function as shown in this section. Which method of graphing do you prefer to use for each function? Explain.

(a) $f(x) = 3x^2 - 4x + 1$

(b) $f(x) = 2(x - 1)^2 - 6$

77. Predicting Graphical Relationships Use a graphing utility to graph f, g, and h in the same viewing window. Before looking at the graphs, try to predict how the graphs of g and h relate to the graph of f.

(a) $f(x) = x^2$, $g(x) = (x - 4)^2$,
   $h(x) = (x - 4)^2 + 3$

(b) $f(x) = x^2$, $g(x) = (x + 1)^2$,
   $h(x) = (x + 1)^2 - 2$

(c) $f(x) = x^2$, $g(x) = (x + 4)^2$,
   $h(x) = (x + 4)^2 + 2$

78. HOW DO YOU SEE IT? Use the graph of $y = f(x)$ to find the intervals on which each of the graphs in (a)–(e) is increasing and decreasing. If not possible, then state the reason.

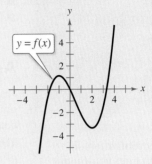

(a) $y = f(-x)$ (b) $y = -f(x)$ (c) $y = \frac{1}{2}f(x)$
(d) $y = -f(x - 1)$ (e) $y = f(x - 2) + 1$

79. Describing Profits Management originally predicted that the profits from the sales of a new product would be approximated by the graph of the function f shown. The actual profits are shown by the function g along with a verbal description. Use the concepts of transformations of graphs to write g in terms of f.

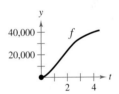

(a) The profits were only three-fourths as large as expected.

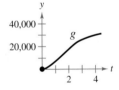

(b) The profits were consistently $10,000 greater than predicted.

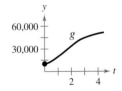

(c) There was a two-year delay in the introduction of the product. After sales began, profits grew as expected.

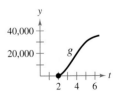

80. Reversing the Order of Transformations Reverse the order of transformations in Example 2(a). Do you obtain the same graph? Do the same for Example 2(b). Do you obtain the same graph? Explain.

P.9 Combinations of Functions: Composite Functions

Arithmetic combinations of functions can help you model and solve real-life problems. For instance, in Exercise 57 on page 100, you will use arithmetic combinations of functions to analyze numbers of pets in the United States.

- Add, subtract, multiply, and divide functions.
- Find the composition of one function with another function.
- Use combinations and compositions of functions to model and solve real-life problems.

Arithmetic Combinations of Functions

Just as two real numbers can be combined by the operations of addition, subtraction, multiplication, and division to form other real numbers, two *functions* can be combined to create new functions. For example, the functions $f(x) = 2x - 3$ and $g(x) = x^2 - 1$ can be combined to form the sum, difference, product, and quotient of f and g.

$$f(x) + g(x) = (2x - 3) + (x^2 - 1) = x^2 + 2x - 4 \qquad \text{Sum}$$

$$f(x) - g(x) = (2x - 3) - (x^2 - 1) = -x^2 + 2x - 2 \qquad \text{Difference}$$

$$f(x)g(x) = (2x - 3)(x^2 - 1) = 2x^3 - 3x^2 - 2x + 3 \qquad \text{Product}$$

$$\frac{f(x)}{g(x)} = \frac{2x - 3}{x^2 - 1}, \quad x \neq \pm 1 \qquad \text{Quotient}$$

The domain of an **arithmetic combination** of functions f and g consists of all real numbers that are common to the domains of f and g. In the case of the quotient $f(x)/g(x)$, there is the further restriction that $g(x) \neq 0$.

Sum, Difference, Product, and Quotient of Functions

Let f and g be two functions with overlapping domains. Then, for all x common to both domains, the *sum, difference, product,* and *quotient* of f and g are defined as follows.

1. Sum: $\quad (f + g)(x) = f(x) + g(x)$

2. Difference: $(f - g)(x) = f(x) - g(x)$

3. Product: $\quad (fg)(x) = f(x) \cdot g(x)$

4. Quotient: $\left(\dfrac{f}{g}\right)(x) = \dfrac{f(x)}{g(x)}, \quad g(x) \neq 0$

EXAMPLE 1 Finding the Sum of Two Functions

Given $f(x) = 2x + 1$ and $g(x) = x^2 + 2x - 1$, find $(f + g)(x)$. Then evaluate the sum when $x = 3$.

Solution The sum of f and g is

$$(f + g)(x) = f(x) + g(x) = (2x + 1) + (x^2 + 2x - 1) = x^2 + 4x.$$

When $x = 3$, the value of this sum is

$$(f + g)(3) = 3^2 + 4(3) = 21.$$

✓ **Checkpoint** ◀))) *Audio-video solution in English & Spanish at LarsonPrecalculus.com.*

Given $f(x) = x^2$ and $g(x) = 1 - x$, find $(f + g)(x)$. Then evaluate the sum when $x = 2$.

EXAMPLE 2 **Finding the Difference of Two Functions**

Given $f(x) = 2x + 1$ and $g(x) = x^2 + 2x - 1$, find $(f - g)(x)$. Then evaluate the difference when $x = 2$.

Solution The difference of f and g is

$$(f - g)(x) = f(x) - g(x) = (2x + 1) - (x^2 + 2x - 1) = -x^2 + 2.$$

When $x = 2$, the value of this difference is

$$(f - g)(2) = -(2)^2 + 2 = -2.$$

✓ *Checkpoint* ◀))) *Audio-video solution in English & Spanish at LarsonPrecalculus.com.*

Given $f(x) = x^2$ and $g(x) = 1 - x$, find $(f - g)(x)$. Then evaluate the difference when $x = 3$.

EXAMPLE 3 **Finding the Product of Two Functions**

Given $f(x) = x^2$ and $g(x) = x - 3$, find $(fg)(x)$. Then evaluate the product when $x = 4$.

Solution The product of f and g is

$$(fg)(x) = f(x)g(x) = (x^2)(x - 3) = x^3 - 3x^2.$$

When $x = 4$, the value of this product is

$$(fg)(4) = 4^3 - 3(4)^2 = 16.$$

✓ *Checkpoint* ◀))) *Audio-video solution in English & Spanish at LarsonPrecalculus.com.*

Given $f(x) = x^2$ and $g(x) = 1 - x$, find $(fg)(x)$. Then evaluate the product when $x = 3$.

In Examples 1–3, both f and g have domains that consist of all real numbers. So, the domains of $f + g$, $f - g$, and fg are also the set of all real numbers. Remember to consider any restrictions on the domains of f and g when forming the sum, difference, product, or quotient of f and g.

EXAMPLE 4 **Finding the Quotients of Two Functions**

Find $(f/g)(x)$ and $(g/f)(x)$ for the functions $f(x) = \sqrt{x}$ and $g(x) = \sqrt{4 - x^2}$. Then find the domains of f/g and g/f.

Solution The quotient of f and g is

$$\left(\frac{f}{g}\right)(x) = \frac{f(x)}{g(x)} = \frac{\sqrt{x}}{\sqrt{4 - x^2}}$$

and the quotient of g and f is

$$\left(\frac{g}{f}\right)(x) = \frac{g(x)}{f(x)} = \frac{\sqrt{4 - x^2}}{\sqrt{x}}.$$

•• **REMARK** Note that the domain of f/g includes $x = 0$, but not $x = 2$, because $x = 2$ yields a zero in the denominator, whereas the domain of g/f includes $x = 2$, but not $x = 0$, because $x = 0$ yields a zero in the denominator.

The domain of f is $[0, \infty)$ and the domain of g is $[-2, 2]$. The intersection of these domains is $[0, 2]$. So, the domains of f/g and g/f are as follows.

Domain of f/g: $[0, 2)$ Domain of g/f: $(0, 2]$

✓ *Checkpoint* ◀))) *Audio-video solution in English & Spanish at LarsonPrecalculus.com.*

Find $(f/g)(x)$ and $(g/f)(x)$ for the functions $f(x) = \sqrt{x - 3}$ and $g(x) = \sqrt{16 - x^2}$. Then find the domains of f/g and g/f.

Composition of Functions

Another way of combining two functions is to form the **composition** of one with the other. For instance, if $f(x) = x^2$ and $g(x) = x + 1,$ then the composition of f with g is

$$f(g(x)) = f(x + 1)$$
$$= (x + 1)^2.$$

This composition is denoted as $f \circ g$ and reads as "f composed with g."

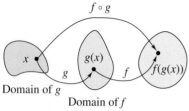

$f \circ g$

x $g(x)$ $f(g(x))$

g f

Domain of g

Domain of f

Figure P.61

Definition of Composition of Two Functions

The **composition** of the function f with the function g is

$$(f \circ g)(x) = f(g(x)).$$

The domain of $f \circ g$ is the set of all x in the domain of g such that $g(x)$ is in the domain of $f.$ (See Figure P.61.)

EXAMPLE 5   **Composition of Functions**

Given $f(x) = x + 2$ and $g(x) = 4 - x^2,$ find the following.

a. $(f \circ g)(x)$

b. $(g \circ f)(x)$

c. $(g \circ f)(-2)$

Solution

a. The composition of f with g is as follows.

$$
\begin{aligned}
(f \circ g)(x) &= f(g(x)) && \text{Definition of } f \circ g \\
&= f(4 - x^2) && \text{Definition of } g(x) \\
&= (4 - x^2) + 2 && \text{Definition of } f(x) \\
&= -x^2 + 6 && \text{Simplify.}
\end{aligned}
$$

b. The composition of g with f is as follows.

$$
\begin{aligned}
(g \circ f)(x) &= g(f(x)) && \text{Definition of } g \circ f \\
&= g(x + 2) && \text{Definition of } f(x) \\
&= 4 - (x + 2)^2 && \text{Definition of } g(x) \\
&= 4 - (x^2 + 4x + 4) && \text{Expand.} \\
&= -x^2 - 4x && \text{Simplify.}
\end{aligned}
$$

Note that, in this case, $(f \circ g)(x) \neq (g \circ f)(x).$

c. Using the result of part (b), write the following.

$$
\begin{aligned}
(g \circ f)(-2) &= -(-2)^2 - 4(-2) && \text{Substitute.} \\
&= -4 + 8 && \text{Simplify.} \\
&= 4 && \text{Simplify.}
\end{aligned}
$$

> **• • REMARK** The following tables of values help illustrate the composition $(f \circ g)(x)$ in Example 5(a).
>
> | x | 0 | 1 | 2 | 3 |
> |---|---|---|---|---|
> | $g(x)$ | 4 | 3 | 0 | -5 |
>
> | $g(x)$ | 4 | 3 | 0 | -5 |
> |---|---|---|---|---|
> | $f(g(x))$ | 6 | 5 | 2 | -3 |
>
> | x | 0 | 1 | 2 | 3 |
> |---|---|---|---|---|
> | $f(g(x))$ | 6 | 5 | 2 | -3 |
>
> Note that the first two tables can be combined (or "composed") to produce the values in the third table.

✓ **Checkpoint** 🔊))) *Audio-video solution in English & Spanish at LarsonPrecalculus.com.*

Given $f(x) = 2x + 5$ and $g(x) = 4x^2 + 1,$ find the following.

a. $(f \circ g)(x)$ **b.** $(g \circ f)(x)$ **c.** $(f \circ g)\left(-\frac{1}{2}\right)$

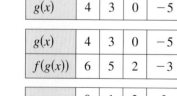

EXAMPLE 6   **Finding the Domain of a Composite Function**

Find the domain of $(f \circ g)(x)$ for the functions

$$f(x) = x^2 - 9 \quad \text{and} \quad g(x) = \sqrt{9 - x^2}.$$

Algebraic Solution

The composition of the functions is as follows.

$$
\begin{aligned}
(f \circ g)(x) &= f(g(x)) \\
&= f\left(\sqrt{9 - x^2}\right) \\
&= \left(\sqrt{9 - x^2}\right)^2 - 9 \\
&= 9 - x^2 - 9 \\
&= -x^2
\end{aligned}
$$

From this, it might appear that the domain of the composition is the set of all real numbers. This, however, is not true. Because the domain of f is the set of all real numbers and the domain of g is $[-3, 3]$, the domain of $f \circ g$ is $[-3, 3]$.

Graphical Solution

The x-coordinates of the points on the graph extend from -3 to 3. So, the domain of $f \circ g$ is $[-3, 3]$.

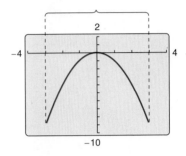

 Checkpoint  Audio-video solution in English & Spanish at LarsonPrecalculus.com.

Find the domain of $(f \circ g)(x)$ for the functions $f(x) = \sqrt{x}$ and $g(x) = x^2 + 4$.

In Examples 5 and 6, you formed the composition of two given functions. In calculus, it is also important to be able to identify two functions that make up a given composite function. For instance, the function $h(x) = (3x - 5)^3$ is the composition of $f(x) = x^3$ and $g(x) = 3x - 5$. That is,

$$h(x) = (3x - 5)^3 = [g(x)]^3 = f(g(x)).$$

Basically, to "decompose" a composite function, look for an "inner" function and an "outer" function. In the function h above, $g(x) = 3x - 5$ is the inner function and $f(x) = x^3$ is the outer function.

EXAMPLE 7   **Decomposing a Composite Function**

Write the function $h(x) = \dfrac{1}{(x - 2)^2}$ as a composition of two functions.

Solution   One way to write h as a composition of two functions is to take the inner function to be $g(x) = x - 2$ and the outer function to be

$$f(x) = \frac{1}{x^2} = x^{-2}.$$

Then write

$$
\begin{aligned}
h(x) &= \frac{1}{(x - 2)^2} \\
&= (x - 2)^{-2} \\
&= f(x - 2) \\
&= f(g(x)).
\end{aligned}
$$

 **Checkpoint** Audio-video solution in English & Spanish at LarsonPrecalculus.com.

Write the function $h(x) = \dfrac{\sqrt[3]{8 - x}}{5}$ as a composition of two functions.

Application

EXAMPLE 8   **Bacteria Count**

The number N of bacteria in a refrigerated food is given by

$$N(T) = 20T^2 - 80T + 500, \quad 2 \le T \le 14$$

where T is the temperature of the food in degrees Celsius. When the food is removed from refrigeration, the temperature of the food is given by

$$T(t) = 4t + 2, \quad 0 \le t \le 3$$

where t is the time in hours.

a. Find the composition $(N \circ T)(t)$ and interpret its meaning in context.

b. Find the time when the bacteria count reaches 2000.

Solution

a. $(N \circ T)(t) = N(T(t))$

$$= 20(4t + 2)^2 - 80(4t + 2) + 500$$

$$= 20(16t^2 + 16t + 4) - 320t - 160 + 500$$

$$= 320t^2 + 320t + 80 - 320t - 160 + 500$$

$$= 320t^2 + 420$$

The composite function $(N \circ T)(t)$ represents the number of bacteria in the food as a function of the amount of time the food has been out of refrigeration.

b. The bacteria count will reach 2000 when $320t^2 + 420 = 2000$. Solve this equation to find that the count will reach 2000 when $t \approx 2.2$ hours. Note that when you solve this equation, you reject the negative value because it is not in the domain of the composite function.

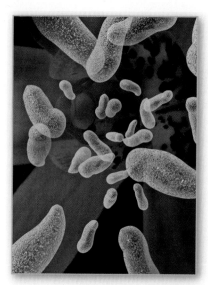

Refrigerated foods can have two types of bacteria: pathogenic bacteria, which can cause foodborne illness, and spoilage bacteria, which give foods an unpleasant look, smell, taste, or texture.

✔ **Checkpoint**  *Audio-video solution in English & Spanish at LarsonPrecalculus.com.*

The number N of bacteria in a refrigerated food is given by

$$N(T) = 8T^2 - 14T + 200, \quad 2 \le T \le 12$$

where T is the temperature of the food in degrees Celsius. When the food is removed from refrigeration, the temperature of the food is given by

$$T(t) = 2t + 2, \quad 0 \le t \le 5$$

where t is the time in hours.

a. Find the composition $(N \circ T)(t)$.

b. Find the time when the bacteria count reaches 1000.

Summarize **(Section P.9)**

1. Explain how to add, subtract, multiply, and divide functions *(page 94)*. For examples of finding arithmetic combinations of functions, see Examples 1–4.

2. Explain how to find the composition of one function with another function *(page 96)*. For examples that use compositions of functions, see Examples 5–7.

3. Describe a real-life example that uses a composition of functions *(page 98, Example 8)*.

P.9   Exercises

Vocabulary: Fill in the blanks.

1. Two functions f and g can be combined by the arithmetic operations of _____, _____, _____, and _____ to create new functions.

2. The _____ of the function f with g is $(f \circ g)(x) = f(g(x))$.

Skills and Applications

Graphing the Sum of Two Functions In Exercises 3 and 4, use the graphs of f and g to graph $h(x) = (f + g)(x)$. To print an enlarged copy of the graph, go to *MathGraphs.com*.

3.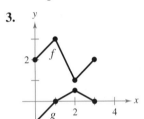

4.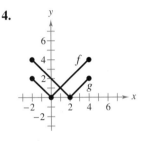

Finding Arithmetic Combinations of Functions In Exercises 5–12, find (a) $(f + g)(x)$, (b) $(f - g)(x)$, (c) $(fg)(x)$, and (d) $(f/g)(x)$. What is the domain of f/g?

5. $f(x) = x + 2$, $g(x) = x - 2$

6. $f(x) = 2x - 5$, $g(x) = 2 - x$

7. $f(x) = x^2$, $g(x) = 4x - 5$

8. $f(x) = 3x + 1$, $g(x) = 5x - 4$

9. $f(x) = x^2 + 6$, $g(x) = \sqrt{1 - x}$

10. $f(x) = \sqrt{x^2 - 4}$, $g(x) = \dfrac{x^2}{x^2 + 1}$

11. $f(x) = \dfrac{1}{x}$, $g(x) = \dfrac{1}{x^2}$

12. $f(x) = \dfrac{x}{x + 1}$, $g(x) = x^3$

Evaluating an Arithmetic Combination of Functions In Exercises 13–24, evaluate the indicated function for $f(x) = x^2 + 1$ and $g(x) = x - 4$.

13. $(f + g)(2)$

14. $(f - g)(-1)$

15. $(f - g)(0)$

16. $(f + g)(1)$

17. $(f - g)(3t)$

18. $(f + g)(t - 2)$

19. $(fg)(6)$

20. $(fg)(-6)$

21. $(f/g)(5)$

22. $(f/g)(0)$

23. $(f/g)(-1) - g(3)$

24. $(fg)(5) + f(4)$

Graphing Two Functions and Their Sum In Exercises 25 and 26, graph the functions f, g, and $f + g$ on the same set of coordinate axes.

25. $f(x) = \frac{1}{2}x$, $g(x) = x - 1$

26. $f(x) = 4 - x^2$, $g(x) = x$

Graphical Reasoning In Exercises 27–30, use a graphing utility to graph f, g, and $f + g$ in the same viewing window. Which function contributes most to the magnitude of the sum when $0 \le x \le 2$? Which function contributes most to the magnitude of the sum when $x > 6$?

27. $f(x) = 3x$, $g(x) = -\dfrac{x^3}{10}$

28. $f(x) = \dfrac{x}{2}$, $g(x) = \sqrt{x}$

29. $f(x) = 3x + 2$, $g(x) = -\sqrt{x + 5}$

30. $f(x) = x^2 - \frac{1}{2}$, $g(x) = -3x^2 - 1$

Finding Compositions of Functions In Exercises 31–34, find (a) $f \circ g$, (b) $g \circ f$, and (c) $g \circ g$.

31. $f(x) = x^2$, $g(x) = x - 1$

32. $f(x) = 3x + 5$, $g(x) = 5 - x$

33. $f(x) = \sqrt[3]{x - 1}$, $g(x) = x^3 + 1$

34. $f(x) = x^3$, $g(x) = \dfrac{1}{x}$

Finding Domains of Functions and Composite Functions In Exercises 35–42, find (a) $f \circ g$ and (b) $g \circ f$. Find the domain of each function and each composite function.

35. $f(x) = \sqrt{x + 4}$, $g(x) = x^2$

36. $f(x) = \sqrt[3]{x - 5}$, $g(x) = x^3 + 1$

37. $f(x) = x^2 + 1$, $g(x) = \sqrt{x}$

38. $f(x) = x^{2/3}$, $g(x) = x^6$

39. $f(x) = |x|$, $g(x) = x + 6$

40. $f(x) = |x - 4|$, $g(x) = 3 - x$

41. $f(x) = \dfrac{1}{x}$, $g(x) = x + 3$

42. $f(x) = \dfrac{3}{x^2 - 1}$, $g(x) = x + 1$

Evaluating Combinations of Functions In Exercises 43–46, use the graphs of f and g to evaluate the functions.

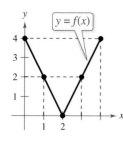
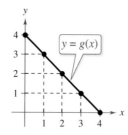

43. (a) $(f + g)(3)$ (b) $(f/g)(2)$
44. (a) $(f - g)(1)$ (b) $(fg)(4)$
45. (a) $(f \circ g)(2)$ (b) $(g \circ f)(2)$
46. (a) $(f \circ g)(1)$ (b) $(g \circ f)(3)$

Decomposing a Composite Function In Exercises 47–54, find two functions f and g such that $(f \circ g)(x) = h(x)$. (There are many correct answers.)

47. $h(x) = (2x + 1)^2$ 48. $h(x) = (1 - x)^3$
49. $h(x) = \sqrt[3]{x^2 - 4}$ 50. $h(x) = \sqrt{9 - x}$

51. $h(x) = \dfrac{1}{x + 2}$ 52. $h(x) = \dfrac{4}{(5x + 2)^2}$

53. $h(x) = \dfrac{-x^2 + 3}{4 - x^2}$

54. $h(x) = \dfrac{27x^3 + 6x}{10 - 27x^3}$

55. **Stopping Distance**   The research and development department of an automobile manufacturer has determined that when a driver is required to stop quickly to avoid an accident, the distance (in feet) the car travels during the driver's reaction time is given by $R(x) = \frac{3}{4}x$, where x is the speed of the car in miles per hour. The distance (in feet) traveled while the driver is braking is given by $B(x) = \frac{1}{15}x^2$.

(a) Find the function that represents the total stopping distance T.

(b) Graph the functions R, B, and T on the same set of coordinate axes for $0 \le x \le 60$.

(c) Which function contributes most to the magnitude of the sum at higher speeds? Explain.

56. **Vital Statistics**   Let $b(t)$ be the number of births in the United States in year t, and let $d(t)$ represent the number of deaths in the United States in year t, where $t = 10$ corresponds to 2010.

(a) If $p(t)$ is the population of the United States in year t, then find the function $c(t)$ that represents the percent change in the population of the United States.

(b) Interpret the value of $c(13)$.

57. **Pets**

Let $d(t)$ be the number of dogs in the United States in year t, and let $c(t)$ be the number of cats in the United States in year t, where $t = 10$ corresponds to 2010.

(a) Find the function $p(t)$ that represents the total number of dogs and cats in the United States.

(b) Interpret the value of $p(13)$.

(c) Let $n(t)$ represent the population of the United States in year t, where $t = 10$ corresponds to 2010. Find and interpret

$$h(t) = \frac{p(t)}{n(t)}.$$

58. **Graphical Reasoning**   An electronically controlled thermostat in a home lowers the temperature automatically during the night. The temperature in the house T (in degrees Fahrenheit) is given in terms of t, the time in hours on a 24-hour clock (see figure).

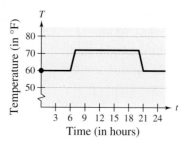

(a) Explain why T is a function of t.

(b) Approximate $T(4)$ and $T(15)$.

(c) The thermostat is reprogrammed to produce a temperature H for which $H(t) = T(t - 1)$. How does this change the temperature?

(d) The thermostat is reprogrammed to produce a temperature H for which $H(t) = T(t) - 1$. How does this change the temperature?

(e) Write a piecewise-defined function that represents the graph.

59. **Geometry**   A square concrete foundation is a base for a cylindrical tank (see figure).

(a) Write the radius r of the tank as a function of the length x of the sides of the square.

(b) Write the area A of the circular base of the tank as a function of the radius r.

(c) Find and interpret $(A \circ r)(x)$.

60. Bacteria Count   The number N of bacteria in a refrigerated food is given by

$$N(T) = 10T^2 - 20T + 600, \quad 2 \le T \le 20$$

where T is the temperature of the food in degrees Celsius. When the food is removed from refrigeration, the temperature of the food is given by

$$T(t) = 3t + 2, \quad 0 \le t \le 6$$

where t is the time in hours.

(a) Find the composition $(N \circ T)(t)$ and interpret its meaning in context.

(b) Find the bacteria count after 0.5 hour.

(c) Find the time when the bacteria count reaches 1500.

61. Salary   You are a sales representative for a clothing manufacturer. You are paid an annual salary, plus a bonus of 3% of your sales over $500,000. Consider the two functions $f(x) = x - 500,000$ and $g(x) = 0.03x$. When x is greater than $500,000, which of the following represents your bonus? Explain your reasoning.

(a) $f(g(x))$

(b) $g(f(x))$

62. Consumer Awareness   The suggested retail price of a new hybrid car is p dollars. The dealership advertises a factory rebate of $2000 and a 10% discount.

(a) Write a function R in terms of p giving the cost of the hybrid car after receiving the rebate from the factory.

(b) Write a function S in terms of p giving the cost of the hybrid car after receiving the dealership discount.

(c) Form the composite functions $(R \circ S)(p)$ and $(S \circ R)(p)$ and interpret each.

(d) Find $(R \circ S)(25{,}795)$ and $(S \circ R)(25{,}795)$. Which yields the lower cost for the hybrid car? Explain.

Exploration

Siblings   In Exercises 63 and 64, three siblings are three different ages. The oldest is twice the age of the middle sibling, and the middle sibling is six years older than one-half the age of the youngest.

63. (a) Write a composite function that gives the oldest sibling's age in terms of the youngest. Explain how you arrived at your answer.

(b) If the oldest sibling is 16 years old, then find the ages of the other two siblings.

64. (a) Write a composite function that gives the youngest sibling's age in terms of the oldest. Explain how you arrived at your answer.

(b) If the youngest sibling is two years old, then find the ages of the other two siblings.

True or False?   In Exercises 65 and 66, determine whether the statement is true or false. Justify your answer.

65. If $f(x) = x + 1$ and $g(x) = 6x$, then

$$(f \circ g)(x) = (g \circ f)(x).$$

66. When you are given two functions $f(x)$ and $g(x)$, you can calculate $(f \circ g)(x)$ if and only if the range of g is a subset of the domain of f.

67. Proof   Prove that the product of two odd functions is an even function, and that the product of two even functions is an even function.

68. **HOW DO YOU SEE IT?**   The graphs labeled L_1, L_2, L_3, and L_4 represent four different pricing discounts, where p is the original price (in dollars) and S is the sale price (in dollars). Match each function with its graph. Describe the situations in parts (c) and (d).

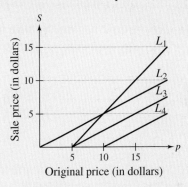

(a) $f(p)$: A 50% discount is applied.

(b) $g(p)$: A $5 discount is applied.

(c) $(g \circ f)(p)$

(d) $(f \circ g)(p)$

69. Conjecture   Use examples to hypothesize whether the product of an odd function and an even function is even or odd. Then prove your hypothesis.

70. Proof

(a) Given a function f, prove that $g(x)$ is even and $h(x)$ is odd, where $g(x) = \frac{1}{2}[f(x) + f(-x)]$ and

$$h(x) = \frac{1}{2}[f(x) - f(-x)].$$

(b) Use the result of part (a) to prove that any function can be written as a sum of even and odd functions. [*Hint:* Add the two equations in part (a).]

(c) Use the result of part (b) to write each function as a sum of even and odd functions.

$$f(x) = x^2 - 2x + 1, \quad k(x) = \frac{1}{x + 1}$$

P.10   Inverse Functions

Inverse functions can help you model and solve real-life problems. For instance, in Exercise 94 on page 110, an inverse function can help you determine the percent load interval for a diesel engine.

- Find inverse functions informally and verify that two functions are inverse functions of each other.
- Use graphs of functions to determine whether functions have inverse functions.
- Use the Horizontal Line Test to determine whether functions are one-to-one.
- Find inverse functions algebraically.

Inverse Functions

Recall from Section P.5 that a set of ordered pairs can represent a function. For instance, the function $f(x) = x + 4$ from the set $A = \{1, 2, 3, 4\}$ to the set $B = \{5, 6, 7, 8\}$ can be written as follows.

$$f(x) = x + 4: \ \{(1, 5), (2, 6), (3, 7), (4, 8)\}$$

In this case, by interchanging the first and second coordinates of each of these ordered pairs, you can form the **inverse function** of f, which is denoted by f^{-1}. It is a function from the set B to the set A, and can be written as follows.

$$f^{-1}(x) = x - 4: \ \{(5, 1), (6, 2), (7, 3), (8, 4)\}$$

Note that the domain of f is equal to the range of f^{-1}, and vice versa, as shown in the figure below. Also note that the functions f and f^{-1} have the effect of "undoing" each other. In other words, when you form the composition of f with f^{-1} or the composition of f^{-1} with f, you obtain the identity function.

$$f(f^{-1}(x)) = f(x - 4) = (x - 4) + 4 = x$$
$$f^{-1}(f(x)) = f^{-1}(x + 4) = (x + 4) - 4 = x$$

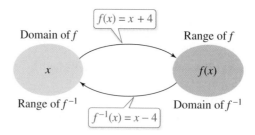

$f(x) = x + 4$

Domain of f Range of f

x $f(x)$

Range of f^{-1} Domain of f^{-1}

$f^{-1}(x) = x - 4$

EXAMPLE 1   **Finding an Inverse Function Informally**

Find the inverse function of $f(x) = 4x$. Then verify that both $f(f^{-1}(x))$ and $f^{-1}(f(x))$ are equal to the identity function.

Solution   The function f *multiplies* each input by 4. To "undo" this function, you need to *divide* each input by 4. So, the inverse function of $f(x) = 4x$ is

$$f^{-1}(x) = \frac{x}{4}.$$

Verify that $f(f^{-1}(x)) = x$ and $f^{-1}(f(x)) = x$ as follows.

$$f(f^{-1}(x)) = f\left(\frac{x}{4}\right) = 4\left(\frac{x}{4}\right) = x \qquad f^{-1}(f(x)) = f^{-1}(4x) = \frac{4x}{4} = x$$

✓ **Checkpoint** 🔊))) *Audio-video solution in English & Spanish at LarsonPrecalculus.com.*

Find the inverse function of $f(x) = \frac{1}{5}x$. Then verify that both $f(f^{-1}(x))$ and $f^{-1}(f(x))$ are equal to the identity function. ∎

Definition of Inverse Function

Let f and g be two functions such that

$$f(g(x)) = x \qquad \text{for every } x \text{ in the domain of } g$$

and

$$g(f(x)) = x \qquad \text{for every } x \text{ in the domain of } f.$$

Under these conditions, the function g is the **inverse function** of the function f. The function g is denoted by f^{-1} (read "f-inverse"). So,

$$f(f^{-1}(x)) = x \quad \text{and} \quad f^{-1}(f(x)) = x.$$

The domain of f must be equal to the range of f^{-1}, and the range of f must be equal to the domain of f^{-1}.

Do not be confused by the use of -1 to denote the inverse function f^{-1}. In this text, whenever f^{-1} is written, it *always* refers to the inverse function of the function f and *not* to the reciprocal of $f(x)$.

If the function g is the inverse function of the function f, then it must also be true that the function f is the inverse function of the function g. For this reason, you can say that the functions f and g are *inverse functions of each other.*

EXAMPLE 2   **Verifying Inverse Functions**

Which of the functions is the inverse function of $f(x) = \dfrac{5}{x - 2}$?

$$g(x) = \frac{x - 2}{5} \qquad h(x) = \frac{5}{x} + 2$$

Solution   By forming the composition of f with g, you have

$$f(g(x)) = f\left(\frac{x - 2}{5}\right) = \frac{5}{\left(\dfrac{x - 2}{5}\right) - 2} = \frac{25}{x - 12} \neq x.$$

Because this composition is not equal to the identity function x, it follows that g *is not* the inverse function of f. By forming the composition of f with h, you have

$$f(h(x)) = f\left(\frac{5}{x} + 2\right) = \frac{5}{\left(\dfrac{5}{x} + 2\right) - 2} = \frac{5}{\left(\dfrac{5}{x}\right)} = x.$$

So, it appears that h *is* the inverse function of f. Confirm this by showing that the composition of h with f is also equal to the identity function, as follows.

$$h(f(x)) = h\left(\frac{5}{x - 2}\right) = \frac{5}{\left(\dfrac{5}{x - 2}\right)} + 2 = x - 2 + 2 = x$$

✓ *Checkpoint* ◀))) Audio-video solution in English & Spanish at LarsonPrecalculus.com.

Which of the functions is the inverse function of $f(x) = \dfrac{x - 4}{7}$?

$$g(x) = 7x + 4 \qquad h(x) = \frac{7}{x - 4}$$

The Graph of an Inverse Function

The graphs of a function f and its inverse function f^{-1} are related to each other in the following way. If the point (a, b) lies on the graph of f, then the point (b, a) must lie on the graph of f^{-1}, and vice versa. This means that the graph of f^{-1} is a *reflection* of the graph of f in the line $y = x$, as shown in Figure P.62.

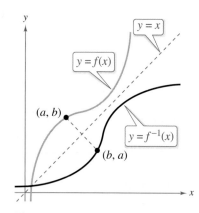

Figure P.62

EXAMPLE 3 **Verifying Inverse Functions Graphically**

Sketch the graphs of the inverse functions $f(x) = 2x - 3$ and $f^{-1}(x) = \frac{1}{2}(x + 3)$ on the same rectangular coordinate system and show that the graphs are reflections of each other in the line $y = x$.

Solution The graphs of f and f^{-1} are shown in Figure P.63. It appears that the graphs are reflections of each other in the line $y = x$. You can further verify this reflective property by testing a few points on each graph. Note in the following list that if the point (a, b) is on the graph of f, then the point (b, a) is on the graph of f^{-1}.

Graph of $f(x) = 2x - 3$	Graph of $f^{-1}(x) = \frac{1}{2}(x + 3)$
$(-1, -5)$	$(-5, -1)$
$(0, -3)$	$(-3, 0)$
$(1, -1)$	$(-1, 1)$
$(2, 1)$	$(1, 2)$
$(3, 3)$	$(3, 3)$

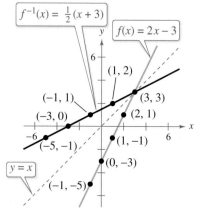

Figure P.63

✓ **Checkpoint** 🔊)) *Audio-video solution in English & Spanish at LarsonPrecalculus.com.*

Sketch the graphs of the inverse functions $f(x) = 4x - 1$ and $f^{-1}(x) = \frac{1}{4}(x + 1)$ on the same rectangular coordinate system and show that the graphs are reflections of each other in the line $y = x$.

EXAMPLE 4 **Verifying Inverse Functions Graphically**

Sketch the graphs of the inverse functions $f(x) = x^2$ $(x \geq 0)$ and $f^{-1}(x) = \sqrt{x}$ on the same rectangular coordinate system and show that the graphs are reflections of each other in the line $y = x$.

Solution The graphs of f and f^{-1} are shown in Figure P.64. It appears that the graphs are reflections of each other in the line $y = x$. You can further verify this reflective property by testing a few points on each graph. Note in the following list that if the point (a, b) is on the graph of f, then the point (b, a) is on the graph of f^{-1}.

Graph of $f(x) = x^2$, $x \geq 0$	Graph of $f^{-1}(x) = \sqrt{x}$
$(0, 0)$	$(0, 0)$
$(1, 1)$	$(1, 1)$
$(2, 4)$	$(4, 2)$
$(3, 9)$	$(9, 3)$

Try showing that $f(f^{-1}(x)) = x$ and $f^{-1}(f(x)) = x$.

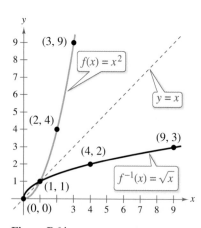

Figure P.64

✓ **Checkpoint** 🔊)) *Audio-video solution in English & Spanish at LarsonPrecalculus.com.*

Sketch the graphs of the inverse functions $f(x) = x^2 + 1$ $(x \geq 0)$ and $f^{-1}(x) = \sqrt{x - 1}$ on the same rectangular coordinate system and show that the graphs are reflections of each other in the line $y = x$.

One-to-One Functions

The reflective property of the graphs of inverse functions gives you a *geometric* test for determining whether a function has an inverse function. This test is called the **Horizontal Line Test** for inverse functions.

Horizontal Line Test for Inverse Functions

A function f has an inverse function if and only if no *horizontal* line intersects the graph of f at more than one point.

If no horizontal line intersects the graph of f at more than one point, then no y-value matches with more than one x-value. This is the essential characteristic of what are called **one-to-one functions.**

One-to-One Functions

A function f is **one-to-one** when each value of the dependent variable corresponds to exactly one value of the independent variable. A function f has an inverse function if and only if f is one-to-one.

Consider the function $f(x) = x^2$. The table on the left is a table of values for $f(x) = x^2$. The table on the right is the same as the table on the left but with the values in the columns interchanged. The table on the right does not represent a function because the input $x = 4$, for instance, matches with two different outputs: $y = -2$ and $y = 2$. So, $f(x) = x^2$ is not one-to-one and does not have an inverse function.

x	$f(x) = x^2$
-2	4
-1	1
0	0
1	1
2	4
3	9

x	y
4	-2
1	-1
0	0
1	1
4	2
9	3

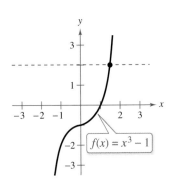

Figure P.65

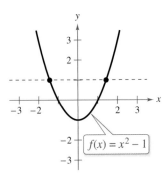

Figure P.66

EXAMPLE 5   **Applying the Horizontal Line Test**

a. The graph of the function $f(x) = x^3 - 1$ is shown in Figure P.65. Because no horizontal line intersects the graph of f at more than one point, f *is* a one-to-one function and *does* have an inverse function.

b. The graph of the function $f(x) = x^2 - 1$ is shown in Figure P.66. Because it is possible to find a horizontal line that intersects the graph of f at more than one point, f *is not* a one-to-one function and *does not* have an inverse function.

✓ *Checkpoint* 🔊 *Audio-video solution in English & Spanish at LarsonPrecalculus.com.*

Use the graph of f to determine whether the function has an inverse function.

a. $f(x) = \frac{1}{2}(3 - x)$

b. $f(x) = |x|$

Finding Inverse Functions Algebraically

For relatively simple functions (such as the one in Example 1), you can find inverse functions by inspection. For more complicated functions, however, it is best to use the following guidelines. The key step in these guidelines is Step 3—interchanging the roles of x and y. This step corresponds to the fact that inverse functions have ordered pairs with the coordinates reversed.

•• **REMARK** Note what happens when you try to find the inverse function of a function that is not one-to-one.

$$f(x) = x^2 + 1 \qquad \text{Original function}$$

$$y = x^2 + 1 \qquad \text{Replace } f(x) \text{ with } y.$$

$$x = y^2 + 1 \qquad \text{Interchange } x \text{ and } y.$$

$$x - 1 = y^2 \qquad \text{Isolate } y\text{-term.}$$

$$y = \pm\sqrt{x - 1} \qquad \text{Solve for } y.$$

You obtain two y-values for each x.

Finding an Inverse Function

1. Use the Horizontal Line Test to decide whether f has an inverse function.

2. In the equation for $f(x)$, replace $f(x)$ with y.

3. Interchange the roles of x and y, and solve for y.

4. Replace y with $f^{-1}(x)$ in the new equation.

5. Verify that f and f^{-1} are inverse functions of each other by showing that the domain of f is equal to the range of f^{-1}, the range of f is equal to the domain of f^{-1}, and $f(f^{-1}(x)) = x$ and $f^{-1}(f(x)) = x$.

EXAMPLE 6   Finding an Inverse Function Algebraically

Find the inverse function of

$$f(x) = \frac{5 - x}{3x + 2}.$$

Solution   The graph of f is shown in Figure P.67. This graph passes the Horizontal Line Test. So, you know that f is one-to-one and has an inverse function.

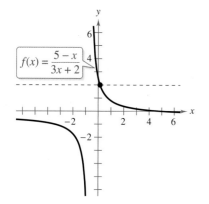

$$f(x) = \frac{5 - x}{3x + 2}$$

Figure P.67

$$f(x) = \frac{5 - x}{3x + 2} \qquad \text{Write original function.}$$

$$y = \frac{5 - x}{3x + 2} \qquad \text{Replace } f(x) \text{ with } y.$$

$$x = \frac{5 - y}{3y + 2} \qquad \text{Interchange } x \text{ and } y.$$

$$x(3y + 2) = 5 - y \qquad \text{Multiply each side by } 3y + 2.$$

$$3xy + 2x = 5 - y \qquad \text{Distributive Property}$$

$$3xy + y = 5 - 2x \qquad \text{Collect terms with } y.$$

$$y(3x + 1) = 5 - 2x \qquad \text{Factor.}$$

$$y = \frac{5 - 2x}{3x + 1} \qquad \text{Solve for } y.$$

$$f^{-1}(x) = \frac{5 - 2x}{3x + 1} \qquad \text{Replace } y \text{ with } f^{-1}(x).$$

Check that $f(f^{-1}(x)) = x$ and $f^{-1}(f(x)) = x$.

✓ **Checkpoint**  Audio-video solution in English & Spanish at LarsonPrecalculus.com.

Find the inverse function of

$$f(x) = \frac{5 - 3x}{x + 2}.$$

EXAMPLE 7   **Finding an Inverse Function Algebraically**

Find the inverse function of

$$f(x) = \sqrt{2x - 3}.$$

Solution The graph of f is a curve, as shown in the figure below. Because this graph passes the Horizontal Line Test, you know that f is one-to-one and has an inverse function.

$f(x) = \sqrt{2x - 3}$	Write original function.
$y = \sqrt{2x - 3}$	Replace $f(x)$ with y.
$x = \sqrt{2y - 3}$	Interchange x and y.
$x^2 = 2y - 3$	Square each side.
$2y = x^2 + 3$	Isolate y-term.
$y = \dfrac{x^2 + 3}{2}$	Solve for y.
$f^{-1}(x) = \dfrac{x^2 + 3}{2}, \quad x \geq 0$	Replace y with $f^{-1}(x)$.

The graph of f^{-1} in the figure is the reflection of the graph of f in the line $y = x$. Note that the range of f is the interval $[0, \infty)$, which implies that the domain of f^{-1} is the interval $[0, \infty)$. Moreover, the domain of f is the interval $\left[\frac{3}{2}, \infty\right)$, which implies that the range of f^{-1} is the interval $\left[\frac{3}{2}, \infty\right)$. Verify that $f(f^{-1}(x)) = x$ and $f^{-1}(f(x)) = x$.

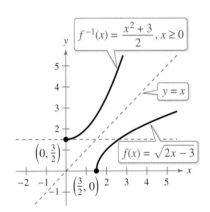

$f^{-1}(x) = \dfrac{x^2 + 3}{2}, x \geq 0$

$y = x$

$f(x) = \sqrt{2x - 3}$

$\left(0, \frac{3}{2}\right)$

$\left(\frac{3}{2}, 0\right)$

✓ *Checkpoint*   *Audio-video solution in English & Spanish at LarsonPrecalculus.com.*

Find the inverse function of

$$f(x) = \sqrt[3]{10 + x}.$$

Summarize   (Section P.10)

1. State the definition of an inverse function *(page 103)*. For examples of finding inverse functions informally and verifying inverse functions, see Examples 1 and 2.

2. Explain how to use the graph of a function to determine whether the function has an inverse function *(page 104)*. For examples of verifying inverse functions graphically, see Examples 3 and 4.

3. Explain how to use the Horizontal Line Test to determine whether a function is one-to-one *(page 105)*. For an example of applying the Horizontal Line Test, see Example 5.

4. Explain how to find an inverse function algebraically *(page 106)*. For examples of finding inverse functions algebraically, see Examples 6 and 7.

P.10 Exercises

See CalcChat.com for tutorial help and worked-out solutions to odd-numbered exercises.

Vocabulary: Fill in the blanks.

1. If the composite functions $f(g(x))$ and $g(f(x))$ both equal x, then the function g is the _____ function of f.
2. The inverse function of f is denoted by _____.
3. The domain of f is the _____ of f^{-1}, and the _____ of f^{-1} is the range of f.
4. The graphs of f and f^{-1} are reflections of each other in the line _____.
5. A function f is _____ when each value of the dependent variable corresponds to exactly one value of the independent variable.
6. A graphical test for the existence of an inverse function of f is called the _____ Line Test.

Skills and Applications

Finding an Inverse Function Informally In Exercises 7–12, find the inverse function of f informally. Verify that $f(f^{-1}(x)) = x$ and $f^{-1}(f(x)) = x$.

7. $f(x) = 6x$
8. $f(x) = \frac{1}{3}x$
9. $f(x) = 3x + 1$
10. $f(x) = \frac{x-1}{5}$
11. $f(x) = \sqrt[3]{x}$
12. $f(x) = x^5$

Verifying Inverse Functions In Exercises 13–16, verify that f and g are inverse functions.

13. $f(x) = -\frac{7}{2}x - 3$, $g(x) = -\frac{2x+6}{7}$
14. $f(x) = \frac{x-9}{4}$, $g(x) = 4x + 9$
15. $f(x) = x^3 + 5$, $g(x) = \sqrt[3]{x-5}$
16. $f(x) = \frac{x^3}{2}$, $g(x) = \sqrt[3]{2x}$

Sketching the Graph of an Inverse Function In Exercises 17–20, use the graph of the function to sketch the graph of its inverse function $y = f^{-1}(x)$.

17.
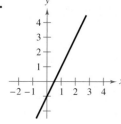

18.
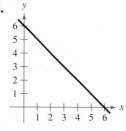

19.
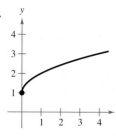

20.
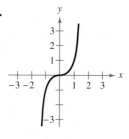

Verifying Inverse Functions In Exercises 21–32, verify that f and g are inverse functions (a) algebraically and (b) graphically.

21. $f(x) = 2x$, $g(x) = \dfrac{x}{2}$
22. $f(x) = x - 5$, $g(x) = x + 5$
23. $f(x) = 7x + 1$, $g(x) = \dfrac{x-1}{7}$
24. $f(x) = 3 - 4x$, $g(x) = \dfrac{3-x}{4}$
25. $f(x) = \dfrac{x^3}{8}$, $g(x) = \sqrt[3]{8x}$
26. $f(x) = \dfrac{1}{x}$, $g(x) = \dfrac{1}{x}$
27. $f(x) = \sqrt{x-4}$, $g(x) = x^2 + 4$, $x \geq 0$
28. $f(x) = 1 - x^3$, $g(x) = \sqrt[3]{1-x}$
29. $f(x) = 9 - x^2$, $x \geq 0$, $g(x) = \sqrt{9-x}$, $x \leq 9$
30. $f(x) = \dfrac{1}{1+x}$, $x \geq 0$, $g(x) = \dfrac{1-x}{x}$, $0 < x \leq 1$
31. $f(x) = \dfrac{x-1}{x+5}$, $g(x) = -\dfrac{5x+1}{x-1}$
32. $f(x) = \dfrac{x+3}{x-2}$, $g(x) = \dfrac{2x+3}{x-1}$

Using a Table to Determine an Inverse Function In Exercises 33 and 34, does the function have an inverse function?

33.

x	-1	0	1	2	3	4
$f(x)$	-2	1	2	1	-2	-6

34.

x	-3	-2	-1	0	2	3
$f(x)$	10	6	4	1	-3	-10

Using a Table to Find an Inverse Function In Exercises 35 and 36, use the table of values for $y = f(x)$ to complete a table for $y = f^{-1}(x)$.

35.

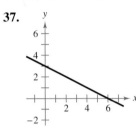

x	-2	-1	0	1	2	3
$f(x)$	-2	0	2	4	6	8

36.

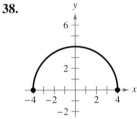

x	-3	-2	-1	0	1	2
$f(x)$	-10	-7	-4	-1	2	5

Applying the Horizontal Line Test In Exercises 37–40, does the function have an inverse function?

37.   **38.**

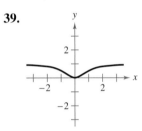

39.   **40.**

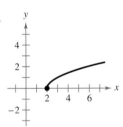

Applying the Horizontal Line Test In Exercises 41–44, use a graphing utility to graph the function, and use the Horizontal Line Test to determine whether the function has an inverse function.

41. $g(x) = (x + 5)^3$ **42.** $f(x) = \frac{1}{8}(x + 2)^2 - 1$

43. $f(x) = -2x\sqrt{16 - x^2}$

44. $h(x) = |x + 4| - |x - 4|$

Finding and Analyzing Inverse Functions In Exercises 45–56, (a) find the inverse function of f, (b) graph both f and f^{-1} on the same set of coordinate axes, (c) describe the relationship between the graphs of f and f^{-1}, and (d) state the domains and ranges of f and f^{-1}.

45. $f(x) = 2x - 3$ **46.** $f(x) = 3x + 1$

47. $f(x) = x^5 - 2$ **48.** $f(x) = x^3 + 1$

49. $f(x) = \sqrt{4 - x^2}$, $0 \le x \le 2$

50. $f(x) = x^2 - 2$, $x \le 0$

51. $f(x) = \dfrac{4}{x}$ **52.** $f(x) = -\dfrac{2}{x}$

53. $f(x) = \dfrac{x + 1}{x - 2}$ **54.** $f(x) = \dfrac{x - 3}{x + 2}$

55. $f(x) = \sqrt[3]{x - 1}$ **56.** $f(x) = x^{3/5}$

Finding an Inverse Function In Exercises 57–72, determine whether the function has an inverse function. If it does, then find the inverse function.

57. $f(x) = x^4$ **58.** $f(x) = \dfrac{1}{x^2}$

59. $g(x) = \dfrac{x}{8}$ **60.** $f(x) = 3x + 5$

61. $p(x) = -4$ **62.** $f(x) = \dfrac{3x + 4}{5}$

63. $f(x) = (x + 3)^2,\quad x \ge -3$

64. $q(x) = (x - 5)^2$

65. $f(x) = \begin{cases} x + 3, & x < 0 \\ 6 - x, & x \ge 0 \end{cases}$

66. $f(x) = \begin{cases} -x, & x \le 0 \\ x^2 - 3x, & x > 0 \end{cases}$

67. $h(x) = -\dfrac{4}{x^2}$

68. $f(x) = |x - 2|,\quad x \le 2$

69. $f(x) = \sqrt{2x + 3}$

70. $f(x) = \sqrt{x - 2}$

71. $f(x) = \dfrac{6x + 4}{4x + 5}$

72. $f(x) = \dfrac{5x - 3}{2x + 5}$

Restricting the Domain In Exercises 73–82, restrict the domain of the function f so that the function is one-to-one and has an inverse function. Then find the inverse function f^{-1}. State the domains and ranges of f and f^{-1}. Explain your results. (There are many correct answers.)

73. $f(x) = (x - 2)^2$ **74.** $f(x) = 1 - x^4$

75. $f(x) = |x + 2|$ **76.** $f(x) = |x - 5|$

77. $f(x) = (x + 6)^2$ **78.** $f(x) = (x - 4)^2$

79. $f(x) = -2x^2 + 5$ **80.** $f(x) = \frac{1}{2}x^2 - 1$

81. $f(x) = |x - 4| + 1$ **82.** $f(x) = -|x - 1| - 2$

Composition with Inverses In Exercises 83–88, use the functions $f(x) = \frac{1}{8}x - 3$ and $g(x) = x^3$ to find the indicated value or function.

83. $(f^{-1} \circ g^{-1})(1)$ **84.** $(g^{-1} \circ f^{-1})(-3)$

85. $(f^{-1} \circ f^{-1})(6)$ **86.** $(g^{-1} \circ g^{-1})(-4)$

87. $(f \circ g)^{-1}$ **88.** $g^{-1} \circ f^{-1}$

Composition with Inverses In Exercises 89–92, use the functions $f(x) = x + 4$ and $g(x) = 2x - 5$ to find the specified function.

89. $g^{-1} \circ f^{-1}$ **90.** $f^{-1} \circ g^{-1}$

91. $(f \circ g)^{-1}$ **92.** $(g \circ f)^{-1}$

93. Hourly Wage   Your wage is $10.00 per hour plus $0.75 for each unit produced per hour. So, your hourly wage y in terms of the number of units produced x is $y = 10 + 0.75x$.

(a) Find the inverse function. What does each variable represent in the inverse function?

(b) Determine the number of units produced when your hourly wage is $24.25.

94. Diesel Mechanics

The function

$$y = 0.03x^2 + 245.50, \quad 0 < x < 100$$

approximates the exhaust temperature y in degrees Fahrenheit, where x is the percent load for a diesel engine.

(a) Find the inverse function. What does each variable represent in the inverse function?

(b) Use a graphing utility to graph the inverse function.

(c) The exhaust temperature of the engine must not exceed 500 degrees Fahrenheit. What is the percent load interval?

Exploration

True or False?   **In Exercises 95 and 96, determine whether the statement is true or false. Justify your answer.**

95. If f is an even function, then f^{-1} exists.

96. If the inverse function of f exists and the graph of f has a y-intercept, then the y-intercept of f is an x-intercept of f^{-1}.

Graphical Analysis   **In Exercises 97 and 98, use the graph of the function f to create a table of values for the given points. Then create a second table that can be used to find f^{-1}, and sketch the graph of f^{-1} if possible.**

97.   **98.**

   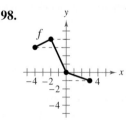

99. Proof   Prove that if f and g are one-to-one functions, then $(f \circ g)^{-1}(x) = (g^{-1} \circ f^{-1})(x)$.

100. Proof   Prove that if f is a one-to-one odd function, then f^{-1} is an odd function.

101. Think About It   The function $f(x) = k(2 - x - x^3)$ has an inverse function, and $f^{-1}(3) = -2$. Find k.

102. Think About It   Consider the functions $f(x) = x + 2$ and $f^{-1}(x) = x - 2$. Evaluate $f(f^{-1}(x))$ and $f^{-1}(f(x))$ for the indicated values of x. What can you conclude about the functions?

x	-10	0	7	45
$f(f^{-1}(x))$				
$f^{-1}(f(x))$				

103. Think About It   Restrict the domain of $f(x) = x^2 + 1$ to $x \geq 0$. Use a graphing utility to graph the function. Does the restricted function have an inverse function? Explain.

104.   **HOW DO YOU SEE IT?**   The cost C for a business to make personalized T-shirts is given by

$$C(x) = 7.50x + 1500$$

where x represents the number of T-shirts.

(a) The graphs of C and C^{-1} are shown below. Match each function with its graph.

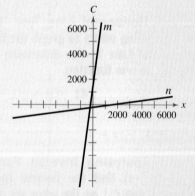

(b) Explain what $C(x)$ and $C^{-1}(x)$ represent in the context of the problem.

One-to-One Function Representation   **In Exercises 105 and 106, determine whether the situation could be represented by a one-to-one function. If so, then write a statement that best describes the inverse function.**

105. The number of miles n a marathon runner has completed in terms of the time t in hours

106. The depth of the tide d at a beach in terms of the time t over a 24-hour period

Chapter Summary

	What Did You Learn?	Explanation/Examples	Review Exercises
Section P.1	Represent and classify real numbers *(p. 2)*.	Real numbers include both rational and irrational numbers. Real numbers are represented graphically on the real number line.	1, 2
	Order real numbers and use inequalities *(p. 4)*.	$a < b$: a is less than b. $a > b$: a is greater than b. $a \leq b$: a is less than or equal to b. $a \geq b$: a is greater than or equal to b.	3, 4
	Find the absolute values of real numbers and find the distance between two real numbers *(p. 6)*.	**Absolute value of a:** $\|a\| = \begin{cases} a, & \text{if } a \geq 0 \\ -a, & \text{if } a < 0 \end{cases}$ **Distance between a and b:** $d(a, b) = \|b - a\| = \|a - b\|$	5–8
	Evaluate algebraic expressions *(p. 8)*.	To evaluate an algebraic expression, substitute numerical values for each of the variables in the expression.	9, 10
	Use the basic rules and properties of algebra *(p. 9)*.	The basic rules of algebra, the properties of negation and equality, the properties of zero, and the properties and operations of fractions can be used to perform operations.	11–22
Section P.2	Identify different types of equations *(p. 14)*, and solve linear equations in one variable and rational equations *(p. 15)*.	**Identity:** true for *every* real number in the domain **Conditional equation:** true for just some (but not all) of the real numbers in the domain **Contradiction:** false for *every* real number in the domain	23–26
	Solve quadratic equations *(p. 17)*, polynomial equations of degree three or greater *(p. 21)*, radical equations *(p. 22)*, and absolute value equations *(p. 23)*.	Four methods of solving quadratic equations are factoring, extracting square roots, completing the square, and the Quadratic Formula. These methods can sometimes be extended to solve polynomial equations of higher degree. When solving equations involving radicals or absolute values, be sure to check for extraneous solutions.	27–38
Section P.3	Plot points in the Cartesian plane *(p. 26)*, and use the Distance Formula *(p. 28)* and the Midpoint Formula *(p. 29)*.	For an ordered pair (x, y), the x-coordinate is the directed distance from the y-axis to the point, and the y-coordinate is the directed distance from the x-axis to the point.	39, 40
	Use a coordinate plane to model and solve real-life problems *(p. 30)*.	The coordinate plane can be used to find the length of a football pass. (See Example 6.)	41, 42
	Sketch graphs of equations *(p. 31)*, and find x- and y-intercepts of graphs of equations *(p. 32)*.	To graph an equation, construct a table of values, plot the points, and connect the points with a smooth curve or line. To find x-intercepts, let y be zero and solve for x. To find y-intercepts, let x be zero and solve for y.	43–48
	Use symmetry to sketch graphs of equations *(p. 33)*.	Graphs can have symmetry with respect to one of the coordinate axes or with respect to the origin. You can test for symmetry algebraically and graphically.	49–52
	Write equations of and sketch graphs of circles *(p. 34)*.	A point (x, y) lies on the circle of radius r and center (h, k) if and only if $(x - h)^2 + (y - k)^2 = r^2$.	53–56

What Did You Learn?	Explanation/Examples	Review Exercises

Section P.4

Use slope to graph linear equations in two variables *(p. 40)*.	**The Slope-Intercept Form of the Equation of a Line** The graph of the equation $y = mx + b$ is a line whose slope is m and whose y-intercept is $(0, b)$.	57–60
Find the slope of a line given two points on the line *(p. 42)*.	The slope m of the nonvertical line through (x_1, y_1) and (x_2, y_2) is $m = (y_2 - y_1)/(x_2 - x_1)$, where $x_1 \neq x_2$.	61, 62
Write linear equations in two variables *(p. 44)*.	**Point-Slope Form of the Equation of a Line** The equation of the line with slope m passing through the point (x_1, y_1) is $y - y_1 = m(x - x_1)$.	63–66
Use slope to identify parallel and perpendicular lines *(p. 45)*.	**Parallel lines:** Slopes are equal. **Perpendicular lines:** Slopes are negative reciprocals of each other.	67, 68
Use slope and linear equations in two variables to model and solve real-life problems *(p. 46)*.	A linear equation in two variables can be used to describe the book value of exercise equipment in a given year. (See Example 7.)	69, 70

Section P.5

Determine whether relations between two variables are functions, and use function notation *(p. 53)*.	A function f from a set A (domain) to a set B (range) is a relation that assigns to each element x in the set A exactly one element y in the set B. **Equation:** $f(x) = 5 - x^2$ $f(2)$: $\quad f(2) = 5 - 2^2 - 1$	71–76
Find the domains of functions *(p. 58)*.	**Domain of $f(x) = 5 - x^2$:** All real numbers	77, 78
Use functions to model and solve real-life problems *(p. 59)*.	A function can be used to model the number of alternative-fueled vehicles in the United States. (See Example 10.)	79
Evaluate difference quotients *(p. 60)*.	**Difference quotient:** $\dfrac{f(x + h) - f(x)}{h}, h \neq 0$	80

Section P.6

Use the Vertical Line Test for functions *(p. 68)*.	A set of points in a coordinate plane is the graph of y as a function of x if and only if no *vertical* line intersects the graph at more than one point.	81, 82
Find the zeros of functions *(p. 69)*.	**Zeros of $f(x)$:** x-values for which $f(x) = 0$	83, 84
Determine intervals on which functions are increasing or decreasing *(p. 70)*, determine relative minimum and relative maximum values of functions *(p. 71)*, and determine the average rates of change of functions *(p. 72)*.	To determine whether a function is increasing, decreasing, or constant on an interval, evaluate the function for several values of x. The points at which the behavior of a function changes can help determine a relative minimum or relative maximum. The average rate of change between any two points is the slope of the line (secant line) through the two points.	85–90
Identify even and odd functions *(p. 73)*.	**Even:** For each x in the domain of f, $f(-x) = f(x)$. **Odd:** For each x in the domain of f, $f(-x) = -f(x)$.	91–94

What Did You Learn?	**Explanation/Examples**	**Review Exercises**

Section P.7

Identify and graph linear (*p. 78*), squaring (*p. 79*), cubic, square root, reciprocal (*p. 80*), step, and other piecewise-defined functions (*p. 81*), and recognize graphs of parent functions (*p. 82*).	**Linear:** $f(x) = ax + b$ $f(x) = -x + 4$ **Squaring:** $f(x) = x^2$ $f(x) = x^2$ $(0, 0)$ **Square Root:** $f(x) = \sqrt{x}$ $f(x) = \sqrt{x}$ $(0, 0)$ **Step:** $f(x) = [\![x]\!]$ $f(x) = [\![x]\!]$ Eight of the most commonly used functions in algebra are shown on page 82.	95–102

Section P.8

Use vertical and horizontal shifts (*p. 85*), reflections (*p. 87*), and nonrigid transformations (*p. 89*) to sketch graphs of functions.	**Vertical shifts:** $h(x) = f(x) + c$ or $h(x) = f(x) - c$ **Horizontal shifts:** $h(x) = f(x - c)$ or $h(x) = f(x + c)$ **Reflection in x-axis:** $h(x) = -f(x)$ **Reflection in y-axis:** $h(x) = f(-x)$ **Nonrigid transformations:** $h(x) = cf(x)$ or $h(x) = f(cx)$	103–112

Section P.9

Add, subtract, multiply, and divide functions (*p. 94*).	$(f + g)(x) = f(x) + g(x)$ $(f - g)(x) = f(x) - g(x)$ $(fg)(x) = f(x) \cdot g(x)$ $(f/g)(x) = f(x)/g(x), \, g(x) \neq 0$	113, 114
Find the composition of one function with another function (*p. 96*).	The composition of the function f with the function g is $(f \circ g)(x) = f(g(x))$.	115, 116
Use combinations and compositions of functions to model and solve real-life problems (*p. 98*).	A composite function can be used to represent the number of bacteria in food as a function of the amount of time the food has been out of refrigeration. (See Example 8.)	117, 118

Section P.10

Find inverse functions informally and verify that two functions are inverse functions of each other (*p. 102*).	Let f and g be two functions such that $f(g(x)) = x$ for every x in the domain of g and $g(f(x)) = x$ for every x in the domain of f. Under these conditions, the function g is the inverse function of the function f.	119, 120
Use graphs of functions to determine whether functions have inverse functions (*p. 104*), use the Horizontal Line Test to determine whether functions are one-to-one (*p. 105*), and find inverse functions algebraically (*p. 106*).	If the point (a, b) lies on the graph of f, then the point (b, a) must lie on the graph of f^{-1}, and vice versa. In short, the graph of f^{-1} is a reflection of the graph of f in the line $y = x$. **Horizontal Line Test for Inverse Functions** A function f has an inverse function if and only if no *horizontal* line intersects the graph of f at more than one point. To find an inverse function, replace $f(x)$ with y, interchange the roles of x and y, solve for y, and then replace y with $f^{-1}(x)$.	121–126

Review Exercises

See CalcChat.com for tutorial help and worked-out solutions to odd-numbered exercises.

P.1 **Classifying Real Numbers** In Exercises 1 and 2, determine which numbers in the set are (a) natural numbers, (b) whole numbers, (c) integers, (d) rational numbers, and (e) irrational numbers.

1. $\left\{ 11, -\frac{8}{9}, \frac{5}{2}, \sqrt{6}, 0.4 \right\}$ 2. $\left\{ \sqrt{15}, -22, 0, 5.2, \frac{3}{7} \right\}$

Plotting and Ordering Real Numbers In Exercises 3 and 4, plot the two real numbers on the real number line. Then place the appropriate inequality symbol (< or >) between them.

3. (a) $\frac{5}{4}$ (b) $\frac{7}{8}$ 4. (a) $\frac{9}{25}$ (b) $\frac{5}{7}$

Finding a Distance In Exercises 5 and 6, find the distance between a and b.

5. $a = -74, b = 48$ 6. $a = -112, b = -6$

Using Absolute Value Notation In Exercises 7 and 8, use absolute value notation to describe the situation.

7. The distance between x and 7 is at least 4.
8. The distance between x and 25 is no more than 10.

Evaluating an Algebraic Expression In Exercises 9 and 10, evaluate the expression for each value of x. (If not possible, then state the reason.)

Expression	Values
9. $-x^2 + x - 1$	(a) $x = 1$ (b) $x = -1$
10. $\dfrac{x}{x-3}$	(a) $x = -3$ (b) $x = 3$

Identifying Rules of Algebra In Exercises 11–16, identify the rule of algebra illustrated by the statement.

11. $2x + (3x - 10) = (2x + 3x) - 10$
12. $4(t + 2) = 4 \cdot t + 4 \cdot 2$
13. $0 + (a - 5) = a - 5$
14. $\dfrac{2}{y+4} \cdot \dfrac{y+4}{2} = 1, \quad y \neq -4$
15. $(t^2 + 1) + 3 = 3 + (t^2 + 1)$
16. $1 \cdot (3x + 4) = 3x + 4$

Performing Operations In Exercises 17–22, perform the operation(s). (Write fractional answers in simplest form.)

17. $|-3| + 4(-2) - 6$ 18. $\dfrac{|-10|}{-10}$

19. $\frac{5}{18} \div \frac{10}{3}$ 20. $(16 - 8) \div 4$

21. $6[4 - 2(6 + 8)]$ 22. $-4[16 - 3(7 - 10)]$

P.2 **Solving an Equation** In Exercises 23–26, solve the equation and check your solution. (If not possible, then explain why.)

23. $3x - 2(x + 5) = 10$ 24. $4x + 2(7 - x) = 5$

25. $\dfrac{x}{5} - 3 = \dfrac{x}{3} + 1$ 26. $\dfrac{18}{x} = \dfrac{10}{x-4}$

Choosing a Method In Exercises 27–30, solve the equation using any convenient method.

27. $2x^2 + 5x + 3 = 0$
28. $16x^2 = 25$
29. $(x + 4)^2 = 18$
30. $x^2 + 6x - 3 = 0$

Solving an Equation In Exercises 31–38, solve the equation. Check your solutions.

31. $5x^4 - 12x^3 = 0$ 32. $x^4 - 5x^2 + 6 = 0$
33. $\sqrt{x + 4} = 3$ 34. $5\sqrt{x} - \sqrt{x - 1} = 6$
35. $(x - 1)^{2/3} - 25 = 0$ 36. $(x + 2)^{3/4} = 27$
37. $|x - 5| = 10$ 38. $|x^2 - 3| = 2x$

P.3 **Plotting, Distance, and Midpoint** In Exercises 39 and 40, (a) plot the points, (b) find the distance between the points, and (c) find the midpoint of the line segment joining the points.

39. $(5, 1), (1, 4)$ 40. $(6, -2), (5, 3)$

Meteorology In Exercises 41 and 42, use the following information. The apparent temperature is a measure of relative discomfort to a person from heat and high humidity. The table shows the actual temperatures x (in degrees Fahrenheit) versus the apparent temperatures y (in degrees Fahrenheit) for a relative humidity of 75%.

x	70	75	80	85	90	95	100
y	70	77	85	95	109	130	150

41. Sketch a scatter plot of the data shown in the table.
42. Find the change in the apparent temperature when the actual temperature changes from 70°F to 100°F.

Sketching the Graph of an Equation In Exercises 43–46, construct a table of values. Use the resulting solution points to sketch the graph of the equation.

43. $y = 2x - 6$ 44. $y = -\frac{1}{2}x + 2$
45. $y = x^2 + 2x$ 46. $y = 2x^2 - x - 9$

Finding *x*- and *y*-Intercepts In Exercises 47 and 48, find the *x*- and *y*-intercepts of the graph of the equation.

47. $y = (x - 3)^2 - 4$ **48.** $y = |x + 1| - 3$

Testing for Symmetry In Exercises 49–52, use the algebraic tests to check for symmetry with respect to both axes and the origin. Then sketch the graph of the equation.

49. $y = -4x + 2$ **50.** $y = 7 - x^2$

51. $y = x^3 + 3$ **52.** $y = |x| + 9$

Writing the Equation of a Circle In Exercises 53 and 54, write the standard form of the equation of the circle for which the endpoints of a diameter are given.

53. $(0, 0), (4, -6)$ **54.** $(-2, -3), (4, -10)$

Sketching the Graph of a Circle In Exercises 55 and 56, find the center and radius of the circle. Then sketch the graph of the circle.

55. $x^2 + y^2 = 9$
56. $(x + 4)^2 + \left(y - \frac{3}{2}\right)^2 = 100$

P.4 **Graphing a Linear Equation** In Exercises 57–60, find the slope and *y*-intercept (if possible) of the equation of the line. Sketch the line.

57. $y = -2x - 7$ **58.** $10x + 2y = 9$

59. $y = 6$ **60.** $x = -3$

Finding the Slope of a Line Through Two Points In Exercises 61 and 62, plot the points and find the slope of the line passing through the pair of points.

61. $(6, 4), (-3, -4)$ **62.** $(-3, 2), (8, 2)$

Finding an Equation of a Line In Exercises 63 and 64, find an equation of the line that passes through the given point and has the indicated slope *m*. Sketch the line.

63. $(10, -3)$, $m = -\frac{1}{2}$ **64.** $(-8, 5)$, $m = 0$

Finding an Equation of a Line In Exercises 65 and 66, find an equation of the line passing through the points.

65. $(-1, 0), (6, 2)$ **66.** $(11, -2), (6, -1)$

Finding Parallel and Perpendicular Lines In Exercises 67 and 68, write the slope-intercept form of the equations of the lines through the given point (a) parallel to and (b) perpendicular to the given line.

67. $5x - 4y = 8$, $(3, -2)$ **68.** $2x + 3y = 5$, $(-8, 3)$

69. Hourly Wage A microchip manufacturer pays its assembly line workers $12.25 per hour. In addition, workers receive a piecework rate of $0.75 per unit produced. Write a linear equation for the hourly wage *W* in terms of the number of units *x* produced per hour.

70. Sales A discount outlet is offering a 20% discount on all items. Write a linear equation giving the sale price *S* for an item with a list price *L*.

P.5 **Testing for Functions Represented Algebraically** In Exercises 71–74, determine whether the equation represents *y* as a function of *x*.

71. $16x - y^4 = 0$ **72.** $2x - y - 3 = 0$
73. $y = \sqrt{1 - x}$ **74.** $|y| = x + 2$

Evaluating a Function In Exercises 75 and 76, evaluate the function at each specified value of the independent variable and simplify.

75. $g(x) = x^{4/3}$

   (a) $g(8)$ (b) $g(t + 1)$ (c) $g(-27)$ (d) $g(-x)$

76. $h(x) = \begin{cases} 2x + 1, & x \le -1 \\ x^2 + 2, & x > -1 \end{cases}$

   (a) $h(-2)$ (b) $h(-1)$ (c) $h(0)$ (d) $h(2)$

Finding the Domain of a Function In Exercises 77 and 78, find the domain of the function. Verify your result with a graph.

77. $f(x) = \sqrt{25 - x^2}$ **78.** $h(x) = \dfrac{x}{x^2 - x - 6}$

79. Physics The velocity of a ball projected upward from ground level is given by $v(t) = -32t + 48$, where *t* is the time in seconds and *v* is the velocity in feet per second.

   (a) Find the velocity when $t = 1$.

   (b) Find the time when the ball reaches its maximum height. [*Hint:* Find the time when $v(t) = 0$.]

80. Evaluating a Difference Quotient Find the difference quotient and simplify your answer.

$$f(x) = 2x^2 + 3x - 1, \quad \frac{f(x + h) - f(x)}{h}, \quad h \ne 0$$

P.6 **Vertical Line Test for Functions** In Exercises 81 and 82, use the Vertical Line Test to determine whether *y* is a function of *x*.

81. $y = (x - 3)^2$ **82.** $x = -|4 - y|$

Finding the Zeros of a Function In Exercises 83 and 84, find the zeros of the function algebraically.

83. $f(x) = \sqrt{2x + 1}$ **84.** $f(x) = x^3 - x^2$

Describing Function Behavior In Exercises 85 and 86, use a graphing utility to graph the function and visually determine the intervals on which the function is increasing, decreasing, or constant.

85. $f(x) = |x| + |x + 1|$ **86.** $f(x) = (x^2 - 4)^2$

Approximating Relative Minima or Maxima In Exercises 87 and 88, use a graphing utility to graph the function and approximate (to two decimal places) any relative minima or maxima.

87. $f(x) = -x^2 + 2x + 1$ **88.** $f(x) = x^3 - 4x^2 - 1$

Average Rate of Change of a Function In Exercises 89 and 90, find the average rate of change of the function from x_1 to x_2.

89. $f(x) = -x^2 + 8x - 4$, $x_1 = 0, x_2 = 4$

90. $f(x) = 2 - \sqrt{x + 1}$, $x_1 = 3, x_2 = 7$

Even, Odd, or Neither? In Exercises 91–94, determine whether the function is even, odd, or neither. Then describe the symmetry.

91. $f(x) = x^5 + 4x - 7$ **92.** $f(x) = x^4 - 20x^2$

93. $f(x) = 2x\sqrt{x^2 + 3}$ **94.** $f(x) = \sqrt[5]{6x^2}$

P.7 **Writing a Linear Function** In Exercises 95 and 96, (a) write the linear function f such that it has the indicated function values, and (b) sketch the graph of the function.

95. $f(2) = -6$, $f(-1) = 3$

96. $f(0) = -5$, $f(4) = -8$

Graphing a Function In Exercises 97–102, sketch the graph of the function.

97. $f(x) = x^2 + 5$ **98.** $g(x) = -3x^3$

99. $f(x) = \sqrt{x + 1}$ **100.** $g(x) = \dfrac{1}{x + 5}$

101. $g(x) = [\![x + 4]\!]$

102. $f(x) = \begin{cases} 5x - 3, & x \geq -1 \\ -4x + 5, & x < -1 \end{cases}$

P.8 **Identifying a Parent Function** In Exercises 103–112, h is related to one of the parent functions described in this chapter. (a) Identify the parent function f. (b) Describe the sequence of transformations from f to h. (c) Sketch the graph of h. (d) Use function notation to write h in terms of f.

103. $h(x) = -(x + 2)^2 + 3$ **104.** $h(x) = \frac{1}{2}(x - 1)^2 - 2$

105. $h(x) = -\frac{1}{3}x^3$ **106.** $h(x) = (x - 2)^3 + 2$

107. $h(x) = -\sqrt{x} + 4$ **108.** $h(x) = -\sqrt{x + 1} + 9$

109. $h(x) = |x + 3| - 5$ **110.** $h(x) = |x - 9|$

111. $h(x) = -[\![x]\!] + 6$ **112.** $h(x) = 5[\![x - 9]\!]$

P.9 **Finding Arithmetic Combinations of Functions** In Exercises 113 and 114, find (a) $(f + g)(x)$, (b) $(f - g)(x)$, (c) $(fg)(x)$, and (d) $(f/g)(x)$. What is the domain of f/g?

113. $f(x) = x^2 + 3$, $g(x) = 2x - 1$

114. $f(x) = x^2 - 4$, $g(x) = \sqrt{3 - x}$

Finding Domains of Functions and Composite Functions In Exercises 115 and 116, find (a) $f \circ g$ and (b) $g \circ f$. Find the domain of each function and each composite function.

115. $f(x) = \frac{1}{3}x - 3$, $g(x) = 3x + 1$

116. $f(x) = \sqrt{x + 1}$, $g(x) = x^2$

Bacteria Count In Exercises 117 and 118, the number N of bacteria in a refrigerated food is given by $N(T) = 25T^2 - 50T + 300$, $1 \leq T \leq 19$, where T is the temperature of the food in degrees Celsius. When the food is removed from refrigeration, the temperature of the food is given by $T(t) = 2t + 1$, $0 \leq t \leq 9$, where t is the time in hours.

117. Find the composition $(N \circ T)(t)$ and interpret its meaning in context.

118. Find the time when the bacteria count reaches 750.

P.10 **Finding an Inverse Function Informally** In Exercises 119 and 120, find the inverse function of f informally. Verify that $f(f^{-1}(x)) = x$ and $f^{-1}(f(x)) = x$.

119. $f(x) = \dfrac{x - 4}{5}$ **120.** $f(x) = x^3 - 1$

Applying the Horizontal Line Test In Exercises 121 and 122, use a graphing utility to graph the function, and use the Horizontal Line Test to determine whether the function has an inverse function.

121. $f(x) = (x - 1)^2$ **122.** $h(t) = \dfrac{2}{t - 3}$

Finding and Analyzing Inverse Functions In Exercises 123 and 124, (a) find the inverse function of f, (b) graph both f and f^{-1} on the same set of coordinate axes, (c) describe the relationship between the graphs of f and f^{-1}, and (d) state the domains and ranges of f and f^{-1}.

123. $f(x) = \frac{1}{2}x - 3$ **124.** $f(x) = \sqrt{x + 1}$

Restricting the Domain In Exercises 125 and 126, restrict the domain of the function f to an interval on which the function is increasing, and determine f^{-1} on that interval.

125. $f(x) = 2(x - 4)^2$ **126.** $f(x) = |x - 2|$

Exploration

True or False? In Exercises 127 and 128, determine whether the statement is true or false. Justify your answer.

127. Relative to the graph of $f(x) = \sqrt{x}$, the function $h(x) = -\sqrt{x + 9} - 13$ is shifted 9 units to the left and 13 units down, then reflected in the x-axis.

128. If f and g are two inverse functions, then the domain of g is equal to the range of f.

Chapter Test

See **CalcChat.com** for tutorial help and worked-out solutions to odd-numbered exercises.

Take this test as you would take a test in class. When you are finished, check your work against the answers given in the back of the book.

1. Place the appropriate inequality symbol ($<$ or $>$) between the real numbers $-\frac{10}{3}$ and $-|-4|$.

2. Find the distance between the real numbers -5.4 and $3\frac{3}{4}$.

3. Identify the rule of algebra illustrated by $(5 - x) + 0 = 5 - x$.

In Exercises 4–7, solve the equation and check your solution. (If not possible, then explain why.)

4. $\frac{2}{3}(x - 1) + \frac{1}{4}x = 10$

5. $(x - 3)(x + 2) = 14$

6. $\dfrac{x - 2}{x + 2} + \dfrac{4}{x + 2} + 4 = 0$

7. $x^4 + x^2 - 6 = 0$

8. Plot the points $(-2, 5)$ and $(6, 0)$. Then find the distance between the points and the midpoint of the line segment joining the points.

In Exercises 9–11, use the algebraic tests to check for symmetry with respect to both axes and the origin. Then sketch the graph of the equation. Identify any x- and y-intercepts.

9. $y = 4 - \frac{3}{4}x$

10. $y = 4 - |x|$

11. $y = x - x^3$

12. Find the center and radius of the circle $(x - 3)^2 + y^2 = 9$. Then sketch its graph.

In Exercises 13 and 14, find an equation of the line passing through the points. Sketch the line.

13. $(2, -3), (-4, 9)$

14. $(3, 0.8), (7, -6)$

15. Write equations of the lines that pass through the point $(0, 4)$ and are (a) parallel to and (b) perpendicular to the line $5x + 2y = 3$.

16. Evaluate the function $f(x) = |x + 2| - 15$ at each specified value of the independent variable and simplify.

   (a) $f(-8)$ (b) $f(14)$ (c) $f(x - 6)$

In Exercises 17–19, (a) use a graphing utility to graph the function, (b) determine the domain of the function, (c) approximate the intervals on which the function is increasing, decreasing, or constant, and (d) determine whether the function is even, odd, or neither.

17. $f(x) = 2x^6 + 5x^4 - x^2$

18. $f(x) = 4x\sqrt{3 - x}$

19. $f(x) = |x + 5|$

In Exercises 20–22, (a) identify the parent function in the transformation, (b) describe the sequence of transformations from f to h, and (c) sketch the graph of h.

20. $h(x) = 3[\![x]\!]$

21. $h(x) = -\sqrt{x + 5} + 8$

22. $h(x) = -2(x - 5)^3 + 3$

In Exercises 23 and 24, find (a) $(f + g)(x)$, (b) $(f - g)(x)$, (c) $(fg)(x)$, (d) $(f/g)(x)$, (e) $(f \circ g)(x)$, and (f) $(g \circ f)(x)$.

23. $f(x) = 3x^2 - 7, g(x) = -x^2 - 4x + 5$

24. $f(x) = 1/x, g(x) = 2\sqrt{x}$

In Exercises 25–27, determine whether the function has an inverse function. If it does, then find the inverse function.

25. $f(x) = x^3 + 8$

26. $f(x) = |x^2 - 3| + 6$

27. $f(x) = 3x\sqrt{x}$

Proofs in Mathematics ■ ■ ■ ■ ■ ■ ■ ■ ■ ■ ■ ■ ■

What does the word *proof* mean to you? In mathematics, the word *proof* means a valid argument. When you are proving a statement or theorem, you must use facts, definitions, and accepted properties in a logical order. You can also use previously proved theorems in your proof. For instance, the proof of the Midpoint Formula below uses the Distance Formula. There are several different proof methods, which you will see in later chapters.

The Midpoint Formula *(p. 29)*

The midpoint of the line segment joining the points (x_1, y_1) and (x_2, y_2) is given by the Midpoint Formula

$$\text{Midpoint} = \left(\frac{x_1 + x_2}{2}, \frac{y_1 + y_2}{2}\right).$$

THE CARTESIAN PLANE

The Cartesian plane was named after the French mathematician René Descartes (1596–1650). While Descartes was lying in bed, he noticed a fly buzzing around on the square ceiling tiles. He discovered that he could describe the position of the fly by the ceiling tile upon which the fly landed. This led to the development of the Cartesian plane. Descartes felt that using a coordinate plane could facilitate descriptions of the positions of objects.

Proof

Using the figure, you must show that $d_1 = d_2$ and $d_1 + d_2 = d_3$.

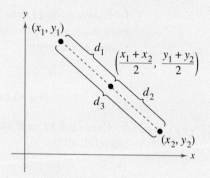

By the Distance Formula, you obtain

$$d_1 = \sqrt{\left(\frac{x_1 + x_2}{2} - x_1\right)^2 + \left(\frac{y_1 + y_2}{2} - y_1\right)^2}$$

$$= \frac{1}{2}\sqrt{(x_2 - x_1)^2 + (y_2 - y_1)^2}$$

$$d_2 = \sqrt{\left(x_2 - \frac{x_1 + x_2}{2}\right)^2 + \left(y_2 - \frac{y_1 + y_2}{2}\right)^2}$$

$$= \frac{1}{2}\sqrt{(x_2 - x_1)^2 + (y_2 - y_1)^2}$$

$$d_3 = \sqrt{(x_2 - x_1)^2 + (y_2 - y_1)^2}$$

So, it follows that $d_1 = d_2$ and $d_1 + d_2 = d_3$. ■

P.S. Problem Solving ▪ ▪ ▪ ▪ ▪ ▪ ▪ ▪ ▪ ▪ ▪ ▪

1. **Monthly Wages** As a salesperson, you receive a monthly salary of $2000, plus a commission of 7% of sales. You receive an offer for a new job at $2300 per month, plus a commission of 5% of sales.

   (a) Write a linear equation for your current monthly wage W_1 in terms of your monthly sales S.

   (b) Write a linear equation for the monthly wage W_2 of your new job offer in terms of the monthly sales S.

   (c) Use a graphing utility to graph both equations in the same viewing window. Find the point of intersection. What does it signify?

   (d) You think you can sell $20,000 per month. Should you change jobs? Explain.

2. **Telephone Keypad** For the numbers 2 through 9 on a telephone keypad (see figure), create two relations: one mapping numbers onto letters, and the other mapping letters onto numbers. Are both relations functions? Explain.

3. **Sums and Differences of Functions** What can be said about the sum and difference of each of the following?

   (a) Two even functions

   (b) Two odd functions

   (c) An odd function and an even function

4. **Inverse Functions** The two functions

   $$f(x) = x \quad \text{and} \quad g(x) = -x$$

   are their own inverse functions. Graph each function and explain why this is true. Graph other linear functions that are their own inverse functions. Find a general formula for a family of linear functions that are their own inverse functions.

5. **Proof** Prove that a function of the following form is even.

   $$y = a_{2n}x^{2n} + a_{2n-2}x^{2n-2} + \cdots + a_2x^2 + a_0$$

6. **Miniature Golf** A miniature golf professional is trying to make a hole-in-one on the miniature golf green shown. The golf ball is at the point $(2.5, 2)$ and the hole is at the point $(9.5, 2)$. The professional wants to bank the ball off the side wall of the green at the point (x, y). Find the coordinates of the point (x, y). Then write an equation for the path of the ball.

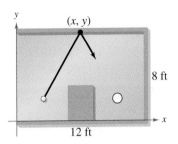

Figure for 6

7. **Titanic** At 2:00 P.M. on April 11, 1912, the *Titanic* left Cobh, Ireland, on her voyage to New York City. At 11:40 P.M. on April 14, the *Titanic* struck an iceberg and sank, having covered only about 2100 miles of the approximately 3400-mile trip.

   (a) What was the total duration of the voyage in hours?

   (b) What was the average speed in miles per hour?

   (c) Write a function relating the distance of the *Titanic* from New York City and the number of hours traveled. Find the domain and range of the function.

   (d) Graph the function from part (c).

8. **Average Rate of Change** Consider the function $f(x) = -x^2 + 4x - 3$. Find the average rate of change of the function from x_1 to x_2.

   (a) $x_1 = 1, x_2 = 2$

   (b) $x_1 = 1, x_2 = 1.5$

   (c) $x_1 = 1, x_2 = 1.25$

   (d) $x_1 = 1, x_2 = 1.125$

   (e) $x_1 = 1, x_2 = 1.0625$

   (f) Does the average rate of change seem to be approaching one value? If so, then state the value.

   (g) Find the equations of the secant lines through the points $(x_1, f(x_1))$ and $(x_2, f(x_2))$ for parts (a)–(e).

   (h) Find the equation of the line through the point $(1, f(1))$ using your answer from part (f) as the slope of the line.

9. **Inverse of a Composition** Consider the functions $f(x) = 4x$ and $g(x) = x + 6$.

   (a) Find $(f \circ g)(x)$.

   (b) Find $(f \circ g)^{-1}(x)$.

   (c) Find $f^{-1}(x)$ and $g^{-1}(x)$.

   (d) Find $(g^{-1} \circ f^{-1})(x)$ and compare the result with that of part (b).

   (e) Repeat parts (a) through (d) for $f(x) = x^3 + 1$ and $g(x) = 2x$.

   (f) Write two one-to-one functions f and g, and repeat parts (a) through (d) for these functions.

   (g) Make a conjecture about $(f \circ g)^{-1}(x)$ and $(g^{-1} \circ f^{-1})(x)$.

10. Trip Time You are in a boat 2 miles from the nearest point on the coast. You are to travel to a point Q, 3 miles down the coast and 1 mile inland (see figure). You row at 2 miles per hour and walk at 4 miles per hour.

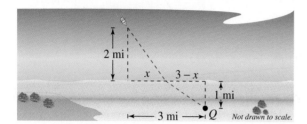

Not drawn to scale.

(a) Write the total time T of the trip as a function of x.

(b) Determine the domain of the function.

(c) Use a graphing utility to graph the function. Be sure to choose an appropriate viewing window.

(d) Find the value of x that minimizes T.

(e) Write a brief paragraph interpreting these values.

11. Heaviside Function The **Heaviside function** $H(x)$ is widely used in engineering applications. (See figure.) To print an enlarged copy of the graph, go to *MathGraphs.com.*

$$H(x) = \begin{cases} 1, & x \geq 0 \\ 0, & x < 0 \end{cases}$$

Sketch the graph of each function by hand.

(a) $H(x) - 2$

(b) $H(x - 2)$

(c) $-H(x)$

(d) $H(-x)$

(e) $\frac{1}{2}H(x)$

(f) $-H(x - 2) + 2$

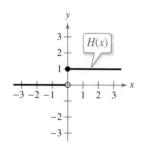

12. Repeated Composition Let $f(x) = \dfrac{1}{1 - x}$.

(a) What are the domain and range of f?

(b) Find $f(f(x))$. What is the domain of this function?

(c) Find $f(f(f(x)))$. Is the graph a line? Why or why not?

13. Associative Property with Compositions Show that the Associative Property holds for compositions of functions—that is,

$$(f \circ (g \circ h))(x) = ((f \circ g) \circ h)(x).$$

14. Graphical Analysis Consider the graph of the function f shown in the figure. Use this graph to sketch the graph of each function. To print an enlarged copy of the graph, go to *MathGraphs.com.*

(a) $f(x + 1)$

(b) $f(x) + 1$

(c) $2f(x)$

(d) $f(-x)$

(e) $-f(x)$

(f) $|f(x)|$

(g) $f(|x|)$

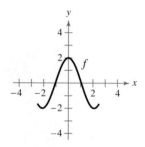

15. Graphical Analysis Use the graphs of f and f^{-1} to complete each table of function values.

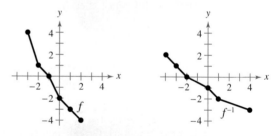

(a)

x	-4	-2	0	4
$(f(f^{-1}(x)))$				

(b)

x	-3	-2	0	1
$(f + f^{-1})(x)$				

(c)

x	-3	-2	0	1
$(f \cdot f^{-1})(x)$				

(d)

x	-4	-3	0	4
$|f^{-1}(x)|$				

1 Trigonometry

Television Coverage *(Exercise 84, page 179)*

Waterslide Design
(Exercise 32, page 197)

Respiratory Cycle *(Exercise 88, page 168)*

Meteorology
(Exercise 99, page 158)

Skateboarding *(Example 10, page 145)*

1.1   Radian and Degree Measure

Angles can help you model and solve real-life problems. For instance, in Exercise 68 on page 131, you will use angles to find the speed of a bicycle.

■ Describe angles.
■ Use radian measure.
■ Use degree measure.
■ Use angles to model and solve real-life problems.

Angles

As derived from the Greek language, the word **trigonometry** means "measurement of triangles." Initially, trigonometry dealt with relationships among the sides and angles of triangles and was used in the development of astronomy, navigation, and surveying. With the development of calculus and the physical sciences in the 17th century, a different perspective arose—one that viewed the classic trigonometric relationships as *functions* with the set of real numbers as their domains. Consequently, the applications of trigonometry expanded to include a vast number of physical phenomena, such as sound waves, planetary orbits, vibrating strings, pendulums, and orbits of atomic particles.

This text incorporates *both* perspectives, starting with angles and their measures.

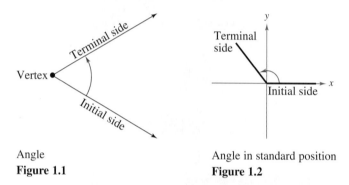

Angle
Figure 1.1

Angle in standard position
Figure 1.2

An **angle** is determined by rotating a ray (half-line) about its endpoint. The starting position of the ray is the **initial side** of the angle, and the position after rotation is the **terminal side,** as shown in Figure 1.1. The endpoint of the ray is the **vertex** of the angle. This perception of an angle fits a coordinate system in which the origin is the vertex and the initial side coincides with the positive x-axis. Such an angle is in **standard position,** as shown in Figure 1.2. Counterclockwise rotation generates **positive angles** and clockwise rotation generates **negative angles,** as shown in Figure 1.3. Angles are labeled with Greek letters such as

α (alpha), β (beta),   and θ (theta)

as well as uppercase letters such as

A, B,   and C.

In Figure 1.4, note that angles α and β have the same initial and terminal sides. Such angles are **coterminal.**

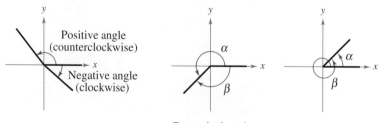

Figure 1.3

Coterminal angles
Figure 1.4

Radian Measure

You determine the **measure of an angle** by the amount of rotation from the initial side to the terminal side. One way to measure angles is in *radians*. This type of measure is especially useful in calculus. To define a radian, you can use a **central angle** of a circle, one whose vertex is the center of the circle, as shown in Figure 1.5.

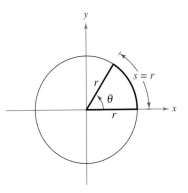

Arc length = radius when $\theta = 1$ radian.
Figure 1.5

> ### Definition of Radian
>
> One **radian** is the measure of a central angle θ that intercepts an arc s equal in length to the radius r of the circle. See Figure 1.5. Algebraically, this means that
>
> $$\theta = \frac{s}{r}$$
>
> where θ is measured in radians. (Note that $\theta = 1$ when $s = r$.)

Because the circumference of a circle is $2\pi r$ units, it follows that a central angle of one full revolution (counterclockwise) corresponds to an arc length of $s = 2\pi r$. Moreover, because $2\pi \approx 6.28$, there are just over six radius lengths in a full circle, as shown in Figure 1.6. Because the units of measure for s and r are the same, the ratio s/r has no units—it is a real number.

Because the measure of an angle of one full revolution is $s/r = 2\pi r/r = 2\pi$ radians, you can obtain the following.

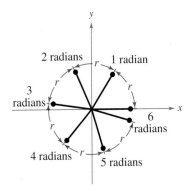

Figure 1.6

$$\frac{1}{2}\text{ revolution} = \frac{2\pi}{2} = \pi \text{ radians} \qquad \frac{1}{4}\text{ revolution} = \frac{2\pi}{4} = \frac{\pi}{2} \text{ radians}$$

$$\frac{1}{6}\text{ revolution} = \frac{2\pi}{6} = \frac{\pi}{3} \text{ radians}$$

These and other common angles are shown below.

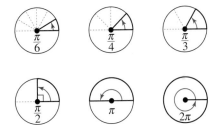

$\frac{\pi}{6}$ $\frac{\pi}{4}$ $\frac{\pi}{3}$ $\frac{\pi}{2}$ π 2π

· · REMARK The phrase "the terminal side of θ lies in a quadrant" is often abbreviated by the phrase "θ lies in a quadrant." The terminal sides of the "quadrant angles" 0, $\pi/2$, π, and $3\pi/2$ do not lie within quadrants.

Recall that the four quadrants in a coordinate system are numbered I, II, III, and IV. The figure below shows which angles between 0 and 2π lie in each of the four quadrants. Note that angles between 0 and $\pi/2$ are **acute** angles and angles between $\pi/2$ and π are **obtuse** angles.

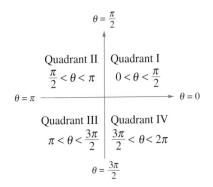

Two angles are coterminal when they have the same initial and terminal sides. For instance, the angles 0 and 2π are coterminal, as are the angles $\pi/6$ and $13\pi/6$. You can find an angle that is coterminal to a given angle θ by adding or subtracting 2π (one revolution), as demonstrated in Example 1. A given angle θ has infinitely many coterminal angles. For instance, $\theta = \pi/6$ is coterminal with $\pi/6 + 2n\pi$, where n is an integer.

EXAMPLE 1   **Finding Coterminal Angles**

▷ **ALGEBRA HELP**   You can review operations involving fractions in Section P.1.

a. For the positive angle $13\pi/6$, subtract 2π to obtain a coterminal angle

$$\frac{13\pi}{6} - 2\pi = \frac{\pi}{6}.$$ See Figure 1.7.

b. For the negative angle $-2\pi/3$, add 2π to obtain a coterminal angle

$$-\frac{2\pi}{3} + 2\pi = \frac{4\pi}{3}.$$ See Figure 1.8.

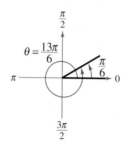

Figure 1.7   **Figure 1.8**

✓ **Checkpoint** ◀))) *Audio-video solution in English & Spanish at LarsonPrecalculus.com.*

Determine two coterminal angles (one positive and one negative) for each angle.

a. $\theta = \dfrac{9\pi}{4}$ **b.** $\theta = -\dfrac{\pi}{3}$

Complementary angles

Supplementary angles
Figure 1.9

Two positive angles α and β are **complementary** (complements of each other) when their sum is $\pi/2$. Two positive angles are **supplementary** (supplements of each other) when their sum is π. See Figure 1.9.

EXAMPLE 2   **Complementary and Supplementary Angles**

a. The complement of $\dfrac{2\pi}{5}$ is $\dfrac{\pi}{2} - \dfrac{2\pi}{5} = \dfrac{5\pi}{10} - \dfrac{4\pi}{10} = \dfrac{\pi}{10}$.

The supplement of $\dfrac{2\pi}{5}$ is $\pi - \dfrac{2\pi}{5} = \dfrac{5\pi}{5} - \dfrac{2\pi}{5} = \dfrac{3\pi}{5}$.

b. Because $4\pi/5$ is greater than $\pi/2$, it has no complement. (Remember that complements are *positive* angles.) The supplement of $4\pi/5$ is

$$\pi - \frac{4\pi}{5} = \frac{5\pi}{5} - \frac{4\pi}{5} = \frac{\pi}{5}.$$

✓ **Checkpoint** ◀))) *Audio-video solution in English & Spanish at LarsonPrecalculus.com.*

If possible, find the complement and the supplement of (a) $\pi/6$ and (b) $5\pi/6$. ■

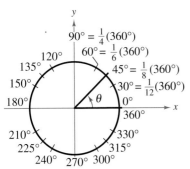

Figure 1.10

$\frac{\pi}{6}$
30°

$\frac{\pi}{4}$
45°

$\frac{\pi}{3}$
60°

$\frac{\pi}{2}$
90°

π
180°

2π
360°

Figure 1.11

Degree Measure

A second way to measure angles is in **degrees,** denoted by the symbol °. A measure of one degree (1°) is equivalent to a rotation of $\frac{1}{360}$ of a complete revolution about the vertex. To measure angles, it is convenient to mark degrees on the circumference of a circle, as shown in Figure 1.10. So, a full revolution (counterclockwise) corresponds to 360°, a half revolution to 180°, a quarter revolution to 90°, and so on.

Because 2π radians corresponds to one complete revolution, degrees and radians are related by the equations

$$360° = 2\pi \text{ rad} \quad \text{and} \quad 180° = \pi \text{ rad.}$$

From the latter equation, you obtain

$$1° = \frac{\pi}{180} \text{ rad} \quad \text{and} \quad 1 \text{ rad} = \left(\frac{180}{\pi}\right)°$$

which lead to the following conversion rules.

Conversions Between Degrees and Radians

1. To convert degrees to radians, multiply degrees by $\dfrac{\pi \text{ rad}}{180°}$.

2. To convert radians to degrees, multiply radians by $\dfrac{180°}{\pi \text{ rad}}$.

To apply these two conversion rules, use the basic relationship π rad = 180°. (See Figure 1.11.)

When no units of angle measure are specified, *radian measure is implied.* For instance, $\theta = 2$ implies that $\theta = 2$ radians.

EXAMPLE 3   **Converting from Degrees to Radians**

a. $135° = (135 \text{ deg})\left(\dfrac{\pi \text{ rad}}{180 \text{ deg}}\right) = \dfrac{3\pi}{4}$ radians   Multiply by $\frac{\pi \text{ rad}}{180°}$.

b. $540° = (540 \text{ deg})\left(\dfrac{\pi \text{ rad}}{180 \text{ deg}}\right) = 3\pi$ radians   Multiply by $\frac{\pi \text{ rad}}{180°}$.

✓ **Checkpoint**))) *Audio-video solution in English & Spanish at LarsonPrecalculus.com.*

Rewrite each angle in radian measure as a multiple of π. (Do not use a calculator.)

a. $\theta = 60°$ **b.** $\theta = 320°$

EXAMPLE 4   **Converting from Radians to Degrees**

a. $-\dfrac{\pi}{2} \text{ rad} = \left(-\dfrac{\pi}{2} \text{ rad}\right)\left(\dfrac{180 \text{ deg}}{\pi \text{ rad}}\right) = -90°$ Multiply by $\frac{180°}{\pi \text{ rad}}$.

b. $2 \text{ rad} = (2 \text{ rad})\left(\dfrac{180 \text{ deg}}{\pi \text{ rad}}\right) = \dfrac{360°}{\pi} \approx 114.59°$ Multiply by $\frac{180°}{\pi \text{ rad}}$.

✓ **Checkpoint**))) *Audio-video solution in English & Spanish at LarsonPrecalculus.com.*

Rewrite each angle in degree measure. (Do not use a calculator.)

a. $\pi/6$ **b.** $5\pi/3$

▷ **TECHNOLOGY**
With calculators, it is convenient to use *decimal* degrees to denote fractional parts of degrees. Historically, however, fractional parts of degrees were expressed in *minutes* and *seconds,* using the prime (′) and double prime (″) notations, respectively. That is,

$$1' = \text{one minute} = \tfrac{1}{60}(1°)$$

$$1'' = \text{one second} = \tfrac{1}{3600}(1°).$$

Consequently, an angle of 64 degrees, 32 minutes, and 47 seconds is represented by $\theta = 64° \, 32' \, 47''$. Many calculators have special keys for converting an angle in degrees, minutes, and seconds (D° M′ S″) to decimal degree form, and vice versa.

Applications

The *radian measure* formula, $\theta = s/r$, can be used to measure arc length along a circle.

> ### Arc Length
>
> For a circle of radius r, a central angle θ intercepts an arc of length s given by
>
> $$s = r\theta \qquad \text{Length of circular arc}$$
>
> where θ is measured in radians. Note that if $r = 1$, then $s = \theta$, and the radian measure of θ equals the arc length.

EXAMPLE 5 **Finding Arc Length**

A circle has a radius of 4 inches. Find the length of the arc intercepted by a central angle of 240°, as shown in Figure 1.12.

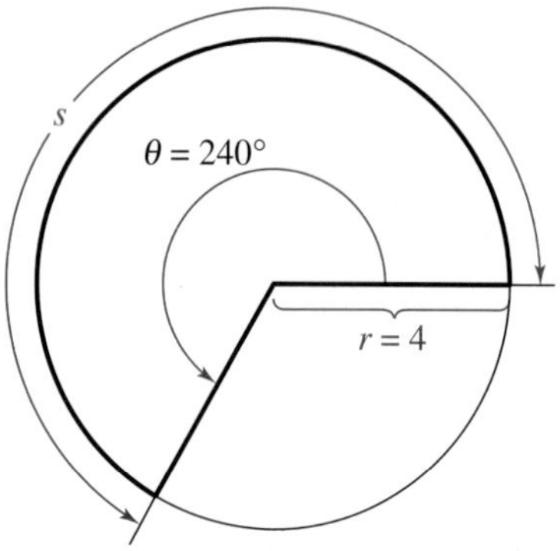

Solution To use the formula $s = r\theta$, first convert 240° to radian measure.

$$240° = (240 \text{ deg})\left(\frac{\pi \text{ rad}}{180 \text{ deg}}\right)$$

$$= \frac{4\pi}{3} \text{ radians}$$

Then, using a radius of $r = 4$ inches, you can find the arc length to be

$$s = r\theta$$

$$= 4\left(\frac{4\pi}{3}\right)$$

$$\approx 16.76 \text{ inches.}$$

Note that the units for r determine the units for $r\theta$ because θ is given in radian measure, which has no units.

✓ *Checkpoint* ◀))) *Audio-video solution in English & Spanish at LarsonPrecalculus.com.*

A circle has a radius of 27 inches. Find the length of the arc intercepted by a central angle of 160°.

Figure 1.12

$\theta = 240°$

$r = 4$

The formula for the length of a circular arc can help you analyze the motion of a particle moving at a *constant speed* along a circular path.

> ### Linear and Angular Speeds
>
> Consider a particle moving at a constant speed along a circular arc of radius r. If s is the length of the arc traveled in time t, then the **linear speed** v of the particle is
>
> $$\text{Linear speed } v = \frac{\text{arc length}}{\text{time}} = \frac{s}{t}.$$
>
> Moreover, if θ is the angle (in radian measure) corresponding to the arc length s, then the **angular speed** ω (the lowercase Greek letter omega) of the particle is
>
> $$\text{Angular speed } \omega = \frac{\text{central angle}}{\text{time}} = \frac{\theta}{t}.$$

•• **REMARK** Linear speed measures how fast the particle moves, and angular speed measures how fast the angle changes. By dividing each side of the formula for arc length by t, you can establish a relationship between linear speed v and angular speed ω, as shown.

$$s = r\theta$$

$$\frac{s}{t} = \frac{r\theta}{t}$$

$$v = r\omega$$

Figure 1.13

EXAMPLE 6   **Finding Linear Speed**

The second hand of a clock is 10.2 centimeters long, as shown in Figure 1.13. Find the linear speed of the tip of this second hand as it passes around the clock face.

Solution   In one revolution, the arc length traveled is

$$s = 2\pi r$$
$$= 2\pi(10.2) \qquad \text{Substitute for } r.$$
$$= 20.4\pi \text{ centimeters.}$$

The time required for the second hand to travel this distance is

$$t = 1 \text{ minute}$$
$$= 60 \text{ seconds.}$$

So, the linear speed of the tip of the second hand is

$$\text{Linear speed} = \frac{s}{t}$$
$$= \frac{20.4\pi \text{ centimeters}}{60 \text{ seconds}}$$
$$\approx 1.068 \text{ centimeters per second.}$$

✓ *Checkpoint* *Audio-video solution in English & Spanish at LarsonPrecalculus.com.*

The second hand of a clock is 8 centimeters long. Find the linear speed of the tip of this second hand as it passes around the clock face.

EXAMPLE 7   **Finding Angular and Linear Speeds**

The blades of a wind turbine are 116 feet long (see Figure 1.14). The propeller rotates at 15 revolutions per minute.

a. Find the angular speed of the propeller in radians per minute.

b. Find the linear speed of the tips of the blades.

Solution

a. Because each revolution generates 2π radians, it follows that the propeller turns

$$(15)(2\pi) = 30\pi \text{ radians per minute.}$$

In other words, the angular speed is

$$\text{Angular speed} = \frac{\theta}{t} = \frac{30\pi \text{ radians}}{1 \text{ minute}} = 30\pi \text{ radians per minute.}$$

Figure 1.14

b. The linear speed is

$$\text{Linear speed} = \frac{s}{t} = \frac{r\theta}{t} = \frac{(116)(30\pi) \text{ feet}}{1 \text{ minute}} \approx 10{,}933 \text{ feet per minute.}$$

✓ *Checkpoint* *Audio-video solution in English & Spanish at LarsonPrecalculus.com.*

The circular blade on a saw rotates at 2400 revolutions per minute.

a. Find the angular speed of the blade in radians per minute.

b. The blade has a radius of 4 inches. Find the linear speed of a blade tip.

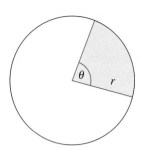

Figure 1.15

A **sector** of a circle is the region bounded by two radii of the circle and their intercepted arc (see Figure 1.15).

Area of a Sector of a Circle

For a circle of radius r, the area A of a sector of the circle with central angle θ is

$$A = \frac{1}{2}r^2\theta$$

where θ is measured in radians.

EXAMPLE 8   **Area of a Sector of a Circle**

A sprinkler on a golf course fairway sprays water over a distance of 70 feet and rotates through an angle of 120° (see Figure 1.16). Find the area of the fairway watered by the sprinkler.

Solution

First convert 120° to radian measure as follows.

$$\theta = 120°$$

$$= (120 \text{ deg})\left(\frac{\pi \text{ rad}}{180 \text{ deg}}\right) \qquad \text{Multiply by } \frac{\pi \text{ rad}}{180°}.$$

$$= \frac{2\pi}{3} \text{ radians}$$

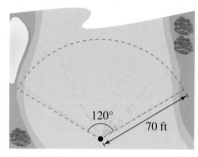

Figure 1.16

Then, using $\theta = 2\pi/3$ and $r = 70$, the area is

$$A = \frac{1}{2}r^2\theta \qquad \text{Formula for the area of a sector of a circle}$$

$$= \frac{1}{2}(70)^2\left(\frac{2\pi}{3}\right) \qquad \text{Substitute for } r \text{ and } \theta.$$

$$= \frac{4900\pi}{3} \qquad \text{Multiply.}$$

$$\approx 5131 \text{ square feet.} \qquad \text{Simplify.}$$

 Checkpoint  *Audio-video solution in English & Spanish at LarsonPrecalculus.com.*

A sprinkler sprays water over a distance of 40 feet and rotates through an angle of 80°. Find the area watered by the sprinkler. ◼

Summarize   (Section 1.1)

1. Describe an angle *(page 122)*.

2. Describe how to determine the measure of an angle using radians *(page 123)*. For examples involving radian measure, see Examples 1 and 2.

3. Describe how to determine the measure of an angle using degrees *(page 125)*. For examples involving degree measure, see Examples 3 and 4.

4. Describe examples of how to use angles to model and solve real-life problems *(pages 126–128, Examples 5–8)*.

1.1 Exercises

See **CalcChat.com** for tutorial help and worked-out solutions to odd-numbered exercises.

Vocabulary: Fill in the blanks.

1. Two angles that have the same initial and terminal sides are _____.
2. One _____ is the measure of a central angle that intercepts an arc equal to the radius of the circle.
3. Two positive angles that have a sum of $\pi/2$ are _____ angles, whereas two positive angles that have a sum of π are _____ angles.
4. The angle measure that is equivalent to a rotation of $\frac{1}{360}$ of a complete revolution about an angle's vertex is one _____.
5. The _____ speed of a particle is the ratio of the arc length to the time traveled, and the _____ speed of a particle is the ratio of the central angle to the time traveled.
6. The area A of a sector of a circle with radius r and central angle θ, where θ is measured in radians, is given by the formula _____.

Skills and Applications

Estimating an Angle   In Exercises 7–10, estimate the angle to the nearest one-half radian.

7.

8.

9.

10.

Determining Quadrants   In Exercises 11 and 12, determine the quadrant in which each angle lies.

11. (a) $\dfrac{\pi}{4}$ (b) $\dfrac{5\pi}{4}$ 12. (a) $-\dfrac{\pi}{6}$ (b) $-\dfrac{11\pi}{9}$

Sketching Angles   In Exercises 13 and 14, sketch each angle in standard position.

13. (a) $\dfrac{\pi}{3}$ (b) $-\dfrac{2\pi}{3}$ 14. (a) $\dfrac{5\pi}{2}$ (b) 4

Finding Coterminal Angles   In Exercises 15 and 16, determine two coterminal angles (one positive and one negative) for each angle. Give your answers in radians.

15. (a) $\dfrac{\pi}{6}$ (b) $\dfrac{7\pi}{6}$ 16. (a) $\dfrac{2\pi}{3}$ (b) $-\dfrac{9\pi}{4}$

Complementary and Supplementary Angles   In Exercises 17–20, find (if possible) the complement and the supplement of each angle.

17. (a) $\dfrac{\pi}{3}$ (b) $\dfrac{\pi}{4}$ 18. (a) $\dfrac{\pi}{12}$ (b) $\dfrac{11\pi}{12}$

19. (a) 1   (b) 2   20. (a) 3   (b) 1.5

Estimating an Angle   In Exercises 21–24, estimate the number of degrees in the angle.

21.    22.

23. 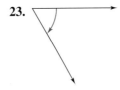   24.

Determining Quadrants   In Exercises 25 and 26, determine the quadrant in which each angle lies.

25. (a) $130°$ (b) $8.3°$
26. (a) $-132°\,50'$ (b) $-3.4°$

Sketching Angles   In Exercises 27 and 28, sketch each angle in standard position.

27. (a) $270°$ (b) $120°$
28. (a) $-135°$ (b) $-750°$

Finding Coterminal Angles   In Exercises 29 and 30, determine two coterminal angles (one positive and one negative) for each angle. Give your answers in degrees.

29. (a) $45°$ (b) $-36°$
30. (a) $120°$ (b) $-420°$

Complementary and Supplementary Angles   In Exercises 31–34, find (if possible) the complement and the supplement of each angle.

31. (a) $18°$ (b) $85°$ 32. (a) $46°$ (b) $93°$
33. (a) $150°$ (b) $79°$ 34. (a) $130°$ (b) $170°$

Converting from Degrees to Radians In Exercises 35 and 36, rewrite each angle in radian measure as a multiple of π. (Do not use a calculator.)

35. (a) $120°$ (b) $-20°$

36. (a) $-60°$ (b) $144°$

Converting from Radians to Degrees In Exercises 37 and 38, rewrite each angle in degree measure. (Do not use a calculator.)

37. (a) $\dfrac{3\pi}{2}$ (b) $\dfrac{7\pi}{6}$

38. (a) $-\dfrac{7\pi}{12}$ (b) $\dfrac{5\pi}{4}$

Converting from Degrees to Radians In Exercises 39–42, convert the angle measure from degrees to radians. Round to three decimal places.

39. $45°$

40. $-48.27°$

41. $0.54°$

42. $345°$

Converting from Radians to Degrees In Exercises 43–46, convert the angle measure from radians to degrees. Round to three decimal places.

43. $\dfrac{5\pi}{11}$

44. $\dfrac{15\pi}{8}$

45. -4.2π

46. -0.57

Converting to Decimal Degree Form In Exercises 47 and 48, convert each angle measure to decimal degree form without using a calculator. Then check your answers using a calculator.

47. (a) $54° \, 45'$ (b) $-128° \, 30'$

48. (a) $-135° \, 36''$ (b) $-408° \, 16' 20''$

Converting to D°M′S″Form In Exercises 49 and 50, convert each angle measure to degrees, minutes, and seconds without using a calculator. Then check your answers using a calculator.

49. (a) $240.6°$ (b) $-145.8°$

50. (a) $-345.12°$ (b) $-3.58°$

Finding Arc Length In Exercises 51 and 52, find the length of the arc on a circle of radius r intercepted by a central angle θ.

51. $r = 15$ inches, $\theta = 120°$

52. $r = 3$ meters, $\theta = 150°$

Finding the Central Angle In Exercises 53 and 54, find the radian measure of the central angle of a circle of radius r that intercepts an arc of length s.

53. $r = 80$ kilometers, $s = 150$ kilometers

54. $r = 14$ feet, $s = 8$ feet

Finding an Angle In Exercises 55 and 56, use the given arc length and radius to find the angle θ (in radians).

55.

56.

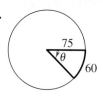

Area of a Sector of a Circle In Exercises 57 and 58, find the area of the sector of a circle of radius r and central angle θ.

57. $r = 12$ millimeters, $\theta = \dfrac{\pi}{4}$

58. $r = 2.5$ feet, $\theta = 225°$

59. Distance Between Cities Find the distance between Dallas, Texas, whose latitude is $32° \, 47' \, 39''$N, and Omaha, Nebraska, whose latitude is $41° \, 15' \, 50''$N. Assume that Earth is a sphere of radius 4000 miles and that the cities are on the same longitude (Omaha is due north of Dallas).

60. Difference in Latitudes Assuming that Earth is a sphere of radius 6378 kilometers, what is the difference in the latitudes of Lynchburg, Virginia, and Myrtle Beach, South Carolina, where Lynchburg is about 400 kilometers due north of Myrtle Beach?

61. Instrumentation The pointer on a voltmeter is 6 centimeters in length (see figure). Find the number of degrees through which the pointer rotates when it moves 2.5 centimeters on the scale.

62. Linear Speed A satellite in a circular orbit 1250 kilometers above Earth makes one complete revolution every 110 minutes. Assuming that Earth is a sphere of radius 6378 kilometers, what is the linear speed (in kilometers per minute) of the satellite?

63. Angular and Linear Speeds The circular blade on a saw rotates at 5000 revolutions per minute.

(a) Find the angular speed of the blade in radians per minute.

(b) The blade has a diameter of $7\frac{1}{4}$ inches. Find the linear speed of a blade tip.

64. Angular and Linear Speeds A carousel with a 50-foot diameter makes 4 revolutions per minute.

(a) Find the angular speed of the carousel in radians per minute.

(b) Find the linear speed (in feet per minute) of the platform rim of the carousel.

65. Angular and Linear Speeds A DVD is approximately 12 centimeters in diameter. The drive motor of the DVD player rotates between 200 and 500 revolutions per minute, depending on what track is being read.

(a) Find an interval for the angular speed of the DVD as it rotates.

(b) Find an interval for the linear speed of a point on the outermost track as the DVD rotates.

66. Angular Speed A car is moving at a rate of 65 miles per hour, and the diameter of its wheels is 2 feet.

(a) Find the number of revolutions per minute the wheels are rotating.

(b) Find the angular speed of the wheels in radians per minute.

67. Linear and Angular Speeds A computerized spin balance machine rotates a 25-inch-diameter tire at 480 revolutions per minute.

(a) Find the road speed (in miles per hour) at which the tire is being balanced.

(b) At what rate should the spin balance machine be set so that the tire is being tested for 55 miles per hour?

68. Speed of a Bicycle

The radii of the pedal sprocket, the wheel sprocket, and the wheel of the bicycle in the figure are 4 inches, 2 inches, and 14 inches, respectively. A cyclist is pedaling at a rate of 1 revolution per second.

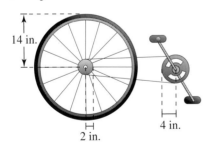

(a) Find the speed of the bicycle in feet per second and miles per hour.

(b) Use your result from part (a) to write a function for the distance d (in miles) a cyclist travels in terms of the number n of revolutions of the pedal sprocket.

(c) Write a function for the distance d (in miles) a cyclist travels in terms of the time t (in seconds). Compare this function with the function from part (b).

69. Area A sprinkler on a golf green sprays water over a distance of 15 meters and rotates through an angle of 140°. Draw a diagram that shows the region that the sprinkler can irrigate. Find the area of the region.

70. Area A car's rear windshield wiper rotates 125°. The total length of the wiper mechanism is 25 inches and wipes the windshield over a distance of 14 inches. Find the area covered by the wiper.

Exploration

True or False? **In Exercises 71–73, determine whether the statement is true or false. Justify your answer.**

71. A measurement of 4 radians corresponds to two complete revolutions from the initial side to the terminal side of an angle.

72. The difference between the measures of two coterminal angles is always a multiple of 360° when expressed in degrees and is always a multiple of 2π radians when expressed in radians.

73. An angle that measures $-1260°$ lies in Quadrant III.

74. HOW DO YOU SEE IT? Determine which angles in the figure are coterminal angles with angle A. Explain your reasoning.

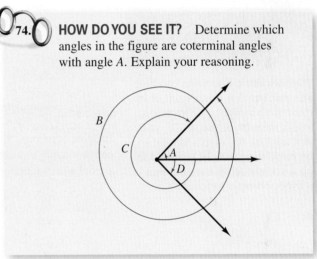

75. Think About It A fan motor turns at a given angular speed. How does the speed of the tips of the blades change when a fan of greater diameter is on the motor? Explain.

76. Think About It Is a degree or a radian the greater unit of measure? Explain.

77. Writing When the radius of a circle increases and the magnitude of a central angle is constant, how does the length of the intercepted arc change? Explain your reasoning.

78. Proof Prove that the area of a circular sector of radius r with central angle θ is

$$A = \frac{1}{2}\theta r^2$$

where θ is measured in radians.

1.2 Trigonometric Functions: The Unit Circle

Trigonometric functions can help you analyze the movement of an oscillating weight. For instance, in Exercise 50 on page 138, you will analyze the displacement of an oscillating weight suspended by a spring using a model that is a trigonometric function.

- Identify a unit circle and describe its relationship to real numbers.
- Evaluate trigonometric functions using the unit circle.
- Use domain and period to evaluate sine and cosine functions, and use a calculator to evaluate trigonometric functions.

The Unit Circle

The two historical perspectives of trigonometry incorporate different methods for introducing the trigonometric functions. One such perspective follows and is based on the unit circle.

Consider the **unit circle** given by

$$x^2 + y^2 = 1 \qquad \text{Unit circle}$$

as shown below.

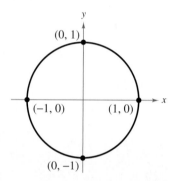

Imagine wrapping the real number line around this circle, with positive numbers corresponding to a counterclockwise wrapping and negative numbers corresponding to a clockwise wrapping, as shown below.

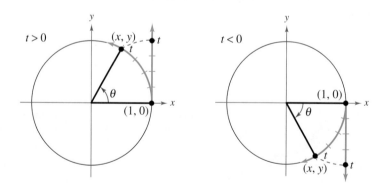

As the real number line wraps around the unit circle, each real number t corresponds to a point (x, y) on the circle. For example, the real number 0 corresponds to the point $(1, 0)$. Moreover, because the unit circle has a circumference of 2π, the real number 2π also corresponds to the point $(1, 0)$.

In general, each real number t also corresponds to a central angle θ (in standard position) whose radian measure is t. With this interpretation of t, the arc length formula

$$s = r\theta \quad \text{(with } r = 1\text{)}$$

indicates that the real number t is the (directional) length of the arc intercepted by the angle θ, given in radians.

Richard Megna/Fundamental Photographs

The Trigonometric Functions

From the preceding discussion, it follows that the coordinates x and y are two functions of the real variable t. You can use these coordinates to define the six trigonometric functions of t.

sine   cosecant   cosine   secant   tangent   cotangent

These six functions are normally abbreviated sin, csc, cos, sec, tan, and cot, respectively.

Definitions of Trigonometric Functions

Let t be a real number and let (x, y) be the point on the unit circle corresponding to t.

$$\sin t = y \qquad\qquad \cos t = x \qquad\qquad \tan t = \frac{y}{x}, \quad x \neq 0$$

$$\csc t = \frac{1}{y}, \quad y \neq 0 \qquad \sec t = \frac{1}{x}, \quad x \neq 0 \qquad \cot t = \frac{x}{y}, \quad y \neq 0$$

•• **REMARK**   Note that the functions in the second row are the *reciprocals* of the corresponding functions in the first row.

In the definitions of the trigonometric functions, note that the tangent and secant are not defined when $x = 0$. For instance, because $t = \pi/2$ corresponds to $(x, y) = (0, 1)$, it follows that $\tan(\pi/2)$ and $\sec(\pi/2)$ are *undefined*. Similarly, the cotangent and cosecant are not defined when $y = 0$. For instance, because $t = 0$ corresponds to $(x, y) = (1, 0)$, cot 0 and csc 0 are *undefined*.

In Figure 1.17, the unit circle is divided into eight equal arcs, corresponding to t-values of

$$0, \frac{\pi}{4}, \frac{\pi}{2}, \frac{3\pi}{4}, \pi, \frac{5\pi}{4}, \frac{3\pi}{2}, \frac{7\pi}{4}, \text{ and } 2\pi.$$

Similarly, in Figure 1.18, the unit circle is divided into 12 equal arcs, corresponding to t-values of

$$0, \frac{\pi}{6}, \frac{\pi}{3}, \frac{\pi}{2}, \frac{2\pi}{3}, \frac{5\pi}{6}, \pi, \frac{7\pi}{6}, \frac{4\pi}{3}, \frac{3\pi}{2}, \frac{5\pi}{3}, \frac{11\pi}{6}, \text{ and } 2\pi.$$

To verify the points on the unit circle in Figure 1.17, note that

$$\left(\frac{\sqrt{2}}{2}, \frac{\sqrt{2}}{2} \right)$$

also lies on the line $y = x$. So, substituting x for y in the equation of the unit circle produces the following.

$$x^2 + x^2 = 1 \implies 2x^2 = 1 \implies x^2 = \frac{1}{2} \implies x = \pm\frac{\sqrt{2}}{2}$$

Because the point is in the first quadrant and $y = x$, you have

$$x = \frac{\sqrt{2}}{2} \quad \text{and} \quad y = \frac{\sqrt{2}}{2}.$$

You can use similar reasoning to verify the rest of the points in Figure 1.17 and the points in Figure 1.18.

Using the (x, y) coordinates in Figures 1.17 and 1.18, you can evaluate the trigonometric functions for common t-values. Examples 1 and 2 demonstrate this procedure. You should study and learn these exact function values for common t-values because they will help you in later sections to perform calculations.

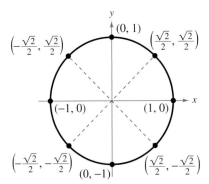

Figure 1.17

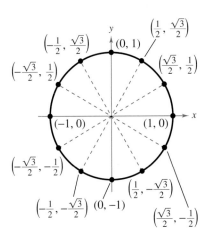

Figure 1.18

| EXAMPLE 1 | **Evaluating Trigonometric Functions** |

▷ **ALGEBRA HELP**   You can review dividing fractions in Section P.1.

Evaluate the six trigonometric functions at each real number.

a. $t = \dfrac{\pi}{6}$ **b.** $t = \dfrac{5\pi}{4}$ **c.** $t = \pi$ **d.** $t = -\dfrac{\pi}{3}$

Solution   For each t-value, begin by finding the corresponding point (x, y) on the unit circle. Then use the definitions of trigonometric functions listed on page 133.

a. $t = \pi/6$ corresponds to the point $(x, y) = \left(\sqrt{3}/2, \, 1/2\right)$.

$$\sin\frac{\pi}{6} = y = \frac{1}{2} \qquad\qquad \csc\frac{\pi}{6} = \frac{1}{y} = \frac{1}{1/2} = 2$$

$$\cos\frac{\pi}{6} = x = \frac{\sqrt{3}}{2} \qquad\qquad \sec\frac{\pi}{6} = \frac{1}{x} = \frac{2}{\sqrt{3}} = \frac{2\sqrt{3}}{3}$$

$$\tan\frac{\pi}{6} = \frac{y}{x} = \frac{1/2}{\sqrt{3}/2} = \frac{1}{\sqrt{3}} = \frac{\sqrt{3}}{3} \qquad\qquad \cot\frac{\pi}{6} = \frac{x}{y} = \frac{\sqrt{3}/2}{1/2} = \sqrt{3}$$

b. $t = 5\pi/4$ corresponds to the point $(x, y) = \left(-\sqrt{2}/2, \, -\sqrt{2}/2\right)$.

$$\sin\frac{5\pi}{4} = y = -\frac{\sqrt{2}}{2} \qquad\qquad \csc\frac{5\pi}{4} = \frac{1}{y} = -\frac{2}{\sqrt{2}} = -\sqrt{2}$$

$$\cos\frac{5\pi}{4} = x = -\frac{\sqrt{2}}{2} \qquad\qquad \sec\frac{5\pi}{4} = \frac{1}{x} = -\frac{2}{\sqrt{2}} = -\sqrt{2}$$

$$\tan\frac{5\pi}{4} = \frac{y}{x} = \frac{-\sqrt{2}/2}{-\sqrt{2}/2} = 1 \qquad\qquad \cot\frac{5\pi}{4} = \frac{x}{y} = \frac{-\sqrt{2}/2}{-\sqrt{2}/2} = 1$$

c. $t = \pi$ corresponds to the point $(x, y) = (-1, 0)$.

$$\sin\pi = y = 0 \qquad\qquad \csc\pi = \frac{1}{y} \text{ is undefined.}$$

$$\cos\pi = x = -1 \qquad\qquad \sec\pi = \frac{1}{x} = \frac{1}{-1} = -1$$

$$\tan\pi = \frac{y}{x} = \frac{0}{-1} = 0 \qquad\qquad \cot\pi = \frac{x}{y} \text{ is undefined.}$$

d. Moving *clockwise* around the unit circle, it follows that $t = -\pi/3$ corresponds to the point $(x, y) = \left(1/2, \, -\sqrt{3}/2\right)$.

$$\sin\left(-\frac{\pi}{3}\right) = y = -\frac{\sqrt{3}}{2} \qquad\qquad \csc\left(-\frac{\pi}{3}\right) = \frac{1}{y} = -\frac{2}{\sqrt{3}} = -\frac{2\sqrt{3}}{3}$$

$$\cos\left(-\frac{\pi}{3}\right) = x = \frac{1}{2} \qquad\qquad \sec\left(-\frac{\pi}{3}\right) = \frac{1}{x} = \frac{1}{1/2} = 2$$

$$\tan\left(-\frac{\pi}{3}\right) = \frac{y}{x} = \frac{-\sqrt{3}/2}{1/2} = -\sqrt{3}$$

$$\cot\left(-\frac{\pi}{3}\right) = \frac{x}{y} = \frac{1/2}{-\sqrt{3}/2} = -\frac{1}{\sqrt{3}} = -\frac{\sqrt{3}}{3}$$

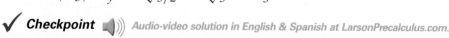 ✓ **Checkpoint**   ))) *Audio-video solution in English & Spanish at LarsonPrecalculus.com.*

Evaluate the six trigonometric functions at each real number.

a. $t = \pi/2$ **b.** $t = 0$ **c.** $t = -5\pi/6$ **d.** $t = -3\pi/4$

Domain and Period of Sine and Cosine

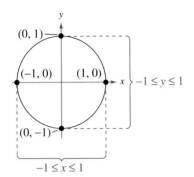

Figure 1.19

The *domain* of the sine and cosine functions is the set of all real numbers. To determine the *range* of these two functions, consider the unit circle shown in Figure 1.19. By definition, $\sin t = y$ and $\cos t = x$. Because (x, y) is on the unit circle, you know that $-1 \le y \le 1$ and $-1 \le x \le 1$. So, the values of sine and cosine also range between -1 and 1.

$$-1 \le y \le 1 \qquad -1 \le x \le 1$$
$$\text{and}$$
$$-1 \le \sin t \le 1 \qquad -1 \le \cos t \le 1$$

Adding 2π to each value of t in the interval $[0, 2\pi]$ results in a revolution around the unit circle, as shown below.

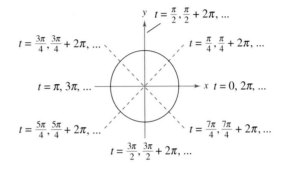

The values of $\sin(t + 2\pi)$ and $\cos(t + 2\pi)$ correspond to those of $\sin t$ and $\cos t$. Similar results can be obtained for repeated revolutions (positive or negative) on the unit circle. This leads to the general result

$$\sin(t + 2\pi n) = \sin t \quad \text{and} \quad \cos(t + 2\pi n) = \cos t$$

for any integer n and real number t. Functions that behave in such a repetitive (or cyclic) manner are called **periodic.**

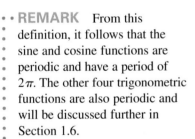

• • REMARK From this definition, it follows that the sine and cosine functions are periodic and have a period of 2π. The other four trigonometric functions are also periodic and will be discussed further in Section 1.6.

Definition of Periodic Function

A function f is **periodic** when there exists a positive real number c such that

$$f(t + c) = f(t)$$

for all t in the domain of f. The smallest number c for which f is periodic is called the **period** of f.

Recall from Section P.6 that a function f is *even* when $f(-t) = f(t)$ and is *odd* when $f(-t) = -f(t)$.

Even and Odd Trigonometric Functions

The cosine and secant functions are *even*.

$$\cos(-t) = \cos t \qquad \sec(-t) = \sec t$$

The sine, cosecant, tangent, and cotangent functions are *odd*.

$$\sin(-t) = -\sin t \qquad \csc(-t) = -\csc t$$
$$\tan(-t) = -\tan t \qquad \cot(-t) = -\cot t$$

EXAMPLE 2   **Evaluating Sine and Cosine**

a. Because $\dfrac{13\pi}{6} = 2\pi + \dfrac{\pi}{6}$, you have $\sin\dfrac{13\pi}{6} = \sin\left(2\pi + \dfrac{\pi}{6}\right) = \sin\dfrac{\pi}{6} = \dfrac{1}{2}$.

b. Because $-\dfrac{7\pi}{2} = -4\pi + \dfrac{\pi}{2}$, you have

$$\cos\left(-\dfrac{7\pi}{2}\right) = \cos\left(-4\pi + \dfrac{\pi}{2}\right) = \cos\dfrac{\pi}{2} = 0.$$

c. For $\sin t = \frac{4}{5}$, $\sin(-t) = -\frac{4}{5}$ because the sine function is odd.

✓ **Checkpoint** ◀))) *Audio-video solution in English & Spanish at LarsonPrecalculus.com.*

a. Use the period of the cosine function to evaluate $\cos(9\pi/2)$.
b. Use the period of the sine function to evaluate $\sin(-7\pi/3)$.
c. Evaluate $\cos t$ given that $\cos(-t) = 0.3$.

▷ **TECHNOLOGY** When evaluating trigonometric functions with a calculator, remember to enclose all fractional angle measures in parentheses. For instance, to evaluate $\sin t$ for $t = \pi/6$, you should enter

(SIN) (() (π) (÷) 6 ()) (ENTER).

These keystrokes yield the correct value of 0.5. Note that some calculators automatically place a left parenthesis after trigonometric functions.

When evaluating a trigonometric function with a calculator, you need to set the calculator to the desired *mode* of measurement (*degree* or *radian*). Most calculators do not have keys for the cosecant, secant, and cotangent functions. To evaluate these functions, you can use the (x⁻¹) key with their respective reciprocal functions: sine, cosine, and tangent. For instance, to evaluate $\csc(\pi/8)$, use the fact that

$$\csc\dfrac{\pi}{8} = \dfrac{1}{\sin(\pi/8)}$$

and enter the following keystroke sequence in *radian* mode.

(() (SIN) (() (π) (÷) 8 ()) ()) (x⁻¹) (ENTER)   Display 2.6131259

EXAMPLE 3   **Using a Calculator**

Function	Mode	Calculator Keystrokes	Display
a. $\sin\dfrac{2\pi}{3}$	Radian	(SIN) (() 2 (π) (÷) 3 ()) (ENTER)	0.8660254
b. $\cot 1.5$	Radian	(() (TAN) (() 1.5 ()) ()) (x⁻¹) (ENTER)	0.0709148

✓ **Checkpoint** ◀))) *Audio-video solution in English & Spanish at LarsonPrecalculus.com.*

Use a calculator to evaluate (a) $\sin(5\pi/7)$ and (b) $\csc 2.0$.

Summarize (Section 1.2)
1. Explain how to identify a unit circle and describe its relationship to real numbers (*page 132*).
2. State the unit circle definitions of the trigonometric functions (*page 133*). For an example of evaluating trigonometric functions using the unit circle, see Example 1.
3. Explain how to use domain and period to evaluate sine and cosine functions (*page 135*) and describe how to use a calculator to evaluate trigonometric functions (*page 136*). For an example of using period and an odd trigonometric function to evaluate sine and cosine functions, see Example 2. For an example of using a calculator to evaluate trigonometric functions, see Example 3.

1.2 Exercises

See **CalcChat.com** for tutorial help and worked-out solutions to odd-numbered exercises.

Vocabulary: Fill in the blanks.

1. Each real number t corresponds to a point (x, y) on the _____ _____.
2. A function f is _____ when there exists a positive real number c such that $f(t + c) = f(t)$ for all t in the domain of f.
3. The smallest number c for which a function f is periodic is called the _____ of f.
4. A function f is _____ when $f(-t) = -f(t)$ and _____ when $f(-t) = f(t)$.

Skills and Applications

Determining Values of Trigonometric Functions
In Exercises 5–8, determine the exact values of the six trigonometric functions of the real number t.

5.
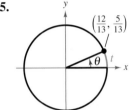

6. $\left(-\frac{8}{17}, \frac{15}{17}\right)$
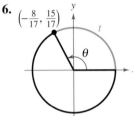

7.
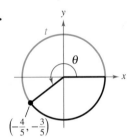
$\left(-\frac{4}{5}, -\frac{3}{5}\right)$

8.
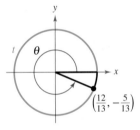
$\left(\frac{12}{13}, -\frac{5}{13}\right)$

Finding a Point on the Unit Circle
In Exercises 9–12, find the point (x, y) on the unit circle that corresponds to the real number t.

9. $t = \pi/2$
10. $t = \pi/4$
11. $t = 5\pi/6$
12. $t = 4\pi/3$

Evaluating Sine, Cosine, and Tangent
In Exercises 13–22, evaluate (if possible) the sine, cosine, and tangent at the real number.

13. $t = \dfrac{\pi}{4}$
14. $t = \dfrac{\pi}{3}$

15. $t = -\dfrac{\pi}{6}$
16. $t = -\dfrac{\pi}{4}$

17. $t = -\dfrac{7\pi}{4}$
18. $t = -\dfrac{4\pi}{3}$

19. $t = \dfrac{11\pi}{6}$
20. $t = \dfrac{5\pi}{3}$

21. $t = -\dfrac{3\pi}{2}$
22. $t = -2\pi$

Evaluating Trigonometric Functions
In Exercises 23–30, evaluate (if possible) the six trigonometric functions at the real number.

23. $t = 2\pi/3$
24. $t = 5\pi/6$
25. $t = 4\pi/3$
26. $t = 7\pi/4$
27. $t = -5\pi/3$
28. $t = 3\pi/2$
29. $t = -\pi/2$
30. $t = -\pi$

Using Period to Evaluate Sine and Cosine
In Exercises 31–36, evaluate the trigonometric function using its period as an aid.

31. $\sin 4\pi$
32. $\cos 3\pi$

33. $\cos \dfrac{7\pi}{3}$
34. $\sin \dfrac{9\pi}{4}$

35. $\sin \dfrac{19\pi}{6}$
36. $\sin\left(-\dfrac{8\pi}{3}\right)$

Using the Value of a Trigonometric Function
In Exercises 37–42, use the value of the trigonometric function to evaluate the indicated functions.

37. $\sin t = \frac{1}{2}$
   (a) $\sin(-t)$
   (b) $\csc(-t)$

38. $\sin(-t) = \frac{3}{8}$
   (a) $\sin t$
   (b) $\csc t$

39. $\cos(-t) = -\frac{1}{5}$
   (a) $\cos t$
   (b) $\sec(-t)$

40. $\cos t = -\frac{3}{4}$
   (a) $\cos(-t)$
   (b) $\sec(-t)$

41. $\sin t = \frac{4}{5}$
   (a) $\sin(\pi - t)$
   (b) $\sin(t + \pi)$

42. $\cos t = \frac{4}{5}$
   (a) $\cos(\pi - t)$
   (b) $\cos(t + \pi)$

Using a Calculator
In Exercises 43–48, use a calculator to evaluate the trigonometric function. Round your answer to four decimal places. (Be sure the calculator is in the correct mode.)

43. $\tan \pi/3$
44. $\csc 2\pi/3$
45. $\csc 0.8$
46. $\cos(-1.7)$
47. $\sec 1.8$
48. $\cot(-0.9)$

49. Harmonic Motion The displacement from equilibrium of an oscillating weight suspended by a spring is given by

$$y(t) = \frac{1}{4}\cos 6t$$

where y is the displacement (in feet) and t is the time (in seconds). Find the displacement when (a) $t = 0$, (b) $t = \frac{1}{4}$, and (c) $t = \frac{1}{2}$.

50. Harmonic Motion

The displacement from equilibrium of an oscillating weight suspended by a spring is given by $y(t) = 3\sin(\pi t/4)$, where y is the displacement (in feet) and t is the time (in seconds).

(a) Complete the table.

t	0	$\frac{1}{2}$	1	$\frac{3}{2}$	2
y					

(b) Use the *table* feature of a graphing utility to determine when the displacement is maximum.

(c) Use the *table* feature of the graphing utility to approximate the time t $(0 < t < 8)$ when the weight reaches equilibrium.

Exploration

True or False? In Exercises 51–54, determine whether the statement is true or false. Justify your answer.

51. Because $\sin(-t) = -\sin t$, the sine of a negative angle is a negative number.

52. The real number 0 corresponds to the point $(0, 1)$ on the unit circle.

53. $\tan a = \tan(a - 6\pi)$

54. $\cos\left(-\frac{7\pi}{2}\right) = \cos\left(\pi + \frac{\pi}{2}\right)$

55. Conjecture Let (x_1, y_1) and (x_2, y_2) be points on the unit circle corresponding to $t = t_1$ and $t = \pi - t_1$, respectively.

(a) Identify the symmetry of the points (x_1, y_1) and (x_2, y_2).

(b) Make a conjecture about any relationship between $\sin t_1$ and $\sin(\pi - t_1)$.

(c) Make a conjecture about any relationship between $\cos t_1$ and $\cos(\pi - t_1)$.

56. Using the Unit Circle Use the unit circle to verify that the cosine and secant functions are even and that the sine, cosecant, tangent, and cotangent functions are odd.

57. Verifying Expressions Are Not Equal Verify that $\cos 2t \neq 2\cos t$ by approximating $\cos 1.5$ and $2\cos 0.75$.

58. Verifying Expressions Are Not Equal Verify that $\sin(t_1 + t_2) \neq \sin t_1 + \sin t_2$ by approximating $\sin 0.25$, $\sin 0.75$, and $\sin 1$.

59. Graphical Analysis With a graphing utility in *radian* and *parametric* modes, enter the equations

$$X_{1T} = \cos T \quad \text{and} \quad Y_{1T} = \sin T$$

and use the following settings.

Tmin = 0, Tmax = 6.3, Tstep = 0.1
Xmin = −1.5, Xmax = 1.5, Xscl = 1
Ymin = −1, Ymax = 1, Yscl = 1

(a) Graph the entered equations and describe the graph.

(b) Use the *trace* feature to move the cursor around the graph. What do the t-values represent? What do the x- and y-values represent?

(c) What are the least and greatest values of x and y?

60. **HOW DO YOU SEE IT?** Use the figure below.

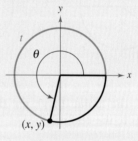

(a) Do all of the trigonometric functions of t exist? Explain your reasoning.

(b) For those trigonometric functions that exist, determine whether the sign of the trigonometric function is positive or negative. Explain your reasoning.

61. Think About It Because $f(t) = \sin t$ is an odd function and $g(t) = \cos t$ is an even function, what can be said about the function $h(t) = f(t)g(t)$?

62. Think About It Because $f(t) = \sin t$ and $g(t) = \tan t$ are odd functions, what can be said about the function $h(t) = f(t)g(t)$?

1.3   Right Triangle Trigonometry

Trigonometric functions can help you analyze real-life situations. For instance, in Exercise 76 on page 149, you will use trigonometric functions to find the height of a helium-filled balloon.

■ Evaluate trigonometric functions of acute angles, and use a calculator to evaluate trigonometric functions.
■ Use the fundamental trigonometric identities.
■ Use trigonometric functions to model and solve real-life problems.

The Six Trigonometric Functions

This section introduces the trigonometric functions from a *right triangle* perspective. Consider a right triangle with one acute angle labeled θ, as shown below. Relative to the angle θ, the three sides of the triangle are the **hypotenuse,** the **opposite side** (the side opposite the angle θ), and the **adjacent side** (the side adjacent to the angle θ).

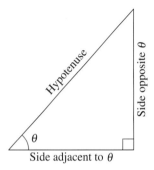

Side adjacent to θ

Using the lengths of these three sides, you can form six ratios that define the six trigonometric functions of the acute angle θ.

sine   cosecant   cosine   secant   tangent   cotangent

In the following definitions, it is important to see that

$$0° < \theta < 90°$$

(θ lies in the first quadrant) and that for such angles the value of each trigonometric function is *positive*.

Right Triangle Definitions of Trigonometric Functions

Let θ be an *acute* angle of a right triangle. The six trigonometric functions of the angle θ are defined as follows. (Note that the functions in the second row are the *reciprocals* of the corresponding functions in the first row.)

$$\sin \theta = \frac{\text{opp}}{\text{hyp}} \qquad \cos \theta = \frac{\text{adj}}{\text{hyp}} \qquad \tan \theta = \frac{\text{opp}}{\text{adj}}$$

$$\csc \theta = \frac{\text{hyp}}{\text{opp}} \qquad \sec \theta = \frac{\text{hyp}}{\text{adj}} \qquad \cot \theta = \frac{\text{adj}}{\text{opp}}$$

The abbreviations

opp,   adj,   and   hyp

represent the lengths of the three sides of a right triangle.

opp = the length of the side *opposite* θ

adj = the length of the side *adjacent to* θ

hyp = the length of the *hypotenuse*

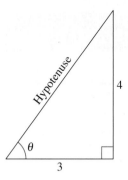

Figure 1.20

EXAMPLE 1   **Evaluating Trigonometric Functions**

Use the triangle in Figure 1.20 to find the values of the six trigonometric functions of θ.

Solution   By the Pythagorean Theorem,

$$(\text{hyp})^2 = (\text{opp})^2 + (\text{adj})^2$$

it follows that

$$\text{hyp} = \sqrt{4^2 + 3^2}$$
$$= \sqrt{25}$$
$$= 5.$$

So, the six trigonometric functions of θ are

$$\sin \theta = \frac{\text{opp}}{\text{hyp}} = \frac{4}{5} \qquad \csc \theta = \frac{\text{hyp}}{\text{opp}} = \frac{5}{4}$$

$$\cos \theta = \frac{\text{adj}}{\text{hyp}} = \frac{3}{5} \qquad \sec \theta = \frac{\text{hyp}}{\text{adj}} = \frac{5}{3}$$

$$\tan \theta = \frac{\text{opp}}{\text{adj}} = \frac{4}{3} \qquad \cot \theta = \frac{\text{adj}}{\text{opp}} = \frac{3}{4}.$$

✓ *Checkpoint* 🔊))) *Audio-video solution in English & Spanish at LarsonPrecalculus.com.*

Use the triangle at the right to find the values of the six trigonometric functions of θ.

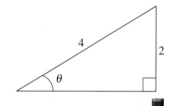

In Example 1, you were given the lengths of two sides of the right triangle, but not the angle θ. Often, you will be asked to find the trigonometric functions of a *given* acute angle θ. To do this, construct a right triangle having θ as one of its angles.

EXAMPLE 2   **Evaluating Trigonometric Functions of 45°**

Find the values of sin 45°, cos 45°, and tan 45°.

Solution   Construct a right triangle having 45° as one of its acute angles, as shown in Figure 1.21. Choose 1 as the length of the adjacent side. From geometry, you know that the other acute angle is also 45°. So, the triangle is isosceles and the length of the opposite side is also 1. Using the Pythagorean Theorem, you find the length of the hypotenuse to be $\sqrt{2}$.

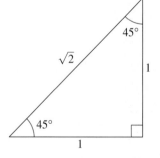

Figure 1.21

$$\sin 45° = \frac{\text{opp}}{\text{hyp}} = \frac{1}{\sqrt{2}} = \frac{\sqrt{2}}{2}$$

$$\cos 45° = \frac{\text{adj}}{\text{hyp}} = \frac{1}{\sqrt{2}} = \frac{\sqrt{2}}{2}$$

$$\tan 45° = \frac{\text{opp}}{\text{adj}} = \frac{1}{1} = 1$$

✓ *Checkpoint* 🔊))) *Audio-video solution in English & Spanish at LarsonPrecalculus.com.*

Find the value of sec 45°.

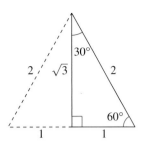

Figure 1.22

▷

▷

EXAMPLE 3 **Trigonometric Functions of 30° and 60°**

Use the equilateral triangle shown in Figure 1.22 to find the values of sin 60°, cos 60°, sin 30°, and cos 30°.

Solution For $\theta = 60°$, you have adj = 1, opp = $\sqrt{3}$, and hyp = 2. So,

$$\sin 60° = \frac{\text{opp}}{\text{hyp}} = \frac{\sqrt{3}}{2} \quad \text{and} \quad \cos 60° = \frac{\text{adj}}{\text{hyp}} = \frac{1}{2}.$$

For $\theta = 30°$, adj = $\sqrt{3}$, opp = 1, and hyp = 2. So,

$$\sin 30° = \frac{\text{opp}}{\text{hyp}} = \frac{1}{2} \quad \text{and} \quad \cos 30° = \frac{\text{adj}}{\text{hyp}} = \frac{\sqrt{3}}{2}.$$

✓ **Checkpoint**))) Audio-video solution in English & Spanish at LarsonPrecalculus.com.

Use the equilateral triangle shown in Figure 1.22 to find the values of tan 60° and tan 30°. ∎

Sines, Cosines, and Tangents of Special Angles

$$\sin 30° = \sin \frac{\pi}{6} = \frac{1}{2} \qquad \cos 30° = \cos \frac{\pi}{6} = \frac{\sqrt{3}}{2} \qquad \tan 30° = \tan \frac{\pi}{6} = \frac{\sqrt{3}}{3}$$

$$\sin 45° = \sin \frac{\pi}{4} = \frac{\sqrt{2}}{2} \qquad \cos 45° = \cos \frac{\pi}{4} = \frac{\sqrt{2}}{2} \qquad \tan 45° = \tan \frac{\pi}{4} = 1$$

$$\sin 60° = \sin \frac{\pi}{3} = \frac{\sqrt{3}}{2} \qquad \cos 60° = \cos \frac{\pi}{3} = \frac{1}{2} \qquad \tan 60° = \tan \frac{\pi}{3} = \sqrt{3}$$

In the box, note that $\sin 30° = \frac{1}{2} = \cos 60°$. This occurs because 30° and 60° are complementary angles. In general, it can be shown from the right triangle definitions that *cofunctions of complementary angles are equal*. That is, if θ is an acute angle, then the following relationships are true.

$$\sin(90° - \theta) = \cos \theta \qquad \cos(90° - \theta) = \sin \theta \qquad \tan(90° - \theta) = \cot \theta$$

$$\cot(90° - \theta) = \tan \theta \qquad \sec(90° - \theta) = \csc \theta \qquad \csc(90° - \theta) = \sec \theta$$

To use a calculator to evaluate trigonometric functions of angles measured in degrees, first set the calculator to *degree* mode and then proceed as demonstrated in Section 1.2.

EXAMPLE 4 **Using a Calculator**

Use a calculator to evaluate sec 5° 40′ 12″.

Solution Begin by converting to decimal degree form. [Recall that $1' = \frac{1}{60}(1°)$ and $1'' = \frac{1}{3600}(1°)$.]

$$5° \, 40' \, 12'' = 5° + \left(\frac{40}{60}\right)° + \left(\frac{12}{3600}\right)° = 5.67°$$

Then, use a calculator to evaluate sec 5.67°.

Function	Calculator Keystrokes	Display
sec 5° 40′ 12″ = sec 5.67°	(COS (5.67)) x⁻¹ ENTER	1.0049166

✓ **Checkpoint**))) Audio-video solution in English & Spanish at LarsonPrecalculus.com.

Use a calculator to evaluate csc 34° 30′ 36″. ∎

Trigonometric Identities

In trigonometry, a great deal of time is spent studying relationships between trigonometric functions (identities).

Fundamental Trigonometric Identities

Reciprocal Identities

$$\sin \theta = \frac{1}{\csc \theta} \qquad \cos \theta = \frac{1}{\sec \theta} \qquad \tan \theta = \frac{1}{\cot \theta}$$

$$\csc \theta = \frac{1}{\sin \theta} \qquad \sec \theta = \frac{1}{\cos \theta} \qquad \cot \theta = \frac{1}{\tan \theta}$$

Quotient Identities

$$\tan \theta = \frac{\sin \theta}{\cos \theta} \qquad \cot \theta = \frac{\cos \theta}{\sin \theta}$$

Pythagorean Identities

$$\sin^2 \theta + \cos^2 \theta = 1$$

$$1 + \tan^2 \theta = \sec^2 \theta$$

$$1 + \cot^2 \theta = \csc^2 \theta$$

Note that $\sin^2 \theta$ represents $(\sin \theta)^2$, $\cos^2 \theta$ represents $(\cos \theta)^2$, and so on.

EXAMPLE 5   Applying Trigonometric Identities

Let θ be an acute angle such that $\sin \theta = 0.6$. Find the values of (a) $\cos \theta$ and (b) $\tan \theta$ using trigonometric identities.

Solution

a. To find the value of $\cos \theta$, use the Pythagorean identity

$$\sin^2 \theta + \cos^2 \theta = 1.$$

So, you have

$(0.6)^2 + \cos^2 \theta = 1$	Substitute 0.6 for $\sin \theta$.
$\cos^2 \theta = 1 - (0.6)^2$	Subtract $(0.6)^2$ from each side.
$\cos^2 \theta = 0.64$	Simplify.
$\cos \theta = \sqrt{0.64}$	Extract positive square root.
$\cos \theta = 0.8.$	Simplify.

b. Now, knowing the sine and cosine of θ, you can find the tangent of θ to be

$$\tan \theta = \frac{\sin \theta}{\cos \theta} = \frac{0.6}{0.8} = 0.75.$$

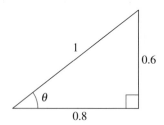

Figure 1.23

Use the definitions of $\cos \theta$ and $\tan \theta$ and the triangle shown in Figure 1.23 to check these results.

✓ **Checkpoint**   *Audio-video solution in English & Spanish at LarsonPrecalculus.com.*

Let θ be an acute angle such that $\cos \theta = 0.25$. Find the values of (a) $\sin \theta$ and (b) $\tan \theta$ using trigonometric identities.

EXAMPLE 6   Applying Trigonometric Identities

Let θ be an acute angle such that $\tan \theta = \frac{1}{3}$. Find the values of (a) $\cot \theta$ and (b) $\sec \theta$ using trigonometric identities.

Solution

a. $\cot \theta = \dfrac{1}{\tan \theta}$ Reciprocal identity

$= \dfrac{1}{1/3}$

$= 3$

b. $\sec^2 \theta = 1 + \tan^2 \theta$ Pythagorean identity

$= 1 + (1/3)^2$

$= 10/9$

$\sec \theta = \sqrt{10}/3$

Use the definitions of $\cot \theta$ and $\sec \theta$ and the triangle shown below to check these results.

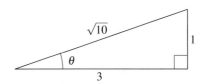

✓ **Checkpoint** *Audio-video solution in English & Spanish at LarsonPrecalculus.com.*

Let β be an acute angle such that $\tan \beta = 2$. Find the values of (a) $\cot \beta$ and (b) $\sec \beta$ using trigonometric identities.

EXAMPLE 7   Using Trigonometric Identities

Use trigonometric identities to transform the left side of the equation into the right side $(0 < \theta < \pi/2)$.

a. $\sin \theta \csc \theta = 1$ **b.** $(\csc \theta + \cot \theta)(\csc \theta - \cot \theta) = 1$

Solution

a. $\sin \theta \csc \theta = \left(\dfrac{1}{\csc \theta}\right)\csc \theta = 1$ Use a reciprocal identity and simplify.

b. $(\csc \theta + \cot \theta)(\csc \theta - \cot \theta)$

$= \csc^2 \theta - \csc \theta \cot \theta + \csc \theta \cot \theta - \cot^2 \theta$ FOIL Method

$= \csc^2 \theta - \cot^2 \theta$ Simplify.

$= 1$ Pythagorean identity

✓ **Checkpoint** *Audio-video solution in English & Spanish at LarsonPrecalculus.com.*

Use trigonometric identities to transform the left side of the equation into the right side $(0 < \theta < \pi/2)$.

a. $\tan \theta \csc \theta = \sec \theta$

b. $(\csc \theta + 1)(\csc \theta - 1) = \cot^2 \theta$

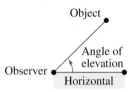

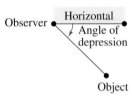

Object

Angle of
elevation

Observer

Horizontal

Observer

Horizontal

Angle of
depression

Object

Figure 1.24

Applications Involving Right Triangles

Many applications of trigonometry involve a process called **solving right triangles.** In this type of application, you are usually given one side of a right triangle and one of the acute angles and are asked to find one of the other sides, *or* you are given two sides and are asked to find one of the acute angles.

In Example 8, the angle you are given is the **angle of elevation,** which represents the angle from the horizontal upward to an object. In other applications you may be given the **angle of depression,** which represents the angle from the horizontal downward to an object. (See Figure 1.24.)

EXAMPLE 8   Using Trigonometry to Solve a Right Triangle

A surveyor is standing 115 feet from the base of the Washington Monument, as shown in the figure at the right. The surveyor measures the angle of elevation to the top of the monument as 78.3°. How tall is the Washington Monument?

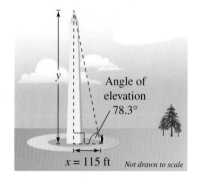

Angle of
elevation
78.3°

$x = 115$ ft   *Not drawn to scale*

Solution   From the figure, you can see that

$$\tan 78.3° = \frac{\text{opp}}{\text{adj}} = \frac{y}{x}$$

where $x = 115$ and y is the height of the monument. So, the height of the Washington Monument is

$$y = x \tan 78.3°$$

$$\approx 115(4.82882)$$

$$\approx 555 \text{ feet.}$$

✓ *Checkpoint*  *Audio-video solution in English & Spanish at LarsonPrecalculus.com.*

How tall is the flagpole in Figure 1.25?

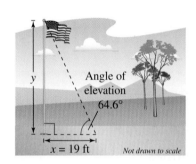

y

Angle of
elevation
64.6°

$x = 19$ ft   *Not drawn to scale*

Figure 1.25

EXAMPLE 9   Using Trigonometry to Solve a Right Triangle

A historic lighthouse is 200 yards from a bike path along the edge of a lake. A walkway to the lighthouse is 400 yards long. Find the acute angle θ between the bike path and the walkway, as illustrated in the figure at the right.

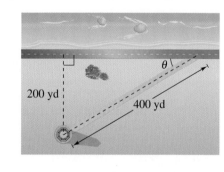

θ

200 yd

400 yd

Solution   From the figure, you can see that the sine of the angle θ is

$$\sin \theta = \frac{\text{opp}}{\text{hyp}} = \frac{200}{400} = \frac{1}{2}.$$

Now you should recognize that $\theta = 30°$.

✓ *Checkpoint*  *Audio-video solution in English & Spanish at LarsonPrecalculus.com.*

Find the acute angle θ between the two paths, as illustrated in Figure 1.26. ■

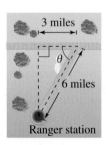

3 miles

θ

6 miles

Ranger station

Figure 1.26

In Example 9, you were able to recognize that $\theta = 30°$ is the acute angle that satisfies the equation $\sin \theta = \frac{1}{2}$. Suppose, however, that you were given the equation $\sin \theta = 0.6$ and were asked to find the acute angle θ. Because $\sin 30° = \frac{1}{2} = 0.5000$ and $\sin 45° = 1/\sqrt{2} \approx 0.7071$, you might guess that θ lies somewhere between $30°$ and $45°$. In a later section, you will study a method by which a more precise value of θ can be determined.

EXAMPLE 10   Solving a Right Triangle

Find the length c of the skateboard ramp shown in the figure below. Find the horizontal length a of the ramp.

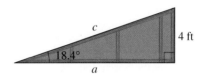

Skateboarders can go to a skatepark, which is a recreational environment built with many different types of ramps and rails.

Solution   From the figure, you can see that

$$\sin 18.4° = \frac{\text{opp}}{\text{hyp}} = \frac{4}{c}.$$

So, the length of the skateboard ramp is

$$c = \frac{4}{\sin 18.4°} \approx \frac{4}{0.3156} \approx 12.7 \text{ feet.}$$

Also from the figure, you can see that

$$\tan 18.4° = \frac{\text{opp}}{\text{adj}} = \frac{4}{a}.$$

So, the horizontal length is

$$a = \frac{4}{\tan 18.4°} \approx 12.0 \text{ feet.}$$

✓ *Checkpoint* ◀))) *Audio-video solution in English & Spanish at LarsonPrecalculus.com.*

Find the length c of the loading ramp shown in the figure below. Find the horizontal length a of the ramp.

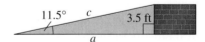

Summarize   (Section 1.3)

1. State the right triangle definitions of the trigonometric functions *(page 139)* and describe how to use a calculator to evaluate trigonometric functions *(page 141)*. For examples of evaluating trigonometric functions of acute angles, see Examples 1–3. For an example of using a calculator to evaluate a trigonometric function, see Example 4.

2. List the fundamental trigonometric identities *(page 142)*. For examples of using the fundamental trigonometric identities, see Examples 5–7.

3. Describe examples of how to use trigonometric functions to model and solve real-life problems *(pages 144 and 145, Examples 8–10)*.

1.3 Exercises

See **CalcChat.com** for tutorial help and worked-out solutions to odd-numbered exercises.

Vocabulary

1. Match each trigonometric function with its right triangle definition.

    (a) sine   (b) cosine   (c) tangent   (d) cosecant   (e) secant   (f) cotangent

    (i) $\dfrac{\text{hypotenuse}}{\text{adjacent}}$ (ii) $\dfrac{\text{adjacent}}{\text{opposite}}$ (iii) $\dfrac{\text{hypotenuse}}{\text{opposite}}$ (iv) $\dfrac{\text{adjacent}}{\text{hypotenuse}}$ (v) $\dfrac{\text{opposite}}{\text{hypotenuse}}$ (vi) $\dfrac{\text{opposite}}{\text{adjacent}}$

In Exercises 2–4, fill in the blanks.

2. Relative to the acute angle θ, the three sides of a right triangle are the _____ side, the _____ side, and the _____.

3. Cofunctions of _____ angles are equal.

4. An angle that measures from the horizontal upward to an object is called the angle of _____, whereas an angle that measures from the horizontal downward to an object is called the angle of _____.

Skills and Applications

Evaluating Trigonometric Functions **In Exercises 5–8, find the exact values of the six trigonometric functions of the angle θ shown in the figure. (Use the Pythagorean Theorem to find the third side of the triangle.)**

5.

6.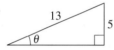

7.

8.

Evaluating Trigonometric Functions **In Exercises 9–12, find the exact values of the six trigonometric functions of the angle θ for each of the two triangles. Explain why the function values are the same.**

9.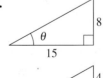

10.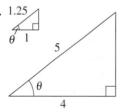

11.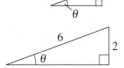

12.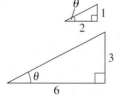

Evaluating Trigonometric Functions **In Exercises 13–20, sketch a right triangle corresponding to the trigonometric function of the acute angle θ. Use the Pythagorean Theorem to determine the third side and then find the other five trigonometric functions of θ.**

13. $\tan \theta = \frac{3}{4}$ **14.** $\cos \theta = \frac{5}{6}$

15. $\sec \theta = \frac{3}{2}$ **16.** $\tan \theta = \frac{4}{5}$

17. $\sin \theta = \frac{1}{5}$ **18.** $\sec \theta = \frac{17}{7}$

19. $\cot \theta = 3$ **20.** $\csc \theta = 9$

Evaluating Trigonometric Functions of 30°, 45°, and 60° **In Exercises 21–30, construct an appropriate triangle to find the missing values. ($0° \le \theta \le 90°, 0 \le \theta \le \pi/2$)**

	Function	θ (deg)	θ (rad)	Function Value
21.	sin	30°		
22.	cos	45°		
23.	sec		$\dfrac{\pi}{4}$	
24.	tan		$\dfrac{\pi}{3}$	
25.	cot			$\dfrac{\sqrt{3}}{3}$
26.	csc			$\sqrt{2}$
27.	csc		$\dfrac{\pi}{6}$	
28.	sin		$\dfrac{\pi}{4}$	
29.	cot			1
30.	tan			$\dfrac{\sqrt{3}}{3}$

Using a Calculator   **In Exercises 31–40, use a calculator to evaluate each function. Round your answers to four decimal places. (Be sure the calculator is in the correct mode.)**

31. (a) $\sin 10°$ (b) $\cos 80°$

32. (a) $\tan 23.5°$ (b) $\cot 66.5°$

33. (a) $\sin 16.35°$ (b) $\csc 16.35°$

34. (a) $\cot 79.56°$ (b) $\sec 79.56°$

35. (a) $\cos 4° 50' 15''$ (b) $\sec 4° 50' 15''$

36. (a) $\sec 42° 12'$ (b) $\csc 48° 7'$

37. (a) $\cot 11° 15'$ (b) $\tan 11° 15'$

38. (a) $\sec 56° 8' 10''$ (b) $\cos 56° 8' 10''$

39. (a) $\csc 32° 40' 3''$ (b) $\tan 44° 28' 16''$

40. (a) $\sec\left(\frac{9}{5} \cdot 20 + 32\right)°$ (b) $\cot\left(\frac{9}{5} \cdot 30 + 32\right)°$

Applying Trigonometric Identities   **In Exercises 41–46, use the given function value(s) and the trigonometric identities to find the indicated trigonometric functions.**

41. $\sin 60° = \dfrac{\sqrt{3}}{2}, \quad \cos 60° = \dfrac{1}{2}$

  (a) $\sin 30°$ (b) $\cos 30°$

  (c) $\tan 60°$ (d) $\cot 60°$

42. $\sin 30° = \dfrac{1}{2}, \quad \tan 30° = \dfrac{\sqrt{3}}{3}$

  (a) $\csc 30°$ (b) $\cot 60°$

  (c) $\cos 30°$ (d) $\cot 30°$

43. $\cos \theta = \frac{1}{3}$

  (a) $\sin \theta$ (b) $\tan \theta$

  (c) $\sec \theta$ (d) $\csc(90° - \theta)$

44. $\sec \theta = 5$

  (a) $\cos \theta$ (b) $\cot \theta$

  (c) $\cot(90° - \theta)$ (d) $\sin \theta$

45. $\cot \alpha = 5$

  (a) $\tan \alpha$ (b) $\csc \alpha$

  (c) $\cot(90° - \alpha)$ (d) $\cos \alpha$

46. $\cos \beta = \dfrac{\sqrt{7}}{4}$

  (a) $\sec \beta$ (b) $\sin \beta$

  (c) $\cot \beta$ (d) $\sin(90° - \beta)$

Using Trigonometric Identities   **In Exercises 47–56, use trigonometric identities to transform the left side of the equation into the right side $(0 < \theta < \pi/2)$.**

47. $\tan \theta \cot \theta = 1$

48. $\cos \theta \sec \theta = 1$

49. $\tan \alpha \cos \alpha = \sin \alpha$

50. $\cot \alpha \sin \alpha = \cos \alpha$

51. $(1 + \sin \theta)(1 - \sin \theta) = \cos^2 \theta$

52. $(1 + \cos \theta)(1 - \cos \theta) = \sin^2 \theta$

53. $(\sec \theta + \tan \theta)(\sec \theta - \tan \theta) = 1$

54. $\sin^2 \theta - \cos^2 \theta = 2 \sin^2 \theta - 1$

55. $\dfrac{\sin \theta}{\cos \theta} + \dfrac{\cos \theta}{\sin \theta} = \csc \theta \sec \theta$

56. $\dfrac{\tan \beta + \cot \beta}{\tan \beta} = \csc^2 \beta$

Evaluating Trigonometric Functions   **In Exercises 57–62, find each value of θ in degrees $(0° < \theta < 90°)$ and radians $(0 < \theta < \pi/2)$ without using a calculator.**

57. (a) $\sin \theta = \frac{1}{2}$ (b) $\csc \theta = 2$

58. (a) $\cos \theta = \dfrac{\sqrt{2}}{2}$ (b) $\tan \theta = 1$

59. (a) $\sec \theta = 2$ (b) $\cot \theta = 1$

60. (a) $\tan \theta = \sqrt{3}$ (b) $\cos \theta = \frac{1}{2}$

61. (a) $\csc \theta = \dfrac{2\sqrt{3}}{3}$ (b) $\sin \theta = \dfrac{\sqrt{2}}{2}$

62. (a) $\cot \theta = \dfrac{\sqrt{3}}{3}$ (b) $\sec \theta = \sqrt{2}$

Finding Side Lengths of a Triangle   **In Exercises 63–66, find the exact values of the indicated variables.**

63. Find x and y.

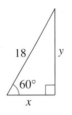

64. Find x and r.

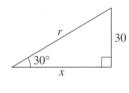

65. Find x and r.

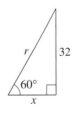

66. Find x and r.

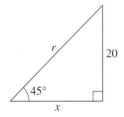

67. Empire State Building   You are standing 45 meters from the base of the Empire State Building. You estimate that the angle of elevation to the top of the 86th floor (the observatory) is 82°. The total height of the building is another 123 meters above the 86th floor. What is the approximate height of the building? One of your friends is on the 86th floor. What is the distance between you and your friend?

68. Height A six-foot-tall person walks from the base of a broadcasting tower directly toward the tip of the shadow cast by the tower. When the person is 132 feet from the tower and 3 feet from the tip of the shadow, the person's shadow starts to appear beyond the tower's shadow.

(a) Draw a right triangle that gives a visual representation of the problem. Show the known quantities of the triangle and use a variable to indicate the height of the tower.

(b) Use a trigonometric function to write an equation involving the unknown quantity.

(c) What is the height of the tower?

69. Angle of Elevation You are skiing down a mountain with a vertical height of 1500 feet. The distance from the top of the mountain to the base is 3000 feet. What is the angle of elevation from the base to the top of the mountain?

70. Width of a River A biologist wants to know the width w of a river in order to properly set instruments for studying the pollutants in the water. From point A, the biologist walks downstream 100 feet and sights to point C (see figure). From this sighting, the biologist determines that $\theta = 54°$. How wide is the river?

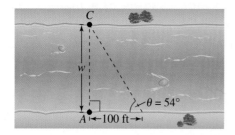

71. Length A guy wire runs from the ground to a cell tower. The wire is attached to the cell tower 150 feet above the ground. The angle formed between the wire and the ground is 43° (see figure).

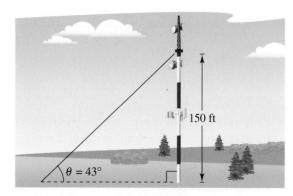

(a) How long is the guy wire?

(b) How far from the base of the tower is the guy wire anchored to the ground?

72. Height of a Mountain In traveling across flat land, you notice a mountain directly in front of you. Its angle of elevation (to the peak) is 3.5°. After you drive 13 miles closer to the mountain, the angle of elevation is 9° (see figure). Approximate the height of the mountain.

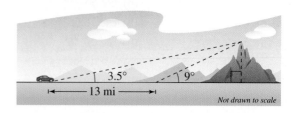

Not drawn to scale

73. Machine Shop Calculations A steel plate has the form of one-fourth of a circle with a radius of 60 centimeters. Two two-centimeter holes are to be drilled in the plate, positioned as shown in the figure. Find the coordinates of the center of each hole.

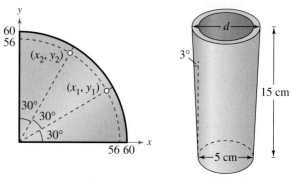

Figure for 73   Figure for 74

74. Machine Shop Calculations A tapered shaft has a diameter of 5 centimeters at the small end and is 15 centimeters long (see figure). The taper is 3°. Find the diameter d of the large end of the shaft.

75. Geometry Use a compass to sketch a quarter of a circle of radius 10 centimeters. Using a protractor, construct an angle of 20° in standard position (see figure). Drop a perpendicular line from the point of intersection of the terminal side of the angle and the arc of the circle. By actual measurement, calculate the coordinates (x, y) of the point of intersection and use these measurements to approximate the six trigonometric functions of a 20° angle.

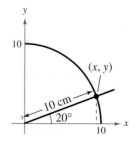

76. Height

A 20-meter line is a tether for a helium-filled balloon. Because of a breeze, the line makes an angle of approximately 85° with the ground.

(a) Draw a right triangle that gives a visual representation of the problem. Show the known quantities of the triangle and use a variable to indicate the height of the balloon.

(b) Use a trigonometric function to write and solve an equation for the height of the balloon.

(c) The breeze becomes stronger and the angle the line makes with the ground decreases. How does this affect the triangle you drew in part (a)?

(d) Complete the table, which shows the heights (in meters) of the balloon for decreasing angle measures θ.

Angle, θ	80°	70°	60°	50°
Height				

Angle, θ	40°	30°	20°	10°
Height				

(e) As θ approaches 0°, how does this affect the height of the balloon? Draw a right triangle to explain your reasoning.

77. Johnstown Inclined Plane The Johnstown Inclined Plane in Pennsylvania is one of the longest and steepest hoists in the world. The railway cars travel a distance of 896.5 feet at an angle of approximately 35.4°, rising to a height of 1693.5 feet above sea level.

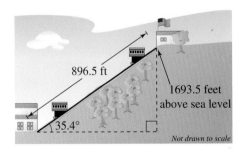

896.5 ft

1693.5 feet above sea level

35.4°

Not drawn to scale

(a) Find the vertical rise of the inclined plane.

(b) Find the elevation of the lower end of the inclined plane.

(c) The cars move up the mountain at a rate of 300 feet per minute. Find the rate at which they rise vertically.

Exploration

78. Writing In right triangle trigonometry, explain why $\sin 30° = \frac{1}{2}$ regardless of the size of the triangle.

True or False? In Exercises 79–84, determine whether the statement is true or false. Justify your answer.

79. $\sin 60° \csc 60° = 1$ **80.** $\sec 30° = \csc 60°$

81. $\sin 45° + \cos 45° = 1$ **82.** $\cot^2 10° - \csc^2 10° = -1$

83. $\dfrac{\sin 60°}{\sin 30°} = \sin 2°$ **84.** $\tan[(5°)^2] = \tan^2 5°$

85. Think About It You are given the value of $\tan \theta$. Is it possible to find the value of $\sec \theta$ without finding the measure of θ? Explain.

86. Think About It

(a) Complete the table.

θ	0.1	0.2	0.3	0.4	0.5
$\sin \theta$					

(b) Is θ or $\sin \theta$ greater for θ in the interval $(0, 0.5]$?

(c) As θ approaches 0, how do θ and $\sin \theta$ compare? Explain.

87. Think About It

(a) Complete the table.

θ	0°	18°	36°	54°	72°	90°
$\sin \theta$						
$\cos \theta$						

(b) Discuss the behavior of the sine function for θ in the range from 0° to 90°.

(c) Discuss the behavior of the cosine function for θ in the range from 0° to 90°.

(d) Use the definitions of the sine and cosine functions to explain the results of parts (b) and (c).

88. HOW DO YOU SEE IT? Use the figure below.

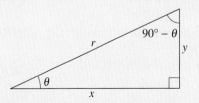

$90° - \theta$

r

y

θ

x

(a) Which side is opposite θ?

(b) Which side is adjacent to $90° - \theta$?

(c) Explain why $\sin \theta = \cos(90° - \theta)$.

1.4 Trigonometric Functions of Any Angle

Trigonometric functions can help you model and solve real-life problems. For instance, in Exercise 99 on page 158, you will use trigonometric functions to model the monthly normal temperatures in New York City and Fairbanks, Alaska.

■ Evaluate trigonometric functions of any angle.
■ Find reference angles.
■ Evaluate trigonometric functions of real numbers.

Introduction

In Section 1.3, the definitions of trigonometric functions were restricted to acute angles. In this section, the definitions are extended to cover *any* angle. When θ is an *acute* angle, the definitions here coincide with those given in the preceding section.

Definitions of Trigonometric Functions of Any Angle

Let θ be an angle in standard position with (x, y) a point on the terminal side of θ and $r = \sqrt{x^2 + y^2} \neq 0$.

$$\sin \theta = \frac{y}{r} \qquad\qquad \cos \theta = \frac{x}{r}$$

$$\tan \theta = \frac{y}{x}, \quad x \neq 0 \qquad \cot \theta = \frac{x}{y}, \quad y \neq 0$$

$$\sec \theta = \frac{r}{x}, \quad x \neq 0 \qquad \csc \theta = \frac{r}{y}, \quad y \neq 0$$

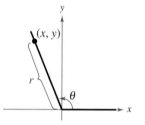

Because $r = \sqrt{x^2 + y^2}$ *cannot* be zero, it follows that the sine and cosine functions are defined for any real value of θ. However, when $x = 0$, the tangent and secant of θ are undefined. For example, the tangent of $90°$ is undefined. Similarly, when $y = 0$, the cotangent and cosecant of θ are undefined.

EXAMPLE 1   **Evaluating Trigonometric Functions**

Let $(-3, 4)$ be a point on the terminal side of θ. Find the sine, cosine, and tangent of θ.

Solution   Referring to Figure 1.27, you can see that $x = -3$, $y = 4$, and

$$\begin{aligned} r &= \sqrt{x^2 + y^2} \\ &= \sqrt{(-3)^2 + 4^2} \\ &= \sqrt{25} \\ &= 5. \end{aligned}$$

So, you have the following.

$$\sin \theta = \frac{y}{r} = \frac{4}{5}$$

$$\cos \theta = \frac{x}{r} = -\frac{3}{5}$$

$$\tan \theta = \frac{y}{x} = -\frac{4}{3}$$

Figure 1.27

✓ **Checkpoint** 🔊)) *Audio-video solution in English & Spanish at LarsonPrecalculus.com.*

Let $(-2, 3)$ be a point on the terminal side of θ. Find the sine, cosine, and tangent of θ.

▷ **ALGEBRA HELP** The formula $r = \sqrt{x^2 + y^2}$ is a result of the Distance Formula. You can review the Distance Formula in Section P.3.

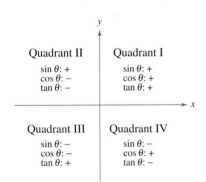

Figure 1.28

The *signs* of the trigonometric functions in the four quadrants can be determined from the definitions of the functions. For instance, because $\cos \theta = x/r$, it follows that $\cos \theta$ is positive wherever $x > 0$, which is in Quadrants I and IV. (Remember, r is always positive.) In a similar manner, you can verify the results shown in Figure 1.28.

EXAMPLE 2 **Evaluating Trigonometric Functions**

Given $\tan \theta = -\frac{5}{4}$ and $\cos \theta > 0$, find $\sin \theta$ and $\sec \theta$.

Solution Note that θ lies in Quadrant IV because that is the only quadrant in which the tangent is negative and the cosine is positive. Moreover, using

$$\tan \theta = \frac{y}{x} = -\frac{5}{4}$$

and the fact that y is negative in Quadrant IV, you can let $y = -5$ and $x = 4$. So, $r = \sqrt{16 + 25} = \sqrt{41}$ and you have the following.

$$\sin \theta = \frac{y}{r}$$

$$= \frac{-5}{\sqrt{41}} \qquad \text{Exact value}$$

$$\approx -0.7809 \qquad \text{Approximate value}$$

$$\sec \theta = \frac{r}{x}$$

$$= \frac{\sqrt{41}}{4} \qquad \text{Exact value}$$

$$\approx 1.6008 \qquad \text{Approximate value}$$

✓ *Checkpoint*  Audio-video solution in English & Spanish at LarsonPrecalculus.com.

Given $\sin \theta = \frac{4}{5}$ and $\tan \theta < 0$, find $\cos \theta$.

EXAMPLE 3 **Trigonometric Functions of Quadrant Angles**

Evaluate the cosine and tangent functions at the four quadrant angles 0, $\frac{\pi}{2}$, π, and $\frac{3\pi}{2}$.

Solution To begin, choose a point on the terminal side of each angle, as shown in Figure 1.29. For each of the four points, $r = 1$ and you have the following.

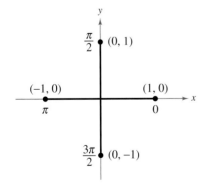

Figure 1.29

$$\cos 0 = \frac{x}{r} = \frac{1}{1} = 1 \qquad\qquad \tan 0 = \frac{y}{x} = \frac{0}{1} = 0 \qquad\qquad (x, y) = (1, 0)$$

$$\cos \frac{\pi}{2} = \frac{x}{r} = \frac{0}{1} = 0 \qquad\qquad \tan \frac{\pi}{2} = \frac{y}{x} = \frac{1}{0} \implies \text{undefined} \qquad (x, y) = (0, 1)$$

$$\cos \pi = \frac{x}{r} = \frac{-1}{1} = -1 \qquad \tan \pi = \frac{y}{x} = \frac{0}{-1} = 0 \qquad\qquad (x, y) = (-1, 0)$$

$$\cos \frac{3\pi}{2} = \frac{x}{r} = \frac{0}{1} = 0 \qquad \tan \frac{3\pi}{2} = \frac{y}{x} = \frac{-1}{0} \implies \text{undefined} \qquad (x, y) = (0, -1)$$

✓ *Checkpoint*  Audio-video solution in English & Spanish at LarsonPrecalculus.com.

Evaluate the sine and cotangent functions at the quadrant angle $\frac{3\pi}{2}$.

Reference Angles

The values of the trigonometric functions of angles greater than $90°$ (or less than $0°$) can be determined from their values at corresponding acute angles called **reference angles.**

> ### Definition of Reference Angle
> Let θ be an angle in standard position. Its **reference angle** is the acute angle θ' formed by the terminal side of θ and the horizontal axis.

The reference angles for θ in Quadrants II, III, and IV are shown below.

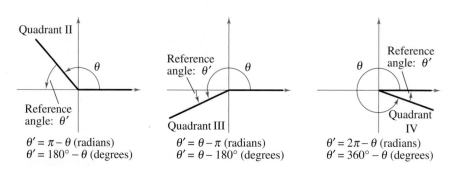

Quadrant II

Reference angle: θ'

$\theta' = \pi - \theta$ (radians)
$\theta' = 180° - \theta$ (degrees)

Reference angle: θ'

Quadrant III

$\theta' = \theta - \pi$ (radians)
$\theta' = \theta - 180°$ (degrees)

Reference angle: θ'

Quadrant IV

$\theta' = 2\pi - \theta$ (radians)
$\theta' = 360° - \theta$ (degrees)

EXAMPLE 4   Finding Reference Angles

Find the reference angle θ'.

a. $\theta = 300°$ **b.** $\theta = 2.3$ **c.** $\theta = -135°$

Solution

a. Because $300°$ lies in Quadrant IV, the angle it makes with the x-axis is

$$\theta' = 360° - 300°$$
$$= 60°. \qquad \text{Degrees}$$

Figure 1.30 shows the angle $\theta = 300°$ and its reference angle $\theta' = 60°$.

b. Because 2.3 lies between $\pi/2 \approx 1.5708$ and $\pi \approx 3.1416$, it follows that it is in Quadrant II and its reference angle is

$$\theta' = \pi - 2.3$$
$$\approx 0.8416. \qquad \text{Radians}$$

Figure 1.31 shows the angle $\theta = 2.3$ and its reference angle $\theta' = \pi - 2.3$.

c. First, determine that $-135°$ is coterminal with $225°$, which lies in Quadrant III. So, the reference angle is

$$\theta' = 225° - 180°$$
$$= 45°. \qquad \text{Degrees}$$

Figure 1.32 shows the angle $\theta = -135°$ and its reference angle $\theta' = 45°$.

✓ Checkpoint  *Audio-video solution in English & Spanish at LarsonPrecalculus.com.*

Find the reference angle θ'.

a. $213°$ **b.** $\dfrac{14\pi}{9}$ **c.** $\dfrac{4\pi}{5}$

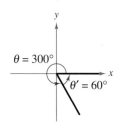

$\theta = 300°$

$\theta' = 60°$

Figure 1.30

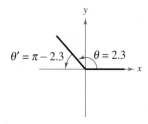

$\theta' = \pi - 2.3$ $\theta = 2.3$

Figure 1.31

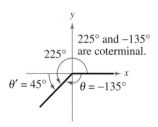

$225°$ and $-135°$ are coterminal.

$225°$

$\theta' = 45°$ $\theta = -135°$

Figure 1.32

Trigonometric Functions of Real Numbers

To see how a reference angle is used to evaluate a trigonometric function, consider the point (x, y) on the terminal side of θ, as shown at the right. By definition, you know that

$$\sin \theta = \frac{y}{r}$$

and

$$\tan \theta = \frac{y}{x}.$$

For the right triangle with acute angle θ' and sides of lengths $|x|$ and $|y|$, you have

$$\sin \theta' = \frac{\text{opp}}{\text{hyp}} = \frac{|y|}{r}$$

and

$$\tan \theta' = \frac{\text{opp}}{\text{adj}} = \frac{|y|}{|x|}.$$

$\text{opp} = |y|, \text{adj} = |x|$

So, it follows that $\sin \theta$ and $\sin \theta'$ are equal, *except possibly in sign*. The same is true for $\tan \theta$ and $\tan \theta'$ *and* for the other four trigonometric functions. In all cases, the quadrant in which θ lies determines the sign of the function value.

Evaluating Trigonometric Functions of Any Angle

To find the value of a trigonometric function of any angle θ:

1. Determine the function value of the associated reference angle θ'.

2. Depending on the quadrant in which θ lies, affix the appropriate sign to the function value.

• • REMARK Learning the table of values at the right is worth the effort because doing so will increase both your efficiency and your confidence. Here is a pattern for the sine function that may help you remember the values.

θ	0°	30°	45°	60°	90°
$\sin \theta$	$\dfrac{\sqrt{0}}{2}$	$\dfrac{\sqrt{1}}{2}$	$\dfrac{\sqrt{2}}{2}$	$\dfrac{\sqrt{3}}{2}$	$\dfrac{\sqrt{4}}{2}$

Reverse the order to get cosine values of the same angles.

By using reference angles and the special angles discussed in the preceding section, you can greatly extend the scope of *exact* trigonometric values. For instance, knowing the function values of 30° means that you know the function values of all angles for which 30° is a reference angle. For convenience, the table below shows the exact values of the sine, cosine, and tangent functions of special angles and quadrant angles.

Trigonometric Values of Common Angles

θ (degrees)	0°	30°	45°	60°	90°	180°	270°
θ (radians)	0	$\dfrac{\pi}{6}$	$\dfrac{\pi}{4}$	$\dfrac{\pi}{3}$	$\dfrac{\pi}{2}$	π	$\dfrac{3\pi}{2}$
$\sin \theta$	0	$\dfrac{1}{2}$	$\dfrac{\sqrt{2}}{2}$	$\dfrac{\sqrt{3}}{2}$	1	0	-1
$\cos \theta$	1	$\dfrac{\sqrt{3}}{2}$	$\dfrac{\sqrt{2}}{2}$	$\dfrac{1}{2}$	0	-1	0
$\tan \theta$	0	$\dfrac{\sqrt{3}}{3}$	1	$\sqrt{3}$	Undef.	0	Undef.

EXAMPLE 5 Using Reference Angles

Evaluate each trigonometric function.

a. $\cos \dfrac{4\pi}{3}$ **b.** $\tan(-210°)$ **c.** $\csc \dfrac{11\pi}{4}$

Solution

a. Because $\theta = 4\pi/3$ lies in Quadrant III, the reference angle is

$$\theta' = \frac{4\pi}{3} - \pi = \frac{\pi}{3}$$

as shown in Figure 1.33. Moreover, the cosine is negative in Quadrant III, so

$$\cos \frac{4\pi}{3} = (-) \cos \frac{\pi}{3}$$

$$= -\frac{1}{2}.$$

b. Because $-210° + 360° = 150°$, it follows that $-210°$ is coterminal with the second-quadrant angle $150°$. So, the reference angle is $\theta' = 180° - 150° = 30°$, as shown in Figure 1.34. Finally, because the tangent is negative in Quadrant II, you have

$$\tan(-210°) = (-) \tan 30°$$

$$= -\frac{\sqrt{3}}{3}.$$

c. Because $(11\pi/4) - 2\pi = 3\pi/4$, it follows that $11\pi/4$ is coterminal with the second-quadrant angle $3\pi/4$. So, the reference angle is $\theta' = \pi - (3\pi/4) = \pi/4$, as shown in Figure 1.35. Because the cosecant is positive in Quadrant II, you have

$$\csc \frac{11\pi}{4} = (+) \csc \frac{\pi}{4}$$

$$= \frac{1}{\sin(\pi/4)}$$

$$= \sqrt{2}.$$

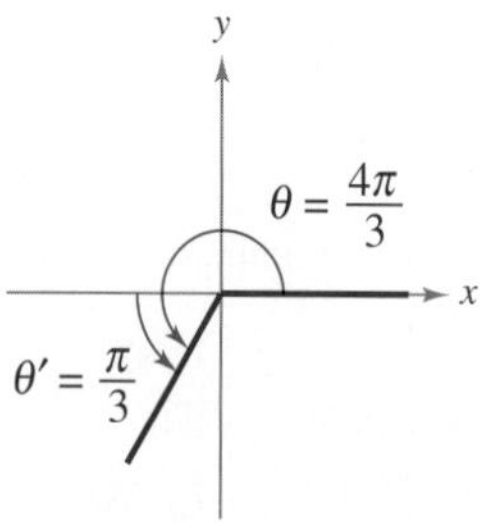

Figure 1.33

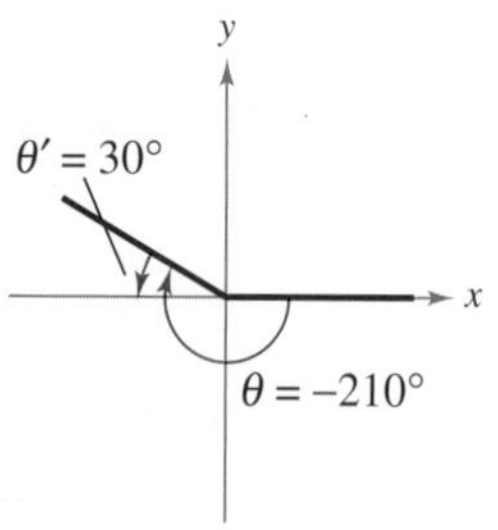

Figure 1.34

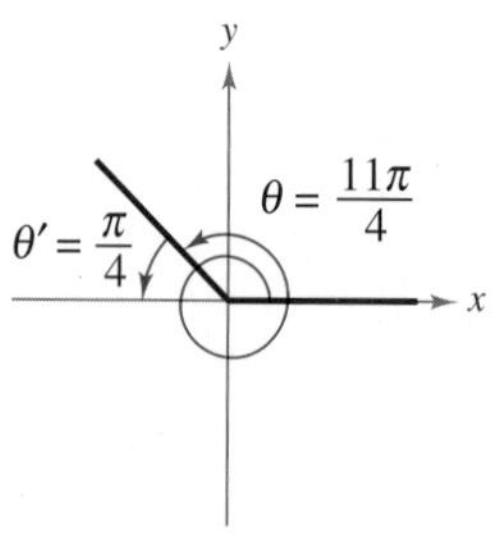

Figure 1.35

✓ **Checkpoint** Audio-video solution in English & Spanish at LarsonPrecalculus.com.

Evaluate each trigonometric function.

a. $\sin \dfrac{7\pi}{4}$ **b.** $\cos(-120°)$ **c.** $\tan \dfrac{11\pi}{6}$

EXAMPLE 6   **Using Trigonometric Identities**

Let θ be an angle in Quadrant II such that $\sin \theta = \frac{1}{3}$. Find (a) $\cos \theta$ and (b) $\tan \theta$ by using trigonometric identities.

Solution

a. Using the Pythagorean identity $\sin^2 \theta + \cos^2 \theta = 1$, you obtain

$$\left(\tfrac{1}{3}\right)^2 + \cos^2 \theta = 1 \quad \Longrightarrow \quad \cos^2 \theta = 1 - \tfrac{1}{9} = \tfrac{8}{9}.$$

Because $\cos \theta < 0$ in Quadrant II, use the negative root to obtain

$$\cos \theta = -\frac{\sqrt{8}}{\sqrt{9}} = -\frac{2\sqrt{2}}{3}.$$

b. Using the trigonometric identity $\tan \theta = \dfrac{\sin \theta}{\cos \theta}$, you obtain

$$\tan \theta = \frac{1/3}{-2\sqrt{2}/3} = -\frac{1}{2\sqrt{2}} = -\frac{\sqrt{2}}{4}.$$

✓ *Checkpoint* ◀))) *Audio-video solution in English & Spanish at LarsonPrecalculus.com.*

Let θ be an angle in Quadrant III such that $\sin \theta = -\frac{4}{5}$. Find (a) $\cos \theta$ and (b) $\tan \theta$ by using trigonometric identities.

EXAMPLE 7   **Using a Calculator**

Use a calculator to evaluate each trigonometric function.

a. $\cot 410°$ b. $\sin(-7)$ c. $\sec \dfrac{\pi}{9}$

Solution

Function	Mode	Calculator Keystrokes	Display
a. $\cot 410°$	Degree	(TAN (410)) x⁻¹ ENTER	0.8390996
b. $\sin(-7)$	Radian	SIN ((−) 7) ENTER	−0.6569866
c. $\sec(\pi/9)$	Radian	(COS (π ÷ 9)) x⁻¹ ENTER	1.0641778

✓ *Checkpoint* ◀))) *Audio-video solution in English & Spanish at LarsonPrecalculus.com.*

Use a calculator to evaluate each trigonometric function.

a. $\tan 119°$ b. $\csc 5$ c. $\cos \dfrac{\pi}{5}$

Summarize   (Section 1.4)

1. State the definitions of the trigonometric functions of any angle *(page 150)*. For examples of evaluating trigonometric functions, see Examples 1–3.

2. Explain how to find a reference angle *(page 152)*. For an example of finding reference angles, see Example 4.

3. Explain how to evaluate a trigonometric function of a real number *(page 153)*. For examples of evaluating trigonometric functions of real numbers, see Examples 5–7.

1.4 Exercises

See CalcChat.com for tutorial help and worked-out solutions to odd-numbered exercises.

Vocabulary: Fill in the blanks.

In Exercises 1–6, let θ be an angle in standard position with (x, y) a point on the terminal side of θ and $r = \sqrt{x^2 + y^2} \neq 0$.

1. $\sin \theta = $ _____

2. $\dfrac{r}{y} = $ _____

3. $\tan \theta = $ _____

4. $\sec \theta = $ _____

5. $\dfrac{x}{r} = $ _____

6. $\dfrac{x}{y} = $ _____

7. Because $r = \sqrt{x^2 + y^2}$ cannot be _____, the sine and cosine functions are _____ for any real value of θ.

8. The acute positive angle formed by the terminal side of an angle θ and the horizontal axis is called the _____ angle of θ and is denoted by θ'.

Skills and Applications

Evaluating Trigonometric Functions In Exercises 9–12, determine the exact values of the six trigonometric functions of each angle θ.

9. (a)   (b)

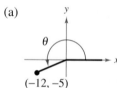   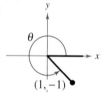

10. (a)   (b)

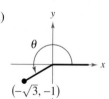   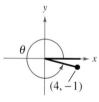

11. (a)   (b)

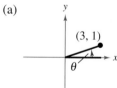   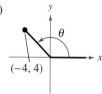

12. (a)   (b)

Evaluating Trigonometric Functions In Exercises 13–18, the point is on the terminal side of an angle in standard position. Determine the exact values of the six trigonometric functions of the angle.

13. $(5, 12)$

14. $(8, 15)$

15. $(-5, -2)$

16. $(-4, 10)$

17. $(-5.4, 7.2)$

18. $\left(3\frac{1}{2}, -7\frac{3}{4}\right)$

Determining a Quadrant In Exercises 19–22, state the quadrant in which θ lies.

19. $\sin \theta > 0$ and $\cos \theta > 0$

20. $\sin \theta < 0$ and $\cos \theta < 0$

21. $\sin \theta > 0$ and $\cos \theta < 0$

22. $\sec \theta > 0$ and $\cot \theta < 0$

Evaluating Trigonometric Functions In Exercises 23–32, find the values of the six trigonometric functions of θ with the given constraint.

Function Value	Constraint
23. $\tan \theta = -\frac{15}{8}$	$\sin \theta > 0$
24. $\cos \theta = \frac{8}{17}$	$\tan \theta < 0$
25. $\sin \theta = \frac{3}{5}$	θ lies in Quadrant II.
26. $\cos \theta = -\frac{4}{5}$	θ lies in Quadrant III.
27. $\cot \theta = -3$	$\cos \theta > 0$
28. $\csc \theta = 4$	$\cot \theta < 0$
29. $\sec \theta = -2$	$\sin \theta < 0$
30. $\sin \theta = 0$	$\sec \theta = -1$
31. $\cot \theta$ is undefined.	$\pi/2 \leq \theta \leq 3\pi/2$
32. $\tan \theta$ is undefined.	$\pi \leq \theta \leq 2\pi$

An Angle Formed by a Line Through the Origin In Exercises 33–36, the terminal side of θ lies on the given line in the specified quadrant. Find the values of the six trigonometric functions of θ by finding a point on the line.

Line	Quadrant
33. $y = -x$	II
34. $y = \frac{1}{3}x$	III
35. $2x - y = 0$	III
36. $4x + 3y = 0$	IV

Trigonometric Function of a Quadrant Angle In Exercises 37–44, evaluate the trigonometric function of the quadrant angle, if possible.

37. $\sin \pi$

38. $\csc \dfrac{3\pi}{2}$

39. $\sec \dfrac{3\pi}{2}$

40. $\sec \pi$

41. $\sin \dfrac{\pi}{2}$

42. $\cot \pi$

43. $\csc \pi$

44. $\cot \dfrac{\pi}{2}$

Finding a Reference Angle In Exercises 45–52, find the reference angle θ' and sketch θ and θ' in standard position.

45. $\theta = 160°$

46. $\theta = 309°$

47. $\theta = -125°$

48. $\theta = -215°$

49. $\theta = \dfrac{2\pi}{3}$

50. $\theta = \dfrac{7\pi}{6}$

51. $\theta = 4.8$

52. $\theta = 11.6$

Using a Reference Angle In Exercises 53–68, evaluate the sine, cosine, and tangent of the angle without using a calculator.

53. $225°$

54. $300°$

55. $750°$

56. $-405°$

57. $-840°$

58. $510°$

59. $\dfrac{2\pi}{3}$

60. $\dfrac{3\pi}{4}$

61. $\dfrac{5\pi}{4}$

62. $\dfrac{7\pi}{6}$

63. $-\dfrac{\pi}{6}$

64. $-\dfrac{\pi}{2}$

65. $\dfrac{9\pi}{4}$

66. $\dfrac{10\pi}{3}$

67. $-\dfrac{3\pi}{2}$

68. $-\dfrac{23\pi}{4}$

Using Trigonometric Identities In Exercises 69–74, use a trigonometric identity to find the indicated value in the specified quadrant.

Function Value	Quadrant	Value
69. $\sin \theta = -\frac{3}{5}$	IV	$\cos \theta$
70. $\cot \theta = -3$	II	$\sin \theta$
71. $\tan \theta = \frac{3}{2}$	III	$\sec \theta$
72. $\csc \theta = -2$	IV	$\cot \theta$
73. $\cos \theta = \frac{5}{8}$	I	$\sec \theta$
74. $\sec \theta = -\frac{9}{4}$	III	$\tan \theta$

Using a Calculator In Exercises 75–90, use a calculator to evaluate the trigonometric function. Round your answer to four decimal places. (Be sure the calculator is in the correct mode.)

75. $\sin 10°$

76. $\sec 225°$

77. $\cos(-110°)$

78. $\csc(-330°)$

79. $\tan 304°$

80. $\cot 178°$

81. $\sec 72°$

82. $\tan(-188°)$

83. $\tan 4.5$

84. $\cot 1.35$

85. $\tan \dfrac{\pi}{9}$

86. $\tan\left(-\dfrac{\pi}{9}\right)$

87. $\sin(-0.65)$

88. $\sec 0.29$

89. $\cot\left(-\dfrac{11\pi}{8}\right)$

90. $\csc\left(-\dfrac{15\pi}{14}\right)$

Solving for θ In Exercises 91–96, find two solutions of each equation. Give your answers in degrees $(0° \le \theta < 360°)$ and in radians $(0 \le \theta < 2\pi)$. Do not use a calculator.

91. (a) $\sin \theta = \frac{1}{2}$ (b) $\sin \theta = -\frac{1}{2}$

92. (a) $\cos \theta = \dfrac{\sqrt{2}}{2}$ (b) $\cos \theta = -\dfrac{\sqrt{2}}{2}$

93. (a) $\csc \theta = \dfrac{2\sqrt{3}}{3}$ (b) $\cot \theta = -1$

94. (a) $\sec \theta = 2$ (b) $\sec \theta = -2$

95. (a) $\tan \theta = 1$ (b) $\cot \theta = -\sqrt{3}$

96. (a) $\sin \theta = \dfrac{\sqrt{3}}{2}$ (b) $\sin \theta = -\dfrac{\sqrt{3}}{2}$

97. Distance An airplane, flying at an altitude of 6 miles, is on a flight path that passes directly over an observer (see figure). Let θ be the angle of elevation from the observer to the plane. Find the distance d from the observer to the plane when (a) $\theta = 30°$, (b) $\theta = 90°$, and (c) $\theta = 120°$.

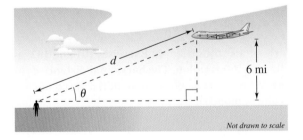

Not drawn to scale

98. Harmonic Motion The displacement from equilibrium of an oscillating weight suspended by a spring is given by $y(t) = 2 \cos 6t$, where y is the displacement (in centimeters) and t is the time (in seconds). Find the displacement when (a) $t = 0$, (b) $t = \frac{1}{4}$, and (c) $t = \frac{1}{2}$.

•• 99. **Data Analysis: Meteorology** •••••••••

The table shows the monthly normal temperatures (in degrees Fahrenheit) for selected months in New York City (*N*) and Fairbanks, Alaska (*F*). *(Source: National Climatic Data Center)*

DATA	Month	New York City, *N*	Fairbanks, *F*
	January	33	−10
	April	52	32
	July	77	62
	October	58	24
	December	38	−6

Spreadsheet at LarsonPrecalculus.com

(a) Use the *regression* feature of a graphing utility to find a model of the form $y = a\sin(bt + c) + d$ for each city. Let *t* represent the month, with $t = 1$ corresponding to January.

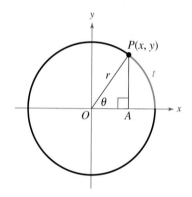

(b) Use the models from part (a) to find the monthly normal temperatures for the two cities in February, March, May, June, August, September, and November.

(c) Compare the models for the two cities.

100. **Sales** A company that produces snowboards forecasts monthly sales over the next 2 years to be

$$S = 23.1 + 0.442t + 4.3\cos\frac{\pi t}{6}$$

where *S* is measured in thousands of units and *t* is the time in months, with $t = 1$ representing January 2014. Predict sales for each of the following months.

(a) February 2014   (b) February 2015

(c) June 2014   (d) June 2015

Path of a Projectile In Exercises 101 and 102, use the following information. The horizontal distance *d* (in feet) traveled by a projectile with an initial speed of *v* feet per second is modeled by

$$d = \frac{v^2}{32}\sin 2\theta$$

where *θ* is the angle at which the projectile is launched.

101. Find the horizontal distance traveled by a golf ball that is hit with an initial speed of 100 feet per second when the golf ball is hit at an angle of (a) $\theta = 30°$, (b) $\theta = 50°$, and (c) $\theta = 60°$.

102. Find the horizontal distance traveled by a model rocket that is launched with an initial speed of 120 feet per second when the model rocket is launched at an angle of (a) $\theta = 60°$, (b) $\theta = 70°$, and (c) $\theta = 80°$.

Exploration

True or False? In Exercises 103 and 104, determine whether the statement is true or false. Justify your answer.

103. In each of the four quadrants, the signs of the secant function and sine function are the same.

104. To find the reference angle for an angle *θ* (given in degrees), find the integer *n* such that $0 \le 360°n - \theta \le 360°$. The difference $360°n - \theta$ is the reference angle.

105. **Think About It** The figure shows point $P(x, y)$ on a unit circle and right triangle *OAP*.

(a) Find sin *t* and cos *t* using the unit circle definitions of sine and cosine (from Section 1.2).

(b) What is the value of *r*? Explain.

(c) Use the definitions of sine and cosine given in this section to find sin *θ* and cos *θ*. Write your answers in terms of *x* and *y*.

(d) Based on your answers to parts (a) and (c), what can you conclude?

106. **HOW DO YOU SEE IT?** Consider an angle in standard position with $r = 12$ centimeters, as shown in the figure. Describe the changes in the values of *x*, *y*, sin *θ*, cos *θ*, and tan *θ* as *θ* increases continuously from 0° to 90°.

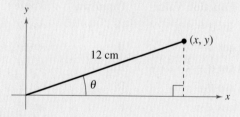

1.5 Graphs of Sine and Cosine Functions

You can use sine and cosine functions in scientific calculations. For instance, in Exercise 88 on page 168, you will use a trigonometric function to model the airflow of your respiratory cycle.

- Sketch the graphs of basic sine and cosine functions.
- Use amplitude and period to help sketch the graphs of sine and cosine functions.
- Sketch translations of the graphs of sine and cosine functions.
- Use sine and cosine functions to model real-life data.

Basic Sine and Cosine Curves

In this section, you will study techniques for sketching the graphs of the sine and cosine functions. The graph of the sine function is a **sine curve.** In Figure 1.36, the black portion of the graph represents one period of the function and is called **one cycle** of the sine curve. The gray portion of the graph indicates that the basic sine curve repeats indefinitely to the left and right. The graph of the cosine function is shown in Figure 1.37.

Recall from Section 1.2 that the domain of the sine and cosine functions is the set of all real numbers. Moreover, the range of each function is the interval $[-1, 1]$, and each function has a period of 2π. Do you see how this information is consistent with the basic graphs shown in Figures 1.36 and 1.37?

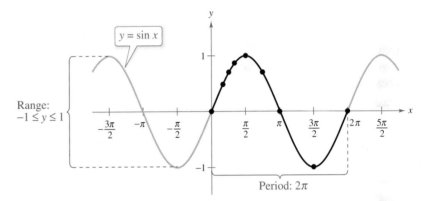

Figure 1.36

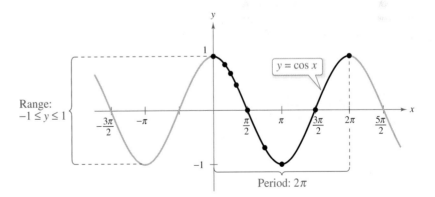

Figure 1.37

Note in Figures 1.36 and 1.37 that the sine curve is symmetric with respect to the *origin*, whereas the cosine curve is symmetric with respect to the *y-axis*. These properties of symmetry follow from the fact that the sine function is odd and the cosine function is even.

To sketch the graphs of the basic sine and cosine functions by hand, it helps to note five **key points** in one period of each graph: the *intercepts, maximum points,* and *minimum points* (see below).

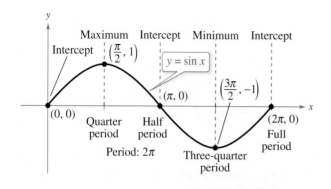 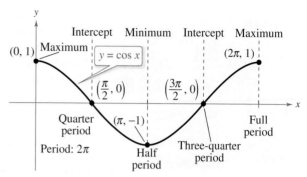

EXAMPLE 1   **Using Key Points to Sketch a Sine Curve**

Sketch the graph of

$$y = 2 \sin x$$

on the interval $[-\pi, 4\pi]$.

Solution   Note that

$$y = 2 \sin x$$
$$= 2(\sin x)$$

indicates that the y-values for the key points will have twice the magnitude of those on the graph of $y = \sin x$. Divide the period 2π into four equal parts to get the key points

Intercept	**Maximum**	**Intercept**	**Minimum**	**Intercept**
$(0, 0),$	$\left(\dfrac{\pi}{2}, 2\right),$	$(\pi, 0),$	$\left(\dfrac{3\pi}{2}, -2\right),$ and	$(2\pi, 0).$

By connecting these key points with a smooth curve and extending the curve in both directions over the interval $[-\pi, 4\pi]$, you obtain the graph shown below.

▷ **TECHNOLOGY** When using a graphing utility to graph trigonometric functions, pay special attention to the viewing window you use. For instance, try graphing $y = [\sin(10x)]/10$ in the standard viewing window in *radian* mode. What do you observe? Use the *zoom* feature to find a viewing window that displays a good view of the graph.

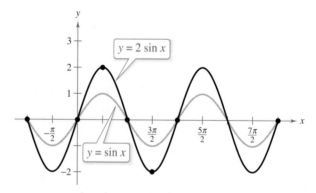

✓ *Checkpoint* ◀))) *Audio-video solution in English & Spanish at LarsonPrecalculus.com.*

Sketch the graph of

$$y = 2 \cos x$$

on the interval $\left[-\dfrac{\pi}{2}, \dfrac{9\pi}{2}\right]$.

Amplitude and Period

In the rest of this section, you will study the graphic effect of each of the constants *a*, *b*, *c*, and *d* in equations of the forms

$$y = d + a \sin(bx - c)$$

and

$$y = d + a \cos(bx - c).$$

A quick review of the transformations you studied in Section P.8 should help in this investigation.

The constant factor *a* in $y = a \sin x$ acts as a *scaling factor*—a *vertical stretch* or *vertical shrink* of the basic sine curve. When $|a| > 1$, the basic sine curve is stretched, and when $|a| < 1$, the basic sine curve is shrunk. The result is that the graph of $y = a \sin x$ ranges between $-a$ and a instead of between -1 and 1. The absolute value of *a* is the **amplitude** of the function $y = a \sin x$. The range of the function $y = a \sin x$ for $a > 0$ is $-a \leq y \leq a$.

Definition of Amplitude of Sine and Cosine Curves

The **amplitude** of $y = a \sin x$ and $y = a \cos x$ represents half the distance between the maximum and minimum values of the function and is given by

$$\text{Amplitude} = |a|.$$

EXAMPLE 2   **Scaling: Vertical Shrinking and Stretching**

In the same coordinate plane, sketch the graph of each function.

a. $y = \frac{1}{2} \cos x$

b. $y = 3 \cos x$

Solution

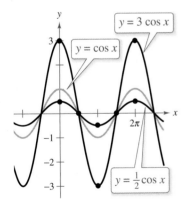

Figure 1.38

a. Because the amplitude of $y = \frac{1}{2} \cos x$ is $\frac{1}{2}$, the maximum value is $\frac{1}{2}$ and the minimum value is $-\frac{1}{2}$. Divide one cycle, $0 \leq x \leq 2\pi$, into four equal parts to get the key points

Maximum	Intercept	Minimum	Intercept	Maximum
$\left(0, \frac{1}{2}\right),$	$\left(\frac{\pi}{2}, 0\right),$	$\left(\pi, -\frac{1}{2}\right),$	$\left(\frac{3\pi}{2}, 0\right),$ and	$\left(2\pi, \frac{1}{2}\right).$

b. A similar analysis shows that the amplitude of $y = 3 \cos x$ is 3, and the key points are

Maximum	Intercept	Minimum	Intercept	Maximum
$(0, 3),$	$\left(\frac{\pi}{2}, 0\right),$	$(\pi, -3),$	$\left(\frac{3\pi}{2}, 0\right),$ and	$(2\pi, 3).$

The graphs of these two functions are shown in Figure 1.38. Notice that the graph of $y = \frac{1}{2} \cos x$ is a vertical *shrink* of the graph of $y = \cos x$ and the graph of $y = 3 \cos x$ is a vertical *stretch* of the graph of $y = \cos x$.

✓ *Checkpoint* 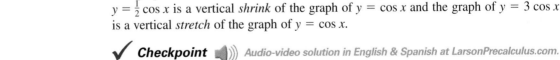 Audio-video solution in English & Spanish at LarsonPrecalculus.com.

In the same coordinate plane, sketch the graph of each function.

a. $y = \frac{1}{3} \sin x$

b. $y = 3 \sin x$

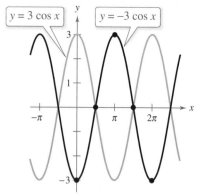

Figure 1.39

You know from Section P.8 that the graph of $y = -f(x)$ is a **reflection** in the x-axis of the graph of $y = f(x)$. For instance, the graph of $y = -3 \cos x$ is a reflection of the graph of $y = 3 \cos x$, as shown in Figure 1.39.

Because $y = a \sin x$ completes one cycle from $x = 0$ to $x = 2\pi$, it follows that $y = a \sin bx$ completes one cycle from $x = 0$ to $x = 2\pi/b$, where b is a positive real number.

Period of Sine and Cosine Functions

Let b be a positive real number. The **period** of $y = a \sin bx$ and $y = a \cos bx$ is given by

$$\text{Period} = \frac{2\pi}{b}.$$

Note that when $0 < b < 1$, the period of $y = a \sin bx$ is greater than 2π and represents a *horizontal stretching* of the graph of $y = a \sin x$. Similarly, when $b > 1$, the period of $y = a \sin bx$ is less than 2π and represents a *horizontal shrinking* of the graph of $y = a \sin x$. When b is negative, the identities $\sin(-x) = -\sin x$ and $\cos(-x) = \cos x$ are used to rewrite the function.

EXAMPLE 3   Scaling: Horizontal Stretching

Sketch the graph of

$$y = \sin \frac{x}{2}.$$

Solution The amplitude is 1. Moreover, because $b = \frac{1}{2}$, the period is

$$\frac{2\pi}{b} = \frac{2\pi}{\frac{1}{2}} = 4\pi. \qquad \text{Substitute for } b.$$

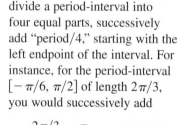

REMARK In general, to divide a period-interval into four equal parts, successively add "period/4," starting with the left endpoint of the interval. For instance, for the period-interval $[-\pi/6, \pi/2]$ of length $2\pi/3$, you would successively add

$$\frac{2\pi/3}{4} = \frac{\pi}{6}$$

to get $-\pi/6, 0, \pi/6, \pi/3$, and $\pi/2$ as the x-values for the key points on the graph.

Now, divide the period-interval $[0, 4\pi]$ into four equal parts using the values $\pi, 2\pi$, and 3π to obtain the key points

Intercept	Maximum	Intercept	Minimum	Intercept
$(0, 0)$,	$(\pi, 1)$,	$(2\pi, 0)$,	$(3\pi, -1)$, and	$(4\pi, 0)$.

The graph is shown below.

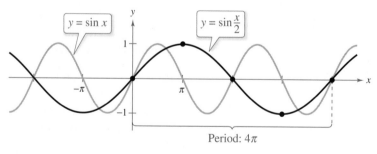

Period: 4π

✓ **Checkpoint**))) *Audio-video solution in English & Spanish at LarsonPrecalculus.com.*

Sketch the graph of

$$y = \cos \frac{x}{3}.$$

Translations of Sine and Cosine Curves

▷ **ALGEBRA HELP** You can review the techniques for shifting, reflecting, and stretching graphs in Section P.8.

The constant c in the general equations

$$y = a\sin(bx - c) \quad \text{and} \quad y = a\cos(bx - c)$$

creates *horizontal translations* (shifts) of the basic sine and cosine curves. Comparing $y = a\sin bx$ with $y = a\sin(bx - c)$, you find that the graph of $y = a\sin(bx - c)$ completes one cycle from $bx - c = 0$ to $bx - c = 2\pi$. By solving for x, you can find the interval for one cycle to be

Left endpoint   Right endpoint

$$\frac{c}{b} \leq x \leq \frac{c}{b} + \frac{2\pi}{b}.$$

Period

This implies that the period of $y = a\sin(bx - c)$ is $2\pi/b$, and the graph of $y = a\sin bx$ is shifted by an amount c/b. The number c/b is the **phase shift.**

Graphs of Sine and Cosine Functions

The graphs of $y = a\sin(bx - c)$ and $y = a\cos(bx - c)$ have the following characteristics. (Assume $b > 0$.)

$$\text{Amplitude} = |a| \qquad \text{Period} = \frac{2\pi}{b}$$

The left and right endpoints of a one-cycle interval can be determined by solving the equations $bx - c = 0$ and $bx - c = 2\pi$.

EXAMPLE 4   Horizontal Translation

Analyze the graph of $y = \dfrac{1}{2}\sin\left(x - \dfrac{\pi}{3}\right)$.

Algebraic Solution

The amplitude is $\frac{1}{2}$ and the period is 2π. By solving the equations

$$x - \frac{\pi}{3} = 0 \implies x = \frac{\pi}{3}$$

and

$$x - \frac{\pi}{3} = 2\pi \implies x = \frac{7\pi}{3}$$

you see that the interval $[\pi/3, 7\pi/3]$ corresponds to one cycle of the graph. Dividing this interval into four equal parts produces the key points

Intercept	Maximum	Intercept	Minimum	Intercept
$\left(\dfrac{\pi}{3}, 0\right)$,	$\left(\dfrac{5\pi}{6}, \dfrac{1}{2}\right)$,	$\left(\dfrac{4\pi}{3}, 0\right)$,	$\left(\dfrac{11\pi}{6}, -\dfrac{1}{2}\right)$,	and $\left(\dfrac{7\pi}{3}, 0\right)$.

Graphical Solution

Use a graphing utility set in *radian* mode to graph $y = (1/2)\sin(x - \pi/3)$, as shown below. Use the *minimum, maximum,* and *zero* or *root* features of the graphing utility to approximate the key points $(1.05, 0)$, $(2.62, 0.5)$, $(4.19, 0)$, $(5.76, -0.5)$, and $(7.33, 0)$.

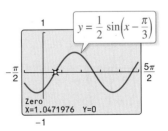

$$y = \frac{1}{2}\sin\left(x - \frac{\pi}{3}\right)$$

Zero
X=1.0471976 Y=0

✓ *Checkpoint*  Audio-video solution in English & Spanish at LarsonPrecalculus.com.

Analyze the graph of $y = 2\cos\left(x - \dfrac{\pi}{2}\right)$.

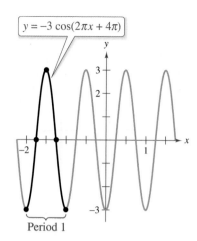

$y = -3\cos(2\pi x + 4\pi)$

Period 1

Figure 1.40

EXAMPLE 5 Horizontal Translation

Sketch the graph of

$$y = -3\cos(2\pi x + 4\pi).$$

Solution The amplitude is 3 and the period is $2\pi/2\pi = 1$. By solving the equations

$$2\pi x + 4\pi = 0$$

$$2\pi x = -4\pi$$

$$x = -2$$

and

$$2\pi x + 4\pi = 2\pi$$

$$2\pi x = -2\pi$$

$$x = -1$$

you see that the interval $[-2, -1]$ corresponds to one cycle of the graph. Dividing this interval into four equal parts produces the key points

Minimum	Intercept	Maximum	Intercept	Minimum
$(-2, -3)$,	$\left(-\frac{7}{4}, 0\right)$,	$\left(-\frac{3}{2}, 3\right)$,	$\left(-\frac{5}{4}, 0\right)$, and	$(-1, -3)$.

The graph is shown in Figure 1.40.

✓ *Checkpoint* Audio-video solution in English & Spanish at LarsonPrecalculus.com.

Sketch the graph of

$$y = -\frac{1}{2}\sin(\pi x + \pi).$$

The final type of transformation is the *vertical translation* caused by the constant d in the equations

$$y = d + a\sin(bx - c) \quad \text{and} \quad y = d + a\cos(bx - c).$$

The shift is d units up for $d > 0$ and d units down for $d < 0$. In other words, the graph oscillates about the horizontal line $y = d$ instead of about the x-axis.

EXAMPLE 6 Vertical Translation

Sketch the graph of

$$y = 2 + 3\cos 2x.$$

Solution The amplitude is 3 and the period is π. The key points over the interval $[0, \pi]$ are

$$(0, 5), \quad \left(\frac{\pi}{4}, 2\right), \quad \left(\frac{\pi}{2}, -1\right), \quad \left(\frac{3\pi}{4}, 2\right), \quad \text{and} \quad (\pi, 5).$$

The graph is shown in Figure 1.41. Compared with the graph of $f(x) = 3\cos 2x$, the graph of $y = 2 + 3\cos 2x$ is shifted up two units.

✓ *Checkpoint* Audio-video solution in English & Spanish at LarsonPrecalculus.com.

Sketch the graph of

$$y = 2\cos x - 5.$$

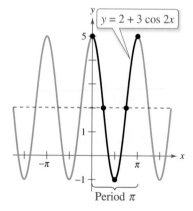

$y = 2 + 3\cos 2x$

Period π

Figure 1.41

Mathematical Modeling

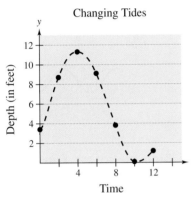

DATA	Time, t	Depth, y
	0	3.4
	2	8.7
	4	11.3
	6	9.1
	8	3.8
	10	0.1
	12	1.2

Spreadsheet at LarsonPrecalculus.com

EXAMPLE 7   **Finding a Trigonometric Model**

The table shows the depths (in feet) of the water at the end of a dock at various times during the morning, where $t = 0$ corresponds to midnight.

a. Use a trigonometric function to model the data.

b. Find the depths at 9 A.M. and 3 P.M.

c. A boat needs at least 10 feet of water to moor at the dock. During what times in the afternoon can it safely dock?

Solution

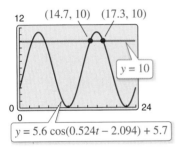

Changing Tides

Depth (in feet)

Time

Figure 1.42

a. Begin by graphing the data, as shown in Figure 1.42. You can use either a sine or cosine model. Suppose you use a cosine model of the form $y = a\cos(bt - c) + d$. The difference between the maximum value and minimum value is twice the amplitude of the function. So, the amplitude is

$$a = \tfrac{1}{2}[(\text{maximum depth}) - (\text{minimum depth})] = \tfrac{1}{2}(11.3 - 0.1) = 5.6.$$

The cosine function completes one half of a cycle between the times at which the maximum and minimum depths occur. So, the period p is

$$p = 2[(\text{time of min. depth}) - (\text{time of max. depth})] = 2(10 - 4) = 12$$

which implies that $b = 2\pi/p \approx 0.524$. Because high tide occurs 4 hours after midnight, consider the left endpoint to be $c/b = 4$, so $c \approx 2.094$. Moreover, because the average depth is $\tfrac{1}{2}(11.3 + 0.1) = 5.7$, it follows that $d = 5.7$. So, you can model the depth with the function $y = 5.6\cos(0.524t - 2.094) + 5.7$.

b. The depths at 9 A.M. and 3 P.M. are as follows.

$$y = 5.6\cos(0.524 \cdot 9 - 2.094) + 5.7 \approx 0.84 \text{ foot} \qquad \text{9 A.M.}$$

$$y = 5.6\cos(0.524 \cdot 15 - 2.094) + 5.7 \approx 10.57 \text{ feet} \qquad \text{3 P.M.}$$

12   (14.7, 10) (17.3, 10)

$y = 10$

$y = 5.6\cos(0.524t - 2.094) + 5.7$

Figure 1.43

c. Using a graphing utility, graph the model with the line $y = 10$. Using the *intersect* feature, you can determine that the depth is at least 10 feet between 2:42 P.M. ($t \approx 14.7$) and 5:18 P.M. ($t \approx 17.3$), as shown in Figure 1.43.

✓ **Checkpoint** 🔊)) *Audio-video solution in English & Spanish at LarsonPrecalculus.com.*

Find a sine model for the data in Example 7.   ■

Summarize   (Section 1.5)

1. Describe how to sketch the graphs of basic sine and cosine functions (*pages 159 and 160*). For an example of sketching the graph of a sine function, see Example 1.

2. Describe how you can use amplitude and period to help sketch the graphs of sine and cosine functions (*pages 161 and 162*). For examples of using amplitude and period to sketch graphs of sine and cosine functions, see Examples 2 and 3.

3. Describe how to sketch translations of the graphs of sine and cosine functions (*pages 163 and 164*). For examples of translating the graphs of sine and cosine functions, see Examples 4–6.

4. Give an example of how to use sine and cosine functions to model real-life data (*page 165, Example 7*).

1.5 Exercises

See CalcChat.com for tutorial help and worked-out solutions to odd-numbered exercises.

Vocabulary: Fill in the blanks.

1. One period of a sine or cosine function is called one _____ of the sine or cosine curve.

2. The _____ of a sine or cosine curve represents half the distance between the maximum and minimum values of the function.

3. For the function $y = a \sin(bx - c)$, $\dfrac{c}{b}$ represents the _____ _____ of the graph of the function.

4. For the function $y = d + a \cos(bx - c)$, d represents a _____ _____ of the graph of the function.

Skills and Applications

Finding the Period and Amplitude In Exercises 5–18, find the period and amplitude.

5. $y = 2 \sin 5x$

6. $y = 3 \cos 2x$

7. $y = \dfrac{3}{4} \cos \dfrac{x}{2}$

8. $y = -3 \sin \dfrac{x}{3}$

9. $y = \dfrac{1}{2} \sin \dfrac{\pi x}{3}$

10. $y = \dfrac{3}{2} \cos \dfrac{\pi x}{2}$

11. $y = -4 \sin x$

12. $y = -\cos \dfrac{2x}{3}$

13. $y = 3 \sin 10x$

14. $y = \tfrac{1}{5} \sin 6x$

15. $y = \dfrac{5}{3} \cos \dfrac{4x}{5}$

16. $y = \dfrac{5}{2} \cos \dfrac{x}{4}$

17. $y = \dfrac{1}{4} \sin 2\pi x$

18. $y = \dfrac{2}{3} \cos \dfrac{\pi x}{10}$

Describing the Relationship Between Graphs In Exercises 19–30, describe the relationship between the graphs of f and g. Consider amplitude, period, and shifts.

19. $f(x) = \sin x$
    $g(x) = \sin(x - \pi)$

20. $f(x) = \cos x$
    $g(x) = \cos(x + \pi)$

21. $f(x) = \cos 2x$
    $g(x) = -\cos 2x$

22. $f(x) = \sin 3x$
    $g(x) = \sin(-3x)$

23. $f(x) = \cos x$
    $g(x) = \cos 2x$

24. $f(x) = \sin x$
    $g(x) = \sin 3x$

25. $f(x) = \sin 2x$
    $g(x) = 3 + \sin 2x$

26. $f(x) = \cos 4x$
    $g(x) = -2 + \cos 4x$

27.
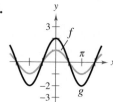

28.
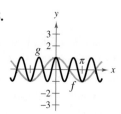

29.
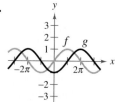

30.
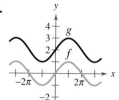

Sketching Graphs of Sine or Cosine Functions In Exercises 31–38, sketch the graphs of f and g in the same coordinate plane. (Include two full periods.)

31. $f(x) = -2 \sin x$
    $g(x) = 4 \sin x$

32. $f(x) = \sin x$
    $g(x) = \sin \dfrac{x}{3}$

33. $f(x) = \cos x$
    $g(x) = 2 + \cos x$

34. $f(x) = 2 \cos 2x$
    $g(x) = -\cos 4x$

35. $f(x) = -\dfrac{1}{2} \sin \dfrac{x}{2}$
    $g(x) = 3 - \dfrac{1}{2} \sin \dfrac{x}{2}$

36. $f(x) = 4 \sin \pi x$
    $g(x) = 4 \sin \pi x - 3$

37. $f(x) = 2 \cos x$
    $g(x) = 2 \cos(x + \pi)$

38. $f(x) = -\cos x$
    $g(x) = -\cos(x - \pi)$

Sketching the Graph of a Sine or Cosine Function In Exercises 39–60, sketch the graph of the function. (Include two full periods.)

39. $y = 5 \sin x$

40. $y = \dfrac{1}{4} \sin x$

41. $y = \dfrac{1}{3} \cos x$

42. $y = 4 \cos x$

43. $y = \cos \dfrac{x}{2}$

44. $y = \sin 4x$

45. $y = \cos 2\pi x$

46. $y = \sin \dfrac{\pi x}{4}$

47. $y = -\sin \dfrac{2\pi x}{3}$

48. $y = -10 \cos \dfrac{\pi x}{6}$

49. $y = 3 \cos(x + \pi)$

50. $y = \sin(x - 2\pi)$

51. $y = \sin\left(x - \dfrac{\pi}{2}\right)$ **52.** $y = 4\cos\left(x + \dfrac{\pi}{4}\right)$

53. $y = 2 - \sin\dfrac{2\pi x}{3}$ **54.** $y = -3 + 5\cos\dfrac{\pi t}{12}$

55. $y = 2 + \frac{1}{10}\cos 60\pi x$ **56.** $y = 2\cos x - 3$

57. $y = 3\cos(x + \pi) - 3$ **58.** $y = -3\cos(6x + \pi)$

59. $y = \dfrac{2}{3}\cos\left(\dfrac{x}{2} - \dfrac{\pi}{4}\right)$ **60.** $y = 4\cos\left(x + \dfrac{\pi}{4}\right) + 4$

Describing a Transformation   In Exercises 61–66, g is related to a parent function $f(x) = \sin(x)$ or $f(x) = \cos(x)$. (a) Describe the sequence of transformations from f to g. (b) Sketch the graph of g. (c) Use function notation to write g in terms of f.

61. $g(x) = \sin(4x - \pi)$ **62.** $g(x) = \sin(2x + \pi)$

63. $g(x) = \cos(x - \pi) + 2$ **64.** $g(x) = 1 + \cos(x + \pi)$

65. $g(x) = 2\sin(4x - \pi) - 3$ **66.** $g(x) = 4 - \sin(2x + \pi)$

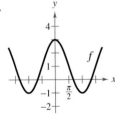 **Graphing a Sine or Cosine Function**   In Exercises 67–72, use a graphing utility to graph the function. (Include two full periods.) Be sure to choose an appropriate viewing window.

67. $y = -2\sin(4x + \pi)$ **68.** $y = -4\sin\left(\dfrac{2}{3}x - \dfrac{\pi}{3}\right)$

69. $y = \cos\left(2\pi x - \dfrac{\pi}{2}\right) + 1$

70. $y = 3\cos\left(\dfrac{\pi x}{2} + \dfrac{\pi}{2}\right) - 2$

71. $y = -0.1\sin\left(\dfrac{\pi x}{10} + \pi\right)$ **72.** $y = \dfrac{1}{100}\sin 120\pi t$

Graphical Reasoning   In Exercises 73–76, find a and d for the function $f(x) = a\cos x + d$ such that the graph of f matches the figure.

73.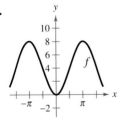

74.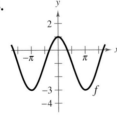

75.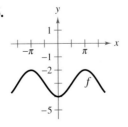

76.

Graphical Reasoning   In Exercises 77–80, find a, b, and c for the function $f(x) = a\sin(bx - c)$ such that the graph of f matches the figure.

77.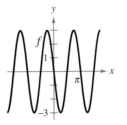

78.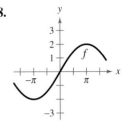

79.

80.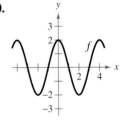

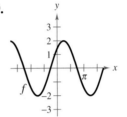 **Graphical Analysis**   In Exercises 81 and 82, use a graphing utility to graph y_1 and y_2 in the interval $[-2\pi, 2\pi]$. Use the graphs to find real numbers x such that $y_1 = y_2$.

81. $y_1 = \sin x$ **82.** $y_1 = \cos x$

   $y_2 = -\dfrac{1}{2}$ $y_2 = -1$

Writing an Equation   In Exercises 83–86, write an equation for the function that is described by the given characteristics.

83. A sine curve with a period of π, an amplitude of 2, a right phase shift of $\pi/2$, and a vertical translation up 1 unit

84. A sine curve with a period of 4π, an amplitude of 3, a left phase shift of $\pi/4$, and a vertical translation down 1 unit

85. A cosine curve with a period of π, an amplitude of 1, a left phase shift of π, and a vertical translation down $\frac{3}{2}$ units

86. A cosine curve with a period of 4π, an amplitude of 3, a right phase shift of $\pi/2$, and a vertical translation up 2 units

87. Respiratory Cycle   After exercising for a few minutes, a person has a respiratory cycle for which the velocity of airflow is approximated by

$$v = 1.75\sin\dfrac{\pi t}{2}$$

where t is the time (in seconds). (Inhalation occurs when $v > 0$, and exhalation occurs when $v < 0$.)

(a) Find the time for one full respiratory cycle.

(b) Find the number of cycles per minute.

(c) Sketch the graph of the velocity function.

• • **88. Respiratory Cycle** • • • • • • • • • • • •

For a person at rest, the velocity v (in liters per second) of airflow during a respiratory cycle (the time from the beginning of one breath to the beginning of the next) is given by

$$v = 0.85 \sin \frac{\pi t}{3}$$

where t is the time (in seconds).

(a) Find the time for one full respiratory cycle.

(b) Find the number of cycles per minute.

(c) Sketch the graph of the velocity function.

89. Data Analysis: Meteorology The table shows the maximum daily high temperatures in Las Vegas L and International Falls I (in degrees Fahrenheit) for month t, with $t = 1$ corresponding to January. *(Source: National Climatic Data Center)*

DATA	Month, t	Las Vegas, L	International Falls, I
	1	57.1	13.8
	2	63.0	22.4
	3	69.5	34.9
	4	78.1	51.5
	5	87.8	66.6
	6	98.9	74.2
	7	104.1	78.6
	8	101.8	76.3
	9	93.8	64.7
	10	80.8	51.7
	11	66.0	32.5
	12	57.3	18.1

Spreadsheet at LarsonPrecalculus.com

(a) A model for the temperatures in Las Vegas is

$$L(t) = 80.60 + 23.50 \cos\left(\frac{\pi t}{6} - 3.67\right).$$

Find a trigonometric model for International Falls.

(b) Use a graphing utility to graph the data points and the model for the temperatures in Las Vegas. How well does the model fit the data?

(c) Use the graphing utility to graph the data points and the model for the temperatures in International Falls. How well does the model fit the data?

(d) Use the models to estimate the average maximum temperature in each city. Which term of the models did you use? Explain.

(e) What is the period of each model? Are the periods what you expected? Explain.

(f) Which city has the greater variability in temperature throughout the year? Which factor of the models determines this variability? Explain.

90. Health The function

$$P = 100 - 20 \cos \frac{5\pi t}{3}$$

approximates the blood pressure P (in millimeters of mercury) at time t (in seconds) for a person at rest.

(a) Find the period of the function.

(b) Find the number of heartbeats per minute.

91. Piano Tuning When tuning a piano, a technician strikes a tuning fork for the A above middle C and sets up a wave motion that can be approximated by $y = 0.001 \sin 880\pi t$, where t is the time (in seconds).

(a) What is the period of the function?

(b) The frequency f is given by $f = 1/p$. What is the frequency of the note?

92. Data Analysis: Astronomy The percent y (in decimal form) of the moon's face illuminated on day x in the year 2014, where $x = 1$ represents January 1, is shown in the table. *(Source: U.S. Naval Observatory)*

DATA	x	y
	1	0.0
	8	0.5
	16	1.0
	24	0.5
	30	0.0
	37	0.5

Spreadsheet at LarsonPrecalculus.com

(a) Create a scatter plot of the data.

(b) Find a trigonometric model that fits the data.

(c) Add the graph of your model in part (b) to the scatter plot. How well does the model fit the data?

(d) What is the period of the model?

(e) Estimate the percent of the moon's face illuminated on March 12, 2014.

93. Ferris Wheel A Ferris wheel is built such that the height h (in feet) above ground of a seat on the wheel at time t (in seconds) can be modeled by

$$h(t) = 53 + 50 \sin\left(\frac{\pi}{10}t - \frac{\pi}{2}\right).$$

(a) Find the period of the model. What does the period tell you about the ride?

(b) Find the amplitude of the model. What does the amplitude tell you about the ride?

(c) Use a graphing utility to graph one cycle of the model.

94. Fuel Consumption The daily consumption C (in gallons) of diesel fuel on a farm is modeled by

$$C = 30.3 + 21.6 \sin\left(\frac{2\pi t}{365} + 10.9\right)$$

where t is the time (in days), with $t = 1$ corresponding to January 1.

(a) What is the period of the model? Is it what you expected? Explain.

(b) What is the average daily fuel consumption? Which term of the model did you use? Explain.

(c) Use a graphing utility to graph the model. Use the graph to approximate the time of the year when consumption exceeds 40 gallons per day.

Exploration

True or False? **In Exercises 95 and 96, determine whether the statement is true or false. Justify your answer.**

95. The graph of the function $f(x) = \sin(x + 2\pi)$ translates the graph of $f(x) = \sin x$ exactly one period to the right so that the two graphs look identical.

96. The function $y = \frac{1}{2}\cos 2x$ has an amplitude that is twice that of the function $y = \cos x$.

Conjecture **In Exercises 97 and 98, graph f and g in the same coordinate plane. Include two full periods. Make a conjecture about the functions.**

97. $f(x) = \sin x$, $g(x) = \cos\left(x - \frac{\pi}{2}\right)$

98. $f(x) = \sin x$, $g(x) = -\cos\left(x + \frac{\pi}{2}\right)$

99. Writing Sketch the graph of $y = \cos bx$ for $b = \frac{1}{2}$, 2, and 3. How does the value of b affect the graph? How many complete cycles of the graph of y occur between 0 and 2π for each value of b?

100. Polynomial Approximations Using calculus, it can be shown that the sine and cosine functions can be approximated by the polynomials

$$\sin x \approx x - \frac{x^3}{3!} + \frac{x^5}{5!}$$

and

$$\cos x \approx 1 - \frac{x^2}{2!} + \frac{x^4}{4!}$$

where x is in radians.

(a) Use a graphing utility to graph the sine function and its polynomial approximation in the same viewing window. How do the graphs compare?

(b) Use the graphing utility to graph the cosine function and its polynomial approximation in the same viewing window. How do the graphs compare?

(c) Study the patterns in the polynomial approximations of the sine and cosine functions and predict the next term in each. Then repeat parts (a) and (b). How did the accuracy of the approximations change when an additional term was added?

101. Polynomial Approximations Use the polynomial approximations of the sine and cosine functions in Exercise 100 to approximate the following function values. Compare the results with those given by a calculator. Is the error in the approximation the same in each case? Explain.

(a) $\sin \dfrac{1}{2}$ (b) $\sin 1$ (c) $\sin \dfrac{\pi}{6}$

(d) $\cos(-0.5)$ (e) $\cos 1$ (f) $\cos \dfrac{\pi}{4}$

102. HOW DO YOU SEE IT? The figure below shows the graph of $y = \sin(x - c)$ for

$$c = -\frac{\pi}{4}, \quad 0, \quad \text{and} \quad \frac{\pi}{4}.$$

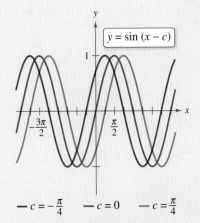

$y = \sin(x - c)$

$-c = -\frac{\pi}{4}$ $-c = 0$ $-c = \frac{\pi}{4}$

(a) How does the value of c affect the graph?

(b) Which graph is equivalent to that of

$$y = -\cos\left(x + \frac{\pi}{4}\right)?$$

Project: Meteorology To work an extended application analyzing the mean monthly temperature and mean monthly precipitation for Honolulu, Hawaii, visit this text's website at *LarsonPrecalculus.com*. (Source: National Climatic Data Center)

1.6 Graphs of Other Trigonometric Functions

You can use graphs of trigonometric functions to model real-life situations such as the distance from a television camera to a unit in a parade, as in Exercise 84 on page 179.

■ Sketch the graphs of tangent functions.
■ Sketch the graphs of cotangent functions.
■ Sketch the graphs of secant and cosecant functions.
■ Sketch the graphs of damped trigonometric functions.

Graph of the Tangent Function

Recall that the tangent function is odd. That is, $\tan(-x) = -\tan x$. Consequently, the graph of $y = \tan x$ is symmetric with respect to the origin. You also know from the identity $\tan x = \sin x / \cos x$ that the tangent is undefined for values at which $\cos x = 0$. Two such values are $x = \pm\pi/2 \approx \pm 1.5708$.

x	$-\dfrac{\pi}{2}$	-1.57	-1.5	$-\dfrac{\pi}{4}$	0	$\dfrac{\pi}{4}$	1.5	1.57	$\dfrac{\pi}{2}$
$\tan x$	Undef.	-1255.8	-14.1	-1	0	1	14.1	1255.8	Undef.

As indicated in the table, $\tan x$ increases without bound as x approaches $\pi/2$ from the left and decreases without bound as x approaches $-\pi/2$ from the right. So, the graph of $y = \tan x$ has *vertical asymptotes* at $x = \pi/2$ and $x = -\pi/2$, as shown below. Moreover, because the period of the tangent function is π, vertical asymptotes also occur at $x = \pi/2 + n\pi$, where n is an integer. The domain of the tangent function is the set of all real numbers other than $x = \pi/2 + n\pi$, and the range is the set of all real numbers.

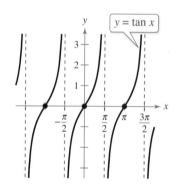

Period: π

Domain: all $x \neq \dfrac{\pi}{2} + n\pi$

Range: $(-\infty, \infty)$

Vertical asymptotes: $x = \dfrac{\pi}{2} + n\pi$

Symmetry: origin

ariadna de raadt/Shutterstock.com

▷ **ALGEBRA HELP**
• You can review odd and even functions in Section P.6.
• You can review symmetry of a graph in Section P.3.
• You can review trigonometric identities in Section 1.3.
• You can review domain and range of a function in Section P.5.
• You can review intercepts of a graph in Section P.3.

Sketching the graph of $y = a\tan(bx - c)$ is similar to sketching the graph of $y = a\sin(bx - c)$ in that you locate key points that identify the intercepts and asymptotes. Two consecutive vertical asymptotes can be found by solving the equations

$$bx - c = -\frac{\pi}{2} \quad \text{and} \quad bx - c = \frac{\pi}{2}.$$

The midpoint between two consecutive vertical asymptotes is an x-intercept of the graph. The period of the function $y = a\tan(bx - c)$ is the distance between two consecutive vertical asymptotes. The amplitude of a tangent function is not defined. After plotting the asymptotes and the x-intercept, plot a few additional points between the two asymptotes and sketch one cycle. Finally, sketch one or two additional cycles to the left and right.

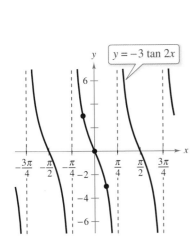

Figure 1.44

EXAMPLE 1 **Sketching the Graph of a Tangent Function**

Sketch the graph of $y = \tan \dfrac{x}{2}$.

Solution

By solving the equations

$$\frac{x}{2} = -\frac{\pi}{2} \quad \text{and} \quad \frac{x}{2} = \frac{\pi}{2}$$

$$x = -\pi \qquad\qquad x = \pi$$

you can see that two consecutive vertical asymptotes occur at $x = -\pi$ and $x = \pi$. Between these two asymptotes, plot a few points, including the x-intercept, as shown in the table. Three cycles of the graph are shown in Figure 1.44.

x	$-\pi$	$-\dfrac{\pi}{2}$	0	$\dfrac{\pi}{2}$	π
$\tan\dfrac{x}{2}$	Undef.	-1	0	1	Undef.

✓ **Checkpoint** *Audio-video solution in English & Spanish at LarsonPrecalculus.com.*

Sketch the graph of $y = \tan \dfrac{x}{4}$.

EXAMPLE 2 **Sketching the Graph of a Tangent Function**

Sketch the graph of $y = -3 \tan 2x$.

Solution

By solving the equations

$$2x = -\frac{\pi}{2} \quad \text{and} \quad 2x = \frac{\pi}{2}$$

$$x = -\frac{\pi}{4} \qquad\qquad x = \frac{\pi}{4}$$

you can see that two consecutive vertical asymptotes occur at $x = -\pi/4$ and $x = \pi/4$. Between these two asymptotes, plot a few points, including the x-intercept, as shown in the table. Three cycles of the graph are shown in Figure 1.45.

x	$-\dfrac{\pi}{4}$	$-\dfrac{\pi}{8}$	0	$\dfrac{\pi}{8}$	$\dfrac{\pi}{4}$
$-3\tan 2x$	Undef.	3	0	-3	Undef.

By comparing the graphs in Examples 1 and 2, you can see that the graph of $y = a\tan(bx - c)$ increases between consecutive vertical asymptotes when $a > 0$ and decreases between consecutive vertical asymptotes when $a < 0$. In other words, the graph for $a < 0$ is a reflection in the x-axis of the graph for $a > 0$.

✓ **Checkpoint** *Audio-video solution in English & Spanish at LarsonPrecalculus.com.*

Sketch the graph of $y = \tan 2x$.

Graph of the Cotangent Function

The graph of the cotangent function is similar to the graph of the tangent function. It also has a period of π. However, from the identity

$$y = \cot x = \frac{\cos x}{\sin x}$$

you can see that the cotangent function has vertical asymptotes when $\sin x$ is zero, which occurs at $x = n\pi$, where n is an integer. The graph of the cotangent function is shown below. Note that two consecutive vertical asymptotes of the graph of $y = a\cot(bx - c)$ can be found by solving the equations $bx - c = 0$ and $bx - c = \pi$.

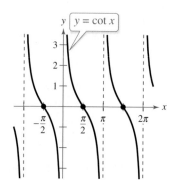

Period: π
Domain: all $x \neq n\pi$
Range: $(-\infty, \infty)$
Vertical asymptotes: $x = n\pi$
Symmetry: origin

EXAMPLE 3   **Sketching the Graph of a Cotangent Function**

Sketch the graph of

$$y = 2\cot\frac{x}{3}.$$

Solution

By solving the equations

$$\frac{x}{3} = 0 \quad \text{and} \quad \frac{x}{3} = \pi$$

$$x = 0 \qquad\qquad x = 3\pi$$

you can see that two consecutive vertical asymptotes occur at $x = 0$ and $x = 3\pi$. Between these two asymptotes, plot a few points, including the x-intercept, as shown in the table. Three cycles of the graph are shown in Figure 1.46. Note that the period is 3π, the distance between consecutive asymptotes.

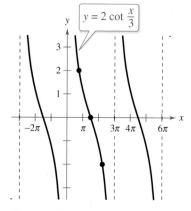

Figure 1.46

x	0	$\dfrac{3\pi}{4}$	$\dfrac{3\pi}{2}$	$\dfrac{9\pi}{4}$	3π
$2\cot\dfrac{x}{3}$	Undef.	2	0	-2	Undef.

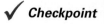 **Checkpoint** ◀))) *Audio-video solution in English & Spanish at LarsonPrecalculus.com.*

Sketch the graph of

$$y = \cot\frac{x}{4}.$$

Graphs of the Reciprocal Functions

You can obtain the graphs of the two remaining trigonometric functions from the graphs of the sine and cosine functions using the reciprocal identities

$$\csc x = \frac{1}{\sin x} \quad \text{and} \quad \sec x = \frac{1}{\cos x}.$$

For instance, at a given value of x, the y-coordinate of sec x is the reciprocal of the y-coordinate of cos x. Of course, when cos $x = 0$, the reciprocal does not exist. Near such values of x, the behavior of the secant function is similar to that of the tangent function. In other words, the graphs of

$$\tan x = \frac{\sin x}{\cos x} \quad \text{and} \quad \sec x = \frac{1}{\cos x}$$

have vertical asymptotes where cos $x = 0$—that is, at $x = \pi/2 + n\pi$, where n is an integer. Similarly,

$$\cot x = \frac{\cos x}{\sin x} \quad \text{and} \quad \csc x = \frac{1}{\sin x}$$

have vertical asymptotes where sin $x = 0$—that is, at $x = n\pi$, where n is an integer.

To sketch the graph of a secant or cosecant function, you should first make a sketch of its reciprocal function. For instance, to sketch the graph of $y = \csc x$, first sketch the graph of $y = \sin x$. Then take reciprocals of the y-coordinates to obtain points on the graph of $y = \csc x$. You can use this procedure to obtain the graphs shown below.

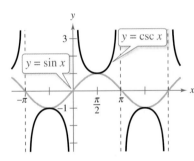

Period: 2π
Domain: all $x \neq n\pi$
Range: $(-\infty, -1] \cup [1, \infty)$
Vertical asymptotes: $x = n\pi$
Symmetry: origin

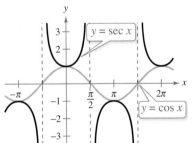

Period: 2π

Domain: all $x \neq \dfrac{\pi}{2} + n\pi$

Range: $(-\infty, -1] \cup [1, \infty)$

Vertical asymptotes: $x = \dfrac{\pi}{2} + n\pi$

Symmetry: y-axis

In comparing the graphs of the cosecant and secant functions with those of the sine and cosine functions, respectively, note that the "hills" and "valleys" are interchanged. For instance, a hill (or maximum point) on the sine curve corresponds to a valley (a relative minimum) on the cosecant curve, and a valley (or minimum point) on the sine curve corresponds to a hill (a relative maximum) on the cosecant curve, as shown in Figure 1.47. Additionally, x-intercepts of the sine and cosine functions become vertical asymptotes of the cosecant and secant functions, respectively (see Figure 1.47).

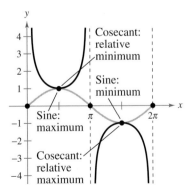

Figure 1.47

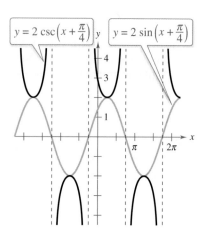

Figure 1.48

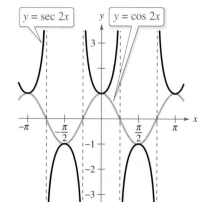

Figure 1.49

EXAMPLE 4   **Sketching the Graph of a Cosecant Function**

Sketch the graph of $y = 2 \csc\left(x + \dfrac{\pi}{4}\right)$.

Solution

Begin by sketching the graph of

$$y = 2 \sin\left(x + \frac{\pi}{4}\right).$$

For this function, the amplitude is 2 and the period is 2π. By solving the equations

$$x + \frac{\pi}{4} = 0 \quad \text{and} \quad x + \frac{\pi}{4} = 2\pi$$

$$x = -\frac{\pi}{4} \qquad\qquad x = \frac{7\pi}{4}$$

you can see that one cycle of the sine function corresponds to the interval from $x = -\pi/4$ to $x = 7\pi/4$. The graph of this sine function is represented by the gray curve in Figure 1.48. Because the sine function is zero at the midpoint and endpoints of this interval, the corresponding cosecant function

$$y = 2 \csc\left(x + \frac{\pi}{4}\right)$$

$$= 2\left(\frac{1}{\sin[x + (\pi/4)]}\right)$$

has vertical asymptotes at $x = -\pi/4$, $x = 3\pi/4$, $x = 7\pi/4$, and so on. The graph of the cosecant function is represented by the black curve in Figure 1.48.

✓ *Checkpoint*  Audio-video solution in English & Spanish at LarsonPrecalculus.com.

Sketch the graph of $y = 2 \csc\left(x + \dfrac{\pi}{2}\right)$.

EXAMPLE 5   **Sketching the Graph of a Secant Function**

Sketch the graph of $y = \sec 2x$.

Solution

Begin by sketching the graph of $y = \cos 2x$, as indicated by the gray curve in Figure 1.49. Then, form the graph of $y = \sec 2x$ as the black curve in the figure. Note that the x-intercepts of $y = \cos 2x$

$$\left(-\frac{\pi}{4}, 0\right), \quad \left(\frac{\pi}{4}, 0\right), \quad \left(\frac{3\pi}{4}, 0\right), \ldots$$

correspond to the vertical asymptotes

$$x = -\frac{\pi}{4}, \quad x = \frac{\pi}{4}, \quad x = \frac{3\pi}{4}, \ldots$$

of the graph of $y = \sec 2x$. Moreover, notice that the period of $y = \cos 2x$ and $y = \sec 2x$ is π.

✓ *Checkpoint*  Audio-video solution in English & Spanish at LarsonPrecalculus.com.

Sketch the graph of $y = \sec \dfrac{x}{2}$.

Damped Trigonometric Graphs

You can graph a *product* of two functions using properties of the individual functions. For instance, consider the function

$$f(x) = x \sin x$$

as the product of the functions $y = x$ and $y = \sin x$. Using properties of absolute value and the fact that $|\sin x| \leq 1$, you have

$$0 \leq |x||\sin x| \leq |x|.$$

Consequently,

$$-|x| \leq x \sin x \leq |x|$$

which means that the graph of $f(x) = x \sin x$ lies between the lines $y = -x$ and $y = x$. Furthermore, because

$$f(x) = x \sin x = \pm x \quad \text{at} \quad x = \frac{\pi}{2} + n\pi$$

and

$$f(x) = x \sin x = 0 \quad \text{at} \quad x = n\pi$$

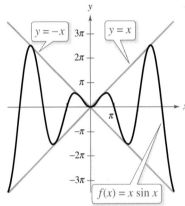

where n is an integer, the graph of f touches the line $y = -x$ or the line $y = x$ at $x = \pi/2 + n\pi$ and has x-intercepts at $x = n\pi$. A sketch of f is shown at the right. In the function $f(x) = x \sin x$, the factor x is called the **damping factor**.

• • • • • • • • • • • • • • • ▷
:
• • REMARK Do you see why the graph of $f(x) = x \sin x$ touches the lines $y = \pm x$ at $x = \pi/2 + n\pi$ and why the graph has x-intercepts at $x = n\pi$? Recall that the sine function is equal to 1 at $\ldots, -3\pi/2, \pi/2, 5\pi/2, \ldots$ $(x = \pi/2 + 2n\pi)$ and -1 at $\ldots, -\pi/2, 3\pi/2, 7\pi/2, \ldots$ $(x = -\pi/2 + 2n\pi)$ and is equal to 0 at $\ldots, -\pi, 0, \pi, 2\pi, 3\pi, \ldots (x = n\pi)$.

EXAMPLE 6   Damped Sine Wave

Sketch the graph of $f(x) = x^2 \sin 3x$.

Solution

Consider $f(x)$ as the product of the two functions

$$y = x^2 \quad \text{and} \quad y = \sin 3x$$

each of which has the set of real numbers as its domain. For any real number x, you know that $x^2 \geq 0$ and $|\sin 3x| \leq 1$. So,

$$x^2|\sin 3x| \leq x^2$$

which means that

$$-x^2 \leq x^2 \sin 3x \leq x^2.$$

Furthermore, because

$$f(x) = x^2 \sin 3x = \pm x^2 \quad \text{at} \quad x = \frac{\pi}{6} + \frac{n\pi}{3}$$

and

$$f(x) = x^2 \sin 3x = 0 \quad \text{at} \quad x = \frac{n\pi}{3}$$

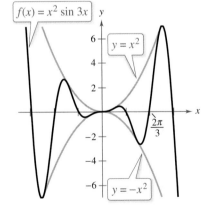

Figure 1.50

the graph of f touches the curve $y = -x^2$ or the curve $y = x^2$ at $x = \pi/6 + n\pi/3$ and has intercepts at $x = n\pi/3$. A sketch of f is shown in Figure 1.50.

✓ **Checkpoint**))) *Audio-video solution in English & Spanish at LarsonPrecalculus.com.*

Sketch the graph of $f(x) = x^2 \sin 4x$.

∎

Below is a summary of the characteristics of the six basic trigonometric functions.

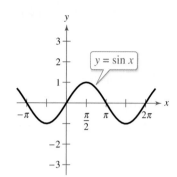

Domain: $(-\infty, \infty)$
Range: $[-1, 1]$
Period: 2π

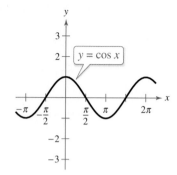

Domain: $(-\infty, \infty)$
Range: $[-1, 1]$
Period: 2π

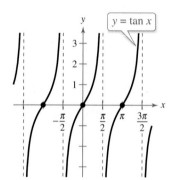

Domain: all $x \neq \dfrac{\pi}{2} + n\pi$
Range: $(-\infty, \infty)$
Period: π

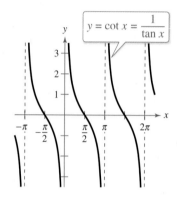

Domain: all $x \neq n\pi$
Range: $(-\infty, \infty)$
Period: π

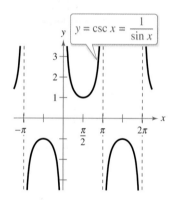

Domain: all $x \neq n\pi$
Range:
$(-\infty, -1] \cup [1, \infty)$
Period: 2π

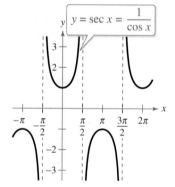

Domain: all $x \neq \dfrac{\pi}{2} + n\pi$
Range:
$(-\infty, -1] \cup [1, \infty)$
Period: 2π

Summarize (Section 1.6)

1. Describe how to sketch the graph of $y = a \tan(bx - c)$ *(page 170)*. For examples of sketching the graphs of tangent functions, see Examples 1 and 2.

2. Describe how to sketch the graph of $y = a \cot(bx - c)$ *(page 172)*. For an example of sketching the graph of a cotangent function, see Example 3.

3. Describe how to sketch the graphs of $y = a \csc(bx - c)$ and $y = a \sec(bx - c)$ *(page 173)*. For examples of sketching the graphs of cosecant and secant functions, see Examples 4 and 5.

4. Describe how to sketch the graph of a damped trigonometric function *(page 175)*. For an example of sketching the graph of a damped trigonometric function, see Example 6.

1.6 Exercises

See **CalcChat.com** for tutorial help and worked-out solutions to odd-numbered exercises.

Vocabulary: Fill in the blanks.

1. The tangent, cotangent, and cosecant functions are _____ , so the graphs of these functions have symmetry with respect to the _____ .

2. The graphs of the tangent, cotangent, secant, and cosecant functions have _____ asymptotes.

3. To sketch the graph of a secant or cosecant function, first make a sketch of its _____ function.

4. For the function $f(x) = g(x) \cdot \sin x$, $g(x)$ is called the _____ factor of the function $f(x)$.

5. The period of $y = \tan x$ is _____ .

6. The domain of $y = \cot x$ is all real numbers such that _____ .

7. The range of $y = \sec x$ is _____ .

8. The period of $y = \csc x$ is _____ .

Skills and Applications

Matching In Exercises 9–14, match the function with its graph. State the period of the function. [The graphs are labeled (a), (b), (c), (d), (e), and (f).]

(a)
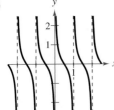

(b)
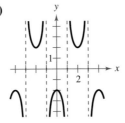

(c)
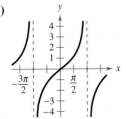

(d)
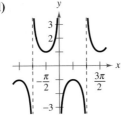

(e)
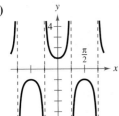

(f)
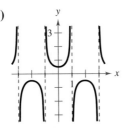

9. $y = \sec 2x$

10. $y = \tan \dfrac{x}{2}$

11. $y = \dfrac{1}{2} \cot \pi x$

12. $y = -\csc x$

13. $y = \dfrac{1}{2} \sec \dfrac{\pi x}{2}$

14. $y = -2 \sec \dfrac{\pi x}{2}$

Sketching the Graph of a Trigonometric Function In Exercises 15–38, sketch the graph of the function. (Include two full periods.)

15. $y = \frac{1}{3} \tan x$

16. $y = \tan 4x$

17. $y = -2 \tan 3x$

18. $y = -3 \tan \pi x$

19. $y = -\frac{1}{2} \sec x$

20. $y = \frac{1}{4} \sec x$

21. $y = \csc \pi x$

22. $y = 3 \csc 4x$

23. $y = \frac{1}{2} \sec \pi x$

24. $y = -2 \sec 4x + 2$

25. $y = \csc \dfrac{x}{2}$

26. $y = \csc \dfrac{x}{3}$

27. $y = 3 \cot 2x$

28. $y = 3 \cot \dfrac{\pi x}{2}$

29. $y = 2 \sec 3x$

30. $y = -\frac{1}{2} \tan x$

31. $y = \tan \dfrac{\pi x}{4}$

32. $y = \tan(x + \pi)$

33. $y = 2 \csc(x - \pi)$

34. $y = \csc(2x - \pi)$

35. $y = 2 \sec(x + \pi)$

36. $y = -\sec \pi x + 1$

37. $y = \dfrac{1}{4} \csc\left(x + \dfrac{\pi}{4}\right)$

38. $y = 2 \cot\left(x + \dfrac{\pi}{2}\right)$

Graphing a Trigonometric Function In Exercises 39–48, use a graphing utility to graph the function. (Include two full periods.)

39. $y = \tan \dfrac{x}{3}$

40. $y = -\tan 2x$

41. $y = -2 \sec 4x$

42. $y = \sec \pi x$

43. $y = \tan\left(x - \dfrac{\pi}{4}\right)$

44. $y = \dfrac{1}{4} \cot\left(x - \dfrac{\pi}{2}\right)$

45. $y = -\csc(4x - \pi)$

46. $y = 2 \sec(2x - \pi)$

47. $y = 0.1 \tan\left(\dfrac{\pi x}{4} + \dfrac{\pi}{4}\right)$

48. $y = \dfrac{1}{3} \sec\left(\dfrac{\pi x}{2} + \dfrac{\pi}{2}\right)$

Solving a Trigonometric Equation Graphically
In Exercises 49–56, use a graph to solve the equation on the interval $[-2\pi, 2\pi]$.

49. $\tan x = 1$

50. $\tan x = \sqrt{3}$

51. $\cot x = -\dfrac{\sqrt{3}}{3}$

52. $\cot x = 1$

53. $\sec x = -2$

54. $\sec x = 2$

55. $\csc x = \sqrt{2}$

56. $\csc x = -\dfrac{2\sqrt{3}}{3}$

Even and Odd Trigonometric Functions　In Exercises 57–64, use the graph of the function to determine whether the function is even, odd, or neither. Verify your answer algebraically.

57. $f(x) = \sec x$

58. $f(x) = \tan x$

59. $g(x) = \cot x$

60. $g(x) = \csc x$

61. $f(x) = x + \tan x$

62. $f(x) = x^2 - \sec x$

63. $g(x) = x \csc x$

64. $g(x) = x^2 \cot x$

Identifying Damped Trigonometric Functions　In Exercises 65–68, match the function with its graph. Describe the behavior of the function as x approaches zero. [The graphs are labeled (a), (b), (c), and (d).]

(a)
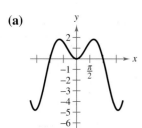

(b)
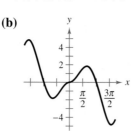

(c)
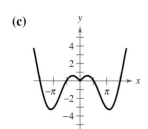

(d)
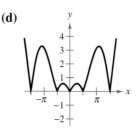

65. $f(x) = |x \cos x|$

66. $f(x) = x \sin x$

67. $g(x) = |x| \sin x$

68. $g(x) = |x| \cos x$

Conjecture　In Exercises 69–72, graph the functions f and g. Use the graphs to make a conjecture about the relationship between the functions.

69. $f(x) = \sin x + \cos\left(x + \dfrac{\pi}{2}\right)$, $\quad g(x) = 0$

70. $f(x) = \sin x - \cos\left(x + \dfrac{\pi}{2}\right)$, $\quad g(x) = 2 \sin x$

71. $f(x) = \sin^2 x$, $\quad g(x) = \frac{1}{2}(1 - \cos 2x)$

72. $f(x) = \cos^2 \dfrac{\pi x}{2}$, $\quad g(x) = \dfrac{1}{2}(1 + \cos \pi x)$

Analyzing a Damped Trigonometric Graph　In Exercises 73–76, use a graphing utility to graph the function and the damping factor of the function in the same viewing window. Describe the behavior of the function as x increases without bound.

73. $g(x) = x \cos \pi x$

74. $f(x) = x^2 \cos x$

75. $f(x) = x^3 \sin x$

76. $h(x) = x^3 \cos x$

Analyzing a Trigonometric Graph　In Exercises 77–82, use a graphing utility to graph the function. Describe the behavior of the function as x approaches zero.

77. $y = \dfrac{6}{x} + \cos x$, $\quad x > 0$

78. $y = \dfrac{4}{x} + \sin 2x$, $\quad x > 0$

79. $g(x) = \dfrac{\sin x}{x}$

80. $f(x) = \dfrac{1 - \cos x}{x}$

81. $f(x) = \sin \dfrac{1}{x}$

82. $h(x) = x \sin \dfrac{1}{x}$

83. Meteorology　The normal monthly high temperatures H (in degrees Fahrenheit) in Erie, Pennsylvania, are approximated by

$$H(t) = 56.94 - 20.86 \cos\left(\frac{\pi t}{6}\right) - 11.58 \sin\left(\frac{\pi t}{6}\right)$$

and the normal monthly low temperatures L are approximated by

$$L(t) = 41.80 - 17.13 \cos\left(\frac{\pi t}{6}\right) - 13.39 \sin\left(\frac{\pi t}{6}\right)$$

where t is the time (in months), with $t = 1$ corresponding to January (see figure). (*Source: National Climatic Data Center*)

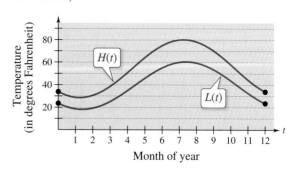

(a) What is the period of each function?

(b) During what part of the year is the difference between the normal high and normal low temperatures greatest? When is it smallest?

(c) The sun is northernmost in the sky around June 21, but the graph shows the warmest temperatures at a later date. Approximate the lag time of the temperatures relative to the position of the sun.

84. Television Coverage

A television camera is on a reviewing platform 27 meters from the street on which a parade will be passing from left to right (see figure). Write the distance d from the camera to a particular unit in the parade as a function of the angle x, and graph the function over the interval $-\pi/2 < x < \pi/2$. (Consider x as negative when a unit in the parade approaches from the left.)

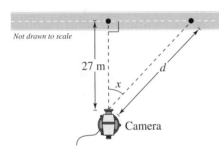

Not drawn to scale

27 m

x

Camera

85. Distance A plane flying at an altitude of 7 miles above a radar antenna will pass directly over the radar antenna (see figure). Let d be the ground distance from the antenna to the point directly under the plane and let x be the angle of elevation to the plane from the antenna. (d is positive as the plane approaches the antenna.) Write d as a function of x and graph the function over the interval $0 < x < \pi$.

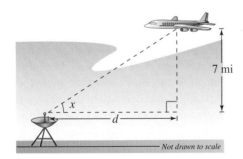

7 mi

x

d

Not drawn to scale

Exploration

True or False? In Exercises 86 and 87, determine whether the statement is true or false. Justify your answer.

86. You can obtain the graph of $y = \csc x$ on a calculator by graphing the reciprocal of $y = \sin x$.

87. You can obtain the graph of $y = \sec x$ on a calculator by graphing a translation of the reciprocal of $y = \sin x$.

Graphical Analysis In Exercises 88 and 89, use a graphing utility to graph the function. Use the graph to determine the behavior of the function as $x \to c$.

(a) As $x \to 0^+$, the value of $f(x) \to$ ▨.

(b) As $x \to 0^-$, the value of $f(x) \to$ ▨.

(c) As $x \to \pi^+$, the value of $f(x) \to$ ▨.

(d) As $x \to \pi^-$, the value of $f(x) \to$ ▨.

88. $f(x) = \cot x$

89. $f(x) = \csc x$

Graphical Analysis In Exercises 90 and 91, use a graphing utility to graph the function. Use the graph to determine the behavior of the function as $x \to c$.

(a) $x \to \left(\dfrac{\pi}{2}\right)^+$

(b) $x \to \left(\dfrac{\pi}{2}\right)^-$

(c) $x \to \left(-\dfrac{\pi}{2}\right)^+$

(d) $x \to \left(-\dfrac{\pi}{2}\right)^-$

90. $f(x) = \tan x$

91. $f(x) = \sec x$

92. **HOW DO YOU SEE IT?** Determine which function is represented by the graph. Do not use a calculator. Explain your reasoning.

(a)

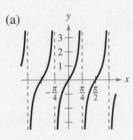

(b)
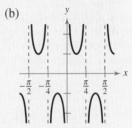

(i) $f(x) = \tan 2x$

(ii) $f(x) = \tan(x/2)$

(iii) $f(x) = 2 \tan x$

(iv) $f(x) = -\tan 2x$

(v) $f(x) = -\tan(x/2)$

(i) $f(x) = \sec 4x$

(ii) $f(x) = \csc 4x$

(iii) $f(x) = \csc(x/4)$

(iv) $f(x) = \sec(x/4)$

(v) $f(x) = \csc(4x - \pi)$

93. Think About It Consider the function $f(x) = x - \cos x$.

(a) Use a graphing utility to graph the function and verify that there exists a zero between 0 and 1. Use the graph to approximate the zero.

(b) Starting with $x_0 = 1$, generate a sequence $x_1, x_2, x_3, \ldots$, where $x_n = \cos(x_{n-1})$. For example,

$x_0 = 1$

$x_1 = \cos(x_0)$

$x_2 = \cos(x_1)$

$x_3 = \cos(x_2)$

$\vdots$

What value does the sequence approach?

1.7 Inverse Trigonometric Functions

You can use inverse trigonometric functions to model and solve real-life problems. For instance, in Exercise 104 on page 188, you will use an inverse trigonometric function to model the angle of elevation from a television camera to a space shuttle launch.

- Evaluate and graph the inverse sine function.
- Evaluate and graph the other inverse trigonometric functions.
- Evaluate the compositions of trigonometric functions.

Inverse Sine Function

Recall from Section P.10 that for a function to have an inverse function, it must be one-to-one—that is, it must pass the Horizontal Line Test. From Figure 1.51, you can see that $y = \sin x$ does not pass the test because different values of x yield the same y-value.

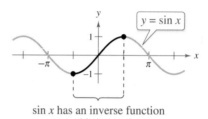

$\sin x$ has an inverse function on this interval.

Figure 1.51

However, when you restrict the domain to the interval $-\pi/2 \le x \le \pi/2$ (corresponding to the black portion of the graph in Figure 1.51), the following properties hold.

1. On the interval $[-\pi/2, \pi/2]$, the function $y = \sin x$ is increasing.
2. On the interval $[-\pi/2, \pi/2]$, $y = \sin x$ takes on its full range of values, $-1 \le \sin x \le 1$.
3. On the interval $[-\pi/2, \pi/2]$, $y = \sin x$ is one-to-one.

So, on the restricted domain $-\pi/2 \le x \le \pi/2$, $y = \sin x$ has a unique inverse function called the **inverse sine function.** It is denoted by

$$y = \arcsin x \quad \text{or} \quad y = \sin^{-1} x.$$

The notation $\sin^{-1} x$ is consistent with the inverse function notation $f^{-1}(x)$. The $\arcsin x$ notation (read as "the arcsine of x") comes from the association of a central angle with its intercepted *arc length* on a unit circle. So, $\arcsin x$ means the angle (or arc) whose sine is x. Both notations, $\arcsin x$ and $\sin^{-1} x$, are commonly used in mathematics, so remember that $\sin^{-1} x$ denotes the *inverse* sine function rather than $1/\sin x$. The values of $\arcsin x$ lie in the interval

$$-\frac{\pi}{2} \le \arcsin x \le \frac{\pi}{2}.$$

The graph of $y = \arcsin x$ is shown in Example 2.

· · REMARK When evaluating the inverse sine function, it helps to remember the phrase "the arcsine of x is the angle (or number) whose sine is x."

Definition of Inverse Sine Function

The **inverse sine function** is defined by

$$y = \arcsin x \quad \text{if and only if} \quad \sin y = x$$

where $-1 \le x \le 1$ and $-\pi/2 \le y \le \pi/2$. The domain of $y = \arcsin x$ is $[-1, 1]$, and the range is $[-\pi/2, \pi/2]$.

▷

REMARK As with the trigonometric functions, much of the work with the inverse trigonometric functions can be done by *exact* calculations rather than by calculator approximations. Exact calculations help to increase your understanding of the inverse functions by relating them to the right triangle definitions of the trigonometric functions.

EXAMPLE 1   **Evaluating the Inverse Sine Function**

If possible, find the exact value.

a. $\arcsin\left(-\dfrac{1}{2}\right)$ **b.** $\sin^{-1}\dfrac{\sqrt{3}}{2}$ **c.** $\sin^{-1}2$

Solution

a. Because $\sin\left(-\dfrac{\pi}{6}\right) = -\dfrac{1}{2}$ and $-\dfrac{\pi}{6}$ lies in $\left[-\dfrac{\pi}{2}, \dfrac{\pi}{2}\right]$, it follows that

$$\arcsin\left(-\dfrac{1}{2}\right) = -\dfrac{\pi}{6}. \qquad \text{Angle whose sine is } -\tfrac{1}{2}$$

b. Because $\sin\dfrac{\pi}{3} = \dfrac{\sqrt{3}}{2}$ and $\dfrac{\pi}{3}$ lies in $\left[-\dfrac{\pi}{2}, \dfrac{\pi}{2}\right]$, it follows that

$$\sin^{-1}\dfrac{\sqrt{3}}{2} = \dfrac{\pi}{3}. \qquad \text{Angle whose sine is } \sqrt{3}/2$$

c. It is not possible to evaluate $y = \sin^{-1}x$ when $x = 2$ because there is no angle whose sine is 2. Remember that the domain of the inverse sine function is $[-1, 1]$.

 Checkpoint  Audio-video solution in English & Spanish at LarsonPrecalculus.com.

If possible, find the exact value.

a. $\arcsin 1$ **b.** $\sin^{-1}(-2)$

EXAMPLE 2   **Graphing the Arcsine Function**

Sketch a graph of

$$y = \arcsin x.$$

Solution

By definition, the equations $y = \arcsin x$ and $\sin y = x$ are equivalent for $-\pi/2 \le y \le \pi/2$. So, their graphs are the same. From the interval $[-\pi/2, \pi/2]$, you can assign values to y in the equation $\sin y = x$ to make a table of values. Then plot the points and connect them with a smooth curve.

y	$-\dfrac{\pi}{2}$	$-\dfrac{\pi}{4}$	$-\dfrac{\pi}{6}$	0	$\dfrac{\pi}{6}$	$\dfrac{\pi}{4}$	$\dfrac{\pi}{2}$
$x = \sin y$	-1	$-\dfrac{\sqrt{2}}{2}$	$-\dfrac{1}{2}$	0	$\dfrac{1}{2}$	$\dfrac{\sqrt{2}}{2}$	1

The resulting graph of $y = \arcsin x$ is shown in Figure 1.52. Note that it is the reflection (in the line $y = x$) of the black portion of the graph in Figure 1.51. Be sure you see that Figure 1.52 shows the *entire* graph of the inverse sine function. Remember that the domain of $y = \arcsin x$ is the closed interval $[-1, 1]$ and the range is the closed interval $[-\pi/2, \pi/2]$.

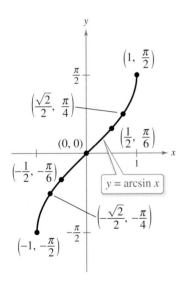

Figure 1.52

 Checkpoint  Audio-video solution in English & Spanish at LarsonPrecalculus.com.

Use a graphing utility to graph $f(x) = \sin x$, $g(x) = \arcsin x$, and $y = x$ in the same viewing window to verify geometrically that g is the inverse function of f. (Be sure to restrict the domain of f properly.)

Other Inverse Trigonometric Functions

The cosine function is decreasing and one-to-one on the interval $0 \leq x \leq \pi$, as shown below.

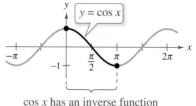

cos x has an inverse function
on this interval.

Consequently, on this interval the cosine function has an inverse function—the **inverse cosine function**—denoted by

$$y = \arccos x \quad \text{or} \quad y = \cos^{-1} x.$$

Similarly, you can define an **inverse tangent function** by restricting the domain of $y = \tan x$ to the interval $(-\pi/2, \pi/2)$. The following list summarizes the definitions of the three most common inverse trigonometric functions. The remaining three are defined in Exercises 115–117.

Definitions of the Inverse Trigonometric Functions

Function	Domain	Range
$y = \arcsin x$ if and only if $\sin y = x$	$-1 \leq x \leq 1$	$-\dfrac{\pi}{2} \leq y \leq \dfrac{\pi}{2}$
$y = \arccos x$ if and only if $\cos y = x$	$-1 \leq x \leq 1$	$0 \leq y \leq \pi$
$y = \arctan x$ if and only if $\tan y = x$	$-\infty < x < \infty$	$-\dfrac{\pi}{2} < y < \dfrac{\pi}{2}$

The graphs of these three inverse trigonometric functions are shown below.

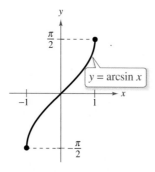

Domain: $[-1, 1]$

Range: $\left[-\dfrac{\pi}{2}, \dfrac{\pi}{2}\right]$

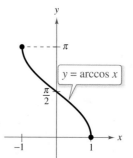

Domain: $[-1, 1]$

Range: $[0, \pi]$

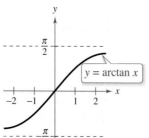

Domain: $(-\infty, \infty)$

Range: $\left(-\dfrac{\pi}{2}, \dfrac{\pi}{2}\right)$

EXAMPLE 3 **Evaluating Inverse Trigonometric Functions**

Find the exact value.

a. $\arccos \dfrac{\sqrt{2}}{2}$

b. $\arctan 0$

c. $\tan^{-1}(-1)$

Solution

a. Because $\cos(\pi/4) = \sqrt{2}/2$ and $\pi/4$ lies in $[0, \pi]$, it follows that

$$\arccos \dfrac{\sqrt{2}}{2} = \dfrac{\pi}{4}. \qquad \text{Angle whose cosine is } \sqrt{2}/2$$

b. Because $\tan 0 = 0$ and 0 lies in $(-\pi/2, \pi/2)$, it follows that

$$\arctan 0 = 0. \qquad \text{Angle whose tangent is } 0$$

c. Because $\tan(-\pi/4) = -1$ and $-\pi/4$ lies in $(-\pi/2, \pi/2)$, it follows that

$$\tan^{-1}(-1) = -\dfrac{\pi}{4}. \qquad \text{Angle whose tangent is } -1$$

✔ *Checkpoint* *Audio-video solution in English & Spanish at LarsonPrecalculus.com.*

Find the exact value of $\cos^{-1}(-1)$.

EXAMPLE 4 **Calculators and Inverse Trigonometric Functions**

Use a calculator to approximate the value, if possible.

a. $\arctan(-8.45)$

b. $\sin^{-1} 0.2447$

c. $\arccos 2$

Solution

Function	Mode	Calculator Keystrokes
a. $\arctan(-8.45)$	Radian	TAN⁻¹ ((−) 8.45) ENTER

From the display, it follows that $\arctan(-8.45) \approx -1.453001$.

b. $\sin^{-1} 0.2447$	Radian	SIN⁻¹ (0.2447) ENTER

From the display, it follows that $\sin^{-1} 0.2447 \approx 0.2472103$.

▷ **c.** $\arccos 2$ Radian   COS⁻¹ (2) ENTER

· ·REMARK Remember that the domain of the inverse sine function and the inverse cosine function is $[-1, 1]$, as indicated in Example 4(c).

In *radian* mode, the calculator should display an *error message* because the domain of the inverse cosine function is $[-1, 1]$.

✔ *Checkpoint* *Audio-video solution in English & Spanish at LarsonPrecalculus.com.*

Use a calculator to approximate the value, if possible.

a. $\arctan 4.84$

b. $\arcsin(-1.1)$

c. $\arccos(-0.349)$

In Example 4, had you set the calculator to *degree* mode, the displays would have been in degrees rather than in radians. This convention is peculiar to calculators. By definition, the values of inverse trigonometric functions are *always in radians*.

Compositions of Functions

▷ **ALGEBRA HELP** You can review the composition of functions in Section P.9.

Recall from Section P.10 that for all x in the domains of f and f^{-1}, inverse functions have the properties

$$f(f^{-1}(x)) = x \quad \text{and} \quad f^{-1}(f(x)) = x.$$

Inverse Properties of Trigonometric Functions

If $-1 \le x \le 1$ and $-\pi/2 \le y \le \pi/2$, then

$$\sin(\arcsin x) = x \quad \text{and} \quad \arcsin(\sin y) = y.$$

If $-1 \le x \le 1$ and $0 \le y \le \pi$, then

$$\cos(\arccos x) = x \quad \text{and} \quad \arccos(\cos y) = y.$$

If x is a real number and $-\pi/2 < y < \pi/2$, then

$$\tan(\arctan x) = x \quad \text{and} \quad \arctan(\tan y) = y.$$

Keep in mind that these inverse properties do not apply for arbitrary values of x and y. For instance,

$$\arcsin\left(\sin \frac{3\pi}{2}\right) = \arcsin(-1) = -\frac{\pi}{2} \ne \frac{3\pi}{2}.$$

In other words, the property $\arcsin(\sin y) = y$ is not valid for values of y outside the interval $[-\pi/2, \pi/2]$.

EXAMPLE 5 Using Inverse Properties

If possible, find the exact value.

a. $\tan[\arctan(-5)]$　　**b.** $\arcsin\left(\sin \dfrac{5\pi}{3}\right)$　　**c.** $\cos(\cos^{-1} \pi)$

Solution

a. Because -5 lies in the domain of the arctangent function, the inverse property applies, and you have

$$\tan[\arctan(-5)] = -5.$$

b. In this case, $5\pi/3$ does not lie in the range of the arcsine function, $-\pi/2 \le y \le \pi/2$. However, $5\pi/3$ is coterminal with

$$\frac{5\pi}{3} - 2\pi = -\frac{\pi}{3}$$

which does lie in the range of the arcsine function, and you have

$$\arcsin\left(\sin \frac{5\pi}{3}\right) = \arcsin\left[\sin\left(-\frac{\pi}{3}\right)\right] = -\frac{\pi}{3}.$$

c. The expression $\cos(\cos^{-1} \pi)$ is not defined because $\cos^{-1} \pi$ is not defined. Remember that the domain of the inverse cosine function is $[-1, 1]$.

✓ **Checkpoint** ◀))) *Audio-video solution in English & Spanish at LarsonPrecalculus.com.*

If possible, find the exact value.

a. $\tan[\tan^{-1}(-14)]$　　**b.** $\sin^{-1}\left(\sin \dfrac{7\pi}{4}\right)$　　**c.** $\cos(\arccos 0.54)$ ∎

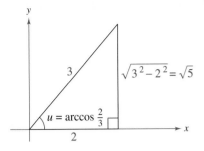

Angle whose cosine is $\frac{2}{3}$
Figure 1.53

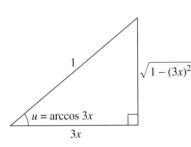

Angle whose sine is $-\frac{3}{5}$
Figure 1.54

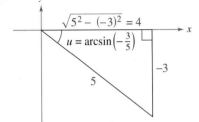

Angle whose cosine is $3x$
Figure 1.55

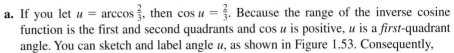

EXAMPLE 6 **Evaluating Compositions of Functions**

Find the exact value.

a. $\tan\left(\arccos \frac{2}{3}\right)$ **b.** $\cos\left[\arcsin\left(-\frac{3}{5}\right)\right]$

Solution

a. If you let $u = \arccos \frac{2}{3}$, then $\cos u = \frac{2}{3}$. Because the range of the inverse cosine function is the first and second quadrants and $\cos u$ is positive, u is a *first*-quadrant angle. You can sketch and label angle u, as shown in Figure 1.53. Consequently,

$$\tan\left(\arccos \frac{2}{3}\right) = \tan u = \frac{\text{opp}}{\text{adj}} = \frac{\sqrt{5}}{2}.$$

b. If you let $u = \arcsin\left(-\frac{3}{5}\right)$, then $\sin u = -\frac{3}{5}$. Because the range of the inverse sine function is the first and fourth quadrants and $\sin u$ is negative, u is a *fourth*-quadrant angle. You can sketch and label angle u, as shown in Figure 1.54. Consequently,

$$\cos\left[\arcsin\left(-\frac{3}{5}\right)\right] = \cos u = \frac{\text{adj}}{\text{hyp}} = \frac{4}{5}.$$

✓ **Checkpoint** *Audio-video solution in English & Spanish at LarsonPrecalculus.com.*

Find the exact value of $\cos\left[\arctan\left(-\frac{3}{4}\right)\right]$.

EXAMPLE 7 **Some Problems from Calculus** ∫

Write each of the following as an algebraic expression in x.

a. $\sin(\arccos 3x)$, $0 \le x \le \frac{1}{3}$ **b.** $\cot(\arccos 3x)$, $0 \le x < \frac{1}{3}$

Solution

If you let $u = \arccos 3x$, then $\cos u = 3x$, where $-1 \le 3x \le 1$. Because

$$\cos u = \frac{\text{adj}}{\text{hyp}} = \frac{3x}{1}$$

you can sketch a right triangle with acute angle u, as shown in Figure 1.55. From this triangle, you can easily convert each expression to algebraic form.

a. $\sin(\arccos 3x) = \sin u = \dfrac{\text{opp}}{\text{hyp}} = \sqrt{1 - 9x^2}$, $0 \le x \le \dfrac{1}{3}$

b. $\cot(\arccos 3x) = \cot u = \dfrac{\text{adj}}{\text{opp}} = \dfrac{3x}{\sqrt{1 - 9x^2}}$, $0 \le x < \dfrac{1}{3}$

✓ **Checkpoint** *Audio-video solution in English & Spanish at LarsonPrecalculus.com.*

Write $\sec(\arctan x)$ as an algebraic expression in x. ◼

Summarize   (Section 1.7)

1. State the definition of the inverse sine function (*page 180*). For examples of evaluating and graphing the inverse sine function, see Examples 1 and 2.

2. State the definitions of the inverse cosine and inverse tangent functions (*page 182*). For examples of evaluating and graphing inverse trigonometric functions, see Examples 3 and 4.

3. State the inverse properties of trigonometric functions (*page 184*). For examples involving the compositions of trigonometric functions, see Examples 5–7.

1.7   Exercises

See CalcChat.com for tutorial help and worked-out solutions to odd-numbered exercises.

Vocabulary: Fill in the blanks.

Function	Alternative Notation	Domain	Range
1. $y = \arcsin x$	_____	_____	$-\dfrac{\pi}{2} \le y \le \dfrac{\pi}{2}$
2. _____	$y = \cos^{-1} x$	$-1 \le x \le 1$	_____
3. $y = \arctan x$	_____	_____	_____

4. Without restrictions, no trigonometric function has an _____ function.

Skills and Applications

Evaluating an Inverse Trigonometric Function
In Exercises 5–18, evaluate the expression without using a calculator.

5. $\arcsin \frac{1}{2}$ **6.** $\arcsin 0$

7. $\arccos \frac{1}{2}$ **8.** $\arccos 0$

9. $\arctan \dfrac{\sqrt{3}}{3}$ **10.** $\arctan 1$

11. $\cos^{-1}\left(-\dfrac{\sqrt{3}}{2}\right)$ **12.** $\sin^{-1}\left(-\dfrac{\sqrt{2}}{2}\right)$

13. $\arctan\left(-\sqrt{3}\right)$ **14.** $\arctan \sqrt{3}$

15. $\arccos\left(-\dfrac{1}{2}\right)$ **16.** $\arcsin \dfrac{\sqrt{2}}{2}$

17. $\sin^{-1}\left(-\dfrac{\sqrt{3}}{2}\right)$ **18.** $\tan^{-1}\left(-\dfrac{\sqrt{3}}{3}\right)$

Graphing an Inverse Trigonometric Function   In Exercises 19 and 20, use a graphing utility to graph f, g, and $y = x$ in the same viewing window to verify geometrically that g is the inverse function of f. (Be sure to restrict the domain of f properly.)

19. $f(x) = \cos x$, $g(x) = \arccos x$
20. $f(x) = \tan x$, $g(x) = \arctan x$

Calculators and Inverse Trigonometric Functions
In Exercises 21–38, use a calculator to evaluate the expression. Round your result to two decimal places.

21. $\arccos 0.37$ **22.** $\arcsin 0.65$

23. $\arcsin(-0.75)$ **24.** $\arccos(-0.7)$

25. $\arctan(-3)$ **26.** $\arctan 25$

27. $\sin^{-1} 0.31$ **28.** $\cos^{-1} 0.26$

29. $\arccos(-0.41)$ **30.** $\arcsin(-0.125)$

31. $\arctan 0.92$ **32.** $\arctan 2.8$

33. $\arcsin \frac{7}{8}$ **34.** $\arccos\left(-\frac{1}{3}\right)$

35. $\tan^{-1} \frac{19}{4}$ **36.** $\tan^{-1}\left(-\frac{95}{7}\right)$

37. $\tan^{-1}\left(-\sqrt{372}\right)$ **38.** $\tan^{-1}\left(-\sqrt{2165}\right)$

Finding Missing Coordinates   In Exercises 39 and 40, determine the missing coordinates of the points on the graph of the function.

39. 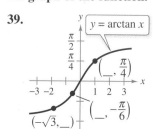   **40.**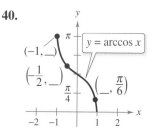

Using an Inverse Trigonometric Function   In Exercises 41–46, use an inverse trigonometric function to write θ as a function of x.

41. 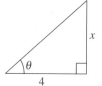   **42.**

43. 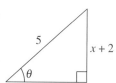   **44.**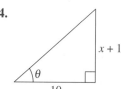

45.   **46.**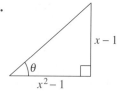

Using Inverse Properties   In Exercises 47–52, use the properties of inverse trigonometric functions to evaluate the expression.

47. $\sin(\arcsin 0.3)$ **48.** $\tan(\arctan 45)$

49. $\cos[\arccos(-0.1)]$ **50.** $\sin[\arcsin(-0.2)]$

51. $\arcsin(\sin 3\pi)$ **52.** $\arccos\left(\cos \dfrac{7\pi}{2}\right)$

Evaluating a Composition of Functions In Exercises 53–64, find the exact value of the expression. (*Hint:* Sketch a right triangle.)

53. $\sin\left(\arctan\frac{3}{4}\right)$

54. $\sec\left(\arcsin\frac{4}{5}\right)$

55. $\cos(\tan^{-1} 2)$

56. $\sin\left(\cos^{-1}\frac{\sqrt{5}}{5}\right)$

57. $\cos\left(\arcsin\frac{5}{13}\right)$

58. $\csc\left[\arctan\left(-\frac{5}{12}\right)\right]$

59. $\sec\left[\arctan\left(-\frac{3}{5}\right)\right]$

60. $\tan\left[\arcsin\left(-\frac{3}{4}\right)\right]$

61. $\sin\left[\arccos\left(-\frac{2}{3}\right)\right]$

62. $\cot\left(\arctan\frac{5}{8}\right)$

63. $\csc\left(\cos^{-1}\frac{\sqrt{3}}{2}\right)$

64. $\sec\left[\sin^{-1}\left(-\frac{\sqrt{2}}{2}\right)\right]$

Writing an Expression In Exercises 65–74, write an algebraic expression that is equivalent to the given expression. (*Hint:* Sketch a right triangle, as demonstrated in Example 7.)

65. $\cot(\arctan x)$

66. $\sin(\arctan x)$

67. $\cos(\arcsin 2x)$

68. $\sec(\arctan 3x)$

69. $\sin(\arccos x)$

70. $\sec[\arcsin(x-1)]$

71. $\tan\left(\arccos\frac{x}{3}\right)$

72. $\cot\left(\arctan\frac{1}{x}\right)$

73. $\csc\left(\arctan\frac{x}{\sqrt{2}}\right)$

74. $\cos\left(\arcsin\frac{x-h}{r}\right)$

Graphical Analysis In Exercises 75 and 76, use a graphing utility to graph f and g in the same viewing window to verify that the two functions are equal. Explain why they are equal. Identify any asymptotes of the graphs.

75. $f(x) = \sin(\arctan 2x)$, $g(x) = \dfrac{2x}{\sqrt{1+4x^2}}$

76. $f(x) = \tan\left(\arccos\dfrac{x}{2}\right)$, $g(x) = \dfrac{\sqrt{4-x^2}}{x}$

Completing an Equation In Exercises 77–80, complete the equation.

77. $\arctan\dfrac{9}{x} = \arcsin(\boxed{})$, $x > 0$

78. $\arcsin\dfrac{\sqrt{36-x^2}}{6} = \arccos(\boxed{})$, $0 \le x \le 6$

79. $\arccos\dfrac{3}{\sqrt{x^2-2x+10}} = \arcsin(\boxed{})$

80. $\arccos\dfrac{x-2}{2} = \arctan(\boxed{})$, $2 < x < 4$

Comparing Graphs In Exercises 81 and 82, sketch a graph of the function and compare the graph of g with the graph of $f(x) = \arcsin x$.

81. $g(x) = \arcsin(x-1)$

82. $g(x) = \arcsin\dfrac{x}{2}$

Sketching the Graph of a Function In Exercises 83–88, sketch a graph of the function.

83. $y = 2\arccos x$

84. $g(t) = \arccos(t+2)$

85. $f(x) = \arctan 2x$

86. $f(x) = \dfrac{\pi}{2} + \arctan x$

87. $h(v) = \arccos\dfrac{v}{2}$

88. $f(x) = \arccos\dfrac{x}{4}$

Graphing an Inverse Trigonometric Function In Exercises 89–94, use a graphing utility to graph the function.

89. $f(x) = 2\arccos(2x)$

90. $f(x) = \pi\arcsin(4x)$

91. $f(x) = \arctan(2x-3)$

92. $f(x) = -3 + \arctan(\pi x)$

93. $f(x) = \pi - \sin^{-1}\left(\dfrac{2}{3}\right)$

94. $f(x) = \dfrac{\pi}{2} + \cos^{-1}\left(\dfrac{1}{\pi}\right)$

Using a Trigonometric Identity In Exercises 95 and 96, write the function in terms of the sine function by using the identity

$$A\cos \omega t + B\sin \omega t = \sqrt{A^2 + B^2}\, \sin\left(\omega t + \arctan\dfrac{A}{B}\right).$$

Use a graphing utility to graph both forms of the function. What does the graph imply?

95. $f(t) = 3\cos 2t + 3\sin 2t$

96. $f(t) = 4\cos \pi t + 3\sin \pi t$

Behavior of an Inverse Trigonometric Function In Exercises 97–102, fill in the blank. If not possible, state the reason.

97. As $x \to 1^-$, the value of $\arcsin x \to \boxed{}$.

98. As $x \to 1^-$, the value of $\arccos x \to \boxed{}$.

99. As $x \to \infty$, the value of $\arctan x \to \boxed{}$.

100. As $x \to -1^+$, the value of $\arcsin x \to \boxed{}$.

101. As $x \to -1^+$, the value of $\arccos x \to \boxed{}$.

102. As $x \to -\infty$, the value of $\arctan x \to \boxed{}$.

103. **Docking a Boat** A boat is pulled in by means of a winch located on a dock 5 feet above the deck of the boat (see figure). Let θ be the angle of elevation from the boat to the winch and let s be the length of the rope from the winch to the boat.

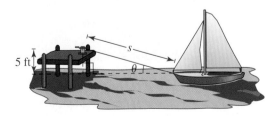

(a) Write θ as a function of s.

(b) Find θ when $s = 40$ feet and $s = 20$ feet.

• • 104. Photography • • • • • • • • • • • • •

A television camera at ground level is filming the lift-off of a space shuttle at a point 750 meters from the launch pad (see figure). Let θ be the angle of elevation to the shuttle and let s be the height of the shuttle.

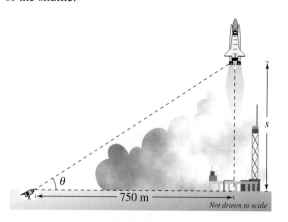

Not drawn to scale

(a) Write θ as a function of s.

(b) Find θ when $s = 300$ meters and $s = 1200$ meters.

105. Photography A photographer is taking a picture of a three-foot-tall painting hung in an art gallery. The camera lens is 1 foot below the lower edge of the painting (see figure). The angle β subtended by the camera lens x feet from the painting is given by

$$\beta = \arctan \frac{3x}{x^2 + 4}, \quad x > 0.$$

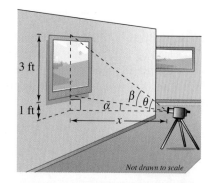

Not drawn to scale

(a) Use a graphing utility to graph β as a function of x.

(b) Move the cursor along the graph to approximate the distance from the picture when β is maximum.

(c) Identify the asymptote of the graph and discuss its meaning in the context of the problem.

106. Granular Angle of Repose Different types of granular substances naturally settle at different angles when stored in cone-shaped piles. This angle θ is called the *angle of repose* (see figure). When rock salt is stored in a cone-shaped pile 11 feet high, the diameter of the pile's base is about 34 feet. *(Source: Bulk-Store Structures, Inc.)*

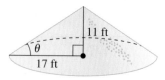

(a) Find the angle of repose for rock salt.

(b) How tall is a pile of rock salt that has a base diameter of 40 feet?

107. Granular Angle of Repose When whole corn is stored in a cone-shaped pile 20 feet high, the diameter of the pile's base is about 82 feet.

(a) Find the angle of repose for whole corn.

(b) How tall is a pile of corn that has a base diameter of 100 feet?

108. Angle of Elevation An airplane flies at an altitude of 6 miles toward a point directly over an observer. Consider θ and x as shown in the figure.

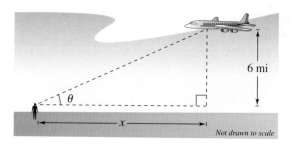

Not drawn to scale

(a) Write θ as a function of x.

(b) Find θ when $x = 7$ miles and $x = 1$ mile.

109. Security Patrol A security car with its spotlight on is parked 20 meters from a warehouse. Consider θ and x as shown in the figure.

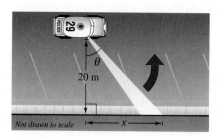

Not drawn to scale

(a) Write θ as a function of x.

(b) Find θ when $x = 5$ meters and $x = 12$ meters.

NASA

Exploration

True or False? In Exercises 110–113, determine whether the statement is true or false. Justify your answer.

110. $\sin \dfrac{5\pi}{6} = \dfrac{1}{2} \implies \arcsin \dfrac{1}{2} = \dfrac{5\pi}{6}$

111. $\tan \dfrac{5\pi}{4} = 1 \implies \arctan 1 = \dfrac{5\pi}{4}$

112. $\arctan x = \dfrac{\arcsin x}{\arccos x}$ **113.** $\sin^{-1} x = \dfrac{1}{\sin x}$

114. HOW DO YOU SEE IT? Use the figure below to determine the value(s) of x for which each statement is true.

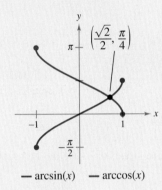

— arcsin(x)   — arccos(x)

(a) $\arcsin x < \arccos x$

(b) $\arcsin x = \arccos x$

(c) $\arcsin x > \arccos x$

115. Inverse Cotangent Function Define the inverse cotangent function by restricting the domain of the cotangent function to the interval $(0, \pi)$, and sketch the graph of the inverse trigonometric function.

116. Inverse Secant Function Define the inverse secant function by restricting the domain of the secant function to the intervals $[0, \pi/2)$ and $(\pi/2, \pi]$, and sketch the graph of the inverse trigonometric function.

117. Inverse Cosecant Function Define the inverse cosecant function by restricting the domain of the cosecant function to the intervals $[-\pi/2, 0)$ and $(0, \pi/2]$, and sketch the graph of the inverse trigonometric function.

118. Writing Use the results of Exercises 115–117 to explain how to graph (a) the inverse cotangent function, (b) the inverse secant function, and (c) the inverse cosecant function on a graphing utility.

Evaluating an Inverse Trigonometric Function In Exercises 119–126, use the results of Exercises 115–117 to evaluate the expression without using a calculator.

119. $\operatorname{arcsec} \sqrt{2}$ **120.** $\operatorname{arcsec} 1$

121. $\operatorname{arccot}(-1)$ **122.** $\operatorname{arccot}\left(-\sqrt{3}\right)$

123. $\operatorname{arccsc} 2$ **124.** $\operatorname{arccsc}(-1)$

125. $\operatorname{arccsc}\left(\dfrac{2\sqrt{3}}{3}\right)$ **126.** $\operatorname{arcsec}\left(-\dfrac{2\sqrt{3}}{3}\right)$

Calculators and Inverse Trigonometric Functions In Exercises 127–134, use the results of Exercises 115–117 and a calculator to approximate the value of the expression. Round your result to two decimal places.

127. $\operatorname{arcsec} 2.54$ **128.** $\operatorname{arcsec}(-1.52)$

129. $\operatorname{arccot} 5.25$ **130.** $\operatorname{arccot}(-10)$

131. $\operatorname{arccot} \frac{5}{3}$ **132.** $\operatorname{arccot}\left(-\frac{16}{7}\right)$

133. $\operatorname{arccsc}\left(-\frac{25}{3}\right)$ **134.** $\operatorname{arccsc}(-12)$

135. Area In calculus, it is shown that the area of the region bounded by the graphs of $y = 0$, $y = 1/(x^2 + 1)$, $x = a$, and $x = b$ is given by

$$\text{Area} = \arctan b - \arctan a$$

(see figure). Find the area for the following values of a and b.

(a) $a = 0, b = 1$ (b) $a = -1, b = 1$

(c) $a = 0, b = 3$ (d) $a = -1, b = 3$

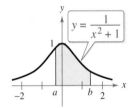

$y = \dfrac{1}{x^2 + 1}$

136. Think About It Use a graphing utility to graph the functions $f(x) = \sqrt{x}$ and $g(x) = 6 \arctan x$. For $x > 0$, it appears that $g > f$. Explain why you know that there exists a positive real number a such that $g < f$ for $x > a$. Approximate the number a.

137. Think About It Consider the functions $f(x) = \sin x$ and $f^{-1}(x) = \arcsin x$.

(a) Use a graphing utility to graph the composite functions $f \circ f^{-1}$ and $f^{-1} \circ f$.

(b) Explain why the graphs in part (a) are not the graph of the line $y = x$. Why do the graphs of $f \circ f^{-1}$ and $f^{-1} \circ f$ differ?

138. Proof Prove each identity.

(a) $\arcsin(-x) = -\arcsin x$

(b) $\arctan(-x) = -\arctan x$

(c) $\arctan x + \arctan \dfrac{1}{x} = \dfrac{\pi}{2}, \quad x > 0$

(d) $\arcsin x + \arccos x = \dfrac{\pi}{2}$

(e) $\arcsin x = \arctan \dfrac{x}{\sqrt{1 - x^2}}$

1.8 Applications and Models

Right triangles often occur in real-life situations. For instance, in Exercise 32 on page 197, you will use right triangles to analyze the design of a new slide at a water park.

- Solve real-life problems involving right triangles.
- Solve real-life problems involving directional bearings.
- Solve real-life problems involving harmonic motion.

Applications Involving Right Triangles

In this section, the three angles of a right triangle are denoted by the letters A, B, and C (where C is the right angle), and the lengths of the sides opposite these angles by the letters a, b, and c, respectively (where c is the hypotenuse).

EXAMPLE 1   Solving a Right Triangle

Solve the right triangle shown at the right for all unknown sides and angles.

Solution   Because $C = 90°$, it follows that

$$A + B = 90° \quad \text{and} \quad B = 90° - 34.2° = 55.8°.$$

To solve for a, use the fact that

$$\tan A = \frac{\text{opp}}{\text{adj}} = \frac{a}{b} \implies a = b \tan A.$$

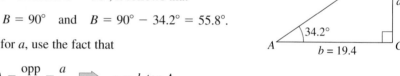

So, $a = 19.4 \tan 34.2° \approx 13.18$. Similarly, to solve for c, use the fact that

$$\cos A = \frac{\text{adj}}{\text{hyp}} = \frac{b}{c} \implies c = \frac{b}{\cos A}.$$

So, $c = \dfrac{19.4}{\cos 34.2°} \approx 23.46$.

✓ **Checkpoint** ◀))) *Audio-video solution in English & Spanish at LarsonPrecalculus.com.*

Solve the right triangle shown in Figure 1.56 for all unknown sides and angles.

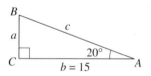

Figure 1.56

EXAMPLE 2   Finding a Side of a Right Triangle

A safety regulation states that the maximum angle of elevation for a rescue ladder is 72°. A fire department's longest ladder is 110 feet. What is the maximum safe rescue height?

Solution   A sketch is shown in Figure 1.57. From the equation $\sin A = a/c$, it follows that

$$a = c \sin A$$

$$= 110 \sin 72°$$

$$\approx 104.6.$$

So, the maximum safe rescue height is about 104.6 feet above the height of the fire truck.

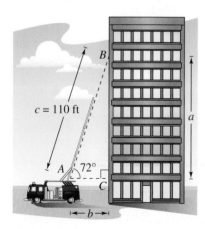

Figure 1.57

✓ **Checkpoint** ◀))) *Audio-video solution in English & Spanish at LarsonPrecalculus.com.*

A ladder that is 16 feet long leans against the side of a house. The angle of elevation of the ladder is 80°. Find the height from the top of the ladder to the ground. ∎

Figure 1.58

Figure 1.59

EXAMPLE 3 Finding a Side of a Right Triangle

At a point 200 feet from the base of a building, the angle of elevation to the *bottom* of a smokestack is 35°, whereas the angle of elevation to the *top* is 53°, as shown in Figure 1.58. Find the height s of the smokestack alone.

Solution

Note from Figure 1.58 that this problem involves two right triangles. For the smaller right triangle, use the fact that

$$\tan 35° = \frac{a}{200}$$

to conclude that the height of the building is

$$a = 200 \tan 35°.$$

For the larger right triangle, use the equation

$$\tan 53° = \frac{a + s}{200}$$

to conclude that $a + s = 200 \tan 53°$. So, the height of the smokestack is

$$s = 200 \tan 53° - a$$

$$= 200 \tan 53° - 200 \tan 35°$$

$$\approx 125.4 \text{ feet.}$$

✓ **Checkpoint** ◀))) *Audio-video solution in English & Spanish at LarsonPrecalculus.com.*

At a point 65 feet from the base of a church, the angles of elevation to the bottom of the steeple and the top of the steeple are 35° and 43°, respectively. Find the height of the steeple.

EXAMPLE 4 Finding an Acute Angle of a Right Triangle

A swimming pool is 20 meters long and 12 meters wide. The bottom of the pool is slanted so that the water depth is 1.3 meters at the shallow end and 4 meters at the deep end, as shown in Figure 1.59. Find the angle of depression (in degrees) of the bottom of the pool.

Solution Using the tangent function, you can see that

$$\tan A = \frac{\text{opp}}{\text{adj}}$$

$$= \frac{2.7}{20}$$

$$= 0.135.$$

So, the angle of depression is

$$A = \arctan 0.135$$

$$\approx 0.13419 \text{ radian}$$

$$\approx 7.69°.$$

✓ **Checkpoint** ◀))) *Audio-video solution in English & Spanish at LarsonPrecalculus.com.*

From the time a small airplane is 100 feet high and 1600 ground feet from its landing runway, the plane descends in a straight line to the runway. Determine the angle of descent (in degrees) of the plane. ◼

Trigonometry and Bearings

· · · · · · · · · · · · · · · ▷

·· REMARK In *air navigation*, bearings are measured in degrees *clockwise* from north. Examples of air navigation bearings are shown below.

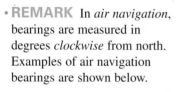

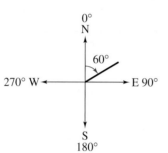

In surveying and navigation, directions can be given in terms of **bearings.** A bearing measures the acute angle that a path or line of sight makes with a fixed north-south line. For instance, the bearing S 35° E, shown below, means 35 degrees east of south.

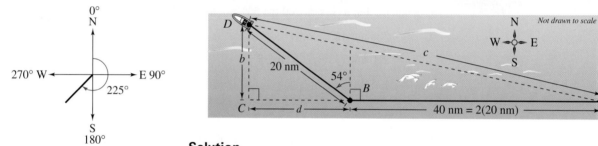

S 35° E   N 80° W   N 45° E

EXAMPLE 5   **Finding Directions in Terms of Bearings**

A ship leaves port at noon and heads due west at 20 knots, or 20 nautical miles (nm) per hour. At 2 P.M. the ship changes course to N 54° W, as shown below. Find the ship's bearing and distance from the port of departure at 3 P.M.

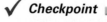

Solution

For triangle BCD, you have

$$B = 90° - 54° = 36°.$$

The two sides of this triangle can be determined to be

$$b = 20 \sin 36° \quad \text{and} \quad d = 20 \cos 36°.$$

For triangle ACD, you can find angle A as follows.

$$\tan A = \frac{b}{d + 40} = \frac{20 \sin 36°}{20 \cos 36° + 40} \approx 0.2092494$$

$$A \approx \arctan 0.2092494 \approx 0.2062732 \text{ radian} \approx 11.82°$$

The angle with the north-south line is $90° - 11.82° = 78.18°$. So, the bearing of the ship is N 78.18° W. Finally, from triangle ACD, you have $\sin A = b/c$, which yields

$$c = \frac{b}{\sin A}$$

$$= \frac{20 \sin 36°}{\sin 11.82°}$$

$$\approx 57.4 \text{ nautical miles.} \qquad \text{Distance from port}$$

✓ **Checkpoint** 🔊))) *Audio-video solution in English & Spanish at LarsonPrecalculus.com.*

A sailboat leaves a pier heading due west at 8 knots. After 15 minutes, the sailboat tacks, changing course to N 16° W at 10 knots. Find the sailboat's distance and bearing from the pier after 12 minutes on this course.

Harmonic Motion

The periodic nature of the trigonometric functions is useful for describing the motion of a point on an object that vibrates, oscillates, rotates, or is moved by wave motion.

For example, consider a ball that is bobbing up and down on the end of a spring, as shown in Figure 1.60. Suppose that 10 centimeters is the maximum distance the ball moves vertically upward or downward from its equilibrium (at rest) position. Suppose further that the time it takes for the ball to move from its maximum displacement above zero to its maximum displacement below zero and back again is $t = 4$ seconds. Assuming the ideal conditions of perfect elasticity and no friction or air resistance, the ball would continue to move up and down in a uniform and regular manner.

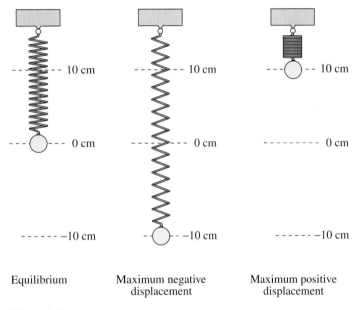

Equilibrium   Maximum negative   Maximum positive
                        displacement   displacement

Figure 1.60

From this spring you can conclude that the period (time for one complete cycle) of the motion is

Period = 4 seconds

its amplitude (maximum displacement from equilibrium) is

Amplitude = 10 centimeters

and its **frequency** (number of cycles per second) is

Frequency = $\frac{1}{4}$ cycle per second.

Motion of this nature can be described by a sine or cosine function and is called **simple harmonic motion.**

Definition of Simple Harmonic Motion

A point that moves on a coordinate line is in **simple harmonic motion** when its distance d from the origin at time t is given by either

$$d = a \sin \omega t \quad \text{or} \quad d = a \cos \omega t$$

where a and ω are real numbers such that $\omega > 0$. The motion has amplitude $|a|$, period $\frac{2\pi}{\omega}$, and frequency $\frac{\omega}{2\pi}$.

EXAMPLE 6 Simple Harmonic Motion

Write an equation for the simple harmonic motion of the ball described in Figure 1.60, where the period is 4 seconds. What is the frequency of this harmonic motion?

Solution

Because the spring is at equilibrium ($d = 0$) when $t = 0$, use the equation

$$d = a \sin \omega t.$$

Moreover, because the maximum displacement from zero is 10 and the period is 4, you have the following.

$$\text{Amplitude} = |a|$$
$$= 10$$

$$\text{Period} = \frac{2\pi}{\omega} = 4 \quad \Longrightarrow \quad \omega = \frac{\pi}{2}$$

Consequently, an equation of motion is

$$d = 10 \sin \frac{\pi}{2}t.$$

Note that the choice of

$$a = 10 \quad \text{or} \quad a = -10$$

depends on whether the ball initially moves up or down. The frequency is

$$\text{Frequency} = \frac{\omega}{2\pi}$$

$$= \frac{\pi/2}{2\pi}$$

$$= \frac{1}{4} \text{ cycle per second.}$$

✓ **Checkpoint** 🔊 *Audio-video solution in English & Spanish at LarsonPrecalculus.com.*

Find a model for simple harmonic motion that satisfies the following conditions: $d = 0$ when $t = 0$, the amplitude is 6 centimeters, and the period is 3 seconds. Then find the frequency. ∎

One illustration of the relationship between sine waves and harmonic motion is in the wave motion that results when a stone is dropped into a calm pool of water. The waves move outward in roughly the shape of sine (or cosine) waves, as shown in Figure 1.61. As an example, suppose you are fishing and your fishing bobber is attached so that it does not move horizontally. As the waves move outward from the dropped stone, your fishing bobber will move up and down in simple harmonic motion, as shown in Figure 1.62.

Figure 1.61

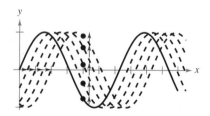

Figure 1.62

EXAMPLE 7 **Simple Harmonic Motion**

Given the equation for simple harmonic motion $d = 6\cos\dfrac{3\pi}{4}t$, find (a) the maximum displacement, (b) the frequency, (c) the value of d when $t = 4$, and (d) the least positive value of t for which $d = 0$.

Algebraic Solution

The given equation has the form $d = a\cos\omega t$, with $a = 6$ and $\omega = 3\pi/4$.

a. The maximum displacement (from the point of equilibrium) is given by the amplitude. So, the maximum displacement is 6.

b. Frequency $= \dfrac{\omega}{2\pi}$

$= \dfrac{3\pi/4}{2\pi}$

$= \dfrac{3}{8}$ cycle per unit of time

c. $d = 6\cos\left[\dfrac{3\pi}{4}(4)\right] = 6\cos 3\pi = 6(-1) = -6$

d. To find the least positive value of t for which $d = 0$, solve the equation

$$6\cos\dfrac{3\pi}{4}t = 0.$$

First divide each side by 6 to obtain

$$\cos\dfrac{3\pi}{4}t = 0.$$

This equation is satisfied when

$$\dfrac{3\pi}{4}t = \dfrac{\pi}{2}, \dfrac{3\pi}{2}, \dfrac{5\pi}{2}, \ldots$$

Multiply these values by $4/(3\pi)$ to obtain

$$t = \dfrac{2}{3}, 2, \dfrac{10}{3}, \ldots$$

So, the least positive value of t is $t = \frac{2}{3}$.

Graphical Solution

Use a graphing utility set in *radian* mode.

a.
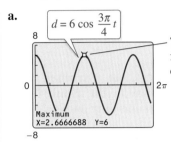
The maximum displacement from the point of equilibrium $(d = 0)$ is 6.

b.
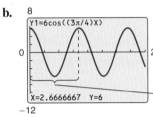
The period is the time for the graph to complete one cycle, which is $t \approx 2.67$. So, the frequency is about $1/2.67 \approx 0.37$ per unit of time.

c.
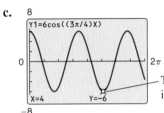
The value of d when $t = 4$ is $d = -6$.

d.
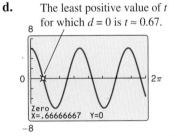
The least positive value of t for which $d = 0$ is $t \approx 0.67$.

✓ **Checkpoint** 🔊))) Audio-video solution in English & Spanish at LarsonPrecalculus.com.

Rework Example 7 for the equation $d = 4\cos 6\pi t$. ■

Summarize (Section 1.8)

1. Describe real-life problems that can be solved using right triangles *(pages 190 and 191, Examples 1–4)*.

2. State the definition of a bearing *(page 192, Example 5)*.

3. State the definition of simple harmonic motion *(page 193, Examples 6 and 7)*.

1.8 Exercises

See CalcChat.com for tutorial help and worked-out solutions to odd-numbered exercises.

Vocabulary: Fill in the blanks.

1. A _____ measures the acute angle that a path or line of sight makes with a fixed north-south line.

2. A point that moves on a coordinate line is said to be in simple _____ _____ when its distance d from the origin at time t is given by either $d = a \sin \omega t$ or $d = a \cos \omega t$.

3. The time for one complete cycle of a point in simple harmonic motion is its _____.

4. The number of cycles per second of a point in simple harmonic motion is its _____.

Skills and Applications

Solving a Right Triangle In Exercises 5–14, solve the right triangle shown in the figure for all unknown sides and angles. Round your answers to two decimal places.

5. $A = 30°, \quad b = 3$
6. $B = 54°, \quad c = 15$
7. $B = 71°, \quad b = 24$
8. $A = 8.4°, \quad a = 40.5$
9. $a = 3, \quad b = 4$
10. $a = 25, \quad c = 35$
11. $b = 16, \quad c = 52$
12. $b = 1.32, \quad c = 9.45$
13. $A = 12°15', \quad c = 430.5$
14. $B = 65°12', \quad a = 14.2$

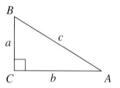

Figure for 5–14

Figure for 15–18

Finding an Altitude In Exercises 15–18, find the altitude of the isosceles triangle shown in the figure. Round your answers to two decimal places.

15. $\theta = 45°, \quad b = 6$
16. $\theta = 18°, \quad b = 10$
17. $\theta = 32°, \quad b = 8$
18. $\theta = 27°, \quad b = 11$

19. **Length** The sun is 25° above the horizon. Find the length of a shadow cast by a building that is 100 feet tall (see figure).

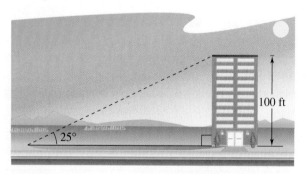

20. **Length** The sun is 20° above the horizon. Find the length of a shadow cast by a park statue that is 12 feet tall.

21. **Height** A ladder that is 20 feet long leans against the side of a house. The angle of elevation of the ladder is 80°. Find the height from the top of the ladder to the ground.

22. **Height** The length of a shadow of a tree is 125 feet when the angle of elevation of the sun is 33°. Approximate the height of the tree.

23. **Height** At a point 50 feet from the base of a church, the angles of elevation to the bottom of the steeple and the top of the steeple are 35° and 47° 40', respectively. Find the height of the steeple.

24. **Distance** An observer in a lighthouse 350 feet above sea level observes two ships directly offshore. The angles of depression to the ships are 4° and 6.5° (see figure). How far apart are the ships?

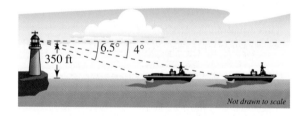

25. **Distance** A passenger in an airplane at an altitude of 10 kilometers sees two towns directly to the east of the plane. The angles of depression to the towns are 28° and 55° (see figure). How far apart are the towns?

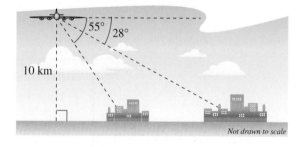

26. Altitude You observe a plane approaching overhead and assume that its speed is 550 miles per hour. The angle of elevation of the plane is 16° at one time and 57° one minute later. Approximate the altitude of the plane.

27. Angle of Elevation An engineer erects a 75-foot cellular telephone tower. Find the angle of elevation to the top of the tower at a point on level ground 50 feet from its base.

28. Angle of Elevation The height of an outdoor basketball backboard is $12\frac{1}{2}$ feet, and the backboard casts a shadow $17\frac{1}{3}$ feet long.

(a) Draw a right triangle that gives a visual representation of the problem. Label the known and unknown quantities.

(b) Use a trigonometric function to write an equation involving the unknown angle of elevation.

(c) Find the angle of elevation of the sun.

29. Angle of Depression A cellular telephone tower that is 150 feet tall is placed on top of a mountain that is 1200 feet above sea level. What is the angle of depression from the top of the tower to a cell phone user who is 5 horizontal miles away and 400 feet above sea level?

30. Angle of Depression A Global Positioning System satellite orbits 12,500 miles above Earth's surface (see figure). Find the angle of depression from the satellite to the horizon. Assume the radius of Earth is 4000 miles.

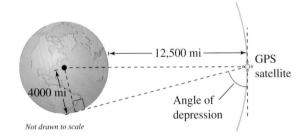

Not drawn to scale

31. Height You are holding one of the tethers attached to the top of a giant character balloon in a parade. Before the start of the parade the balloon is upright and the bottom is floating approximately 20 feet above ground level. You are standing approximately 100 feet ahead of the balloon (see figure).

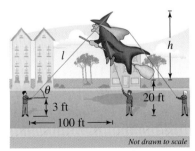

Not drawn to scale

(a) Find an equation for the length l of the tether you are holding in terms of h, the height of the balloon from top to bottom.

(b) Find an equation for the angle of elevation θ from you to the top of the balloon.

(c) The angle of elevation to the top of the balloon is 35°. Find the height h of the balloon.

32. Waterslide Design

The designers of a water park are creating a new slide and have sketched some preliminary drawings. The length of the ladder is 30 feet, and its angle of elevation is 60° (see figure).

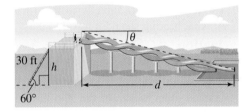

(a) Find the height h of the slide.

(b) Find the angle of depression θ from the top of the slide to the end of the slide at the ground in terms of the horizontal distance d a rider travels.

(c) Safety restrictions require the angle of depression to be no less than 25° and no more than 30°. Find an interval for how far a rider travels horizontally.

33. Speed Enforcement A police department has set up a speed enforcement zone on a straight length of highway. A patrol car is parked parallel to the zone, 200 feet from one end and 150 feet from the other end (see figure).

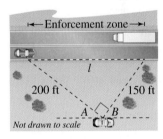

Not drawn to scale

(a) Find the length l of the zone and the measures of the angles A and B (in degrees).

(b) Find the minimum amount of time (in seconds) it takes for a vehicle to pass through the zone without exceeding the posted speed limit of 35 miles per hour.

34. Airplane Ascent During takeoff, an airplane's angle of ascent is 18° and its speed is 275 feet per second.

(a) Find the plane's altitude after 1 minute.

(b) How long will it take for the plane to climb to an altitude of 10,000 feet?

35. Navigation An airplane flying at 600 miles per hour has a bearing of 52°. After flying for 1.5 hours, how far north and how far east will the plane have traveled from its point of departure?

36. Navigation A jet leaves Reno, Nevada, and is headed toward Miami, Florida, at a bearing of 100°. The distance between the two cities is approximately 2472 miles.

(a) How far north and how far west is Reno relative to Miami?

(b) The jet is to return directly to Reno from Miami. At what bearing should it travel?

37. Navigation A ship leaves port at noon and has a bearing of S 29° W. The ship sails at 20 knots.

(a) How many nautical miles south and how many nautical miles west will the ship have traveled by 6:00 P.M.?

(b) At 6:00 P.M., the ship changes course to due west. Find the ship's bearing and distance from the port of departure at 7:00 P.M.

38. Navigation A privately owned yacht leaves a dock in Myrtle Beach, South Carolina, and heads toward Freeport in the Bahamas at a bearing of S 1.4° E. The yacht averages a speed of 20 knots over the 428-nautical-mile trip.

(a) How long will it take the yacht to make the trip?

(b) How far east and south is the yacht after 12 hours?

(c) A plane leaves Myrtle Beach to fly to Freeport. What bearing should be taken?

39. Navigation A ship is 45 miles east and 30 miles south of port. The captain wants to sail directly to port. What bearing should be taken?

40. Navigation An airplane is 160 miles north and 85 miles east of an airport. The pilot wants to fly directly to the airport. What bearing should be taken?

41. Surveying A surveyor wants to find the distance across a pond (see figure). The bearing from A to B is N 32° W. The surveyor walks 50 meters from A to C, and at the point C the bearing to B is N 68° W.

(a) Find the bearing from A to C.

(b) Find the distance from A to B.

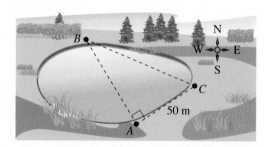

42. Location of a Fire Fire tower A is 30 kilometers due west of fire tower B. A fire is spotted from the towers, and the bearings from A and B are N 76° E and N 56° W, respectively (see figure). Find the distance d of the fire from the line segment AB.

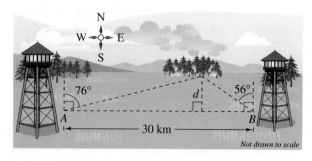

Not drawn to scale

43. Geometry Determine the angle between the diagonal of a cube and the diagonal of its base, as shown in the figure.

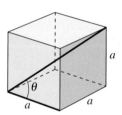

44. Geometry Determine the angle between the diagonal of a cube and its edge, as shown in the figure.

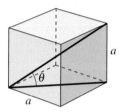

45. Geometry Find the length of the sides of a regular pentagon inscribed in a circle of radius 25 inches.

46. Geometry Find the length of the sides of a regular hexagon inscribed in a circle of radius 25 inches.

Harmonic Motion **In Exercises 47–50, find a model for simple harmonic motion satisfying the specified conditions.**

Displacement ($t = 0$)	Amplitude	Period
47. 0	4 centimeters	2 seconds
48. 0	3 meters	6 seconds
49. 3 inches	3 inches	1.5 seconds
50. 2 feet	2 feet	10 seconds

51. Tuning Fork A point on the end of a tuning fork moves in simple harmonic motion described by $d = a \sin \omega t$. Find ω given that the tuning fork for middle C has a frequency of 264 vibrations per second.

52. Wave Motion A buoy oscillates in simple harmonic motion as waves go past. The buoy moves a total of 3.5 feet from its low point to its high point (see figure), and it returns to its high point every 10 seconds. Write an equation that describes the motion of the buoy where the high point corresponds to the time $t = 0$.

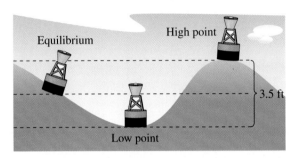

Harmonic Motion **In Exercises 53–56, for the simple harmonic motion described by the trigonometric function, find (a) the maximum displacement, (b) the frequency, (c) the value of d when $t = 5$, and (d) the least positive value of t for which $d = 0$. Use a graphing utility to verify your results.**

53. $d = 9 \cos \dfrac{6\pi}{5} t$

54. $d = \dfrac{1}{2} \cos 20\pi t$

55. $d = \dfrac{1}{4} \sin 6\pi t$

56. $d = \dfrac{1}{64} \sin 792\pi t$

57. Oscillation of a Spring A ball that is bobbing up and down on the end of a spring has a maximum displacement of 3 inches. Its motion (in ideal conditions) is modeled by $y = \frac{1}{4} \cos 16t$, $t > 0$, where y is measured in feet and t is the time in seconds.

(a) Graph the function.

(b) What is the period of the oscillations?

(c) Determine the first time the weight passes the point of equilibrium ($y = 0$).

58. Data Analysis The table shows the average sales S (in millions of dollars) of an outerwear manufacturer for each month t, where $t = 1$ represents January.

Time, t	1	2	3	4	5	6
Sales, S	13.46	11.15	8.00	4.85	2.54	1.70

Time, t	7	8	9	10	11	12
Sales, S	2.54	4.85	8.00	11.15	13.46	14.30

(a) Create a scatter plot of the data.

(b) Find a trigonometric model that fits the data. Graph the model with your scatter plot. How well does the model fit the data?

(c) What is the period of the model? Do you think it is reasonable given the context? Explain your reasoning.

(d) Interpret the meaning of the model's amplitude in the context of the problem.

59. Data Analysis The numbers of hours H of daylight in Denver, Colorado, on the 15th of each month are: 1(9.67), 2(10.72), 3(11.92), 4(13.25), 5(14.37), 6(14.97), 7(14.72), 8(13.77), 9(12.48), 10(11.18), 11(10.00), 12(9.38). The month is represented by t, with $t = 1$ corresponding to January. A model for the data is

$$H(t) = 12.13 + 2.77 \sin\!\left(\frac{\pi t}{6} - 1.60\right).$$

 (a) Use a graphing utility to graph the data points and the model in the same viewing window.

(b) What is the period of the model? Is it what you expected? Explain.

(c) What is the amplitude of the model? What does it represent in the context of the problem? Explain.

Exploration

60. HOW DO YOU SEE IT? The graph below shows the displacement of an object in simple harmonic motion.

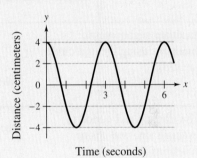

(a) What is the amplitude?

(b) What is the period?

(c) Is the equation of the simple harmonic motion of the form $d = a \sin \omega t$ or $d = a \cos \omega t$?

True or False? **In Exercises 61 and 62, determine whether the statement is true or false. Justify your answer.**

61. The Leaning Tower of Pisa is not vertical, but when you know the angle of elevation θ to the top of the tower as you stand d feet away from it, you can find its height h using the formula $h = d \tan \theta$.

62. The bearing N 24° E means 24 degrees north of east.

Chapter Summary

	What Did You Learn?	Explanation/Examples	Review Exercises
Section 1.1	Describe angles *(p. 122)*.		1–4
	Convert between degrees and radians *(p. 125)*.	To convert degrees to radians, multiply degrees by $(\pi\,\text{rad})/180°$. To convert radians to degrees, multiply radians by $180°/(\pi\,\text{rad})$.	5–14
	Use angles to model and solve real-life problems *(p. 126)*.	Angles can help you find the length of a circular arc and the area of a sector of a circle. (See Examples 5 and 8.)	15–18
Section 1.2	Identify a unit circle and describe its relationship to real numbers *(p. 132)*.		19–22
	Evaluate trigonometric functions using the unit circle *(p. 133)*.	$t = \dfrac{2\pi}{3}$ corresponds to $(x, y) = \left(-\dfrac{1}{2}, \dfrac{\sqrt{3}}{2}\right)$. So $\cos\dfrac{2\pi}{3} = -\dfrac{1}{2}$, $\sin\dfrac{2\pi}{3} = \dfrac{\sqrt{3}}{2}$, and $\tan\dfrac{2\pi}{3} = -\sqrt{3}$.	23, 24
	Use domain and period to evaluate sine and cosine functions *(p. 135)*, and use a calculator to evaluate trigonometric functions *(p. 136)*.	Because $\dfrac{13\pi}{6} = 2\pi + \dfrac{\pi}{6}$, $\sin\dfrac{13\pi}{6} = \sin\dfrac{\pi}{6} = \dfrac{1}{2}$. $\sin\dfrac{3\pi}{8} \approx 0.9239$, $\cot(-1.2) \approx -0.3888$	25–32
Section 1.3	Evaluate trigonometric functions of acute angles *(p. 139)*, and use a calculator to evaluate trigonometric functions *(p. 141)*.	$\sin\theta = \dfrac{\text{opp}}{\text{hyp}}$, $\cos\theta = \dfrac{\text{adj}}{\text{hyp}}$, $\tan\theta = \dfrac{\text{opp}}{\text{adj}}$ $\csc\theta = \dfrac{\text{hyp}}{\text{opp}}$, $\sec\theta = \dfrac{\text{hyp}}{\text{adj}}$, $\cot\theta = \dfrac{\text{adj}}{\text{opp}}$ $\tan 34.7° \approx 0.6924$, $\csc 29°\,15' \approx 2.0466$	33–38
	Use the fundamental trigonometric identities *(p. 142)*.	$\sin\theta = \dfrac{1}{\csc\theta}$, $\tan\theta = \dfrac{\sin\theta}{\cos\theta}$, $\sin^2\theta + \cos^2\theta = 1$	39, 40
	Use trigonometric functions to model and solve real-life problems *(p. 144)*.	Trigonometric functions can help you find the height of a monument, the angle between two paths, and the length of a ramp. (See Examples 8–10.)	41, 42

	What Did You Learn?	**Explanation/Examples**	**Review Exercises**
Section 1.4	Evaluate trigonometric functions of any angle (p. 150).	Let $(3, 4)$ be a point on the terminal side of θ. Then $\sin \theta = \frac{4}{5}$, $\cos \theta = \frac{3}{5}$, and $\tan \theta = \frac{4}{3}$.	43–50
	Find reference angles (p. 152).	Let θ be an angle in standard position. Its reference angle is the acute angle θ' formed by the terminal side of θ and the horizontal axis.	51–54
	Evaluate trigonometric functions of real numbers (p. 153).	$\cos \dfrac{7\pi}{3} = \dfrac{1}{2}$ because $\theta' = \dfrac{7\pi}{3} - 2\pi = \dfrac{\pi}{3}$ and $\cos \dfrac{\pi}{3} = \dfrac{1}{2}$.	55–62
Section 1.5	Sketch the graphs of sine and cosine functions using amplitude and period (p. 159).		63, 64
	Sketch translations of the graphs of sine and cosine functions (p. 163).	For $y = d + a \sin(bx - c)$ and $y = d + a \cos(bx - c)$, the constant c creates a horizontal translation. The constant d creates a vertical translation. (See Examples 4–6.)	65–68
	Use sine and cosine functions to model real-life data (p. 165).	A cosine function can help you model the depth of the water at the end of a dock at various times. (See Example 7.)	69, 70
Section 1.6	Sketch the graphs of tangent (p. 170), cotangent (p. 172), secant (p. 173), and cosecant (p. 173) functions.		71–74
	Sketch the graphs of damped trigonometric functions (p. 175).	In $f(x) = x \cos 2x$, the factor x is called the damping factor.	75, 76
Section 1.7	Evaluate and graph inverse trigonometric functions (p. 180).	$\sin^{-1} \dfrac{\sqrt{3}}{2} = \dfrac{\pi}{3}$, $\quad \cos^{-1}\left(-\dfrac{\sqrt{2}}{2}\right) = \dfrac{3\pi}{4}$, $\quad \tan^{-1}(-1) = -\dfrac{\pi}{4}$	77–86
	Evaluate the compositions of trigonometric functions (p. 184).	$\cos\left[\arctan\left(\frac{5}{12}\right)\right] = \frac{12}{13}$, $\quad \sin(\sin^{-1} 0.4) = 0.4$	87–92
Section 1.8	Solve real-life problems involving right triangles (p. 190).	A trigonometric function can help you find the height of a smokestack on top of a building. (See Example 3.)	93, 94
	Solve real-life problems involving directional bearings (p. 192).	Trigonometric functions can help you find a ship's bearing and distance from a port at a given time. (See Example 5.)	95
	Solve real-life problems involving harmonic motion (p. 193).	Sine or cosine functions can help you describe the motion of an object that vibrates, oscillates, rotates, or is moved by wave motion. (See Examples 6 and 7.)	96

Review Exercises   See CalcChat.com for tutorial help and worked-out solutions to odd-numbered exercises.

1.1 **Using Radian or Degree Measure** In Exercises 1–4, (a) sketch the angle in standard position, (b) determine the quadrant in which the angle lies, and (c) determine one positive and one negative coterminal angle.

1. $\dfrac{15\pi}{4}$

2. $-\dfrac{4\pi}{3}$

3. $-110°$

4. $280°$

Converting from Degrees to Radians In Exercises 5–8, convert the angle measure from degrees to radians. Round to three decimal places.

5. $450°$

6. $-112.5°$

7. $-33°\,45'$

8. $197°\,17'$

Converting from Radians to Degrees In Exercises 9–12, convert the angle measure from radians to degrees. Round to three decimal places.

9. $\dfrac{3\pi}{10}$

10. $-\dfrac{11\pi}{6}$

11. -3.5

12. 5.7

Converting to D° M′ S″ Form In Exercises 13 and 14, convert the angle measure to degrees, minutes, and seconds without using a calculator. Then check your answer using a calculator.

13. $198.4°$

14. $-5.96°$

15. Arc Length Find the length of the arc on a circle of radius 20 inches intercepted by a central angle of $138°$.

16. Phonograph Phonograph records are vinyl discs that rotate on a turntable. A typical record album is 12 inches in diameter and plays at $33\frac{1}{3}$ revolutions per minute.

(a) What is the angular speed of a record album?

(b) What is the linear speed of the outer edge of a record album?

17. Circular Sector Find the area of the sector of a circle of radius 18 inches and central angle $\theta = 120°$.

18. Circular Sector Find the area of the sector of a circle of radius 6.5 millimeters and central angle $\theta = 5\pi/6$.

1.2 **Finding a Point on the Unit Circle** In Exercises 19–22, find the point (x, y) on the unit circle that corresponds to the real number t.

19. $t = \dfrac{2\pi}{3}$

20. $t = \dfrac{7\pi}{4}$

21. $t = \dfrac{7\pi}{6}$

22. $t = -\dfrac{4\pi}{3}$

Evaluating Trigonometric Functions In Exercises 23 and 24, evaluate (if possible) the six trigonometric functions at the real number.

23. $t = \dfrac{3\pi}{4}$

24. $t = -\dfrac{2\pi}{3}$

Using Period to Evaluate Sine and Cosine In Exercises 25–28, evaluate the trigonometric function using its period as an aid.

25. $\sin \dfrac{11\pi}{4}$

26. $\cos 4\pi$

27. $\sin\left(-\dfrac{17\pi}{6}\right)$

28. $\cos\left(-\dfrac{13\pi}{3}\right)$

Using a Calculator In Exercises 29–32, use a calculator to evaluate the trigonometric function. Round your answer to four decimal places. (Be sure the calculator is in the correct mode.)

29. $\tan 33$

30. $\csc 10.5$

31. $\sec \dfrac{12\pi}{5}$

32. $\sin\left(-\dfrac{\pi}{9}\right)$

1.3 **Evaluating Trigonometric Functions** In Exercises 33 and 34, find the exact values of the six trigonometric functions of the angle θ shown in the figure. (Use the Pythagorean Theorem to find the third side of the triangle.)

33.

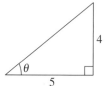

34.

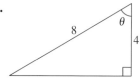

Using a Calculator In Exercises 35–38, use a calculator to evaluate the trigonometric function. Round your answer to four decimal places. (Be sure the calculator is in the correct mode.)

35. $\tan 33°$

36. $\sec 79.3°$

37. $\cot 15°\,14'$

38. $\cos 78°\,11'\,58''$

Applying Trigonometric Identities In Exercises 39 and 40, use the given function value and the trigonometric identities to find the indicated trigonometric functions.

39. $\sin \theta = \frac{1}{3}$

    (a) $\csc \theta$ (b) $\cos \theta$

    (c) $\sec \theta$ (d) $\tan \theta$

40. $\csc \theta = 5$

    (a) $\sin \theta$ (b) $\cot \theta$

    (c) $\tan \theta$ (d) $\sec(90° - \theta)$

41. Railroad Grade   A train travels 3.5 kilometers on a straight track with a grade of $1°\,10'$ (see figure). What is the vertical rise of the train in that distance?

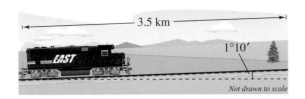

3.5 km

$1°10'$

Not drawn to scale

42. Guy Wire   A guy wire runs from the ground to the top of a 25-foot telephone pole. The angle formed between the wire and the ground is $52°$. How far from the base of the pole is the wire attached to the ground? Assume the pole is perpendicular to the ground.

1.4 Evaluating Trigonometric Functions   In Exercises 43–46, the point is on the terminal side of an angle in standard position. Determine the exact values of the six trigonometric functions of the angle.

43. $(12, 16)$ **44.** $(3, -4)$

45. $(0.3, 0.4)$ **46.** $\left(-\frac{10}{3}, -\frac{2}{3}\right)$

Evaluating Trigonometric Functions   In Exercises 47–50, find the values of the remaining five trigonometric functions of θ with the given constraint.

Function Value	Constraint
47. $\sec \theta = \frac{6}{5}$	$\tan \theta < 0$
48. $\csc \theta = \frac{3}{2}$	$\cos \theta < 0$
49. $\cos \theta = -\frac{2}{5}$	$\sin \theta > 0$
50. $\sin \theta = -\frac{1}{2}$	$\cos \theta > 0$

Finding a Reference Angle   In Exercises 51–54, find the reference angle θ' and sketch θ and θ' in standard position.

51. $\theta = 264°$ **52.** $\theta = 635°$

53. $\theta = -6\pi/5$ **54.** $\theta = 17\pi/3$

Using a Reference Angle   In Exercises 55–58, evaluate the sine, cosine, and tangent of the angle without using a calculator.

55. $\pi/3$ **56.** $-5\pi/4$

57. $-150°$ **58.** $495°$

Using a Calculator   In Exercises 59–62, use a calculator to evaluate the trigonometric function. Round your answer to four decimal places. (Be sure the calculator is in the correct mode.)

59. $\sin 4$

60. $\cot(-4.8)$

61. $\sin(12\pi/5)$

62. $\tan(-25\pi/7)$

1.5 Sketching the Graph of a Sine or Cosine Function   In Exercises 63–68, sketch the graph of the function. (Include two full periods.)

63. $y = \sin 6x$

64. $f(x) = 5 \sin(2x/5)$

65. $y = 5 + \sin x$

66. $y = -4 - \cos \pi x$

67. $g(t) = \frac{5}{2} \sin(t - \pi)$

68. $g(t) = 3 \cos(t + \pi)$

69. Sound Waves   Sine functions of the form $y = a \sin bx$, where x is measured in seconds, can model sine waves.

(a) Write an equation of a sound wave whose amplitude is 2 and whose period is $\frac{1}{264}$ second.

(b) What is the frequency of the sound wave described in part (a)?

70. Data Analysis: Meteorology   The times S of sunset (Greenwich Mean Time) at $40°$ north latitude on the 15th of each month are: 1(16:59), 2(17:35), 3(18:06), 4(18:38), 5(19:08), 6(19:30), 7(19:28), 8(18:57), 9(18:09), 10(17:21), 11(16:44), 12(16:36). The month is represented by t, with $t = 1$ corresponding to January. A model (in which minutes have been converted to the decimal parts of an hour) for the data is $S(t) = 18.09 + 1.41 \sin[(\pi t/6) + 4.60]$.

(a) Use a graphing utility to graph the data points and the model in the same viewing window.

(b) What is the period of the model? Is it what you expected? Explain.

(c) What is the amplitude of the model? What does it represent in the model? Explain.

1.6 Sketching the Graph of a Trigonometric Function   In Exercises 71–74, sketch the graph of the function. (Include two full periods.)

71. $f(t) = \tan\left(t + \frac{\pi}{2}\right)$

72. $f(x) = \frac{1}{2} \cot x$

73. $f(x) = \frac{1}{2} \csc \frac{x}{2}$

74. $h(t) = \sec\left(t - \frac{\pi}{4}\right)$

Analyzing a Damped Trigonometric Graph   In Exercises 75 and 76, use a graphing utility to graph the function and the damping factor of the function in the same viewing window. Describe the behavior of the function as x increases without bound.

75. $f(x) = x \cos x$ **76.** $g(x) = x^4 \cos x$

1.7 Evaluating an Inverse Trigonometric Function In Exercises 77–80, evaluate the expression without using a calculator.

77. $\arcsin(-1)$

78. $\cos^{-1} 1$

79. $\text{arccot } \sqrt{3}$

80. $\text{arcsec}\left(-\sqrt{2}\right)$

Calculators and Inverse Trigonometric Functions In Exercises 81–84, use a calculator to evaluate the expression. Round your result to two decimal places.

81. $\tan^{-1}(-1.5)$

82. $\arccos 0.324$

83. $\text{arccot } 10.5$

84. $\text{arccsc}(-2.01)$

Graphing an Inverse Trigonometric Function In Exercises 85 and 86, use a graphing utility to graph the function.

85. $f(x) = \arctan(x/2)$

86. $f(x) = -\arcsin 2x$

Evaluating a Composition of Functions In Exercises 87–90, find the exact value of the expression. (*Hint:* Sketch a right triangle.)

87. $\cos\left(\arctan \frac{3}{4}\right)$

88. $\sec\left(\tan^{-1} \frac{12}{5}\right)$

89. $\sec\left[\sin^{-1}\left(-\frac{1}{4}\right)\right]$

90. $\cot\left[\arcsin\left(-\frac{12}{13}\right)\right]$

⨍ Writing an Expression In Exercises 91 and 92, write an algebraic expression that is equivalent to the given expression. (*Hint:* Sketch a right triangle.)

91. $\tan\left[\arccos(x/2)\right]$

92. $\sec(\arcsin x)$

1.8

93. **Angle of Elevation** The height of a radio transmission tower is 70 meters, and it casts a shadow of length 30 meters. Draw a right triangle that gives a visual representation of the problem. Label the known and unknown quantities. Then find the angle of elevation of the sun.

94. **Height** Your football has landed at the edge of the roof of your school building. When you are 25 feet from the base of the building, the angle of elevation to your football is 21°. How high off the ground is your football?

95. **Distance** From city A to city B, a plane flies 650 miles at a bearing of 48°. From city B to city C, the plane flies 810 miles at a bearing of 115°. Find the distance from city A to city C and the bearing from city A to city C.

96. **Wave Motion** Your fishing bobber oscillates in simple harmonic motion from the waves in the lake where you fish. Your bobber moves a total of 1.5 inches from its high point to its low point and returns to its high point every 3 seconds. Write an equation modeling the motion of your bobber, where the high point corresponds to the time $t = 0$.

Exploration

True or False? In Exercises 97 and 98, determine whether the statement is true or false. Justify your answer.

97. $y = \sin \theta$ is not a function because $\sin 30° = \sin 150°$.

98. Because $\tan 3\pi/4 = -1$, $\arctan(-1) = 3\pi/4$.

99. **Writing** Describe the behavior of $f(\theta) = \sec \theta$ at the zeros of $g(\theta) = \cos \theta$. Explain your reasoning.

100. **Conjecture**

(a) Use a graphing utility to complete the table.

θ	0.1	0.4	0.7	1.0	1.3
$\tan\left(\theta - \dfrac{\pi}{2}\right)$					
$-\cot \theta$					

(b) Make a conjecture about the relationship between $\tan[\theta - (\pi/2)]$ and $-\cot \theta$.

101. **Writing** When graphing the sine and cosine functions, determining the amplitude is part of the analysis. Explain why this is not true for the other four trigonometric functions.

102. **Graphical Reasoning** The formulas for the area of a circular sector and the arc length are $A = \frac{1}{2}r^2\theta$ and $s = r\theta$, respectively. (r is the radius and θ is the angle measured in radians.)

(a) For $\theta = 0.8$, write the area and arc length as functions of r. What is the domain of each function? Use a graphing utility to graph the functions. Use the graphs to determine which function changes more rapidly as r increases. Explain.

(b) For $r = 10$ centimeters, write the area and arc length as functions of θ. What is the domain of each function? Use the graphing utility to graph and identify the functions.

103. **Writing** Describe a real-life application that can be represented by a simple harmonic motion model and is different from any that you have seen in this chapter. Explain which function you would use to model your application and why. Explain how you would determine the amplitude, period, and frequency of the model for your application.

Chapter Test

See **CalcChat.com** for tutorial help and worked-out solutions to odd-numbered exercises.

Take this test as you would take a test in class. When you are finished, check your work against the answers given in the back of the book.

1. Consider an angle that measures $\dfrac{5\pi}{4}$ radians.

   (a) Sketch the angle in standard position.

   (b) Determine two coterminal angles (one positive and one negative).

   (c) Convert the angle to degree measure.

2. A truck is moving at a rate of 105 kilometers per hour, and the diameter of each of its wheels is 1 meter. Find the angular speed of the wheels in radians per minute.

3. A water sprinkler sprays water on a lawn over a distance of 25 feet and rotates through an angle of 130°. Find the area of the lawn watered by the sprinkler.

4. Find the exact values of the six trigonometric functions of the angle θ shown in the figure.

5. Given that $\tan \theta = \frac{3}{2}$, find the other five trigonometric functions of θ.

6. Determine the reference angle θ' of the angle $\theta = 205°$ and sketch θ and θ' in standard position.

7. Determine the quadrant in which θ lies when $\sec \theta < 0$ and $\tan \theta > 0$.

8. Find two exact values of θ in degrees ($0 \le \theta < 360°$) for which $\cos \theta = -\sqrt{3}/2$. (Do not use a calculator.)

9. Use a calculator to approximate two values of θ in radians ($0 \le \theta < 2\pi$) for which $\csc \theta = 1.030$. Round the results to two decimal places.

In Exercises 10 and 11, find the values of the remaining five trigonometric functions of θ with the given constraint.

10. $\cos \theta = \frac{3}{5}, \quad \tan \theta < 0$

11. $\sec \theta = -\frac{29}{20}, \quad \sin \theta > 0$

In Exercises 12 and 13, sketch the graph of the function. (Include two full periods.)

12. $g(x) = -2 \sin\left(x - \dfrac{\pi}{4}\right)$

13. $f(\alpha) = \dfrac{1}{2} \tan 2\alpha$

In Exercises 14 and 15, use a graphing utility to graph the function. If the function is periodic, then find its period.

14. $y = \sin 2\pi x + 2 \cos \pi x$

15. $y = 6t \cos(0.25t), \quad 0 \le t \le 32$

16. Find a, b, and c for the function $f(x) = a \sin(bx + c)$ such that the graph of f matches the figure.

17. Find the exact value of $\cot\left(\arcsin \frac{3}{8}\right)$ without using a calculator.

18. Graph the function $f(x) = 2 \arcsin\left(\frac{1}{2}x\right)$.

19. A plane is 90 miles south and 110 miles east of London Heathrow Airport. What bearing should be taken to fly directly to the airport?

20. Write the equation for the simple harmonic motion of a ball on a spring that starts at its lowest point of 6 inches below equilibrium, bounces to its maximum height of 6 inches above equilibrium, and returns to its lowest point in a total of 2 seconds.

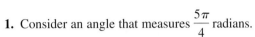

$(-2, 6)$

θ

Figure for 4

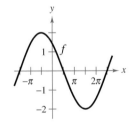

Figure for 16

Proofs in Mathematics ■ ■ ■ ■ ■ ■ ■ ■ ■ ■ ■ ■ ■

The Pythagorean Theorem

The Pythagorean Theorem is one of the most famous theorems in mathematics. More than 100 different proofs now exist. James A. Garfield, the twentieth president of the United States, developed a proof of the Pythagorean Theorem in 1876. His proof, shown below, involves the fact that a trapezoid can be formed from two congruent right triangles and an isosceles right triangle.

The Pythagorean Theorem

In a right triangle, the sum of the squares of the lengths of the legs is equal to the square of the length of the hypotenuse, where a and b are the legs and c is the hypotenuse.

$$a^2 + b^2 = c^2$$

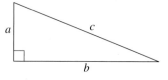

Proof

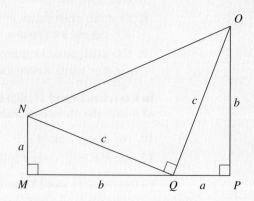

$$\text{Area of trapezoid } MNOP = \text{Area of } \triangle MNQ + \text{Area of } \triangle PQO + \text{Area of } \triangle NOQ$$

$$\frac{1}{2}(a + b)(a + b) = \frac{1}{2}ab + \frac{1}{2}ab + \frac{1}{2}c^2$$

$$\frac{1}{2}(a + b)(a + b) = ab + \frac{1}{2}c^2$$

$$(a + b)(a + b) = 2ab + c^2$$

$$a^2 + 2ab + b^2 = 2ab + c^2$$

$$a^2 + b^2 = c^2$$

P.S. Problem Solving ▪ ▪ ▪ ▪ ▪ ▪ ▪ ▪ ▪ ▪ ▪ ▪ ▪

1. Angle of Rotation The restaurant at the top of the Space Needle in Seattle, Washington, is circular and has a radius of 47.25 feet. The dining part of the restaurant revolves, making about one complete revolution every 48 minutes. A dinner party, seated at the edge of the revolving restaurant at 6:45 P.M., finishes at 8:57 P.M.

(a) Find the angle through which the dinner party rotated.

(b) Find the distance the party traveled during dinner.

2. Bicycle Gears A bicycle's gear ratio is the number of times the freewheel turns for every one turn of the chainwheel (see figure). The table shows the numbers of teeth in the freewheel and chainwheel for the first five gears of an 18-speed touring bicycle. The chainwheel completes one rotation for each gear. Find the angle through which the freewheel turns for each gear. Give your answers in both degrees and radians.

DATA	Gear Number	Number of Teeth in Freewheel	Number of Teeth in Chainwheel
	1	32	24
	2	26	24
	3	22	24
	4	32	40
	5	19	24

Spreadsheet at LarsonPrecalculus.com

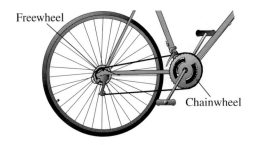

Freewheel

Chainwheel

3. Surveying A surveyor in a helicopter is trying to determine the width of an island, as shown in the figure.

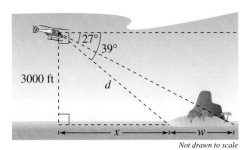

127°

39°

3000 ft

d

x

w

Not drawn to scale

(a) What is the shortest distance d the helicopter would have to travel to land on the island?

(b) What is the horizontal distance x the helicopter would have to travel before it would be directly over the nearer end of the island?

(c) Find the width w of the island. Explain how you found your answer.

4. Similar Triangles and Trigonometric Functions Use the figure below.

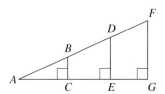

(a) Explain why $\triangle ABC$, $\triangle ADE$, and $\triangle AFG$ are similar triangles.

(b) What does similarity imply about the ratios

$$\frac{BC}{AB}, \quad \frac{DE}{AD}, \quad \text{and} \quad \frac{FG}{AF}?$$

(c) Does the value of sin A depend on which triangle from part (a) you use to calculate it? Would using a different right triangle similar to the three given triangles change the value of sin A?

(d) Do your conclusions from part (c) apply to the other five trigonometric functions? Explain.

5. Graphical Analysis Use a graphing utility to graph h, and use the graph to decide whether h is even, odd, or neither.

(a) $h(x) = \cos^2 x$

(b) $h(x) = \sin^2 x$

6. Squares of Even and Odd Functions Given that f is an even function and g is an odd function, use the results of Exercise 5 to make a conjecture about h, where

(a) $h(x) = [f(x)]^2$

(b) $h(x) = [g(x)]^2$.

7. Height of a Ferris Wheel Car The model for the height h (in feet) of a Ferris wheel car is

$$h = 50 + 50 \sin 8\pi t$$

where t is the time (in minutes). (The Ferris wheel has a radius of 50 feet.) This model yields a height of 50 feet when $t = 0$. Alter the model so that the height of the car is 1 foot when $t = 0$.

8. Periodic Function The function f is periodic, with period c. So, $f(t + c) = f(t)$. Are the following statements true? Explain.

(a) $f(t - 2c) = f(t)$

(b) $f\left(t + \frac{1}{2}c\right) = f\left(\frac{1}{2}t\right)$

(c) $f\left(\frac{1}{2}(t + c)\right) = f\left(\frac{1}{2}t\right)$

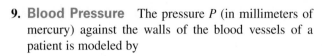

9. Blood Pressure The pressure P (in millimeters of mercury) against the walls of the blood vessels of a patient is modeled by

$$P = 100 - 20 \cos\left(\frac{8\pi}{3}t\right)$$

where t is time (in seconds).

(a) Use a graphing utility to graph the model.

(b) What is the period of the model? What does the period tell you about this situation?

(c) What is the amplitude of the model? What does it tell you about this situation?

(d) If one cycle of this model is equivalent to one heartbeat, what is the pulse of this patient?

(e) A physician wants this patient's pulse rate to be 64 beats per minute or less. What should the period be? What should the coefficient of t be?

10. Biorhythms A popular theory that attempts to explain the ups and downs of everyday life states that each of us has three cycles, called biorhythms, which begin at birth. These three cycles can be modeled by sine waves.

Physical (23 days): $P = \sin \dfrac{2\pi t}{23}, \quad t \geq 0$

Emotional (28 days): $E = \sin \dfrac{2\pi t}{28}, \quad t \geq 0$

Intellectual (33 days): $I = \sin \dfrac{2\pi t}{33}, \quad t \geq 0$

where t is the number of days since birth. Consider a person who was born on July 20, 1990.

(a) Use a graphing utility to graph the three models in the same viewing window for $7300 \leq t \leq 7380$.

(b) Describe the person's biorhythms during the month of September 2010.

(c) Calculate the person's three energy levels on September 22, 2010.

11. (a) **Graphical Reasoning** Use a graphing utility to graph the functions

$f(x) = 2\cos 2x + 3\sin 3x$ and

$g(x) = 2\cos 2x + 3\sin 4x$.

(b) Use the graphs from part (a) to find the period of each function.

(c) Is the function $h(x) = A\cos \alpha x + B\sin \beta x$, where α and β are positive integers, periodic? Explain your reasoning.

12. Analyzing Trigonometric Functions Two trigonometric functions f and g have periods of 2, and their graphs intersect at $x = 5.35$.

(a) Give one positive value of x less than 5.35 and one value of x greater than 5.35 at which the functions have the same value.

(b) Determine one negative value of x at which the graphs intersect.

(c) Is it true that $f(13.35) = g(-4.65)$? Explain your reasoning.

13. Refraction When you stand in shallow water and look at an object below the surface of the water, the object will look farther away from you than it really is. This is because when light rays pass between air and water, the water refracts, or bends, the light rays. The index of refraction for water is 1.333. This is the ratio of the sine of θ_1 and the sine of θ_2 (see figure).

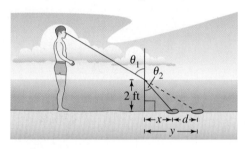

(a) While standing in water that is 2 feet deep, you look at a rock at angle $\theta_1 = 60°$ (measured from a line perpendicular to the surface of the water). Find θ_2.

(b) Find the distances x and y.

(c) Find the distance d between where the rock is and where it appears to be.

(d) What happens to d as you move closer to the rock? Explain your reasoning.

14. Polynomial Approximation In calculus, it can be shown that the arctangent function can be approximated by the polynomial

$$\arctan x \approx x - \frac{x^3}{3} + \frac{x^5}{5} - \frac{x^7}{7}$$

where x is in radians.

(a) Use a graphing utility to graph the arctangent function and its polynomial approximation in the same viewing window. How do the graphs compare?

(b) Study the pattern in the polynomial approximation of the arctangent function and guess the next term. Then repeat part (a). How does the accuracy of the approximation change when you add additional terms?

2 Analytic Trigonometry

Projectile Motion
(Example 10, page 248)

Standing Waves *(page 238)*

Honeycomb Cell *(Example 10, page 230)*

Shadow Length
(Exercise 66, page 223)

Friction *(Exercise 61, page 216)*

2.1   Using Fundamental Identities

Fundamental trigonometric identities can help you simplify trigonometric expressions. For instance, in Exercise 61 on page 216, you will use trigonometric identities to simplify an expression for the coefficient of friction.

- Recognize and write the fundamental trigonometric identities.
- Use the fundamental trigonometric identities to evaluate trigonometric functions, simplify trigonometric expressions, and rewrite trigonometric expressions.

Introduction

In Chapter 1, you studied the basic definitions, properties, graphs, and applications of the individual trigonometric functions. In this chapter, you will learn how to use the fundamental identities to do the following.

1. Evaluate trigonometric functions.
2. Simplify trigonometric expressions.
3. Develop additional trigonometric identities.
4. Solve trigonometric equations.

Fundamental Trigonometric Identities

Reciprocal Identities

$$\sin u = \frac{1}{\csc u} \qquad \cos u = \frac{1}{\sec u} \qquad \tan u = \frac{1}{\cot u}$$

$$\csc u = \frac{1}{\sin u} \qquad \sec u = \frac{1}{\cos u} \qquad \cot u = \frac{1}{\tan u}$$

Quotient Identities

$$\tan u = \frac{\sin u}{\cos u} \qquad \cot u = \frac{\cos u}{\sin u}$$

Pythagorean Identities

$$\sin^2 u + \cos^2 u = 1 \qquad 1 + \tan^2 u = \sec^2 u \qquad 1 + \cot^2 u = \csc^2 u$$

Cofunction Identities

$$\sin\left(\frac{\pi}{2} - u\right) = \cos u \qquad \cos\left(\frac{\pi}{2} - u\right) = \sin u$$

$$\tan\left(\frac{\pi}{2} - u\right) = \cot u \qquad \cot\left(\frac{\pi}{2} - u\right) = \tan u$$

$$\sec\left(\frac{\pi}{2} - u\right) = \csc u \qquad \csc\left(\frac{\pi}{2} - u\right) = \sec u$$

Even/Odd Identities

$$\sin(-u) = -\sin u \qquad \cos(-u) = \cos u \qquad \tan(-u) = -\tan u$$

$$\csc(-u) = -\csc u \qquad \sec(-u) = \sec u \qquad \cot(-u) = -\cot u$$

• • REMARK   You should learn the fundamental trigonometric identities well, because you will use them frequently in trigonometry and they will also appear in calculus. Note that u can be an angle, a real number, or a variable.

Pythagorean identities are sometimes used in radical form such as

$$\sin u = \pm\sqrt{1 - \cos^2 u}$$

or

$$\tan u = \pm\sqrt{\sec^2 u - 1}$$

where the sign depends on the choice of u.

Using the Fundamental Identities

One common application of trigonometric identities is to use given values of trigonometric functions to evaluate other trigonometric functions.

EXAMPLE 1   **Using Identities to Evaluate a Function**

Use the values $\sec u = -\frac{3}{2}$ and $\tan u > 0$ to find the values of all six trigonometric functions.

Solution   Using a reciprocal identity, you have

$$\cos u = \frac{1}{\sec u} = \frac{1}{-3/2} = -\frac{2}{3}.$$

Using a Pythagorean identity, you have

$$\sin^2 u = 1 - \cos^2 u \qquad\qquad \text{Pythagorean identity}$$
$$= 1 - \left(-\tfrac{2}{3}\right)^2 \qquad\qquad \text{Substitute } -\tfrac{2}{3} \text{ for } \cos u.$$
$$= \tfrac{5}{9}. \qquad\qquad\qquad \text{Simplify.}$$

Because $\sec u < 0$ and $\tan u > 0$, it follows that u lies in Quadrant III. Moreover, because $\sin u$ is negative when u is in Quadrant III, choose the negative root and obtain $\sin u = -\sqrt{5}/3$. Knowing the values of the sine and cosine enables you to find the values of all six trigonometric functions.

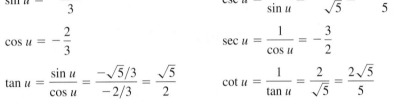

$$\sin u = -\frac{\sqrt{5}}{3} \qquad\qquad \csc u = \frac{1}{\sin u} = -\frac{3}{\sqrt{5}} = -\frac{3\sqrt{5}}{5}$$

$$\cos u = -\frac{2}{3} \qquad\qquad \sec u = \frac{1}{\cos u} = -\frac{3}{2}$$

$$\tan u = \frac{\sin u}{\cos u} = \frac{-\sqrt{5}/3}{-2/3} = \frac{\sqrt{5}}{2} \qquad\qquad \cot u = \frac{1}{\tan u} = \frac{2}{\sqrt{5}} = \frac{2\sqrt{5}}{5}$$

✓ *Checkpoint* 🔊))) *Audio-video solution in English & Spanish at LarsonPrecalculus.com.*

Use the values $\tan x = \frac{1}{3}$ and $\cos x < 0$ to find the values of all six trigonometric functions.

EXAMPLE 2   **Simplifying a Trigonometric Expression**

Simplify

$$\sin x \cos^2 x - \sin x.$$

Solution   First factor out a common monomial factor and then use a fundamental identity.

$$\sin x \cos^2 x - \sin x = \sin x(\cos^2 x - 1) \qquad \text{Factor out common monomial factor.}$$
$$= -\sin x(1 - \cos^2 x) \qquad \text{Factor out } -1.$$
$$= -\sin x(\sin^2 x) \qquad\qquad \text{Pythagorean identity}$$
$$= -\sin^3 x \qquad\qquad\qquad \text{Multiply.}$$

✓ *Checkpoint* 🔊))) *Audio-video solution in English & Spanish at LarsonPrecalculus.com.*

Simplify

$$\cos^2 x \csc x - \csc x.$$

▷ **TECHNOLOGY** To use a graphing utility to check the result of Example 2, graph

$$y_1 = \sin x \cos^2 x - \sin x$$

and

$$y_2 = -\sin^3 x$$

in the same viewing window, as shown below. Because Example 2 shows the equivalence algebraically and the two graphs appear to coincide, you can conclude that the expressions are equivalent.

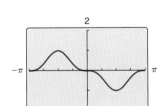

When factoring trigonometric expressions, it is helpful to find a special polynomial factoring form that fits the expression, as shown in Example 3.

EXAMPLE 3 **Factoring Trigonometric Expressions**

Factor each expression.

a. $\sec^2 \theta - 1$ **b.** $4 \tan^2 \theta + \tan \theta - 3$

Solution

a. This expression has the form $u^2 - v^2$, which is the difference of two squares. It factors as

$$\sec^2 \theta - 1 = (\sec \theta + 1)(\sec \theta - 1).$$

b. This expression has the polynomial form $ax^2 + bx + c$, and it factors as

$$4 \tan^2 \theta + \tan \theta - 3 = (4 \tan \theta - 3)(\tan \theta + 1).$$

✓ **Checkpoint**  Audio-video solution in English & Spanish at LarsonPrecalculus.com.

Factor each expression.

a. $1 - \cos^2 \theta$ **b.** $2 \csc^2 \theta - 7 \csc \theta + 6$

On occasion, factoring or simplifying can best be done by first rewriting the expression in terms of just *one* trigonometric function or in terms of *sine and cosine only*. Examples 4 and 5, respectively, show these strategies.

EXAMPLE 4 **Factoring a Trigonometric Expression**

Factor $\csc^2 x - \cot x - 3$.

Solution   Use the identity $\csc^2 x = 1 + \cot^2 x$ to rewrite the expression.

$$\csc^2 x - \cot x - 3 = (1 + \cot^2 x) - \cot x - 3 \qquad \text{Pythagorean identity}$$

$$= \cot^2 x - \cot x - 2 \qquad \text{Combine like terms.}$$

$$= (\cot x - 2)(\cot x + 1) \qquad \text{Factor.}$$

✓ **Checkpoint** Audio-video solution in English & Spanish at LarsonPrecalculus.com.

Factor $\sec^2 x + 3 \tan x + 1$.

· · REMARK   Remember that when adding rational expressions, you must first find the least common denominator (LCD). In Example 5, the LCD is $\sin t$.

EXAMPLE 5 **Simplifying a Trigonometric Expression**

$$\sin t + \cot t \cos t = \sin t + \left(\frac{\cos t}{\sin t}\right) \cos t \qquad \text{Quotient identity}$$

$$= \frac{\sin^2 t + \cos^2 t}{\sin t} \qquad \text{Add fractions.}$$

$$= \frac{1}{\sin t} \qquad \text{Pythagorean identity}$$

$$= \csc t \qquad \text{Reciprocal identity}$$

✓ **Checkpoint** Audio-video solution in English & Spanish at LarsonPrecalculus.com.

Simplify $\csc x - \cos x \cot x$.

 **EXAMPLE 6**   **Adding Trigonometric Expressions**

Perform the addition $\dfrac{\sin \theta}{1 + \cos \theta} + \dfrac{\cos \theta}{\sin \theta}$ and simplify.

Solution

$$\frac{\sin \theta}{1 + \cos \theta} + \frac{\cos \theta}{\sin \theta} = \frac{(\sin \theta)(\sin \theta) + (\cos \theta)(1 + \cos \theta)}{(1 + \cos \theta)(\sin \theta)}$$

$$= \frac{\sin^2 \theta + \cos^2 \theta + \cos \theta}{(1 + \cos \theta)(\sin \theta)} \qquad \text{Multiply.}$$

$$= \frac{1 + \cos \theta}{(1 + \cos \theta)(\sin \theta)} \qquad \begin{array}{l}\text{Pythagorean identity:} \\ \sin^2 \theta + \cos^2 \theta = 1\end{array}$$

$$= \frac{1}{\sin \theta} \qquad \text{Divide out common factor.}$$

$$= \csc \theta \qquad \text{Reciprocal identity}$$

✓ *Checkpoint* ◀))) *Audio-video solution in English & Spanish at LarsonPrecalculus.com.*

Perform the addition $\dfrac{1}{1 + \sin \theta} + \dfrac{1}{1 - \sin \theta}$ and simplify.   ■

The next two examples involve techniques for rewriting expressions in forms that are used in calculus.

EXAMPLE 7   **Rewriting a Trigonometric Expression**

Rewrite $\dfrac{1}{1 + \sin x}$ so that it is *not* in fractional form.

Solution   From the Pythagorean identity

$$\cos^2 x = 1 - \sin^2 x = (1 - \sin x)(1 + \sin x)$$

multiplying both the numerator and the denominator by $(1 - \sin x)$ will produce a monomial denominator.

$$\frac{1}{1 + \sin x} = \frac{1}{1 + \sin x} \cdot \frac{1 - \sin x}{1 - \sin x} \qquad \begin{array}{l}\text{Multiply numerator and} \\ \text{denominator by } (1 - \sin x).\end{array}$$

$$= \frac{1 - \sin x}{1 - \sin^2 x} \qquad \text{Multiply.}$$

$$= \frac{1 - \sin x}{\cos^2 x} \qquad \text{Pythagorean identity}$$

$$= \frac{1}{\cos^2 x} - \frac{\sin x}{\cos^2 x} \qquad \text{Write as separate fractions.}$$

$$= \frac{1}{\cos^2 x} - \frac{\sin x}{\cos x} \cdot \frac{1}{\cos x} \qquad \text{Product of fractions}$$

$$= \sec^2 x - \tan x \sec x \qquad \text{Reciprocal and quotient identities}$$

✓ *Checkpoint* ◀))) *Audio-video solution in English & Spanish at LarsonPrecalculus.com.*

Rewrite $\dfrac{\cos^2 \theta}{1 - \sin \theta}$ so that it is *not* in fractional form.   ■

EXAMPLE 8   **Trigonometric Substitution**

Use the substitution $x = 2 \tan \theta$, $0 < \theta < \pi/2$, to write

$$\sqrt{4 + x^2}$$

as a trigonometric function of θ.

Solution   Begin by letting $x = 2 \tan \theta$. Then, you obtain

$$\sqrt{4 + x^2} = \sqrt{4 + (2 \tan \theta)^2} \qquad \text{Substitute } 2 \tan \theta \text{ for } x.$$

$$= \sqrt{4 + 4 \tan^2 \theta} \qquad \text{Rule of exponents}$$

$$= \sqrt{4(1 + \tan^2 \theta)} \qquad \text{Factor.}$$

$$= \sqrt{4 \sec^2 \theta} \qquad \text{Pythagorean identity}$$

$$= 2 \sec \theta. \qquad \sec \theta > 0 \text{ for } 0 < \theta < \pi/2$$

✓ *Checkpoint* 🔊))) Audio-video solution in English & Spanish at *LarsonPrecalculus.com*.

Use the substitution $x = 3 \sin \theta$, $0 < \theta < \pi/2$, to write

$$\sqrt{9 - x^2}$$

as a trigonometric function of θ.

The figure below shows the right triangle illustration of the trigonometric substitution $x = 2 \tan \theta$ in Example 8.

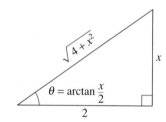

Angle whose tangent is $x/2$

Use this triangle to check the solution of Example 8, as follows. For $0 < \theta < \pi/2$, you have

$$\text{opp} = x, \quad \text{adj} = 2, \quad \text{and} \quad \text{hyp} = \sqrt{4 + x^2}.$$

With these expressions, you can write

$$\sec \theta = \frac{\text{hyp}}{\text{adj}} = \frac{\sqrt{4 + x^2}}{2}.$$

So, $2 \sec \theta = \sqrt{4 + x^2}$, and the solution checks.

Summarize   (Section 2.1)

1. State the fundamental trigonometric identities *(page 210)*.

2. Explain how to use the fundamental trigonometric identities to evaluate trigonometric functions, simplify trigonometric expressions, and rewrite trigonometric expressions *(pages 211–214)*. For examples of these concepts, see Examples 1–8.

2.1 Exercises

Vocabulary: Fill in the blank to complete the trigonometric identity.

1. $\dfrac{\sin u}{\cos u} = $ _____

2. $\dfrac{1}{\csc u} = $ _____

3. $\dfrac{1}{\tan u} = $ _____

4. $\sec\left(\dfrac{\pi}{2} - u\right) = $ _____

5. $1 + $ _____ $= \csc^2 u$

6. $\cot(-u) = $ _____

Skills and Applications

Using Identities to Evaluate a Function In Exercises 7–14, use the given values to find the values (if possible) of all six trigonometric functions.

7. $\sin x = \dfrac{1}{2}, \quad \cos x = \dfrac{\sqrt{3}}{2}$

8. $\csc \theta = \dfrac{25}{7}, \quad \tan \theta = \dfrac{7}{24}$

9. $\cos\left(\dfrac{\pi}{2} - x\right) = \dfrac{3}{5}, \quad \cos x = \dfrac{4}{5}$

10. $\sin(-x) = -\dfrac{1}{3}, \quad \tan x = -\dfrac{\sqrt{2}}{4}$

11. $\sec x = 4, \quad \sin x > 0$

12. $\csc \theta = -5, \quad \cos \theta < 0$

13. $\sin \theta = -1, \quad \cot \theta = 0$

14. $\tan \theta$ is undefined, $\quad \sin \theta > 0$

Matching Trigonometric Expressions In Exercises 15–20, match the trigonometric expression with one of the following.

(a) $\csc x$

(b) -1

(c) 1

(d) $\sin x \tan x$

(e) $\sec^2 x$

(f) $\sec^2 x + \tan^2 x$

15. $\sec x \cos x$

16. $\cot^2 x - \csc^2 x$

17. $\sec^4 x - \tan^4 x$

18. $\cot x \sec x$

19. $\dfrac{\sec^2 x - 1}{\sin^2 x}$

20. $\dfrac{\cos^2[(\pi/2) - x]}{\cos x}$

Factoring a Trigonometric Expression In Exercises 21–28, factor the expression and use the fundamental identities to simplify. There is more than one correct form of each answer.

21. $\tan^2 x - \tan^2 x \sin^2 x$

22. $\sin^2 x \sec^2 x - \sin^2 x$

23. $\dfrac{\sec^2 x - 1}{\sec x - 1}$

24. $\dfrac{\cos x - 2}{\cos^2 x - 4}$

25. $1 - 2\cos^2 x + \cos^4 x$

26. $\sec^4 x - \tan^4 x$

27. $\cot^3 x + \cot^2 x + \cot x + 1$

28. $\sec^3 x - \sec^2 x - \sec x + 1$

Factoring a Trigonometric Expression In Exercises 29–32, factor the trigonometric expression. There is more than one correct form of each answer.

29. $3\sin^2 x - 5\sin x - 2$

30. $6\cos^2 x + 5\cos x - 6$

31. $\cot^2 x + \csc x - 1$

32. $\sin^2 x + 3\cos x + 3$

Multiplying Trigonometric Expressions In Exercises 33 and 34, perform the multiplication and use the fundamental identities to simplify. There is more than one correct form of each answer.

33. $(\sin x + \cos x)^2$

34. $(2\csc x + 2)(2\csc x - 2)$

Simplifying a Trigonometric Expression In Exercises 35–44, use the fundamental identities to simplify the expression. There is more than one correct form of each answer.

35. $\cot \theta \sec \theta$

36. $\tan(-x)\cos x$

37. $\sin \phi(\csc \phi - \sin \phi)$

38. $\cos t(1 + \tan^2 t)$

39. $\dfrac{1 - \sin^2 x}{\csc^2 x - 1}$

40. $\dfrac{\tan \theta \cot \theta}{\sec \theta}$

41. $\cos\left(\dfrac{\pi}{2} - x\right)\sec x$

42. $\dfrac{\cos^2 y}{1 - \sin y}$

43. $\sin \beta \tan \beta + \cos \beta$

44. $\cot u \sin u + \tan u \cos u$

Adding or Subtracting Trigonometric Expressions In Exercises 45–48, perform the addition or subtraction and use the fundamental identities to simplify. There is more than one correct form of each answer.

45. $\dfrac{1}{1 + \cos x} + \dfrac{1}{1 - \cos x}$

46. $\dfrac{1}{\sec x + 1} - \dfrac{1}{\sec x - 1}$

47. $\tan x - \dfrac{\sec^2 x}{\tan x}$

48. $\dfrac{\cos x}{1 + \sin x} + \dfrac{1 + \sin x}{\cos x}$

ƒ Rewriting a Trigonometric Expression In Exercises 49 and 50, rewrite the expression so that it is not in fractional form. There is more than one correct form of each answer.

49. $\dfrac{\sin^2 y}{1 - \cos y}$

50. $\dfrac{5}{\tan x + \sec x}$

Trigonometric Functions and Expressions In Exercises 51 and 52, use a graphing utility to determine which of the six trigonometric functions is equal to the expression. Verify your answer algebraically.

51. $\cos x \cot x + \sin x$

52. $\dfrac{1}{\sin x}\left(\dfrac{1}{\cos x} - \cos x\right)$

Trigonometric Substitution In Exercises 53–56, use the trigonometric substitution to write the algebraic expression as a trigonometric function of θ, where $0 < \theta < \pi/2$.

53. $\sqrt{9 - x^2}, \quad x = 3\cos\theta$

54. $\sqrt{49 - x^2}, \quad x = 7\sin\theta$

55. $\sqrt{x^2 - 4}, \quad x = 2\sec\theta$

56. $\sqrt{9x^2 + 25}, \quad 3x = 5\tan\theta$

Trigonometric Substitution In Exercises 57 and 58, use the trigonometric substitution to write the algebraic equation as a trigonometric equation of θ, where $-\pi/2 < \theta < \pi/2$. Then find $\sin\theta$ and $\cos\theta$.

57. $3 = \sqrt{9 - x^2}, \quad x = 3\sin\theta$

58. $-5\sqrt{3} = \sqrt{100 - x^2}, \quad x = 10\cos\theta$

Solving a Trigonometric Equation In Exercises 59 and 60, use a graphing utility to solve the equation for θ, where $0 \le \theta < 2\pi$.

59. $\sin\theta = \sqrt{1 - \cos^2\theta}$

60. $\sec\theta = \sqrt{1 + \tan^2\theta}$

61. Friction

The forces acting on an object weighing W units on an inclined plane positioned at an angle of θ with the horizontal (see figure) are modeled by

$$\mu W \cos\theta = W \sin\theta$$

where μ is the coefficient of friction. Solve the equation for μ and simplify the result.

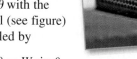

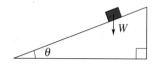

62. Rate of Change The rate of change of the function $f(x) = \sec x + \cos x$ is given by the expression $\sec x \tan x - \sin x$. Show that this expression can also be written as $\sin x \tan^2 x$.

Exploration

True or False? In Exercises 63 and 64, determine whether the statement is true or false. Justify your answer.

63. The even and odd trigonometric identities are helpful for determining whether the value of a trigonometric function is positive or negative.

64. A cofunction identity can transform a tangent function into a cosecant function.

Finding Limits of Trigonometric Functions In Exercises 65 and 66, fill in the blanks.

65. As $x \to \left(\dfrac{\pi}{2}\right)^-$, $\tan x \to \;\rule{1.2em}{0.8em}\;$ and $\cot x \to \;\rule{1.2em}{0.8em}\;$.

66. As $x \to \pi^+$, $\sin x \to \;\rule{1.2em}{0.8em}\;$ and $\csc x \to \;\rule{1.2em}{0.8em}\;$.

Determining Identities In Exercises 67 and 68, determine whether the equation is an identity, and give a reason for your answer.

67. $\dfrac{(\sin k\theta)}{(\cos k\theta)} = \tan\theta, \quad k$ is a constant.

68. $\sin\theta \csc\theta = 1$

69. Trigonometric Substitution Use the trigonometric substitution $u = a\tan\theta$, where $-\pi/2 < \theta < \pi/2$ and $a > 0$, to simplify the expression $\sqrt{a^2 + u^2}$.

70. HOW DO YOU SEE IT?

Explain how to use the figure to derive the Pythagorean identities

$$\sin^2\theta + \cos^2\theta = 1,$$

$$1 + \tan^2\theta = \sec^2\theta,$$

and $\quad 1 + \cot^2\theta = \csc^2\theta.$

Discuss how to remember these identities and other fundamental trigonometric identities.

71. Writing Trigonometric Functions in Terms of Sine Write each of the other trigonometric functions of θ in terms of $\sin\theta$.

72. Rewriting a Trigonometric Expression Rewrite the following expression in terms of $\sin\theta$ and $\cos\theta$.

$$\dfrac{\sec\theta\,(1 + \tan\theta)}{\sec\theta + \csc\theta}$$

2.2   Verifying Trigonometric Identities

Trigonometric identities enable you to rewrite trigonometric equations that model real-life situations. For instance, in Exercise 66 on page 223, trigonometric identities can help you simplify the equation that models the length of a shadow cast by a gnomon (a device used to tell time).

■ **Verify trigonometric identities.**

Introduction

In this section, you will study techniques for verifying trigonometric identities. In the next section, you will study techniques for solving trigonometric equations. The key to verifying identities *and* solving equations is the ability to use the fundamental identities and the rules of algebra to rewrite trigonometric expressions.

Remember that a *conditional equation* is an equation that is true for only some of the values in its domain. For example, the conditional equation

$$\sin x = 0 \qquad\qquad \text{Conditional equation}$$

is true only for

$$x = n\pi$$

where n is an integer. When you find these values, you are *solving* the equation.

On the other hand, an equation that is true for all real values in the domain of the variable is an *identity*. For example, the familiar equation

$$\sin^2 x = 1 - \cos^2 x \qquad \text{Identity}$$

is true for all real numbers x. So, it is an identity.

Verifying Trigonometric Identities

Although there are similarities, verifying that a trigonometric equation is an identity is quite different from solving an equation. There is no well-defined set of rules to follow in verifying trigonometric identities, and it is best to learn the process by practicing.

Guidelines for Verifying Trigonometric Identities

1. Work with one side of the equation at a time. It is often better to work with the more complicated side first.
2. Look for opportunities to factor an expression, add fractions, square a binomial, or create a monomial denominator.
3. Look for opportunities to use the fundamental identities. Note which functions are in the final expression you want. Sines and cosines pair up well, as do secants and tangents, and cosecants and cotangents.
4. If the preceding guidelines do not help, then try converting all terms to sines and cosines.
5. Always try *something*. Even making an attempt that leads to a dead end can provide insight.

Verifying trigonometric identities is a useful process when you need to convert a trigonometric expression into a form that is more useful algebraically. When you verify an identity, you cannot *assume* that the two sides of the equation are equal because you are trying to verify that they *are* equal. As a result, when verifying identities, you cannot use operations such as adding the same quantity to each side of the equation or cross multiplication.

EXAMPLE 1   **Verifying a Trigonometric Identity**

Verify the identity $\dfrac{\sec^2 \theta - 1}{\sec^2 \theta} = \sin^2 \theta$.

•• **REMARK**   Remember that an identity is only true for all real values in the domain of the variable. For instance, in Example 1 the identity is not true when $\theta = \pi/2$ because $\sec^2 \theta$ is not defined when $\theta = \pi/2$.

Solution   Start with the left side because it is more complicated.

$$\dfrac{\sec^2 \theta - 1}{\sec^2 \theta} = \dfrac{(\tan^2 \theta + 1) - 1}{\sec^2 \theta} \qquad \text{Pythagorean identity}$$

$$= \dfrac{\tan^2 \theta}{\sec^2 \theta} \qquad \text{Simplify.}$$

$$= \tan^2 \theta (\cos^2 \theta) \qquad \text{Reciprocal identity}$$

$$= \dfrac{\sin^2 \theta}{(\cos^2 \theta)} (\cos^2 \theta) \qquad \text{Quotient identity}$$

$$= \sin^2 \theta \qquad \text{Simplify.}$$

Notice that you verify the identity by starting with the left side of the equation (the more complicated side) and using the fundamental trigonometric identities to simplify it until you obtain the right side.

✓ **Checkpoint**  *Audio-video solution in English & Spanish at LarsonPrecalculus.com.*

Verify the identity $\dfrac{\sin^2 \theta + \cos^2 \theta}{\cos^2 \theta \sec^2 \theta} = 1$.   ■

There can be more than one way to verify an identity. Here is another way to verify the identity in Example 1.

$$\dfrac{\sec^2 \theta - 1}{\sec^2 \theta} = \dfrac{\sec^2 \theta}{\sec^2 \theta} - \dfrac{1}{\sec^2 \theta} \qquad \text{Write as separate fractions.}$$

$$= 1 - \cos^2 \theta \qquad \text{Reciprocal identity}$$

$$= \sin^2 \theta \qquad \text{Pythagorean identity}$$

EXAMPLE 2   **Verifying a Trigonometric Identity**

Verify the identity $2 \sec^2 \alpha = \dfrac{1}{1 - \sin \alpha} + \dfrac{1}{1 + \sin \alpha}$.

Algebraic Solution

Start with the right side because it is more complicated.

$$\dfrac{1}{1 - \sin \alpha} + \dfrac{1}{1 + \sin \alpha} = \dfrac{1 + \sin \alpha + 1 - \sin \alpha}{(1 - \sin \alpha)(1 + \sin \alpha)} \qquad \text{Add fractions.}$$

$$= \dfrac{2}{1 - \sin^2 \alpha} \qquad \text{Simplify.}$$

$$= \dfrac{2}{\cos^2 \alpha} \qquad \text{Pythagorean identity}$$

$$= 2 \sec^2 \alpha \qquad \text{Reciprocal identity}$$

Numerical Solution

Use a graphing utility to create a table that shows the values of $y_1 = 2/\cos^2 x$ and $y_2 = 1/(1 - \sin x) + 1/(1 + \sin x)$ for different values of x.

X	Y1	Y2
-.5	2.5969	2.5969
-.25	2.1304	2.1304
0	2	2
.25	2.1304	2.1304
.5	2.5969	2.5969
.75	3.7357	3.7357
1	6.851	6.851
X=-.5		

The values for y_1 and y_2 appear to be identical, so the equation appears to be an identity.

✓ **Checkpoint**  *Audio-video solution in English & Spanish at LarsonPrecalculus.com.*

Verify the identity $2 \csc^2 \beta = \dfrac{1}{1 - \cos \beta} + \dfrac{1}{1 + \cos \beta}$.   ■

In Example 2, you needed to write the Pythagorean identity $\sin^2 u + \cos^2 u = 1$ in the equivalent form $\cos^2 u = 1 - \sin^2 u$. When verifying identities, you may find it useful to write the Pythagorean identities in one of these equivalent forms.

Pythagorean Identities

$$\sin^2 u + \cos^2 u = 1$$

$$1 + \tan^2 u = \sec^2 u$$

$$1 + \cot^2 u = \csc^2 u$$

Equivalent Forms

$$\sin^2 u = 1 - \cos^2 u$$

$$\cos^2 u = 1 - \sin^2 u$$

$$1 = \sec^2 u - \tan^2 u$$

$$\tan^2 u = \sec^2 u - 1$$

$$1 = \csc^2 u - \cot^2 u$$

$$\cot^2 u = \csc^2 u - 1$$

EXAMPLE 3 **Verifying a Trigonometric Identity**

Verify the identity $(\tan^2 x + 1)(\cos^2 x - 1) = -\tan^2 x$.

Algebraic Solution

By applying identities before multiplying, you obtain the following.

$$(\tan^2 x + 1)(\cos^2 x - 1) = (\sec^2 x)(-\sin^2 x) \qquad \text{Pythagorean identities}$$

$$= -\frac{\sin^2 x}{\cos^2 x} \qquad \text{Reciprocal identity}$$

$$= -\left(\frac{\sin x}{\cos x}\right)^2 \qquad \text{Property of exponents}$$

$$= -\tan^2 x \qquad \text{Quotient identity}$$

Graphical Solution

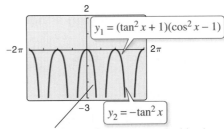

Because the graphs appear to coincide, the given equation appears to be an identity.

 **Checkpoint** Audio-video solution in English & Spanish at LarsonPrecalculus.com.

Verify the identity $(\sec^2 x - 1)(\sin^2 x - 1) = -\sin^2 x$.

• • • • • • • • • • • • • • • ▷

•• **REMARK** Although a graphing utility can be useful in helping to verify an identity, you must use algebraic techniques to produce a *valid* proof.

EXAMPLE 4 **Converting to Sines and Cosines**

Verify the identity $\tan x + \cot x = \sec x \csc x$.

Solution Convert the left side into sines and cosines.

$$\tan x + \cot x = \frac{\sin x}{\cos x} + \frac{\cos x}{\sin x} \qquad \text{Quotient identities}$$

$$= \frac{\sin^2 x + \cos^2 x}{\cos x \sin x} \qquad \text{Add fractions.}$$

$$= \frac{1}{\cos x \sin x} \qquad \text{Pythagorean identity}$$

$$= \frac{1}{\cos x} \cdot \frac{1}{\sin x} \qquad \text{Product of fractions}$$

$$= \sec x \csc x \qquad \text{Reciprocal identities}$$

 Checkpoint  Audio-video solution in English & Spanish at LarsonPrecalculus.com.

Verify the identity $\csc x - \sin x = \cos x \cot x$.

Recall from algebra that *rationalizing the denominator* using conjugates is, on occasion, a powerful simplification technique. A related form of this technique works for simplifying trigonometric expressions as well. For instance, to simplify

$$\frac{1}{1 - \cos x}$$

multiply the numerator and the denominator by $1 + \cos x$.

$$\frac{1}{1 - \cos x} = \frac{1}{1 - \cos x}\left(\frac{1 + \cos x}{1 + \cos x}\right)$$

$$= \frac{1 + \cos x}{1 - \cos^2 x}$$

$$= \frac{1 + \cos x}{\sin^2 x}$$

$$= \csc^2 x(1 + \cos x)$$

The expression $\csc^2 x(1 + \cos x)$ is considered a simplified form of

$$\frac{1}{1 - \cos x}$$

because $\csc^2 x(1 + \cos x)$ does not contain fractions.

EXAMPLE 5   Verifying a Trigonometric Identity

Verify the identity $\sec x + \tan x = \dfrac{\cos x}{1 - \sin x}$.

Algebraic Solution

Begin with the *right* side and create a monomial denominator by multiplying the numerator and the denominator by $1 + \sin x$.

$$\frac{\cos x}{1 - \sin x} = \frac{\cos x}{1 - \sin x}\left(\frac{1 + \sin x}{1 + \sin x}\right) \qquad \text{Multiply numerator and denominator by } 1 + \sin x.$$

$$= \frac{\cos x + \cos x \sin x}{1 - \sin^2 x} \qquad \text{Multiply.}$$

$$= \frac{\cos x + \cos x \sin x}{\cos^2 x} \qquad \text{Pythagorean identity}$$

$$= \frac{\cos x}{\cos^2 x} + \frac{\cos x \sin x}{\cos^2 x} \qquad \text{Write as separate fractions.}$$

$$= \frac{1}{\cos x} + \frac{\sin x}{\cos x} \qquad \text{Simplify.}$$

$$= \sec x + \tan x \qquad \text{Identities}$$

Graphical Solution

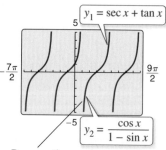

$y_1 = \sec x + \tan x$

$y_2 = \dfrac{\cos x}{1 - \sin x}$

Because the graphs appear to coincide, the given equation appears to be an identity.

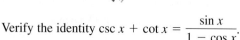 ✓ **Checkpoint**))) *Audio-video solution in English & Spanish at LarsonPrecalculus.com.*

Verify the identity $\csc x + \cot x = \dfrac{\sin x}{1 - \cos x}$.

■

In Examples 1 through 5, you have been verifying trigonometric identities by working with one side of the equation and converting to the form given on the other side. On occasion, it is practical to work with each side *separately*, to obtain one common form that is equivalent to both sides. This is illustrated in Example 6.

EXAMPLE 6   **Working with Each Side Separately**

Verify the identity $\dfrac{\cot^2 \theta}{1 + \csc \theta} = \dfrac{1 - \sin \theta}{\sin \theta}$.

Algebraic Solution

Working with the left side, you have

$$\frac{\cot^2 \theta}{1 + \csc \theta} = \frac{\csc^2 \theta - 1}{1 + \csc \theta} \qquad \text{Pythagorean identity}$$

$$= \frac{(\csc \theta - 1)(\csc \theta + 1)}{1 + \csc \theta} \qquad \text{Factor.}$$

$$= \csc \theta - 1. \qquad \text{Simplify.}$$

Now, simplifying the right side, you have

$$\frac{1 - \sin \theta}{\sin \theta} = \frac{1}{\sin \theta} - \frac{\sin \theta}{\sin \theta} = \csc \theta - 1.$$

This verifies the identity because both sides are equal to $\csc \theta - 1$.

Numerical Solution

Use a graphing utility to create a table that shows the values of

$$y_1 = \frac{\cot^2 x}{1 + \csc x} \quad \text{and} \quad y_2 = \frac{1 - \sin x}{\sin x}$$

for different values of x.

X	Y1	Y2
-.5	-3.086	-3.086
-.25	-5.042	-5.042
0	ERROR	ERROR
.25	3.042	3.042
.5	1.0858	1.0858
.75	.46705	.46705
1	.1884	.1884
X=1		

The values for y_1 and y_2 appear to be identical, so the equation appears to be an identity.

 Checkpoint ◀))) *Audio-video solution in English & Spanish at LarsonPrecalculus.com.*

Verify the identity $\dfrac{\tan^2 \theta}{1 + \sec \theta} = \dfrac{1 - \cos \theta}{\cos \theta}$.

Example 7 shows powers of trigonometric functions rewritten as more complicated sums of products of trigonometric functions. This is a common procedure used in calculus.

EXAMPLE 7   **Two Examples from Calculus** $\displaystyle\int$

Verify each identity.

a. $\tan^4 x = \tan^2 x \sec^2 x - \tan^2 x$ **b.** $\csc^4 x \cot x = \csc^2 x(\cot x + \cot^3 x)$

Solution

a. $\tan^4 x = (\tan^2 x)(\tan^2 x)$ Write as separate factors.

$\qquad\quad = \tan^2 x(\sec^2 x - 1)$ Pythagorean identity

$\qquad\quad = \tan^2 x \sec^2 x - \tan^2 x$ Multiply.

b. $\csc^4 x \cot x = \csc^2 x \csc^2 x \cot x$ Write as separate factors.

$\qquad\qquad\quad = \csc^2 x(1 + \cot^2 x) \cot x$ Pythagorean identity

$\qquad\qquad\quad = \csc^2 x(\cot x + \cot^3 x)$ Multiply.

✓ **Checkpoint** ◀))) *Audio-video solution in English & Spanish at LarsonPrecalculus.com.*

Verify each identity.

a. $\tan^3 x = \tan x \sec^2 x - \tan x$ **b.** $\sin^3 x \cos^4 x = (\cos^4 x - \cos^6 x)\sin x$ ■

Summarize (Section 2.2)

1. State the guidelines for verifying trigonometric identities *(page 217)*. For examples of verifying trigonometric identities, see Examples 1–7.

2.2 Exercises

See **CalcChat.com** for tutorial help and worked-out solutions to odd-numbered exercises.

Vocabulary

In Exercises 1 and 2, fill in the blanks.

1. An equation that is true for all real values in its domain is called an _____.

2. An equation that is true for only some values in its domain is called a _____ _____.

In Exercises 3–8, fill in the blank to complete the fundamental trigonometric identity.

3. $\dfrac{1}{\cot u} = $ _____

4. $\dfrac{\cos u}{\sin u} = $ _____

5. $\sin^2 u + $ _____ $= 1$

6. $\cos\left(\dfrac{\pi}{2} - u\right) = $ _____

7. $\csc(-u) = $ _____

8. $\sec(-u) = $ _____

Skills and Applications

Verifying a Trigonometric Identity In Exercises 9–50, verify the identity.

9. $\tan t \cot t = 1$

10. $\sec y \cos y = 1$

11. $\cot^2 y(\sec^2 y - 1) = 1$

12. $\cos x + \sin x \tan x = \sec x$

13. $(1 + \sin \alpha)(1 - \sin \alpha) = \cos^2 \alpha$

14. $\cos^2 \beta - \sin^2 \beta = 2 \cos^2 \beta - 1$

15. $\cos^2 \beta - \sin^2 \beta = 1 - 2 \sin^2 \beta$

16. $\sin^2 \alpha - \sin^4 \alpha = \cos^2 \alpha - \cos^4 \alpha$

17. $\dfrac{\tan^2 \theta}{\sec \theta} = \sin \theta \tan \theta$

18. $\dfrac{\cot^3 t}{\csc t} = \cos t(\csc^2 t - 1)$

19. $\dfrac{\cot^2 t}{\csc t} = \dfrac{1 - \sin^2 t}{\sin t}$

20. $\dfrac{1}{\tan \beta} + \tan \beta = \dfrac{\sec^2 \beta}{\tan \beta}$

21. $\sin^{1/2} x \cos x - \sin^{5/2} x \cos x = \cos^3 x \sqrt{\sin x}$

22. $\sec^6 x(\sec x \tan x) - \sec^4 x(\sec x \tan x) = \sec^5 x \tan^3 x$

23. $\dfrac{\cot x}{\sec x} = \csc x - \sin x$

24. $\dfrac{\sec \theta - 1}{1 - \cos \theta} = \sec \theta$

25. $\sec x - \cos x = \sin x \tan x$

26. $\sec x(\csc x - 2 \sin x) = \cot x - \tan x$

27. $\dfrac{1}{\tan x} + \dfrac{1}{\cot x} = \tan x + \cot x$

28. $\dfrac{1}{\sin x} - \dfrac{1}{\csc x} = \csc x - \sin x$

29. $\dfrac{1 + \sin \theta}{\cos \theta} + \dfrac{\cos \theta}{1 + \sin \theta} = 2 \sec \theta$

30. $\dfrac{\cos \theta \cot \theta}{1 - \sin \theta} - 1 = \csc \theta$

31. $\dfrac{1}{\cos x + 1} + \dfrac{1}{\cos x - 1} = -2 \csc x \cot x$

32. $\cos x - \dfrac{\cos x}{1 - \tan x} = \dfrac{\sin x \cos x}{\sin x - \cos x}$

33. $\tan\left(\dfrac{\pi}{2} - \theta\right) \tan \theta = 1$

34. $\dfrac{\cos[(\pi/2) - x]}{\sin[(\pi/2) - x]} = \tan x$

35. $\dfrac{\tan x \cot x}{\cos x} = \sec x$

36. $\dfrac{\csc(-x)}{\sec(-x)} = -\cot x$

37. $(1 + \sin y)[1 + \sin(-y)] = \cos^2 y$

38. $\dfrac{\tan x + \tan y}{1 - \tan x \tan y} = \dfrac{\cot x + \cot y}{\cot x \cot y - 1}$

39. $\dfrac{\tan x + \cot y}{\tan x \cot y} = \tan y + \cot x$

40. $\dfrac{\cos x - \cos y}{\sin x + \sin y} + \dfrac{\sin x - \sin y}{\cos x + \cos y} = 0$

41. $\sqrt{\dfrac{1 + \sin \theta}{1 - \sin \theta}} = \dfrac{1 + \sin \theta}{|\cos \theta|}$

42. $\sqrt{\dfrac{1 - \cos \theta}{1 + \cos \theta}} = \dfrac{1 - \cos \theta}{|\sin \theta|}$

43. $\cos^2 \beta + \cos^2\left(\dfrac{\pi}{2} - \beta\right) = 1$

44. $\sec^2 y - \cot^2\left(\dfrac{\pi}{2} - y\right) = 1$

45. $\sin t \csc\left(\dfrac{\pi}{2} - t\right) = \tan t$

46. $\sec^2\left(\dfrac{\pi}{2} - x\right) - 1 = \cot^2 x$

47. $\tan(\sin^{-1} x) = \dfrac{x}{\sqrt{1 - x^2}}$

48. $\cos(\sin^{-1} x) = \sqrt{1 - x^2}$

49. $\tan\left(\sin^{-1} \dfrac{x - 1}{4}\right) = \dfrac{x - 1}{\sqrt{16 - (x - 1)^2}}$

50. $\tan\left(\cos^{-1} \dfrac{x + 1}{2}\right) = \dfrac{\sqrt{4 - (x + 1)^2}}{x + 1}$

Error Analysis   In Exercises 51 and 52, describe the error(s).

51. $(1 + \tan x)[1 + \cot(-x)]$
$$= (1 + \tan x)(1 + \cot x)$$
$$= 1 + \cot x + \tan x + \tan x \cot x$$
$$= 1 + \cot x + \tan x + 1$$
$$= 2 + \cot x + \tan x$$

52. $\dfrac{1 + \sec(-\theta)}{\sin(-\theta) + \tan(-\theta)} = \dfrac{1 - \sec\theta}{\sin\theta - \tan\theta}$
$$= \dfrac{1 - \sec\theta}{(\sin\theta)[1 - (1/\cos\theta)]}$$
$$= \dfrac{1 - \sec\theta}{\sin\theta(1 - \sec\theta)}$$
$$= \dfrac{1}{\sin\theta} = \csc\theta$$

Determining Trigonometric Identities   In Exercises 53–58, (a) use a graphing utility to graph each side of the equation to determine whether the equation is an identity, (b) use the _table_ feature of the graphing utility to determine whether the equation is an identity, and (c) confirm the results of parts (a) and (b) algebraically.

53. $(1 + \cot^2 x)(\cos^2 x) = \cot^2 x$

54. $\csc x(\csc x - \sin x) + \dfrac{\sin x - \cos x}{\sin x} + \cot x = \csc^2 x$

55. $2 + \cos^2 x - 3\cos^4 x = \sin^2 x(3 + 2\cos^2 x)$

56. $\tan^4 x + \tan^2 x - 3 = \sec^2 x(4\tan^2 x - 3)$

57. $\dfrac{1 + \cos x}{\sin x} = \dfrac{\sin x}{1 - \cos x}$

58. $\dfrac{\cot\alpha}{\csc\alpha + 1} = \dfrac{\csc\alpha + 1}{\cot\alpha}$

Verifying a Trigonometric Identity   In Exercises 59–62, verify the identity.

59. $\tan^5 x = \tan^3 x \sec^2 x - \tan^3 x$

60. $\sec^4 x \tan^2 x = (\tan^2 x + \tan^4 x)\sec^2 x$

61. $\cos^3 x \sin^2 x = (\sin^2 x - \sin^4 x)\cos x$

62. $\sin^4 x + \cos^4 x = 1 - 2\cos^2 x + 2\cos^4 x$

Using Cofunction Identities   In Exercises 63 and 64, use the cofunction identities to evaluate the expression without using a calculator.

63. $\sin^2 25° + \sin^2 65°$

64. $\tan^2 63° + \cot^2 16° - \sec^2 74° - \csc^2 27°$

65. Rate of Change   The rate of change of the function $f(x) = \sin x + \csc x$ with respect to change in the variable x is given by the expression $\cos x - \csc x \cot x$. Show that the expression for the rate of change can also be written as $-\cos x \cot^2 x$.

66. Shadow Length
The length s of a shadow cast by a vertical gnomon (a device used to tell time) of height h when the angle of the sun above the horizon is θ can be modeled by the equation

$$s = \dfrac{h\sin(90° - \theta)}{\sin\theta}.$$

(a) Verify that the expression for s is equal to $h\cot\theta$.

(b) Use a graphing utility to complete the table. Let $h = 5$ feet.

θ	15°	30°	45°	60°	75°	90°
s						

(c) Use your table from part (b) to determine the angles of the sun that result in the maximum and minimum lengths of the shadow.

(d) Based on your results from part (c), what time of day do you think it is when the angle of the sun above the horizon is 90°?

Exploration

True or False?   In Exercises 67–69, determine whether the statement is true or false. Justify your answer.

67. There can be more than one way to verify a trigonometric identity.

68. The equation $\sin^2\theta + \cos^2\theta = 1 + \tan^2\theta$ is an identity because $\sin^2(0) + \cos^2(0) = 1$ and $1 + \tan^2(0) = 1$.

69. $\sin x^2 = \sin^2 x$

 70.   HOW DO YOU SEE IT?   Explain how to use the figure to derive the identity
$$\dfrac{\sec^2\theta - 1}{\sec^2\theta} = \sin^2\theta$$
given in Example 1.

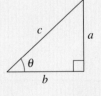

Think About It   In Exercises 71–74, explain why the equation is not an identity and find one value of the variable for which the equation is not true.

71. $\sin\theta = \sqrt{1 - \cos^2\theta}$ **72.** $\tan\theta = \sqrt{\sec^2\theta - 1}$

73. $1 - \cos\theta = \sin\theta$ **74.** $1 + \tan\theta = \sec\theta$

2.3 Solving Trigonometric Equations

Trigonometric equations can help you solve a variety of real-life problems. For instance, in Exercise 94 on page 234, you will solve a trigonometric equation to determine the height above ground of a seat on a Ferris wheel.

- Use standard algebraic techniques to solve trigonometric equations.
- Solve trigonometric equations of quadratic type.
- Solve trigonometric equations involving multiple angles.
- Use inverse trigonometric functions to solve trigonometric equations.

Introduction

To solve a trigonometric equation, use standard algebraic techniques (when possible) such as collecting like terms and factoring. Your preliminary goal in solving a trigonometric equation is to *isolate* the trigonometric function on one side of the equation. For example, to solve the equation $2 \sin x = 1$, divide each side by 2 to obtain

$$\sin x = \frac{1}{2}.$$

To solve for x, note in the figure below that the equation $\sin x = \frac{1}{2}$ has solutions $x = \pi/6$ and $x = 5\pi/6$ in the interval $[0, 2\pi)$. Moreover, because $\sin x$ has a period of 2π, there are infinitely many other solutions, which can be written as

$$x = \frac{\pi}{6} + 2n\pi \quad \text{and} \quad x = \frac{5\pi}{6} + 2n\pi \qquad \text{General solution}$$

where n is an integer, as shown below.

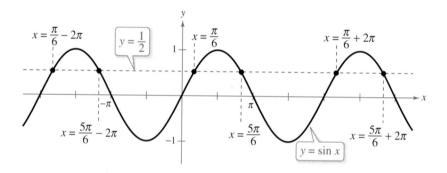

The figure below illustrates another way to show that the equation $\sin x = \frac{1}{2}$ has infinitely many solutions. Any angles that are coterminal with $\pi/6$ or $5\pi/6$ will also be solutions of the equation.

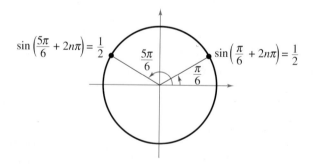

When solving trigonometric equations, you should write your answer(s) using exact values, when possible, rather than decimal approximations.

EXAMPLE 1   Collecting Like Terms

Solve

$$\sin x + \sqrt{2} = -\sin x.$$

Solution   Begin by isolating $\sin x$ on one side of the equation.

$\sin x + \sqrt{2} = -\sin x$	Write original equation.
$\sin x + \sin x + \sqrt{2} = 0$	Add $\sin x$ to each side.
$\sin x + \sin x = -\sqrt{2}$	Subtract $\sqrt{2}$ from each side.
$2 \sin x = -\sqrt{2}$	Combine like terms.
$\sin x = -\dfrac{\sqrt{2}}{2}$	Divide each side by 2.

Because $\sin x$ has a period of 2π, first find all solutions in the interval $[0, 2\pi)$. These solutions are $x = 5\pi/4$ and $x = 7\pi/4$. Finally, add multiples of 2π to each of these solutions to obtain the general form

$$x = \frac{5\pi}{4} + 2n\pi \quad \text{and} \quad x = \frac{7\pi}{4} + 2n\pi \qquad \text{General solution}$$

where n is an integer.

 ✓ **Checkpoint**   ◄))) *Audio-video solution in English & Spanish at LarsonPrecalculus.com.*

Solve $\sin x - \sqrt{2} = -\sin x$.

EXAMPLE 2   Extracting Square Roots

Solve

$$3 \tan^2 x - 1 = 0.$$

Solution   Begin by isolating $\tan x$ on one side of the equation.

$3 \tan^2 x - 1 = 0$	Write original equation.
$3 \tan^2 x = 1$	Add 1 to each side.
$\tan^2 x = \dfrac{1}{3}$	Divide each side by 3.
$\tan x = \pm\dfrac{1}{\sqrt{3}}$	Extract square roots.
$\tan x = \pm\dfrac{\sqrt{3}}{3}$	Rationalize the denominator.

•• **REMARK**   When you extract square roots, make sure you account for both the positive and negative solutions. ▷

Because $\tan x$ has a period of π, first find all solutions in the interval $[0, \pi)$. These solutions are $x = \pi/6$ and $x = 5\pi/6$. Finally, add multiples of π to each of these solutions to obtain the general form

$$x = \frac{\pi}{6} + n\pi \quad \text{and} \quad x = \frac{5\pi}{6} + n\pi \qquad \text{General solution}$$

where n is an integer.

 ✓ **Checkpoint**   ◄))) *Audio-video solution in English & Spanish at LarsonPrecalculus.com.*

Solve $4 \sin^2 x - 3 = 0$.

The equations in Examples 1 and 2 involved only one trigonometric function. When two or more functions occur in the same equation, collect all terms on one side and try to separate the functions by factoring or by using appropriate identities. This may produce factors that yield no solutions, as illustrated in Example 3.

EXAMPLE 3 **Factoring**

Solve $\cot x \cos^2 x = 2 \cot x$.

Solution Begin by collecting all terms on one side of the equation and factoring.

$$\cot x \cos^2 x = 2 \cot x \qquad \text{Write original equation.}$$

$$\cot x \cos^2 x - 2 \cot x = 0 \qquad \text{Subtract 2 cot } x \text{ from each side.}$$

$$\cot x (\cos^2 x - 2) = 0 \qquad \text{Factor.}$$

By setting each of these factors equal to zero, you obtain

$$\cot x = 0 \quad \text{and} \quad \cos^2 x - 2 = 0$$

$$\cos^2 x = 2$$

$$\cos x = \pm \sqrt{2}.$$

In the interval $(0, \pi)$, the equation $\cot x = 0$ has the solution

$$x = \frac{\pi}{2}.$$

No solution exists for $\cos x = \pm \sqrt{2}$ because $\pm \sqrt{2}$ are outside the range of the cosine function. Because $\cot x$ has a period of π, you obtain the general form of the solution by adding multiples of π to $x = \pi/2$ to get

$$x = \frac{\pi}{2} + n\pi \qquad \text{General solution}$$

where n is an integer. Confirm this graphically by sketching the graph of $y = \cot x \cos^2 x - 2 \cot x$, as shown below.

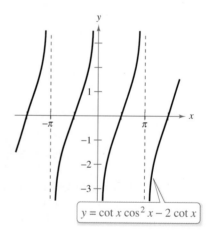

$$y = \cot x \cos^2 x - 2 \cot x$$

Notice that the x-intercepts occur at

$$-\frac{3\pi}{2}, \quad -\frac{\pi}{2}, \quad \frac{\pi}{2}, \quad \frac{3\pi}{2}$$

and so on. These x-intercepts correspond to the solutions of $\cot x \cos^2 x - 2 \cot x = 0$.

✓ **Checkpoint** ◀))) Audio-video solution in English & Spanish at LarsonPrecalculus.com.

Solve $\sin^2 x = 2 \sin x$.

▷ **ALGEBRA HELP** You can review the techniques for solving quadratic equations in Section P.2.

Equations of Quadratic Type

Many trigonometric equations are of quadratic type $ax^2 + bx + c = 0$, as shown below. To solve equations of this type, factor the quadratic or, when this is not possible, use the Quadratic Formula.

Quadratic in sin x	**Quadratic in sec x**
$2 \sin^2 x - \sin x - 1 = 0$	$\sec^2 x - 3 \sec x - 2 = 0$
$2(\sin x)^2 - \sin x - 1 = 0$	$(\sec x)^2 - 3(\sec x) - 2 = 0$

EXAMPLE 4 Factoring an Equation of Quadratic Type

Find all solutions of $2 \sin^2 x - \sin x - 1 = 0$ in the interval $[0, 2\pi)$.

Algebraic Solution

Treat the equation as a quadratic in $\sin x$ and factor.

$2 \sin^2 x - \sin x - 1 = 0$ Write original equation.

$(2 \sin x + 1)(\sin x - 1) = 0$ Factor.

Setting each factor equal to zero, you obtain the following solutions in the interval $[0, 2\pi)$.

$2 \sin x + 1 = 0$ and $\sin x - 1 = 0$

$\sin x = -\dfrac{1}{2}$ $\sin x = 1$

$x = \dfrac{7\pi}{6}, \dfrac{11\pi}{6}$ $x = \dfrac{\pi}{2}$

Graphical Solution

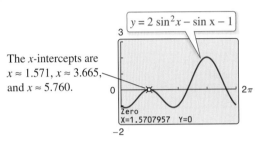

$y = 2 \sin^2 x - \sin x - 1$

The x-intercepts are $x \approx 1.571$, $x \approx 3.665$, and $x \approx 5.760$.

Zero
X=1.5707957 Y=0

From the above figure, you can conclude that the approximate solutions of $2 \sin^2 x - \sin x - 1 = 0$ in the interval $[0, 2\pi)$ are

$x \approx 1.571 \approx \dfrac{\pi}{2}$, $x \approx 3.665 \approx \dfrac{7\pi}{6}$, and $x \approx 5.760 \approx \dfrac{11\pi}{6}$.

✓ **Checkpoint** ◀))) Audio-video solution in English & Spanish at LarsonPrecalculus.com.

Find all solutions of $2 \sin^2 x - 3 \sin x + 1 = 0$ in the interval $[0, 2\pi)$.

EXAMPLE 5 Rewriting with a Single Trigonometric Function

Solve $2 \sin^2 x + 3 \cos x - 3 = 0$.

Solution This equation contains both sine and cosine functions. You can rewrite the equation so that it has only cosine functions by using the identity $\sin^2 x = 1 - \cos^2 x$.

$2 \sin^2 x + 3 \cos x - 3 = 0$	Write original equation.
$2(1 - \cos^2 x) + 3 \cos x - 3 = 0$	Pythagorean identity
$2 \cos^2 x - 3 \cos x + 1 = 0$	Multiply each side by -1.
$(2 \cos x - 1)(\cos x - 1) = 0$	Factor.

By setting each factor equal to zero, you can find the solutions in the interval $[0, 2\pi)$ to be $x = 0$, $x = \pi/3$, and $x = 5\pi/3$. Because $\cos x$ has a period of 2π, the general solution is

$$x = 2n\pi, \qquad x = \dfrac{\pi}{3} + 2n\pi, \qquad x = \dfrac{5\pi}{3} + 2n\pi \qquad \text{General solution}$$

where n is an integer.

✓ **Checkpoint** ◀))) Audio-video solution in English & Spanish at LarsonPrecalculus.com.

Solve $3 \sec^2 x - 2 \tan^2 x - 4 = 0$.

Sometimes you must square each side of an equation to obtain a quadratic, as demonstrated in the next example. Because this procedure can introduce extraneous solutions, you should check any solutions in the original equation to see whether they are valid or extraneous.

•• **REMARK** You square each side of the equation in Example 6 because the squares of the sine and cosine functions are related by a Pythagorean identity. The same is true for the squares of the secant and tangent functions and for the squares of the cosecant and cotangent functions.

EXAMPLE 6 **Squaring and Converting to Quadratic Type**

Find all solutions of $\cos x + 1 = \sin x$ in the interval $[0, 2\pi)$.

Solution It is not clear how to rewrite this equation in terms of a single trigonometric function. Notice what happens when you square each side of the equation.

$$\cos x + 1 = \sin x \qquad \text{Write original equation.}$$
$$\cos^2 x + 2 \cos x + 1 = \sin^2 x \qquad \text{Square each side.}$$
$$\cos^2 x + 2 \cos x + 1 = 1 - \cos^2 x \qquad \text{Pythagorean identity}$$
$$\cos^2 x + \cos^2 x + 2 \cos x + 1 - 1 = 0 \qquad \text{Rewrite equation.}$$
$$2 \cos^2 x + 2 \cos x = 0 \qquad \text{Combine like terms.}$$
$$2 \cos x (\cos x + 1) = 0 \qquad \text{Factor.}$$

Setting each factor equal to zero produces

$$2 \cos x = 0 \qquad \text{and} \qquad \cos x + 1 = 0$$
$$\cos x = 0 \qquad\qquad\qquad \cos x = -1$$
$$x = \frac{\pi}{2}, \frac{3\pi}{2} \qquad\qquad\qquad x = \pi.$$

Because you squared the original equation, check for extraneous solutions.

Check $x = \dfrac{\pi}{2}$

$$\cos \frac{\pi}{2} + 1 \overset{?}{=} \sin \frac{\pi}{2} \qquad \text{Substitute } \frac{\pi}{2} \text{ for } x.$$
$$0 + 1 = 1 \qquad \text{Solution checks.} \checkmark$$

Check $x = \dfrac{3\pi}{2}$

$$\cos \frac{3\pi}{2} + 1 \overset{?}{=} \sin \frac{3\pi}{2} \qquad \text{Substitute } \frac{3\pi}{2} \text{ for } x.$$
$$0 + 1 \neq -1 \qquad \text{Solution does not check.}$$

Check $x = \pi$

$$\cos \pi + 1 \overset{?}{=} \sin \pi \qquad \text{Substitute } \pi \text{ for } x.$$
$$-1 + 1 = 0 \qquad \text{Solution checks.} \checkmark$$

Of the three possible solutions, $x = 3\pi/2$ is extraneous. So, in the interval $[0, 2\pi)$, the only two solutions are

$$x = \frac{\pi}{2} \quad \text{and} \quad x = \pi.$$

✓ *Checkpoint* ◀))) *Audio-video solution in English & Spanish at LarsonPrecalculus.com.*

Find all solutions of $\sin x + 1 = \cos x$ in the interval $[0, 2\pi)$.

Functions Involving Multiple Angles

The next two examples involve trigonometric functions of multiple angles of the forms cos ku and tan ku. To solve equations of these forms, first solve the equation for ku, and then divide your result by k.

EXAMPLE 7   **Solving a Multiple-Angle Equation**

Solve $2 \cos 3t - 1 = 0$.

Solution

$2 \cos 3t - 1 = 0$	Write original equation.
$2 \cos 3t = 1$	Add 1 to each side.
$\cos 3t = \dfrac{1}{2}$	Divide each side by 2.

In the interval $[0, 2\pi)$, you know that $3t = \pi/3$ and $3t = 5\pi/3$ are the only solutions, so, in general, you have

$$3t = \frac{\pi}{3} + 2n\pi \quad \text{and} \quad 3t = \frac{5\pi}{3} + 2n\pi.$$

Dividing these results by 3, you obtain the general solution

$$t = \frac{\pi}{9} + \frac{2n\pi}{3} \quad \text{and} \quad t = \frac{5\pi}{9} + \frac{2n\pi}{3} \qquad \text{General solution}$$

where n is an integer.

✓ *Checkpoint* ◀))) *Audio-video solution in English & Spanish at LarsonPrecalculus.com.*

Solve $2 \sin 2t - \sqrt{3} = 0$.

EXAMPLE 8   **Solving a Multiple-Angle Equation**

$3 \tan \dfrac{x}{2} + 3 = 0$	Original equation
$3 \tan \dfrac{x}{2} = -3$	Subtract 3 from each side.
$\tan \dfrac{x}{2} = -1$	Divide each side by 3.

In the interval $[0, \pi)$, you know that $x/2 = 3\pi/4$ is the only solution, so, in general, you have

$$\frac{x}{2} = \frac{3\pi}{4} + n\pi.$$

Multiplying this result by 2, you obtain the general solution

$$x = \frac{3\pi}{2} + 2n\pi \qquad \text{General solution}$$

where n is an integer.

✓ *Checkpoint* ◀))) *Audio-video solution in English & Spanish at LarsonPrecalculus.com.*

Solve $2 \tan \dfrac{x}{2} - 2 = 0$.

Using Inverse Functions

EXAMPLE 9   **Using Inverse Functions**

$$\sec^2 x - 2\tan x = 4 \qquad \text{Original equation}$$

$$1 + \tan^2 x - 2\tan x - 4 = 0 \qquad \text{Pythagorean identity}$$

$$\tan^2 x - 2\tan x - 3 = 0 \qquad \text{Combine like terms.}$$

$$(\tan x - 3)(\tan x + 1) = 0 \qquad \text{Factor.}$$

Setting each factor equal to zero, you obtain two solutions in the interval $(-\pi/2, \pi/2)$. [Recall that the range of the inverse tangent function is $(-\pi/2, \pi/2)$.]

$$x = \arctan 3 \quad \text{and} \quad x = \arctan(-1) = -\pi/4$$

Finally, because $\tan x$ has a period of π, you add multiples of π to obtain

$$x = \arctan 3 + n\pi \quad \text{and} \quad x = (-\pi/4) + n\pi \qquad \text{General solution}$$

where n is an integer. You can use a calculator to approximate the value of $\arctan 3$.

✓ **Checkpoint** ◀))) *Audio-video solution in English & Spanish at LarsonPrecalculus.com.*

Solve $4\tan^2 x + 5\tan x - 6 = 0$.

EXAMPLE 10   **Surface Area of a Honeycomb Cell**

The surface area S (in square inches) of a honeycomb cell is given by

$$S = 6hs + 1.5s^2\left[\left(\sqrt{3} - \cos\theta\right)/\sin\theta\right], \quad 0 < \theta \le 90°$$

where $h = 2.4$ inches, $s = 0.75$ inch, and θ is the angle shown in Figure 2.1. What value of θ gives the minimum surface area?

Solution   Letting $h = 2.4$ and $s = 0.75$, you obtain

$$S = 10.8 + 0.84375\left[\left(\sqrt{3} - \cos\theta\right)/\sin\theta\right].$$

Graph this function using a graphing utility. The minimum point on the graph, which occurs at $\theta \approx 54.7°$, is shown in Figure 2.2. By using calculus, the exact minimum point on the graph can be shown to occur at $\theta = \arccos\left(1/\sqrt{3}\right) \approx 0.9553 \approx 54.7356°$.

✓ **Checkpoint** ◀))) *Audio-video solution in English & Spanish at LarsonPrecalculus.com.*

In Example 10, for what value(s) of θ is the surface area 12 square inches?

It is possible to find the minimum surface area of a honeycomb cell using a graphing utility or using calculus and the arccosine function.

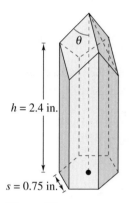

$h = 2.4$ in.

$s = 0.75$ in.

Figure 2.1

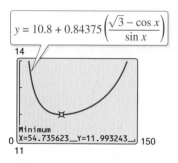

$$y = 10.8 + 0.84375\left(\frac{\sqrt{3} - \cos x}{\sin x}\right)$$

14

Minimum
X=54.735623   Y=11.993243   150

0

11

Figure 2.2

Summarize   (Section 2.3)

1. Describe how to use standard algebraic techniques to solve trigonometric equations *(page 224)*. For examples of using standard algebraic techniques to solve trigonometric equations, see Examples 1–3.

2. Explain how to solve a trigonometric equation of quadratic type *(page 227)*. For examples of solving trigonometric equations of quadratic type, see Examples 4–6.

3. Explain how to solve a trigonometric equation involving multiple angles *(page 229)*. For examples of solving trigonometric equations involving multiple angles, see Examples 7 and 8.

4. Explain how to use inverse trigonometric functions to solve trigonometric equations *(page 230)*. For examples of using inverse trigonometric functions to solve trigonometric equations, see Examples 9 and 10.

2.3   Exercises

See CalcChat.com for tutorial help and worked-out solutions to odd-numbered exercises.

Vocabulary: Fill in the blanks.

1. When solving a trigonometric equation, the preliminary goal is to _____ the trigonometric function involved in the equation.

2. The equation $2 \sin \theta + 1 = 0$ has the solutions $\theta = \dfrac{7\pi}{6} + 2n\pi$ and $\theta = \dfrac{11\pi}{6} + 2n\pi$, which are called _____ solutions.

3. The equation $2 \tan^2 x - 3 \tan x + 1 = 0$ is a trigonometric equation that is of _____ type.

4. A solution of an equation that does not satisfy the original equation is called an _____ solution.

Skills and Applications

Verifying Solutions   In Exercises 5–10, verify that the x-values are solutions of the equation.

5. $\tan x - \sqrt{3} = 0$

    (a) $x = \dfrac{\pi}{3}$

    (b) $x = \dfrac{4\pi}{3}$

6. $\sec x - 2 = 0$

    (a) $x = \dfrac{\pi}{3}$

    (b) $x = \dfrac{5\pi}{3}$

7. $3 \tan^2 2x - 1 = 0$

    (a) $x = \dfrac{\pi}{12}$

    (b) $x = \dfrac{5\pi}{12}$

8. $2 \cos^2 4x - 1 = 0$

    (a) $x = \dfrac{\pi}{16}$

    (b) $x = \dfrac{3\pi}{16}$

9. $2 \sin^2 x - \sin x - 1 = 0$

    (a) $x = \dfrac{\pi}{2}$

    (b) $x = \dfrac{7\pi}{6}$

10. $\csc^4 x - 4 \csc^2 x = 0$

    (a) $x = \dfrac{\pi}{6}$

    (b) $x = \dfrac{5\pi}{6}$

Solving a Trigonometric Equation   In Exercises 11–24, solve the equation.

11. $\sqrt{3} \csc x - 2 = 0$

12. $\tan x + \sqrt{3} = 0$

13. $\cos x + 1 = -\cos x$

14. $3 \sin x + 1 = \sin x$

15. $3 \sec^2 x - 4 = 0$

16. $3 \cot^2 x - 1 = 0$

17. $4 \cos^2 x - 1 = 0$

18. $\sin^2 x = 3 \cos^2 x$

19. $2 \sin^2 2x = 1$

20. $\tan^2 3x = 3$

21. $\tan 3x(\tan x - 1) = 0$

22. $\cos 2x(2 \cos x + 1) = 0$

23. $\sin x(\sin x + 1) = 0$

24. $(2 \sin^2 x - 1)(\tan^2 x - 3) = 0$

Solving a Trigonometric Equation   In Exercises 25–38, find all solutions of the equation in the interval $[0, 2\pi)$.

25. $\cos^3 x = \cos x$

26. $\sec^2 x - 1 = 0$

27. $3 \tan^3 x = \tan x$

28. $2 \sin^2 x = 2 + \cos x$

29. $\sec^2 x - \sec x = 2$

30. $\sec x \csc x = 2 \csc x$

31. $2 \sin x + \csc x = 0$

32. $\sin x - 2 = \cos x - 2$

33. $2 \cos^2 x + \cos x - 1 = 0$

34. $2 \sin^2 x + 3 \sin x + 1 = 0$

35. $2 \sec^2 x + \tan^2 x - 3 = 0$

36. $\cos x + \sin x \tan x = 2$

37. $\csc x + \cot x = 1$

38. $\sec x + \tan x = 1$

Solving a Multiple-Angle Equation   In Exercises 39–44, solve the multiple-angle equation.

39. $2 \cos 2x - 1 = 0$

40. $2 \sin 2x + \sqrt{3} = 0$

41. $\tan 3x - 1 = 0$

42. $\sec 4x - 2 = 0$

43. $2 \cos \dfrac{x}{2} - \sqrt{2} = 0$

44. $2 \sin \dfrac{x}{2} + \sqrt{3} = 0$

Finding x-Intercepts   In Exercises 45–48, find the x-intercepts of the graph.

45. $y = \sin \dfrac{\pi x}{2} + 1$

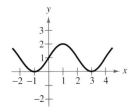

46. $y = \sin \pi x + \cos \pi x$

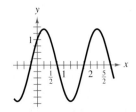

47. $y = \tan^2\left(\dfrac{\pi x}{6}\right) - 3$

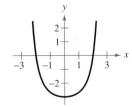

48. $y = \sec^4\left(\dfrac{\pi x}{8}\right) - 4$

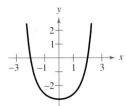

Approximating Solutions   In Exercises 49–58, use a graphing utility to approximate the solutions (to three decimal places) of the equation in the interval $[0, 2\pi)$.

49. $2 \sin x + \cos x = 0$

50. $4 \sin^3 x + 2 \sin^2 x - 2 \sin x - 1 = 0$

51. $\dfrac{1 + \sin x}{\cos x} + \dfrac{\cos x}{1 + \sin x} = 4$

52. $\dfrac{\cos x \cot x}{1 - \sin x} = 3$

53. $x \tan x - 1 = 0$ **54.** $x \cos x - 1 = 0$

55. $\sec^2 x + 0.5 \tan x - 1 = 0$

56. $\csc^2 x + 0.5 \cot x - 5 = 0$

57. $2 \tan^2 x + 7 \tan x - 15 = 0$

58. $6 \sin^2 x - 7 \sin x + 2 = 0$

Using the Quadratic Formula   In Exercises 59–62, use the Quadratic Formula to solve the equation in the interval $[0, 2\pi)$. Then use a graphing utility to approximate the angle x.

59. $12 \sin^2 x - 13 \sin x + 3 = 0$

60. $3 \tan^2 x + 4 \tan x - 4 = 0$

61. $\tan^2 x + 3 \tan x + 1 = 0$

62. $4 \cos^2 x - 4 \cos x - 1 = 0$

Using Inverse Functions   In Exercises 63–74, use inverse functions where needed to find all solutions of the equation in the interval $[0, 2\pi)$.

63. $\tan^2 x + \tan x - 12 = 0$

64. $\tan^2 x - \tan x - 2 = 0$

65. $\sec^2 x - 6 \tan x = -4$

66. $\sec^2 x + \tan x - 3 = 0$

67. $2 \sin^2 x + 5 \cos x = 4$ **68.** $2 \cos^2 x + 7 \sin x = 5$

69. $\cot^2 x - 9 = 0$ **70.** $\cot^2 x - 6 \cot x + 5 = 0$

71. $\sec^2 x - 4 \sec x = 0$

72. $\sec^2 x + 2 \sec x - 8 = 0$

73. $\csc^2 x + 3 \csc x - 4 = 0$

74. $\csc^2 x - 5 \csc x = 0$

Approximating Solutions   In Exercises 75–78, use a graphing utility to approximate the solutions (to three decimal places) of the equation in the given interval.

75. $3 \tan^2 x + 5 \tan x - 4 = 0,\quad \left[-\dfrac{\pi}{2}, \dfrac{\pi}{2}\right]$

76. $\cos^2 x - 2 \cos x - 1 = 0,\quad [0, \pi]$

77. $4 \cos^2 x - 2 \sin x + 1 = 0,\quad \left[-\dfrac{\pi}{2}, \dfrac{\pi}{2}\right]$

78. $2 \sec^2 x + \tan x - 6 = 0,\quad \left[-\dfrac{\pi}{2}, \dfrac{\pi}{2}\right]$

Approximating Maximum and Minimum Points   In Exercises 79–84, (a) use a graphing utility to graph the function and approximate the maximum and minimum points on the graph in the interval $[0, 2\pi)$, and (b) solve the trigonometric equation and demonstrate that its solutions are the x-coordinates of the maximum and minimum points of f. (Calculus is required to find the trigonometric equation.)

Function	Trigonometric Equation
79. $f(x) = \sin^2 x + \cos x$	$2 \sin x \cos x - \sin x = 0$
80. $f(x) = \cos^2 x - \sin x$	$-2 \sin x \cos x - \cos x = 0$
81. $f(x) = \sin x + \cos x$	$\cos x - \sin x = 0$
82. $f(x) = 2 \sin x + \cos 2x$	$2 \cos x - 4 \sin x \cos x = 0$
83. $f(x) = \sin x \cos x$	$-\sin^2 x + \cos^2 x = 0$
84. $f(x) = \sec x + \tan x - x$	$\sec x \tan x + \sec^2 x = 1$

Number of Points of Intersection   In Exercises 85 and 86, use the graph to approximate the number of points of intersection of the graphs of y_1 and y_2.

85. $y_1 = 2 \sin x$ **86.** $y_1 = 2 \sin x$
$\quad\ y_2 = 3x + 1$ $\quad\ y_2 = \frac{1}{2}x + 1$

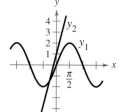
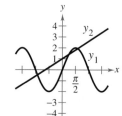

87. Graphical Reasoning   Consider the function $f(x) = (\sin x)/x$ and its graph shown in the figure.

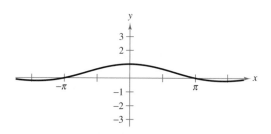

(a) What is the domain of the function?

(b) Identify any symmetry and any asymptotes of the graph.

(c) Describe the behavior of the function as $x \to 0$.

(d) How many solutions does the equation

$$\frac{\sin x}{x} = 0$$

have in the interval $[-8, 8]$? Find the solutions.

88. Graphical Reasoning Consider the function

$$f(x) = \cos\frac{1}{x}$$

and its graph shown in the figure.

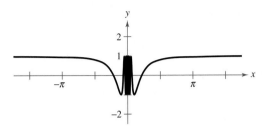

(a) What is the domain of the function?

(b) Identify any symmetry and any asymptotes of the graph.

(c) Describe the behavior of the function as $x \to 0$.

(d) How many solutions does the equation

$$\cos\frac{1}{x} = 0$$

have in the interval $[-1, 1]$? Find the solutions.

(e) Does the equation $\cos(1/x) = 0$ have a greatest solution? If so, then approximate the solution. If not, then explain why.

89. Harmonic Motion A weight is oscillating on the end of a spring (see figure). The position of the weight relative to the point of equilibrium is given by

$$y = \tfrac{1}{12}(\cos 8t - 3\sin 8t)$$

where y is the displacement (in meters) and t is the time (in seconds). Find the times when the weight is at the point of equilibrium $(y = 0)$ for $0 \le t \le 1$.

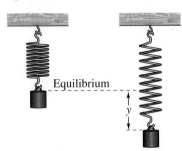

Equilibrium

90. Damped Harmonic Motion The displacement from equilibrium of a weight oscillating on the end of a spring is given by

$$y = 1.56t^{-1/2}\cos 1.9t$$

where y is the displacement (in feet) and t is the time (in seconds). Use a graphing utility to graph the displacement function for $0 \le t \le 10$. Find the time beyond which the displacement does not exceed 1 foot from equilibrium.

91. Sales The monthly sales S (in hundreds of units) of skiing equipment at a sports store are approximated by

$$S = 58.3 + 32.5\cos\frac{\pi t}{6}$$

where t is the time (in months), with $t = 1$ corresponding to January. Determine the months in which sales exceed 7500 units.

92. Projectile Motion A baseball is hit at an angle of θ with the horizontal and with an initial velocity of $v_0 = 100$ feet per second. An outfielder catches the ball 300 feet from home plate (see figure). Find θ when the range r of a projectile is given by

$$r = \frac{1}{32}v_0^2\sin 2\theta.$$

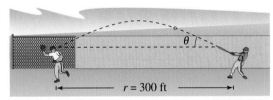

$r = 300$ ft

Not drawn to scale

93. Data Analysis: Meteorology The table shows the normal daily high temperatures in Houston H (in degrees Fahrenheit) for month t, with $t = 1$ corresponding to January. *(Source: NOAA)*

Month, t	Houston, H
1	62.3
2	66.5
3	73.3
4	79.1
5	85.5
6	90.7
7	93.6
8	93.5
9	89.3
10	82.0
11	72.0
12	64.6

Spreadsheet at LarsonPrecalculus.com

(a) Create a scatter plot of the data.

(b) Find a cosine model for the temperatures.

(c) Use a graphing utility to graph the data points and the model for the temperatures. How well does the model fit the data?

(d) What is the overall normal daily high temperature?

(e) Use the graphing utility to describe the months during which the normal daily high temperature is above 86°F and below 86°F.

94. Ferris Wheel

The height h (in feet) above ground of a seat on a Ferris wheel at time t (in minutes) can be modeled by

$$h(t) = 53 + 50 \sin\left(\frac{\pi}{16}t - \frac{\pi}{2}\right).$$

The wheel makes one revolution every 32 seconds. The ride begins when $t = 0$.

(a) During the first 32 seconds of the ride, when will a person on the Ferris wheel be 53 feet above ground?

(b) When will a person be at the top of the Ferris wheel for the first time during the ride? If the ride lasts 160 seconds, then how many times will a person be at the top of the ride, and at what times?

95. Geometry The area of a rectangle (see figure) inscribed in one arc of the graph of $y = \cos x$ is given by

$$A = 2x \cos x, \quad 0 < x < \pi/2.$$

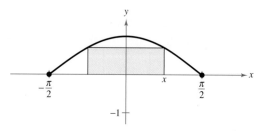

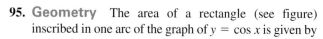 (a) Use a graphing utility to graph the area function, and approximate the area of the largest inscribed rectangle.

(b) Determine the values of x for which $A \geq 1$.

96. Quadratic Approximation Consider the function

$$f(x) = 3 \sin(0.6x - 2).$$

(a) Approximate the zero of the function in the interval $[0, 6]$.

(b) A quadratic approximation agreeing with f at $x = 5$ is

$$g(x) = -0.45x^2 + 5.52x - 13.70.$$

Use a graphing utility to graph f and g in the same viewing window. Describe the result.

(c) Use the Quadratic Formula to find the zeros of g. Compare the zero in the interval $[0, 6]$ with the result of part (a).

Fixed Point In Exercises 97 and 98, find the smallest positive fixed point of the function f. [A *fixed point* of a function f is a real number c such that $f(c) = c$.]

97. $f(x) = \tan(\pi x/4)$ **98.** $f(x) = \cos x$

Exploration

True or False? In Exercises 99 and 100, determine whether the statement is true or false. Justify your answer.

99. The equation $2 \sin 4t - 1 = 0$ has four times the number of solutions in the interval $[0, 2\pi)$ as the equation $2 \sin t - 1 = 0$.

100. If you correctly solve a trigonometric equation to the statement $\sin x = 3.4$, then you can finish solving the equation by using an inverse function.

101. Think About It Explain what happens when you divide each side of the equation $\cot x \cos^2 x = 2 \cot x$ by $\cot x$. Is this a correct method to use when solving equations?

102. HOW DO YOU SEE IT? Explain how to use the figure to solve the equation $2 \cos x - 1 = 0$.

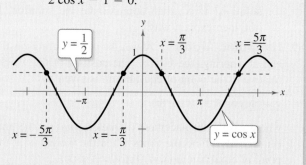

103. Graphical Reasoning Use a graphing utility to confirm the solutions found in Example 6 in two different ways.

(a) Graph both sides of the equation and find the x-coordinates of the points at which the graphs intersect.

Left side: $y = \cos x + 1$

Right side: $y = \sin x$

(b) Graph the equation $y = \cos x + 1 - \sin x$ and find the x-intercepts of the graph. Do both methods produce the same x-values? Which method do you prefer? Explain.

104. Discussion Explain in your own words how knowledge of algebra is important when solving trigonometric equations.

Project: Meteorology To work an extended application analyzing the normal daily high temperatures in Phoenix and in Seattle, visit this text's website at *LarsonPrecalculus.com*. (*Source: NOAA*)

2.4 Sum and Difference Formulas

Trigonometric identities enable you to rewrite trigonometric expressions. For instance, in Exercise 79 on page 240, you will use an identity to rewrite a trigonometric expression in a form that helps you analyze a harmonic motion equation.

■ Use sum and difference formulas to evaluate trigonometric functions, verify identities, and solve trigonometric equations.

Using Sum and Difference Formulas

In this and the following section, you will study the uses of several trigonometric identities and formulas.

Sum and Difference Formulas

$$\sin(u + v) = \sin u \cos v + \cos u \sin v$$

$$\sin(u - v) = \sin u \cos v - \cos u \sin v$$

$$\cos(u + v) = \cos u \cos v - \sin u \sin v$$

$$\cos(u - v) = \cos u \cos v + \sin u \sin v$$

$$\tan(u + v) = \frac{\tan u + \tan v}{1 - \tan u \tan v} \qquad \tan(u - v) = \frac{\tan u - \tan v}{1 + \tan u \tan v}$$

For a proof of the sum and difference formulas for $\cos(u \pm v)$ and $\tan(u \pm v)$, see Proofs in Mathematics on page 256.

Examples 1 and 2 show how **sum and difference formulas** can enable you to find exact values of trigonometric functions involving sums or differences of special angles.

EXAMPLE 1 **Evaluating a Trigonometric Function**

Find the exact value of $\sin \dfrac{\pi}{12}$.

Solution To find the *exact* value of $\sin \pi/12$, use the fact that

$$\frac{\pi}{12} = \frac{\pi}{3} - \frac{\pi}{4}.$$

Consequently, the formula for $\sin(u - v)$ yields

$$\sin \frac{\pi}{12} = \sin\left(\frac{\pi}{3} - \frac{\pi}{4}\right)$$

$$= \sin \frac{\pi}{3} \cos \frac{\pi}{4} - \cos \frac{\pi}{3} \sin \frac{\pi}{4}$$

$$= \frac{\sqrt{3}}{2}\left(\frac{\sqrt{2}}{2}\right) - \frac{1}{2}\left(\frac{\sqrt{2}}{2}\right)$$

$$= \frac{\sqrt{6} - \sqrt{2}}{4}.$$

Try checking this result on your calculator. You will find that $\sin \pi/12 \approx 0.259$.

✓ *Checkpoint* ◀))) *Audio-video solution in English & Spanish at LarsonPrecalculus.com.*

Find the exact value of $\cos \dfrac{\pi}{12}$.   ■

•• **REMARK** Another way to
solve Example 2 is to use the
fact that $75° = 120° - 45°$
together with the formula for
$\cos(u - v)$.

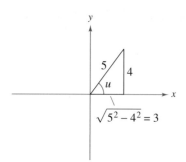

Figure 2.3

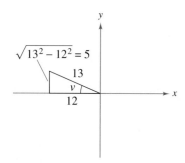

Figure 2.4

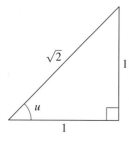

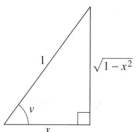

Figure 2.5

EXAMPLE 2 **Evaluating a Trigonometric Function**

Find the exact value of $\cos 75°$.

Solution Using the fact that $75° = 30° + 45°$, together with the formula for $\cos(u + v)$, you obtain

$$\cos 75° = \cos(30° + 45°)$$

$$= \cos 30° \cos 45° - \sin 30° \sin 45°$$

$$= \frac{\sqrt{3}}{2}\left(\frac{\sqrt{2}}{2}\right) - \frac{1}{2}\left(\frac{\sqrt{2}}{2}\right)$$

$$= \frac{\sqrt{6} - \sqrt{2}}{4}.$$

✓ **Checkpoint** 🔊))) *Audio-video solution in English & Spanish at LarsonPrecalculus.com.*

Find the exact value of $\sin 75°$.

EXAMPLE 3 **Evaluating a Trigonometric Expression**

Find the exact value of $\sin(u + v)$ given $\sin u = 4/5$, where $0 < u < \pi/2$, and $\cos v = -12/13$, where $\pi/2 < v < \pi$.

Solution Because $\sin u = 4/5$ and u is in Quadrant I, $\cos u = 3/5$, as shown in Figure 2.3. Because $\cos v = -12/13$ and v is in Quadrant II, $\sin v = 5/13$, as shown in Figure 2.4. You can find $\sin(u + v)$ as follows.

$$\sin(u + v) = \sin u \cos v + \cos u \sin v$$

$$= \frac{4}{5}\left(-\frac{12}{13}\right) + \frac{3}{5}\left(\frac{5}{13}\right)$$

$$= -\frac{33}{65}$$

✓ **Checkpoint** 🔊))) *Audio-video solution in English & Spanish at LarsonPrecalculus.com.*

Find the exact value of $\cos(u + v)$ given $\sin u = 12/13$, where $0 < u < \pi/2$, and $\cos v = -3/5$, where $\pi/2 < v < \pi$.

EXAMPLE 4 **An Application of a Sum Formula**

Write $\cos(\arctan 1 + \arccos x)$ as an algebraic expression.

Solution This expression fits the formula for $\cos(u + v)$. Figure 2.5 shows angles $u = \arctan 1$ and $v = \arccos x$. So,

$$\cos(u + v) = \cos(\arctan 1) \cos(\arccos x) - \sin(\arctan 1) \sin(\arccos x)$$

$$= \frac{1}{\sqrt{2}} \cdot x - \frac{1}{\sqrt{2}} \cdot \sqrt{1 - x^2}$$

$$= \frac{x - \sqrt{1 - x^2}}{\sqrt{2}}.$$

✓ **Checkpoint** 🔊))) *Audio-video solution in English & Spanish at LarsonPrecalculus.com.*

Write $\sin(\arctan 1 + \arccos x)$ as an algebraic expression. ∎

Hipparchus, considered the most eminent of Greek astronomers, was born about 190 B.C. in Nicaea. He is credited with the invention of trigonometry. He also derived the sum and difference formulas for $\sin(A \pm B)$ and $\cos(A \pm B)$.

EXAMPLE 5   **Proving a Cofunction Identity**

Use a difference formula to prove the cofunction identity $\cos\left(\dfrac{\pi}{2} - x\right) = \sin x$.

Solution   Using the formula for $\cos(u - v)$, you have

$$\cos\left(\frac{\pi}{2} - x\right) = \cos\frac{\pi}{2}\cos x + \sin\frac{\pi}{2}\sin x$$

$$= (0)(\cos x) + (1)(\sin x)$$

$$= \sin x.$$

✓ **Checkpoint** ◀))) *Audio-video solution in English & Spanish at LarsonPrecalculus.com.*

Use a difference formula to prove the cofunction identity $\sin\left(x - \dfrac{\pi}{2}\right) = -\cos x$. ■

Sum and difference formulas can be used to rewrite expressions such as

$$\sin\left(\theta + \frac{n\pi}{2}\right) \quad \text{and} \quad \cos\left(\theta + \frac{n\pi}{2}\right), \quad \text{where } n \text{ is an integer}$$

as expressions involving only $\sin\theta$ or $\cos\theta$. The resulting formulas are called **reduction formulas.**

EXAMPLE 6   **Deriving Reduction Formulas**

Simplify each expression.

a. $\cos\left(\theta - \dfrac{3\pi}{2}\right)$

b. $\tan(\theta + 3\pi)$

Solution

a. Using the formula for $\cos(u - v)$, you have

$$\cos\left(\theta - \frac{3\pi}{2}\right) = \cos\theta\cos\frac{3\pi}{2} + \sin\theta\sin\frac{3\pi}{2}$$

$$= (\cos\theta)(0) + (\sin\theta)(-1)$$

$$= -\sin\theta.$$

b. Using the formula for $\tan(u + v)$, you have

$$\tan(\theta + 3\pi) = \frac{\tan\theta + \tan 3\pi}{1 - \tan\theta\tan 3\pi}$$

$$= \frac{\tan\theta + 0}{1 - (\tan\theta)(0)}$$

$$= \tan\theta.$$

✓ **Checkpoint** ◀))) *Audio-video solution in English & Spanish at LarsonPrecalculus.com.*

Simplify each expression.

a. $\sin\left(\dfrac{3\pi}{2} - \theta\right)$ **b.** $\tan\left(\theta - \dfrac{\pi}{4}\right)$ ■

EXAMPLE 7 **Solving a Trigonometric Equation**

Find all solutions of $\sin[x + (\pi/4)] + \sin[x - (\pi/4)] = -1$ in the interval $[0, 2\pi)$.

Algebraic Solution

Using sum and difference formulas, rewrite the equation as

$$\sin x \cos \frac{\pi}{4} + \cos x \sin \frac{\pi}{4} + \sin x \cos \frac{\pi}{4} - \cos x \sin \frac{\pi}{4} = -1$$

$$2 \sin x \cos \frac{\pi}{4} = -1$$

$$2(\sin x)\left(\frac{\sqrt{2}}{2}\right) = -1$$

$$\sin x = -\frac{1}{\sqrt{2}}$$

$$\sin x = -\frac{\sqrt{2}}{2}.$$

So, the only solutions in the interval $[0, 2\pi)$ are $x = 5\pi/4$ and $x = 7\pi/4$.

Graphical Solution

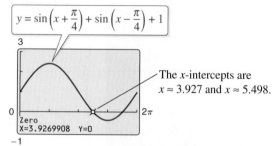

The x-intercepts are $x \approx 3.927$ and $x \approx 5.498$.

From the above figure, you can conclude that the approximate solutions in the interval $[0, 2\pi)$ are

$$x \approx 3.927 \approx \frac{5\pi}{4} \quad \text{and} \quad x \approx 5.498 \approx \frac{7\pi}{4}.$$

 ✓ **Checkpoint** )) *Audio-video solution in English & Spanish at LarsonPrecalculus.com.*

Find all solutions of $\sin[x + (\pi/2)] + \sin[x - (3\pi/2)] = 1$ in the interval $[0, 2\pi)$.

The next example is an application from calculus.

EXAMPLE 8 **An Application from Calculus**

Verify that $\dfrac{\sin(x + h) - \sin x}{h} = (\cos x)\left(\dfrac{\sin h}{h}\right) - (\sin x)\left(\dfrac{1 - \cos h}{h}\right)$, where $h \neq 0$.

Solution Using the formula for $\sin(u + v)$, you have

$$\frac{\sin(x + h) - \sin x}{h} = \frac{\sin x \cos h + \cos x \sin h - \sin x}{h}$$

$$= \frac{\cos x \sin h - \sin x(1 - \cos h)}{h}$$

$$= (\cos x)\left(\frac{\sin h}{h}\right) - (\sin x)\left(\frac{1 - \cos h}{h}\right).$$

✓ **Checkpoint**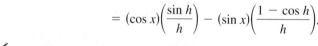

Verify that $\dfrac{\cos(x + h) - \cos x}{h} = (\cos x)\left(\dfrac{\cos h - 1}{h}\right) - (\sin x)\left(\dfrac{\sin h}{h}\right)$, where $h \neq 0$.

One application of the sum and difference formulas is in the analysis of standing waves, such as those that can be produced when plucking a guitar string. You will investigate standing waves in Exercise 80.

Summarize (Section 2.4)
1. State the sum and difference formulas for sine, cosine, and tangent *(page 235)*. For examples of using the sum and difference formulas to evaluate trigonometric functions, verify identities, and solve trigonometric equations, see Examples 1–8.

2.4 Exercises

See **CalcChat.com** for tutorial help and worked-out solutions to odd-numbered exercises.

Vocabulary: Fill in the blank.

1. $\sin(u - v) =$ _____

2. $\cos(u + v) =$ _____

3. $\tan(u + v) =$ _____

4. $\sin(u + v) =$ _____

5. $\cos(u - v) =$ _____

6. $\tan(u - v) =$ _____

Skills and Applications

Evaluating Trigonometric Expressions In Exercises 7–10, find the exact value of each expression.

7. (a) $\cos\left(\dfrac{\pi}{4} + \dfrac{\pi}{3}\right)$ (b) $\cos\dfrac{\pi}{4} + \cos\dfrac{\pi}{3}$

8. (a) $\sin\left(\dfrac{7\pi}{6} - \dfrac{\pi}{3}\right)$ (b) $\sin\dfrac{7\pi}{6} - \sin\dfrac{\pi}{3}$

9. (a) $\sin(135° - 30°)$ (b) $\sin 135° - \cos 30°$

10. (a) $\cos(120° + 45°)$ (b) $\cos 120° + \cos 45°$

Evaluating Trigonometric Functions In Exercises 11–26, find the exact values of the sine, cosine, and tangent of the angle.

11. $\dfrac{11\pi}{12} = \dfrac{3\pi}{4} + \dfrac{\pi}{6}$

12. $\dfrac{7\pi}{12} = \dfrac{\pi}{3} + \dfrac{\pi}{4}$

13. $\dfrac{17\pi}{12} = \dfrac{9\pi}{4} - \dfrac{5\pi}{6}$

14. $-\dfrac{\pi}{12} = \dfrac{\pi}{6} - \dfrac{\pi}{4}$

15. $105° = 60° + 45°$

16. $165° = 135° + 30°$

17. $195° = 225° - 30°$

18. $255° = 300° - 45°$

19. $\dfrac{13\pi}{12}$

20. $-\dfrac{7\pi}{12}$

21. $-\dfrac{13\pi}{12}$

22. $\dfrac{5\pi}{12}$

23. $285°$

24. $-105°$

25. $-165°$

26. $15°$

Rewriting a Trigonometric Expression In Exercises 27–34, write the expression as the sine, cosine, or tangent of an angle.

27. $\sin 3 \cos 1.2 - \cos 3 \sin 1.2$

28. $\cos\dfrac{\pi}{7} \cos\dfrac{\pi}{5} - \sin\dfrac{\pi}{7} \sin\dfrac{\pi}{5}$

29. $\sin 60° \cos 15° + \cos 60° \sin 15°$

30. $\cos 130° \cos 40° - \sin 130° \sin 40°$

31. $\dfrac{\tan 45° - \tan 30°}{1 + \tan 45° \tan 30°}$

32. $\dfrac{\tan 140° - \tan 60°}{1 + \tan 140° \tan 60°}$

33. $\cos 3x \cos 2y + \sin 3x \sin 2y$

34. $\dfrac{\tan 2x + \tan x}{1 - \tan 2x \tan x}$

Evaluating a Trigonometric Expression In Exercises 35–40, find the exact value of the expression.

35. $\sin\dfrac{\pi}{12} \cos\dfrac{\pi}{4} + \cos\dfrac{\pi}{12} \sin\dfrac{\pi}{4}$

36. $\cos\dfrac{\pi}{16} \cos\dfrac{3\pi}{16} - \sin\dfrac{\pi}{16} \sin\dfrac{3\pi}{16}$

37. $\sin 120° \cos 60° - \cos 120° \sin 60°$

38. $\cos 120° \cos 30° + \sin 120° \sin 30°$

39. $\dfrac{\tan(5\pi/6) - \tan(\pi/6)}{1 + \tan(5\pi/6) \tan(\pi/6)}$

40. $\dfrac{\tan 25° + \tan 110°}{1 - \tan 25° \tan 110°}$

Evaluating a Trigonometric Expression In Exercises 41–46, find the exact value of the trigonometric expression given that $\sin u = \frac{5}{13}$ and $\cos v = -\frac{3}{5}$. (Both u and v are in Quadrant II.)

41. $\sin(u + v)$ **42.** $\cos(u - v)$

43. $\tan(u + v)$ **44.** $\csc(u - v)$

45. $\sec(v - u)$ **46.** $\cot(u + v)$

Evaluating a Trigonometric Expression In Exercises 47–52, find the exact value of the trigonometric expression given that $\sin u = -\frac{7}{25}$ and $\cos v = -\frac{4}{5}$. (Both u and v are in Quadrant III.)

47. $\cos(u + v)$ **48.** $\sin(u + v)$

49. $\tan(u - v)$ **50.** $\cot(v - u)$

51. $\csc(u - v)$ **52.** $\sec(v - u)$

An Application of a Sum or Difference Formula In Exercises 53–56, write the trigonometric expression as an algebraic expression.

53. $\sin(\arcsin x + \arccos x)$ **54.** $\sin(\arctan 2x - \arccos x)$

55. $\cos(\arccos x + \arcsin x)$

56. $\cos(\arccos x - \arctan x)$

Proving a Trigonometric Identity In Exercises 57–64, prove the identity.

57. $\sin\left(\dfrac{\pi}{2} - x\right) = \cos x$

58. $\sin\left(\dfrac{\pi}{2} + x\right) = \cos x$

59. $\sin\left(\dfrac{\pi}{6} + x\right) = \dfrac{1}{2}\left(\cos x + \sqrt{3}\sin x\right)$

60. $\cos\left(\dfrac{5\pi}{4} - x\right) = -\dfrac{\sqrt{2}}{2}(\cos x + \sin x)$

61. $\cos(\pi - \theta) + \sin\left(\dfrac{\pi}{2} + \theta\right) = 0$

62. $\tan\left(\dfrac{\pi}{4} - \theta\right) = \dfrac{1 - \tan\theta}{1 + \tan\theta}$

63. $\cos(x + y)\cos(x - y) = \cos^2 x - \sin^2 y$

64. $\sin(x + y) + \sin(x - y) = 2\sin x \cos y$

Deriving a Reduction Formula In Exercises 65–68, simplify the expression algebraically and use a graphing utility to confirm your answer graphically.

65. $\cos\left(\dfrac{3\pi}{2} - x\right)$

66. $\cos(\pi + x)$

67. $\sin\left(\dfrac{3\pi}{2} + \theta\right)$

68. $\tan(\pi + \theta)$

Solving a Trigonometric Equation In Exercises 69–74, find all solutions of the equation in the interval $[0, 2\pi)$.

69. $\sin(x + \pi) - \sin x + 1 = 0$

70. $\cos(x + \pi) - \cos x - 1 = 0$

71. $\cos\left(x + \dfrac{\pi}{4}\right) - \cos\left(x - \dfrac{\pi}{4}\right) = 1$

72. $\sin\left(x + \dfrac{\pi}{6}\right) - \sin\left(x - \dfrac{7\pi}{6}\right) = \dfrac{\sqrt{3}}{2}$

73. $\tan(x + \pi) + 2\sin(x + \pi) = 0$

74. $\sin\left(x + \dfrac{\pi}{2}\right) - \cos^2 x = 0$

Approximating Solutions In Exercises 75–78, use a graphing utility to approximate the solutions of the equation in the interval $[0, 2\pi)$.

75. $\cos\left(x + \dfrac{\pi}{4}\right) + \cos\left(x - \dfrac{\pi}{4}\right) = 1$

76. $\tan(x + \pi) - \cos\left(x + \dfrac{\pi}{2}\right) = 0$

77. $\sin\left(x + \dfrac{\pi}{2}\right) + \cos^2 x = 0$

78. $\cos\left(x - \dfrac{\pi}{2}\right) - \sin^2 x = 0$

79. Harmonic Motion

A weight is attached to a spring suspended vertically from a ceiling. When a driving force is applied to the system, the weight moves vertically from its equilibrium position, and this motion is modeled by

$$y = \dfrac{1}{3}\sin 2t + \dfrac{1}{4}\cos 2t$$

where y is the distance from equilibrium (in feet) and t is the time (in seconds).

(a) Use the identity

$$a\sin B\theta + b\cos B\theta = \sqrt{a^2 + b^2}\,\sin(B\theta + C)$$

where $C = \arctan(b/a)$, $a > 0$, to write the model in the form

$$y = \sqrt{a^2 + b^2}\,\sin(Bt + C).$$

(b) Find the amplitude of the oscillations of the weight.

(c) Find the frequency of the oscillations of the weight.

80. Standing Waves The equation of a standing wave is obtained by adding the displacements of two waves traveling in opposite directions (see figure). Assume that each of the waves has amplitude A, period T, and wavelength λ. If the models for these waves are

$$y_1 = A\cos 2\pi\left(\dfrac{t}{T} - \dfrac{x}{\lambda}\right) \quad \text{and} \quad y_2 = A\cos 2\pi\left(\dfrac{t}{T} + \dfrac{x}{\lambda}\right)$$

then show that

$$y_1 + y_2 = 2A\cos\dfrac{2\pi t}{T}\cos\dfrac{2\pi x}{\lambda}.$$

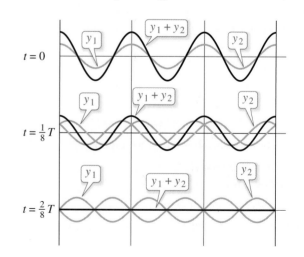

Exploration

True or False? **In Exercises 81–84, determine whether the statement is true or false. Justify your answer.**

81. $\sin(u \pm v) = \sin u \cos v \pm \cos u \sin v$

82. $\cos(u \pm v) = \cos u \cos v \pm \sin u \sin v$

83. $\tan\left(x - \dfrac{\pi}{4}\right) = \dfrac{\tan x + 1}{1 - \tan x}$

84. $\sin\left(x - \dfrac{\pi}{2}\right) = -\cos x$

85. An Application from Calculus Let $x = \pi/3$ in the identity in Example 8 and define the functions f and g as follows.

$$f(h) = \frac{\sin[(\pi/3) + h] - \sin(\pi/3)}{h}$$

$$g(h) = \cos\frac{\pi}{3}\left(\frac{\sin h}{h}\right) - \sin\frac{\pi}{3}\left(\frac{1 - \cos h}{h}\right)$$

(a) What are the domains of the functions f and g?

(b) Use a graphing utility to complete the table.

h	0.5	0.2	0.1	0.05	0.02	0.01
$f(h)$						
$g(h)$						

(c) Use the graphing utility to graph the functions f and g.

(d) Use the table and the graphs to make a conjecture about the values of the functions f and g as $h \to 0^+$.

86. HOW DO YOU SEE IT? Explain how to use the figure to justify each statement.

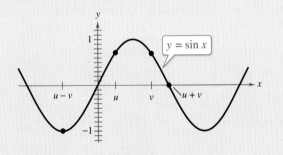

(a) $\sin(u + v) \neq \sin u + \sin v$

(b) $\sin(u - v) \neq \sin u - \sin v$

Verifying an Identity **In Exercises 87–90, verify the identity.**

87. $\cos(n\pi + \theta) = (-1)^n \cos \theta$, n is an integer

88. $\sin(n\pi + \theta) = (-1)^n \sin \theta$, n is an integer

89. $a \sin B\theta + b \cos B\theta = \sqrt{a^2 + b^2} \sin(B\theta + C)$, where $C = \arctan(b/a)$ and $a > 0$

90. $a \sin B\theta + b \cos B\theta = \sqrt{a^2 + b^2} \cos(B\theta - C)$, where $C = \arctan(a/b)$ and $b > 0$

Rewriting a Trigonometric Expression **In Exercises 91–94, use the formulas given in Exercises 89 and 90 to write the trigonometric expression in the following forms.**

(a) $\sqrt{a^2 + b^2} \sin(B\theta + C)$ (b) $\sqrt{a^2 + b^2} \cos(B\theta - C)$

91. $\sin \theta + \cos \theta$ 　　　　　**92.** $3 \sin 2\theta + 4 \cos 2\theta$

93. $12 \sin 3\theta + 5 \cos 3\theta$ 　**94.** $\sin 2\theta + \cos 2\theta$

Rewriting a Trigonometric Expression **In Exercises 95 and 96, use the formulas given in Exercises 89 and 90 to write the trigonometric expression in the form $a \sin B\theta + b \cos B\theta$.**

95. $2 \sin\left(\theta + \dfrac{\pi}{4}\right)$ 　　**96.** $5 \cos\left(\theta - \dfrac{\pi}{4}\right)$

Angle Between Two Lines **In Exercises 97 and 98, use the figure, which shows two lines whose equations are $y_1 = m_1 x + b_1$ and $y_2 = m_2 x + b_2$. Assume that both lines have positive slopes. Derive a formula for the angle between the two lines. Then use your formula to find the angle between the given pair of lines.**

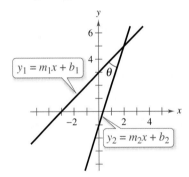

97. $y = x$ and $y = \sqrt{3}x$

98. $y = x$ and $y = \dfrac{1}{\sqrt{3}}x$

Graphical Reasoning **In Exercises 99 and 100, use a graphing utility to graph y_1 and y_2 in the same viewing window. Use the graphs to determine whether $y_1 = y_2$. Explain your reasoning.**

99. $y_1 = \cos(x + 2)$, 　$y_2 = \cos x + \cos 2$

100. $y_1 = \sin(x + 4)$, 　$y_2 = \sin x + \sin 4$

101. Proof

(a) Write a proof of the formula for $\sin(u + v)$.

(b) Write a proof of the formula for $\sin(u - v)$.

2.5 Multiple-Angle and Product-to-Sum Formulas

A variety of trigonometric formulas enable you to rewrite trigonometric functions in more convenient forms. For instance, in Exercise 73 on page 250, you will use a half-angle formula to relate the Mach number of a supersonic airplane to the apex angle of the cone formed by the sound waves behind the airplane.

■ Use multiple-angle formulas to rewrite and evaluate trigonometric functions.
■ Use power-reducing formulas to rewrite and evaluate trigonometric functions.
■ Use half-angle formulas to rewrite and evaluate trigonometric functions.
■ Use product-to-sum and sum-to-product formulas to rewrite and evaluate trigonometric functions.
■ Use trigonometric formulas to rewrite real-life models.

Multiple-Angle Formulas

In this section, you will study four other categories of trigonometric identities.

1. The first category involves *functions of multiple angles* such as $\sin ku$ and $\cos ku$.

2. The second category involves *squares of trigonometric functions* such as $\sin^2 u$.

3. The third category involves *functions of half-angles* such as $\sin(u/2)$.

4. The fourth category involves *products of trigonometric functions* such as $\sin u \cos v$.

You should learn the **double-angle formulas** because they are used often in trigonometry and calculus. For proofs of these formulas, see Proofs in Mathematics on page 257.

Double-Angle Formulas

$$\sin 2u = 2 \sin u \cos u \qquad\qquad \cos 2u = \cos^2 u - \sin^2 u$$

$$\tan 2u = \frac{2 \tan u}{1 - \tan^2 u} \qquad\qquad\qquad = 2 \cos^2 u - 1$$

$$\qquad\qquad\qquad\qquad\qquad\qquad = 1 - 2 \sin^2 u$$

EXAMPLE 1   **Solving a Multiple-Angle Equation**

Solve $2 \cos x + \sin 2x = 0$.

Solution Begin by rewriting the equation so that it involves functions of x (rather than $2x$). Then factor and solve.

$$2 \cos x + \sin 2x = 0 \qquad \text{Write original equation.}$$

$$2 \cos x + 2 \sin x \cos x = 0 \qquad \text{Double-angle formula}$$

$$2 \cos x(1 + \sin x) = 0 \qquad \text{Factor.}$$

$$2 \cos x = 0 \quad \text{and} \quad 1 + \sin x = 0 \qquad \text{Set factors equal to zero.}$$

$$x = \frac{\pi}{2}, \frac{3\pi}{2} \qquad\qquad x = \frac{3\pi}{2} \qquad \text{Solutions in } [0, 2\pi)$$

So, the general solution is

$$x = \frac{\pi}{2} + 2n\pi \quad \text{and} \quad x = \frac{3\pi}{2} + 2n\pi$$

where n is an integer. Try verifying these solutions graphically.

✓ **Checkpoint** ◀))) Audio-video solution in English & Spanish at LarsonPrecalculus.com.

Solve $\cos 2x + \cos x = 0$.

Lukich/Shutterstock.com

EXAMPLE 2   **Evaluating Functions Involving Double Angles**

Use the following to find $\sin 2\theta$, $\cos 2\theta$, and $\tan 2\theta$.

$$\cos \theta = \frac{5}{13}, \quad \frac{3\pi}{2} < \theta < 2\pi$$

Solution   From Figure 2.6,

$$\sin \theta = \frac{y}{r} = -\frac{12}{13} \quad \text{and} \quad \tan \theta = \frac{y}{x} = -\frac{12}{5}.$$

Consequently, using each of the double-angle formulas, you can write

$$\sin 2\theta = 2 \sin \theta \cos \theta = 2\left(-\frac{12}{13}\right)\left(\frac{5}{13}\right) = -\frac{120}{169}$$

$$\cos 2\theta = 2 \cos^2 \theta - 1 = 2\left(\frac{25}{169}\right) - 1 = -\frac{119}{169}$$

$$\tan 2\theta = \frac{2 \tan \theta}{1 - \tan^2 \theta} = \frac{2\left(-\dfrac{12}{5}\right)}{1 - \left(-\dfrac{12}{5}\right)^2} = \frac{120}{119}.$$

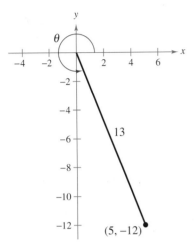

Figure 2.6

✓ *Checkpoint* ◀))) *Audio-video solution in English & Spanish at LarsonPrecalculus.com.*

Use the following to find $\sin 2\theta$, $\cos 2\theta$, and $\tan 2\theta$.

$$\sin \theta = \frac{3}{5}, \quad 0 < \theta < \frac{\pi}{2}$$ ■

The double-angle formulas are not restricted to angles 2θ and θ. Other *double* combinations, such as 4θ and 2θ or 6θ and 3θ, are also valid. Here are two examples.

$$\sin 4\theta = 2 \sin 2\theta \cos 2\theta \quad \text{and} \quad \cos 6\theta = \cos^2 3\theta - \sin^2 3\theta$$

By using double-angle formulas together with the sum formulas given in the preceding section, you can form other multiple-angle formulas.

EXAMPLE 3   **Deriving a Triple-Angle Formula**

Rewrite $\sin 3x$ in terms of $\sin x$.

Solution

$\sin 3x = \sin(2x + x)$	Rewrite as a sum.
$= \sin 2x \cos x + \cos 2x \sin x$	Sum formula
$= 2 \sin x \cos x \cos x + (1 - 2 \sin^2 x) \sin x$	Double-angle formulas
$= 2 \sin x \cos^2 x + \sin x - 2 \sin^3 x$	Distributive Property
$= 2 \sin x(1 - \sin^2 x) + \sin x - 2 \sin^3 x$	Pythagorean identity
$= 2 \sin x - 2 \sin^3 x + \sin x - 2 \sin^3 x$	Distributive Property
$= 3 \sin x - 4 \sin^3 x$	Simplify.

✓ *Checkpoint* ◀))) *Audio-video solution in English & Spanish at LarsonPrecalculus.com.*

Rewrite $\cos 3x$ in terms of $\cos x$.   ■

Power-Reducing Formulas

The double-angle formulas can be used to obtain the following **power-reducing formulas.**

Power-Reducing Formulas

$$\sin^2 u = \frac{1 - \cos 2u}{2}$$

$$\cos^2 u = \frac{1 + \cos 2u}{2}$$

$$\tan^2 u = \frac{1 - \cos 2u}{1 + \cos 2u}$$

For a proof of the power-reducing formulas, see Proofs in Mathematics on page 257. Example 4 shows a typical power reduction used in calculus.

EXAMPLE 4 **Reducing a Power**

Rewrite $\sin^4 x$ in terms of first powers of the cosines of multiple angles.

Solution Note the repeated use of power-reducing formulas.

$$\sin^4 x = (\sin^2 x)^2 \qquad \text{Property of exponents}$$

$$= \left(\frac{1 - \cos 2x}{2}\right)^2 \qquad \text{Power-reducing formula}$$

$$= \frac{1}{4}(1 - 2\cos 2x + \cos^2 2x) \qquad \text{Expand.}$$

$$= \frac{1}{4}\left(1 - 2\cos 2x + \frac{1 + \cos 4x}{2}\right) \qquad \text{Power-reducing formula}$$

$$= \frac{1}{4} - \frac{1}{2}\cos 2x + \frac{1}{8} + \frac{1}{8}\cos 4x \qquad \text{Distributive Property}$$

$$= \frac{3}{8} - \frac{1}{2}\cos 2x + \frac{1}{8}\cos 4x \qquad \text{Simplify.}$$

$$= \frac{1}{8}(3 - 4\cos 2x + \cos 4x) \qquad \text{Factor out common factor.}$$

You can use a graphing utility to check this result, as shown below. Notice that the graphs coincide.

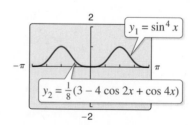

$y_1 = \sin^4 x$

$y_2 = \frac{1}{8}(3 - 4\cos 2x + \cos 4x)$

 Checkpoint Audio-video solution in English & Spanish at *LarsonPrecalculus.com*.

Rewrite $\tan^4 x$ in terms of first powers of the cosines of multiple angles.

Half-Angle Formulas

You can derive some useful alternative forms of the power-reducing formulas by replacing u with $u/2$. The results are called **half-angle formulas.**

•• REMARK To find the exact value of a trigonometric function with an angle measure in D°M′S″ form using a half-angle formula, first convert the angle measure to decimal degree form. Then multiply the resulting angle measure by 2. ▷

> ### Half-Angle Formulas
>
> $$\sin\frac{u}{2} = \pm\sqrt{\frac{1-\cos u}{2}} \qquad \cos\frac{u}{2} = \pm\sqrt{\frac{1+\cos u}{2}}$$
>
> $$\tan\frac{u}{2} = \frac{1-\cos u}{\sin u} = \frac{\sin u}{1+\cos u}$$
>
> The signs of $\sin\dfrac{u}{2}$ and $\cos\dfrac{u}{2}$ depend on the quadrant in which $\dfrac{u}{2}$ lies.

•• REMARK Use your calculator to verify the result obtained in Example 5. That is, evaluate $\sin 105°$ and $\left(\sqrt{2+\sqrt{3}}\right)/2$. Note that both values are approximately 0.9659258. ▷

EXAMPLE 5   **Using a Half-Angle Formula**

Find the exact value of $\sin 105°$.

Solution   Begin by noting that $105°$ is half of $210°$. Then, using the half-angle formula for $\sin(u/2)$ and the fact that $105°$ lies in Quadrant II, you have

$$\sin 105° = \sqrt{\frac{1-\cos 210°}{2}} = \sqrt{\frac{1+\left(\sqrt{3}/2\right)}{2}} = \frac{\sqrt{2+\sqrt{3}}}{2}.$$

The positive square root is chosen because $\sin\theta$ is positive in Quadrant II.

✓ **Checkpoint** ◀))) *Audio-video solution in English & Spanish at LarsonPrecalculus.com.*

Find the exact value of $\cos 105°$.

EXAMPLE 6   **Solving a Trigonometric Equation**

Find all solutions of $1 + \cos^2 x = 2\cos^2\dfrac{x}{2}$ in the interval $[0, 2\pi)$.

Algebraic Solution

$$1 + \cos^2 x = 2\cos^2\frac{x}{2} \qquad \text{Write original equation.}$$

$$1 + \cos^2 x = 2\left(\pm\sqrt{\frac{1+\cos x}{2}}\right)^2 \qquad \text{Half-angle formula}$$

$$1 + \cos^2 x = 1 + \cos x \qquad \text{Simplify.}$$

$$\cos^2 x - \cos x = 0 \qquad \text{Simplify.}$$

$$\cos x(\cos x - 1) = 0 \qquad \text{Factor.}$$

By setting the factors $\cos x$ and $\cos x - 1$ equal to zero, you find that the solutions in the interval $[0, 2\pi)$ are

$$x = \frac{\pi}{2}, \quad x = \frac{3\pi}{2}, \quad \text{and} \quad x = 0.$$

Graphical Solution

The x-intercepts are $x = 0$, $x \approx 1.571$, and $x \approx 4.712$.

From the above figure, you can conclude that the approximate solutions of $1 + \cos^2 x = 2\cos^2 x/2$ in the interval $[0, 2\pi)$ are

$$x = 0, \quad x \approx 1.571 \approx \frac{\pi}{2}, \quad \text{and} \quad x \approx 4.712 \approx \frac{3\pi}{2}.$$

✓ **Checkpoint** ◀))) *Audio-video solution in English & Spanish at LarsonPrecalculus.com.*

Find all solutions of $\cos^2 x = \sin^2\dfrac{x}{2}$ in the interval $[0, 2\pi)$.

Product-to-Sum Formulas

Each of the following **product-to-sum formulas** can be verified using the sum and difference formulas discussed in the preceding section.

Product-to-Sum Formulas

$$\sin u \sin v = \frac{1}{2}[\cos(u - v) - \cos(u + v)]$$

$$\cos u \cos v = \frac{1}{2}[\cos(u - v) + \cos(u + v)]$$

$$\sin u \cos v = \frac{1}{2}[\sin(u + v) + \sin(u - v)]$$

$$\cos u \sin v = \frac{1}{2}[\sin(u + v) - \sin(u - v)]$$

Product-to-sum formulas are used in calculus to solve problems involving the products of sines and cosines of two different angles.

EXAMPLE 7   Writing Products as Sums

Rewrite the product $\cos 5x \sin 4x$ as a sum or difference.

Solution   Using the appropriate product-to-sum formula, you obtain

$$\cos 5x \sin 4x = \frac{1}{2}[\sin(5x + 4x) - \sin(5x - 4x)]$$

$$= \frac{1}{2} \sin 9x - \frac{1}{2} \sin x.$$

✓ *Checkpoint* ◀))) *Audio-video solution in English & Spanish at LarsonPrecalculus.com.*

Rewrite the product $\sin 5x \cos 3x$ as a sum or difference.

Occasionally, it is useful to reverse the procedure and write a sum of trigonometric functions as a product. This can be accomplished with the following **sum-to-product formulas.**

Sum-to-Product Formulas

$$\sin u + \sin v = 2 \sin\left(\frac{u + v}{2}\right) \cos\left(\frac{u - v}{2}\right)$$

$$\sin u - \sin v = 2 \cos\left(\frac{u + v}{2}\right) \sin\left(\frac{u - v}{2}\right)$$

$$\cos u + \cos v = 2 \cos\left(\frac{u + v}{2}\right) \cos\left(\frac{u - v}{2}\right)$$

$$\cos u - \cos v = -2 \sin\left(\frac{u + v}{2}\right) \sin\left(\frac{u - v}{2}\right)$$

For a proof of the sum-to-product formulas, see Proofs in Mathematics on page 258.

EXAMPLE 8   Using a Sum-to-Product Formula

Find the exact value of $\cos 195° + \cos 105°$.

Solution   Using the appropriate sum-to-product formula, you obtain

$$\cos 195° + \cos 105° = 2 \cos\left(\frac{195° + 105°}{2}\right) \cos\left(\frac{195° - 105°}{2}\right)$$

$$= 2 \cos 150° \cos 45°$$

$$= 2\left(-\frac{\sqrt{3}}{2}\right)\left(\frac{\sqrt{2}}{2}\right)$$

$$= -\frac{\sqrt{6}}{2}.$$

✓ *Checkpoint* *Audio-video solution in English & Spanish at LarsonPrecalculus.com.*

Find the exact value of $\sin 195° + \sin 105°$.

EXAMPLE 9   Solving a Trigonometric Equation

Solve $\sin 5x + \sin 3x = 0$.

Solution

$$\sin 5x + \sin 3x = 0 \qquad \text{Write original equation.}$$

$$2 \sin\left(\frac{5x + 3x}{2}\right) \cos\left(\frac{5x - 3x}{2}\right) = 0 \qquad \text{Sum-to-product formula}$$

$$2 \sin 4x \cos x = 0 \qquad \text{Simplify.}$$

By setting the factor $2 \sin 4x$ equal to zero, you can find that the solutions in the interval $[0, 2\pi)$ are

$$x = 0, \frac{\pi}{4}, \frac{\pi}{2}, \frac{3\pi}{4}, \pi, \frac{5\pi}{4}, \frac{3\pi}{2}, \frac{7\pi}{4}.$$

The equation $\cos x = 0$ yields no additional solutions, so you can conclude that the solutions are of the form $x = n\pi/4$ where n is an integer. To confirm this graphically, sketch the graph of $y = \sin 5x + \sin 3x$, as shown below.

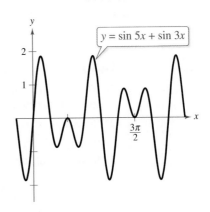

Notice from the graph that the x-intercepts occur at multiples of $\pi/4$.

✓ *Checkpoint* *Audio-video solution in English & Spanish at LarsonPrecalculus.com.*

Solve $\sin 4x - \sin 2x = 0$.

Application

EXAMPLE 10   **Projectile Motion**

Ignoring air resistance, the range of a projectile fired at an angle θ with the horizontal and with an initial velocity of v_0 feet per second is given by

$$r = \frac{1}{16}v_0^2 \sin\theta\cos\theta$$

where r is the horizontal distance (in feet) that the projectile travels. A football player can kick a football from ground level with an initial velocity of 80 feet per second.

a. Write the projectile motion model in a simpler form.

b. At what angle must the player kick the football so that the football travels 200 feet?

Solution

a. You can use a double-angle formula to rewrite the projectile motion model as

$$r = \frac{1}{32}v_0^2(2\sin\theta\cos\theta) \qquad \text{Rewrite original projectile motion model.}$$

$$= \frac{1}{32}v_0^2 \sin 2\theta. \qquad \text{Rewrite model using a double-angle formula.}$$

b. $\quad r = \frac{1}{32}v_0^2 \sin 2\theta \qquad \text{Write projectile motion model.}$

$$200 = \frac{1}{32}(80)^2 \sin 2\theta \qquad \text{Substitute 200 for } r \text{ and 80 for } v_0.$$

$$200 = 200 \sin 2\theta \qquad \text{Simplify.}$$

$$1 = \sin 2\theta \qquad \text{Divide each side by 200.}$$

You know that $2\theta = \pi/2$, so dividing this result by 2 produces $\theta = \pi/4$. Because $\pi/4 = 45°$, the player must kick the football at an angle of 45° so that the football travels 200 feet.

Kicking a football with an initial velocity of 80 feet per second at an angle of 45° with the horizontal results in a distance traveled of 200 feet.

✓ *Checkpoint*  *Audio-video solution in English & Spanish at LarsonPrecalculus.com.*

In Example 10, for what angle is the horizontal distance the football travels a maximum?

Summarize   (Section 2.5)

1. State the double-angle formulas *(page 242)*. For examples of using multiple-angle formulas to rewrite and evaluate trigonometric functions, see Examples 1–3.

2. State the power-reducing formulas *(page 244)*. For an example of using power-reducing formulas to rewrite a trigonometric function, see Example 4.

3. State the half-angle formulas *(page 245)*. For examples of using half-angle formulas to rewrite and evaluate trigonometric functions, see Examples 5 and 6.

4. State the product-to-sum and sum-to-product formulas *(page 246)*. For an example of using a product-to-sum formula to rewrite a trigonometric function, see Example 7. For examples of using sum-to-product formulas to rewrite and evaluate trigonometric functions, see Examples 8 and 9.

5. Describe an example of how to use a trigonometric formula to rewrite a real-life model *(page 248, Example 10)*.

2.5 Exercises

See **CalcChat.com** for tutorial help and worked-out solutions to odd-numbered exercises.

Vocabulary: Fill in the blank to complete the trigonometric formula.

1. $\sin 2u =$ _____

2. $\cos 2u =$ _____

3. $\dfrac{1 - \cos 2u}{1 + \cos 2u} =$ _____

4. $\sin \dfrac{u}{2} =$ _____

5. $\sin u \cos v =$ _____

6. $\cos u - \cos v =$ _____

Skills and Applications

Solving a Multiple-Angle Equation In Exercises 7–14, find the exact solutions of the equation in the interval $[0, 2\pi)$.

7. $\sin 2x - \sin x = 0$

8. $\sin 2x \sin x = \cos x$

9. $\cos 2x - \cos x = 0$

10. $\cos 2x + \sin x = 0$

11. $\sin 4x = -2 \sin 2x$

12. $(\sin 2x + \cos 2x)^2 = 1$

13. $\tan 2x - \cot x = 0$

14. $\tan 2x - 2 \cos x = 0$

Using a Double-Angle Formula In Exercises 15–20, use a double-angle formula to rewrite the expression.

15. $6 \sin x \cos x$

16. $\sin x \cos x$

17. $6 \cos^2 x - 3$

18. $\cos^2 x - \frac{1}{2}$

19. $4 - 8 \sin^2 x$

20. $10 \sin^2 x - 5$

Evaluating Functions Involving Double Angles
In Exercises 21–24, find the exact values of $\sin 2u$, $\cos 2u$, and $\tan 2u$ using the double-angle formulas.

21. $\sin u = -3/5$, $3\pi/2 < u < 2\pi$

22. $\cos u = -4/5$, $\pi/2 < u < \pi$

23. $\tan u = 3/5$, $0 < u < \pi/2$

24. $\sec u = -2$, $\pi < u < 3\pi/2$

25. Deriving a Multiple-Angle Formula Rewrite $\cos 4x$ in terms of $\cos x$.

26. Deriving a Multiple-Angle Formula Rewrite $\tan 3x$ in terms of $\tan x$.

Reducing Powers In Exercises 27–32, use the power-reducing formulas to rewrite the expression in terms of the first power of the cosine.

27. $\cos^4 x$

28. $\sin^4 2x$

29. $\tan^4 2x$

30. $\tan^2 2x \cos^4 2x$

31. $\sin^2 2x \cos^2 2x$

32. $\sin^4 x \cos^2 x$

Using Half-Angle Formulas In Exercises 33–36, use the half-angle formulas to determine the exact values of the sine, cosine, and tangent of the angle.

33. $75°$

34. $67° \, 30'$

35. $\pi/8$

36. $7\pi/12$

Using Half-Angle Formulas In Exercises 37–40, (a) determine the quadrant in which $u/2$ lies, and (b) find the exact values of $\sin(u/2)$, $\cos(u/2)$, and $\tan(u/2)$ using the half-angle formulas.

37. $\cos u = 7/25$, $0 < u < \pi/2$

38. $\sin u = 5/13$, $\pi/2 < u < \pi$

39. $\tan u = -5/12$, $3\pi/2 < u < 2\pi$

40. $\cot u = 3$, $\pi < u < 3\pi/2$

Using Half-Angle Formulas In Exercises 41–44, use the half-angle formulas to simplify the expression.

41. $\sqrt{\dfrac{1 - \cos 6x}{2}}$

42. $\sqrt{\dfrac{1 + \cos 4x}{2}}$

43. $-\sqrt{\dfrac{1 - \cos 8x}{1 + \cos 8x}}$

44. $-\sqrt{\dfrac{1 - \cos(x - 1)}{2}}$

Solving a Trigonometric Equation In Exercises 45–48, find all solutions of the equation in the interval $[0, 2\pi)$. Use a graphing utility to graph the equation and verify the solutions.

45. $\sin \dfrac{x}{2} + \cos x = 0$

46. $\sin \dfrac{x}{2} + \cos x - 1 = 0$

47. $\cos \dfrac{x}{2} - \sin x = 0$

48. $\tan \dfrac{x}{2} - \sin x = 0$

Using Product-to-Sum Formulas In Exercises 49–52, use the product-to-sum formulas to rewrite the product as a sum or difference.

49. $\sin 5\theta \sin 3\theta$

50. $7 \cos(-5\beta) \sin 3\beta$

51. $\cos 2\theta \cos 4\theta$

52. $\sin(x + y) \cos(x - y)$

Using Sum-to-Product Formulas In Exercises 53–56, use the sum-to-product formulas to rewrite the sum or difference as a product.

53. $\sin 5\theta - \sin 3\theta$

54. $\sin 3\theta + \sin \theta$

55. $\cos 6x + \cos 2x$

56. $\cos\left(\theta + \dfrac{\pi}{2}\right) - \cos\left(\theta - \dfrac{\pi}{2}\right)$

Using Sum-to-Product Formulas **In Exercises 57–60, use the sum-to-product formulas to find the exact value of the expression.**

57. $\sin 75° + \sin 15°$

58. $\cos 120° + \cos 60°$

59. $\cos \dfrac{3\pi}{4} - \cos \dfrac{\pi}{4}$

60. $\sin \dfrac{5\pi}{4} - \sin \dfrac{3\pi}{4}$

Solving a Trigonometric Equation **In Exercises 61–64, find all solutions of the equation in the interval $[0, 2\pi)$. Use a graphing utility to graph the equation and verify the solutions.**

61. $\sin 6x + \sin 2x = 0$

62. $\cos 2x - \cos 6x = 0$

63. $\dfrac{\cos 2x}{\sin 3x - \sin x} - 1 = 0$

64. $\sin^2 3x - \sin^2 x = 0$

Verifying a Trigonometric Identity **In Exercises 65–72, verify the identity.**

65. $\csc 2\theta = \dfrac{\csc \theta}{2 \cos \theta}$

66. $\sin \dfrac{\alpha}{3} \cos \dfrac{\alpha}{3} = \dfrac{1}{2} \sin \dfrac{2\alpha}{3}$

67. $1 + \cos 10y = 2 \cos^2 5y$

68. $\cos^4 x - \sin^4 x = \cos 2x$

69. $(\sin x + \cos x)^2 = 1 + \sin 2x$

70. $\tan \dfrac{u}{2} = \csc u - \cot u$

71. $\dfrac{\sin x \pm \sin y}{\cos x + \cos y} = \tan \dfrac{x \pm y}{2}$

72. $\cos\left(\dfrac{\pi}{3} + x\right) + \cos\left(\dfrac{\pi}{3} - x\right) = \cos x$

73. Mach Number • • • • • • • • • • • • • • • •

The Mach number M of a supersonic airplane is the ratio of its speed to the speed of sound. When an airplane travels faster than the speed of sound, the sound waves form a cone behind the airplane. The Mach number is related to the apex angle θ of the cone by $\sin(\theta/2) = 1/M$.

(a) Use a half-angle formula to rewrite the equation in terms of $\cos \theta$.

(b) Find the angle θ that corresponds to a Mach number of 1.

(c) Find the angle θ that corresponds to a Mach number of 4.5.

(d) The speed of sound is about 760 miles per hour. Determine the speed of an object with the Mach numbers from parts (b) and (c).

74. Projectile Motion The range of a projectile fired at an angle θ with the horizontal and with an initial velocity of v_0 feet per second is

$$r = \dfrac{1}{32} v_0^2 \sin 2\theta$$

where r is measured in feet. An athlete throws a javelin at 75 feet per second. At what angle must the athlete throw the javelin so that the javelin travels 130 feet?

75. Railroad Track When two railroad tracks merge, the overlapping portions of the tracks are in the shapes of circular arcs (see figure). The radius of each arc r (in feet) and the angle θ are related by

$$\dfrac{x}{2} = 2r \sin^2 \dfrac{\theta}{2}.$$

Write a formula for x in terms of $\cos \theta$.

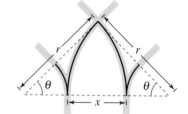

Exploration

 76. **HOW DO YOU SEE IT?** Explain how to use the figure to verify the double-angle formulas (a) $\sin 2u = 2 \sin u \cos u$ and (b) $\cos 2u = \cos^2 u - \sin^2 u$.

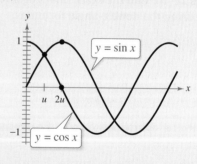

True or False? **In Exercises 77 and 78, determine whether the statement is true or false. Justify your answer.**

77. Because the sine function is an odd function, for a negative number u, $\sin 2u = -2 \sin u \cos u$.

78. $\sin \dfrac{u}{2} = -\sqrt{\dfrac{1 - \cos u}{2}}$ when u is in the second quadrant.

79. Complementary Angles If ϕ and θ are complementary angles, then show that (a) $\sin(\phi - \theta) = \cos 2\theta$ and (b) $\cos(\phi - \theta) = \sin 2\theta$.

Chapter Summary

	What Did You Learn?	Explanation/Examples	Review Exercises
Section 2.1	Recognize and write the fundamental trigonometric identities (p. 210).	**Reciprocal Identities** $\sin u = 1/\csc u \qquad \cos u = 1/\sec u \qquad \tan u = 1/\cot u$ $\csc u = 1/\sin u \qquad \sec u = 1/\cos u \qquad \cot u = 1/\tan u$ **Quotient Identities:** $\tan u = \dfrac{\sin u}{\cos u}, \quad \cot u = \dfrac{\cos u}{\sin u}$ **Pythagorean Identities:** $\sin^2 u + \cos^2 u = 1$, $1 + \tan^2 u = \sec^2 u, \quad 1 + \cot^2 u = \csc^2 u$ **Cofunction Identities** $\sin[(\pi/2) - u] = \cos u \qquad \cos[(\pi/2) - u] = \sin u$ $\tan[(\pi/2) - u] = \cot u \qquad \cot[(\pi/2) - u] = \tan u$ $\sec[(\pi/2) - u] = \csc u \qquad \csc[(\pi/2) - u] = \sec u$ **Even/Odd Identities** $\sin(-u) = -\sin u \qquad \cos(-u) = \cos u \qquad \tan(-u) = -\tan u$ $\csc(-u) = -\csc u \qquad \sec(-u) = \sec u \qquad \cot(-u) = -\cot u$	1–4
	Use the fundamental trigonometric identities to evaluate trigonometric functions, simplify trigonometric expressions, and rewrite trigonometric expressions (p. 211).	In some cases, when factoring or simplifying trigonometric expressions, it is helpful to rewrite the expression in terms of just *one* trigonometric function or in terms of *sine and cosine only*.	5–18
Section 2.2	Verify trigonometric identities (p. 217).	**Guidelines for Verifying Trigonometric Identities** **1.** Work with one side of the equation at a time. **2.** Look to factor an expression, add fractions, square a binomial, or create a monomial denominator. **3.** Look to use the fundamental identities. Note which functions are in the final expression you want. Sines and cosines pair up well, as do secants and tangents, and cosecants and cotangents. **4.** If the preceding guidelines do not help, then try converting all terms to sines and cosines. **5.** Always try *something*.	19–26
Section 2.3	Use standard algebraic techniques to solve trigonometric equations (p. 224).	Use standard algebraic techniques (when possible) such as collecting like terms, extracting square roots, and factoring to solve trigonometric equations.	27–32
	Solve trigonometric equations of quadratic type (p. 227).	To solve trigonometric equations of quadratic type $ax^2 + bx + c = 0$, factor the quadratic or, when this is not possible, use the Quadratic Formula.	33–36
	Solve trigonometric equations involving multiple angles (p. 229).	To solve equations that contain forms such as $\sin ku$ or $\cos ku$, first solve the equation for ku, and then divide your result by k.	37–42
	Use inverse trigonometric functions to solve trigonometric equations (p. 230).	After factoring an equation, you may get an equation such as $(\tan x - 3)(\tan x + 1) = 0$. In such cases, use inverse trigonometric functions to solve. (See Example 9.)	43–46

What Did You Learn?	**Explanation/Examples**	**Review Exercises**
Section 2.4 — Use sum and difference formulas to evaluate trigonometric functions, verify identities, and solve trigonometric equations (*p. 235*).	**Sum and Difference Formulas** $\sin(u + v) = \sin u \cos v + \cos u \sin v$ $\sin(u - v) = \sin u \cos v - \cos u \sin v$ $\cos(u + v) = \cos u \cos v - \sin u \sin v$ $\cos(u - v) = \cos u \cos v + \sin u \sin v$ $\tan(u + v) = \dfrac{\tan u + \tan v}{1 - \tan u \tan v}$ $\tan(u - v) = \dfrac{\tan u - \tan v}{1 + \tan u \tan v}$	47–62
Section 2.5 — Use multiple-angle formulas to rewrite and evaluate trigonometric functions (*p. 242*).	**Double-Angle Formulas** $\sin 2u = 2 \sin u \cos u \qquad \cos 2u = \cos^2 u - \sin^2 u$ $\tan 2u = \dfrac{2 \tan u}{1 - \tan^2 u} \qquad\qquad\quad = 2\cos^2 u - 1$ $\qquad\qquad\qquad\qquad\qquad\qquad\quad = 1 - 2\sin^2 u$	63–66
Use power-reducing formulas to rewrite and evaluate trigonometric functions (*p. 244*).	**Power-Reducing Formulas** $\sin^2 u = \dfrac{1 - \cos 2u}{2}, \quad \cos^2 u = \dfrac{1 + \cos 2u}{2}$ $\tan^2 u = \dfrac{1 - \cos 2u}{1 + \cos 2u}$	67, 68
Use half-angle formulas to rewrite and evaluate trigonometric functions (*p. 245*).	**Half-Angle Formulas** $\sin \dfrac{u}{2} = \pm\sqrt{\dfrac{1 - \cos u}{2}}, \quad \cos \dfrac{u}{2} = \pm\sqrt{\dfrac{1 + \cos u}{2}}$ $\tan \dfrac{u}{2} = \dfrac{1 - \cos u}{\sin u} = \dfrac{\sin u}{1 + \cos u}$ The signs of $\sin(u/2)$ and $\cos(u/2)$ depend on the quadrant in which $u/2$ lies.	69–74
Use product-to-sum and sum-to-product formulas to rewrite and evaluate trigonometric functions (*p. 246*).	**Product-to-Sum Formulas** $\sin u \sin v = (1/2)[\cos(u - v) - \cos(u + v)]$ $\cos u \cos v = (1/2)[\cos(u - v) + \cos(u + v)]$ $\sin u \cos v = (1/2)[\sin(u + v) + \sin(u - v)]$ $\cos u \sin v = (1/2)[\sin(u + v) - \sin(u - v)]$ **Sum-to-Product Formulas** $\sin u + \sin v = 2 \sin\left(\dfrac{u + v}{2}\right) \cos\left(\dfrac{u - v}{2}\right)$ $\sin u - \sin v = 2 \cos\left(\dfrac{u + v}{2}\right) \sin\left(\dfrac{u - v}{2}\right)$ $\cos u + \cos v = 2 \cos\left(\dfrac{u + v}{2}\right) \cos\left(\dfrac{u - v}{2}\right)$ $\cos u - \cos v = -2 \sin\left(\dfrac{u + v}{2}\right) \sin\left(\dfrac{u - v}{2}\right)$	75–78
Use trigonometric formulas to rewrite real-life models (*p. 248*).	A trigonometric formula can be used to rewrite the projectile motion model $r = (1/16)v_0^2 \sin \theta \cos \theta$. (See Example 10.)	79, 80

Review Exercises   See CalcChat.com for tutorial help and worked-out solutions to odd-numbered exercises.

2.1 Recognizing a Fundamental Identity In Exercises 1–4, name the trigonometric function that is equivalent to the expression.

1. $\dfrac{\sin x}{\cos x}$

2. $\dfrac{1}{\sin x}$

3. $\dfrac{1}{\tan x}$

4. $\sqrt{\cot^2 x + 1}$

Using Identities to Evaluate a Function In Exercises 5 and 6, use the given values and fundamental trigonometric identities to find the values (if possible) of all six trigonometric functions.

5. $\tan \theta = \dfrac{2}{3}, \quad \sec \theta = \dfrac{\sqrt{13}}{3}$

6. $\sin\left(\dfrac{\pi}{2} - x\right) = \dfrac{\sqrt{2}}{2}, \quad \sin x = -\dfrac{\sqrt{2}}{2}$

Simplifying a Trigonometric Expression In Exercises 7–16, use the fundamental trigonometric identities to simplify the expression. There is more than one correct form of each answer.

7. $\dfrac{1}{\cot^2 x + 1}$

8. $\dfrac{\tan \theta}{1 - \cos^2 \theta}$

9. $\tan^2 x(\csc^2 x - 1)$

10. $\cot^2 x(\sin^2 x)$

11. $\dfrac{\cot\left(\dfrac{\pi}{2} - u\right)}{\cos u}$

12. $\dfrac{\sec^2(-\theta)}{\csc^2 \theta}$

13. $\cos^2 x + \cos^2 x \cot^2 x$

14. $(\tan x + 1)^2 \cos x$

15. $\dfrac{1}{\csc \theta + 1} - \dfrac{1}{\csc \theta - 1}$

16. $\dfrac{\tan^2 x}{1 + \sec x}$

Trigonometric Substitution In Exercises 17 and 18, use the trigonometric substitution to write the algebraic expression as a trigonometric function of θ, where $0 < \theta < \pi/2$.

17. $\sqrt{25 - x^2}, x = 5 \sin \theta$

18. $\sqrt{x^2 - 16}, x = 4 \sec \theta$

2.2 Verifying a Trigonometric Identity In Exercises 19–26, verify the identity.

19. $\cos x(\tan^2 x + 1) = \sec x$

20. $\sec^2 x \cot x - \cot x = \tan x$

21. $\sec\left(\dfrac{\pi}{2} - \theta\right) = \csc \theta$

22. $\cot\left(\dfrac{\pi}{2} - x\right) = \tan x$

23. $\dfrac{1}{\tan \theta \csc \theta} = \cos \theta$

24. $\dfrac{1}{\tan x \csc x \sin x} = \cot x$

25. $\sin^5 x \cos^2 x = (\cos^2 x - 2 \cos^4 x + \cos^6 x) \sin x$

26. $\cos^3 x \sin^2 x = (\sin^2 x - \sin^4 x) \cos x$

2.3 Solving a Trigonometric Equation In Exercises 27–32, solve the equation.

27. $\sin x = \sqrt{3} - \sin x$

28. $4 \cos \theta = 1 + 2 \cos \theta$

29. $3\sqrt{3} \tan u = 3$

30. $\dfrac{1}{2} \sec x - 1 = 0$

31. $3 \csc^2 x = 4$

32. $4 \tan^2 u - 1 = \tan^2 u$

Solving a Trigonometric Equation In Exercises 33–42, find all solutions of the equation in the interval $[0, 2\pi)$.

33. $2 \cos^2 x - \cos x = 1$

34. $2 \cos^2 x + 3 \cos x = 0$

35. $\cos^2 x + \sin x = 1$

36. $\sin^2 x + 2 \cos x = 2$

37. $2 \sin 2x - \sqrt{2} = 0$

38. $2 \cos \dfrac{x}{2} + 1 = 0$

39. $3 \tan^2\left(\dfrac{x}{3}\right) - 1 = 0$

40. $\sqrt{3} \tan 3x = 0$

41. $\cos 4x(\cos x - 1) = 0$

42. $3 \csc^2 5x = -4$

Using Inverse Functions In Exercises 43–46, use inverse functions where needed to find all solutions of the equation in the interval $[0, 2\pi)$.

43. $\tan^2 x - 2 \tan x = 0$

44. $2 \tan^2 x - 3 \tan x = -1$

45. $\tan^2 \theta + \tan \theta - 6 = 0$

46. $\sec^2 x + 6 \tan x + 4 = 0$

2.4 Evaluating Trigonometric Functions In Exercises 47–50, find the exact values of the sine, cosine, and tangent of the angle.

47. $285° = 315° - 30°$

48. $345° = 300° + 45°$

49. $\dfrac{25\pi}{12} = \dfrac{11\pi}{6} + \dfrac{\pi}{4}$

50. $\dfrac{19\pi}{12} = \dfrac{11\pi}{6} - \dfrac{\pi}{4}$

Rewriting a Trigonometric Expression In Exercises 51 and 52, write the expression as the sine, cosine, or tangent of an angle.

51. $\sin 60° \cos 45° - \cos 60° \sin 45°$

52. $\dfrac{\tan 68° - \tan 115°}{1 + \tan 68° \tan 115°}$

Evaluating a Trigonometric Expression In Exercises 53–56, find the exact value of the trigonometric expression given that $\tan u = \dfrac{3}{4}$ and $\cos v = -\dfrac{4}{5}$. (u is in Quadrant I and v is in Quadrant III.)

53. $\sin(u + v)$

54. $\tan(u + v)$

55. $\cos(u - v)$

56. $\sin(u - v)$

Proving a Trigonometric Identity   In Exercises 57–60, prove the identity.

57. $\cos\left(x + \dfrac{\pi}{2}\right) = -\sin x$ **58.** $\tan\left(x - \dfrac{\pi}{2}\right) = -\cot x$

59. $\tan(\pi - x) = -\tan x$

60. $\cos 3x = 4\cos^3 x - 3\cos x$

Solving a Trigonometric Equation   In Exercises 61 and 62, find all solutions of the equation in the interval $[0, 2\pi)$.

61. $\sin\left(x + \dfrac{\pi}{4}\right) - \sin\left(x - \dfrac{\pi}{4}\right) = 1$

62. $\cos\left(x + \dfrac{\pi}{6}\right) - \cos\left(x - \dfrac{\pi}{6}\right) = 1$

2.5 Evaluating Functions Involving Double Angles
In Exercises 63 and 64, find the exact values of $\sin 2u$, $\cos 2u$, and $\tan 2u$ using the double-angle formulas.

63. $\sin u = -\dfrac{4}{5},\quad \pi < u < 3\pi/2$

64. $\cos u = -2/\sqrt{5},\quad \pi/2 < u < \pi$

Verifying a Trigonometric Identity   In Exercises 65 and 66, use the double-angle formulas to verify the identity algebraically and use a graphing utility to confirm your result graphically.

65. $\sin 4x = 8\cos^3 x \sin x - 4\cos x \sin x$

66. $\tan^2 x = \dfrac{1 - \cos 2x}{1 + \cos 2x}$

♩ Reducing Powers   In Exercises 67 and 68, use the power-reducing formulas to rewrite the expression in terms of the first power of the cosine.

67. $\tan^2 2x$ **68.** $\sin^2 x \tan^2 x$

Using Half-Angle Formulas   In Exercises 69 and 70, use the half-angle formulas to determine the exact values of the sine, cosine, and tangent of the angle.

69. $-75°$ **70.** $\dfrac{19\pi}{12}$

Using Half-Angle Formulas   In Exercises 71 and 72, (a) determine the quadrant in which $u/2$ lies, and (b) find the exact values of $\sin(u/2)$, $\cos(u/2)$, and $\tan(u/2)$ using the half-angle formulas.

71. $\tan u = \dfrac{4}{3},\ \pi < u < 3\pi/2$

72. $\cos u = -\dfrac{2}{7},\ \pi/2 < u < \pi$

Using Half-Angle Formulas   In Exercises 73 and 74, use the half-angle formulas to simplify the expression.

73. $-\sqrt{\dfrac{1 + \cos 10x}{2}}$ **74.** $\dfrac{\sin 6x}{1 + \cos 6x}$

Using Product-to-Sum Formulas   In Exercises 75 and 76, use the product-to-sum formulas to rewrite the product as a sum or difference.

75. $\cos 4\theta \sin 6\theta$ **76.** $2\sin 7\theta \cos 3\theta$

Using Sum-to-Product Formulas   In Exercises 77 and 78, use the sum-to-product formulas to rewrite the sum or difference as a product.

77. $\cos 6\theta + \cos 5\theta$

78. $\sin\left(x + \dfrac{\pi}{4}\right) - \sin\left(x - \dfrac{\pi}{4}\right)$

79. Projectile Motion   A baseball leaves the hand of a player at first base at an angle of θ with the horizontal and at an initial velocity of $v_0 = 80$ feet per second. A player at second base 100 feet away catches the ball. Find θ when the range r of a projectile is

$$r = \dfrac{1}{32}v_0^2 \sin 2\theta.$$

80. Geometry   A trough for feeding cattle is 4 meters long and its cross sections are isosceles triangles with the two equal sides being $\dfrac{1}{2}$ meter (see figure). The angle between the two sides is θ.

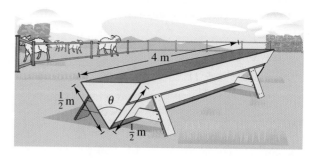

(a) Write the trough's volume as a function of $\theta/2$.

(b) Write the volume of the trough as a function of θ and determine the value of θ such that the volume is maximum.

Exploration

True or False?   In Exercises 81–84, determine whether the statement is true or false. Justify your answer.

81. If $\dfrac{\pi}{2} < \theta < \pi$, then $\cos\dfrac{\theta}{2} < 0$.

82. $\sin(x + y) = \sin x + \sin y$

83. $4\sin(-x)\cos(-x) = -2\sin 2x$

84. $4\sin 45° \cos 15° = 1 + \sqrt{3}$

85. Think About It   When a trigonometric equation has an infinite number of solutions, is it true that the equation is an identity? Explain.

Chapter Test

See **CalcChat.com** for tutorial help and worked-out solutions to odd-numbered exercises.

Take this test as you would take a test in class. When you are finished, check your work against the answers given in the back of the book.

1. When $\tan \theta = \frac{6}{5}$ and $\cos \theta < 0$, evaluate (if possible) all six trigonometric functions of θ.

2. Use the fundamental identities to simplify $\csc^2 \beta (1 - \cos^2 \beta)$.

3. Factor and simplify $\dfrac{\sec^4 x - \tan^4 x}{\sec^2 x + \tan^2 x}$.   4. Add and simplify $\dfrac{\cos \theta}{\sin \theta} + \dfrac{\sin \theta}{\cos \theta}$.

5. Determine the values of θ, $0 \le \theta < 2\pi$, for which $\tan \theta = -\sqrt{\sec^2 \theta - 1}$.

6. Use a graphing utility to graph the functions $y_1 = \cos x + \sin x \tan x$ and $y_2 = \sec x$ in the same viewing window. Make a conjecture about y_1 and y_2. Verify the result algebraically.

In Exercises 7–12, verify the identity.

7. $\sin \theta \sec \theta = \tan \theta$ 8. $\sec^2 x \tan^2 x + \sec^2 x = \sec^4 x$

9. $\dfrac{\csc \alpha + \sec \alpha}{\sin \alpha + \cos \alpha} = \cot \alpha + \tan \alpha$ 10. $\tan\left(x + \dfrac{\pi}{2}\right) = -\cot x$

11. $\sin(n\pi + \theta) = (-1)^n \sin \theta$, n is an integer.

12. $(\sin x + \cos x)^2 = 1 + \sin 2x$

13. Rewrite $\sin^4 \dfrac{x}{2}$ in terms of the first power of the cosine.

14. Use a half-angle formula to simplify the expression $(\sin 4\theta)/(1 + \cos 4\theta)$.

15. Rewrite $4 \sin 3\theta \cos 2\theta$ as a sum or difference.

16. Rewrite $\cos 3\theta - \cos \theta$ as a product.

In Exercises 17–20, find all solutions of the equation in the interval $[0, 2\pi)$.

17. $\tan^2 x + \tan x = 0$ 18. $\sin 2\alpha - \cos \alpha = 0$

19. $4 \cos^2 x - 3 = 0$ 20. $\csc^2 x - \csc x - 2 = 0$

21. Use a graphing utility to approximate the solutions (to three decimal places) of $5 \sin x - x = 0$ in the interval $[0, 2\pi)$.

22. Find the exact value of $\cos 105°$ using the fact that $105° = 135° - 30°$.

23. Use the figure to find the exact values of $\sin 2u$, $\cos 2u$, and $\tan 2u$.

24. Cheyenne, Wyoming, has a latitude of $41°$N. At this latitude, the position of the sun at sunrise can be modeled by

$$D = 31 \sin\left(\dfrac{2\pi}{365}t - 1.4\right)$$

where t is the time (in days), with $t = 1$ representing January 1. In this model, D represents the number of degrees north or south of due east that the sun rises. Use a graphing utility to determine the days on which the sun is more than $20°$ north of due east at sunrise.

25. The heights above ground h_1 and h_2 (in feet) of two people in different seats on a Ferris wheel can be modeled by

$$h_1 = 28 \cos 10t + 38 \quad \text{and} \quad h_2 = 28 \cos\left[10\left(t - \dfrac{\pi}{6}\right)\right] + 38, \; 0 \le t \le 2$$

where t is the time (in minutes). When are the two people at the same height?

$(2, -5)$

Figure for 23

Proofs in Mathematics ■ ■ ■ ■ ■ ■ ■ ■ ■ ■ ■ ■ ■ ■ ■

Proof

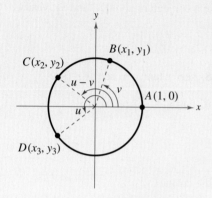

Use the figures at the left for the proofs of the formulas for $\cos(u \pm v)$. In the top figure, let A be the point $(1, 0)$ and then use u and v to locate the points $B(x_1, y_1)$, $C(x_2, y_2)$, and $D(x_3, y_3)$ on the unit circle. So, $x_i^2 + y_i^2 = 1$ for $i = 1, 2$, and 3. For convenience, assume that $0 < v < u < 2\pi$. In the bottom figure, note that arcs AC and BD have the same length. So, line segments AC and BD are also equal in length, which implies that

$$\sqrt{(x_2 - 1)^2 + (y_2 - 0)^2} = \sqrt{(x_3 - x_1)^2 + (y_3 - y_1)^2}$$

$$x_2^2 - 2x_2 + 1 + y_2^2 = x_3^2 - 2x_1x_3 + x_1^2 + y_3^2 - 2y_1y_3 + y_1^2$$

$$(x_2^2 + y_2^2) + 1 - 2x_2 = (x_3^2 + y_3^2) + (x_1^2 + y_1^2) - 2x_1x_3 - 2y_1y_3$$

$$1 + 1 - 2x_2 = 1 + 1 - 2x_1x_3 - 2y_1y_3$$

$$x_2 = x_3x_1 + y_3y_1.$$

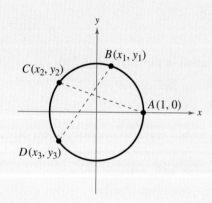

Finally, by substituting the values $x_2 = \cos(u - v)$, $x_3 = \cos u$, $x_1 = \cos v$, $y_3 = \sin u$, and $y_1 = \sin v$, you obtain $\cos(u - v) = \cos u \cos v + \sin u \sin v$. To establish the formula for $\cos(u + v)$, consider $u + v = u - (-v)$ and use the formula just derived to obtain

$$\cos(u + v) = \cos[u - (-v)]$$

$$= \cos u \cos(-v) + \sin u \sin(-v)$$

$$= \cos u \cos v - \sin u \sin v.$$

You can use the sum and difference formulas for sine and cosine to prove the formulas for $\tan(u \pm v)$.

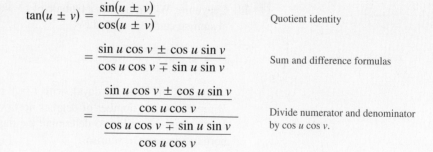

$$\tan(u \pm v) = \frac{\sin(u \pm v)}{\cos(u \pm v)} \qquad \text{Quotient identity}$$

$$= \frac{\sin u \cos v \pm \cos u \sin v}{\cos u \cos v \mp \sin u \sin v} \qquad \text{Sum and difference formulas}$$

$$= \frac{\dfrac{\sin u \cos v \pm \cos u \sin v}{\cos u \cos v}}{\dfrac{\cos u \cos v \mp \sin u \sin v}{\cos u \cos v}} \qquad \begin{array}{l}\text{Divide numerator and denominator}\\ \text{by } \cos u \cos v.\end{array}$$

$$= \frac{\dfrac{\sin u \cos v}{\cos u \cos v} \pm \dfrac{\cos u \sin v}{\cos u \cos v}}{\dfrac{\cos u \cos v}{\cos u \cos v} \mp \dfrac{\sin u \sin v}{\cos u \cos v}}$$ Write as separate fractions.

$$= \frac{\dfrac{\sin u}{\cos u} \pm \dfrac{\sin v}{\cos v}}{1 \mp \dfrac{\sin u}{\cos u} \cdot \dfrac{\sin v}{\cos v}}$$ Simplify.

$$= \frac{\tan u \pm \tan v}{1 \mp \tan u \tan v}$$ Quotient identity

TRIGONOMETRY AND ASTRONOMY

Early astronomers used trigonometry to calculate measurements in the universe. For instance, they used trigonometry to calculate the circumference of Earth and the distance from Earth to the moon. Another major accomplishment in astronomy using trigonometry was computing distances to stars.

Double-Angle Formulas   *(p. 242)*

$$\sin 2u = 2 \sin u \cos u \qquad \cos 2u = \cos^2 u - \sin^2 u$$

$$\tan 2u = \frac{2 \tan u}{1 - \tan^2 u} \qquad\qquad = 2 \cos^2 u - 1$$

$$= 1 - 2 \sin^2 u$$

Proof

To prove all three formulas, let $v = u$ in the corresponding sum formulas.

$$\sin 2u = \sin(u + u) = \sin u \cos u + \cos u \sin u = 2 \sin u \cos u$$

$$\cos 2u = \cos(u + u) = \cos u \cos u - \sin u \sin u = \cos^2 u - \sin^2 u$$

$$\tan 2u = \tan(u + u) = \frac{\tan u + \tan u}{1 - \tan u \tan u} = \frac{2 \tan u}{1 - \tan^2 u}$$

Power-Reducing Formulas   *(p. 244)*

$$\sin^2 u = \frac{1 - \cos 2u}{2} \qquad \cos^2 u = \frac{1 + \cos 2u}{2} \qquad \tan^2 u = \frac{1 - \cos 2u}{1 + \cos 2u}$$

Proof

To prove the first formula, solve for $\sin^2 u$ in the double-angle formula $\cos 2u = 1 - 2 \sin^2 u$, as follows.

$$\cos 2u = 1 - 2 \sin^2 u$$ Write double-angle formula.

$$2 \sin^2 u = 1 - \cos 2u$$ Subtract $\cos 2u$ from, and add $2 \sin^2 u$ to, each side.

$$\sin^2 u = \frac{1 - \cos 2u}{2}$$ Divide each side by 2.

In a similar way, you can prove the second formula by solving for $\cos^2 u$ in the double-angle formula

$$\cos 2u = 2 \cos^2 u - 1.$$

To prove the third formula, use a quotient identity, as follows.

$$\tan^2 u = \frac{\sin^2 u}{\cos^2 u}$$

$$= \frac{\dfrac{1 - \cos 2u}{2}}{\dfrac{1 + \cos 2u}{2}}$$

$$= \frac{1 - \cos 2u}{1 + \cos 2u}$$

Sum-to-Product Formulas   *(p. 246)*

$$\sin u + \sin v = 2 \sin\left(\frac{u + v}{2}\right) \cos\left(\frac{u - v}{2}\right)$$

$$\sin u - \sin v = 2 \cos\left(\frac{u + v}{2}\right) \sin\left(\frac{u - v}{2}\right)$$

$$\cos u + \cos v = 2 \cos\left(\frac{u + v}{2}\right) \cos\left(\frac{u - v}{2}\right)$$

$$\cos u - \cos v = -2 \sin\left(\frac{u + v}{2}\right) \sin\left(\frac{u - v}{2}\right)$$

Proof

To prove the first formula, let $x = u + v$ and $y = u - v$. Then substitute $u = (x + y)/2$ and $v = (x - y)/2$ in the product-to-sum formula.

$$\sin u \cos v = \frac{1}{2}[\sin(u + v) + \sin(u - v)]$$

$$\sin\left(\frac{x + y}{2}\right) \cos\left(\frac{x - y}{2}\right) = \frac{1}{2}(\sin x + \sin y)$$

$$2 \sin\left(\frac{x + y}{2}\right) \cos\left(\frac{x - y}{2}\right) = \sin x + \sin y$$

The other sum-to-product formulas can be proved in a similar manner.

P.S. Problem Solving ■ ■ ■ ■ ■ ■ ■ ■ ■ ■ ■ ■ ■ ■

1. **Writing Trigonometric Functions in Terms of Cosine** Write each of the other trigonometric functions of θ in terms of $\cos \theta$.

2. **Verifying a Trigonometric Identity** Verify that for all integers n,

$$\cos\left[\frac{(2n+1)\pi}{2}\right] = 0.$$

3. **Verifying a Trigonometric Identity** Verify that for all integers n,

$$\sin\left[\frac{(12n+1)\pi}{6}\right] = \frac{1}{2}.$$

4. **Sound Wave** A sound wave is modeled by

$$p(t) = \frac{1}{4\pi}\left[p_1(t) + 30p_2(t) + p_3(t) + p_5(t) + 30p_6(t)\right]$$

where $p_n(t) = \frac{1}{n}\sin(524n\pi t)$, and t is the time (in seconds).

(a) Find the sine components $p_n(t)$ and use a graphing utility to graph the components. Then verify the graph of p shown below.

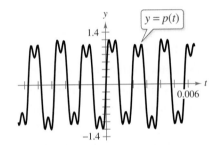

(b) Find the period of each sine component of p. Is p periodic? If so, then what is its period?

(c) Use the graphing utility to find the t-intercepts of the graph of p over one cycle.

(d) Use the graphing utility to approximate the absolute maximum and absolute minimum values of p over one cycle.

5. **Geometry** Three squares of side s are placed side by side (see figure). Make a conjecture about the relationship between the sum $u + v$ and w. Prove your conjecture by using the identity for the tangent of the sum of two angles.

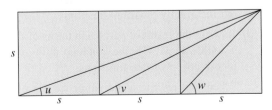

6. **Projectile Motion** The path traveled by an object (neglecting air resistance) that is projected at an initial height of h_0 feet, an initial velocity of v_0 feet per second, and an initial angle θ is given by

$$y = -\frac{16}{v_0^2 \cos^2 \theta}x^2 + (\tan \theta)x + h_0$$

where x and y are measured in feet. Find a formula for the maximum height of an object projected from ground level at velocity v_0 and angle θ. To do this, find half of the horizontal distance

$$\frac{1}{32}v_0^2 \sin 2\theta$$

and then substitute it for x in the general model for the path of a projectile (where $h_0 = 0$).

7. **Geometry** The length of each of the two equal sides of an isosceles triangle is 10 meters (see figure). The angle between the two sides is θ.

(a) Write the area of the triangle as a function of $\theta/2$.

(b) Write the area of the triangle as a function of θ. Determine the value of θ such that the area is a maximum.

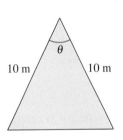

Figure for 7   Figure for 8

8. **Geometry** Use the figure to derive the formulas for

$$\sin\frac{\theta}{2}, \cos\frac{\theta}{2}, \text{ and } \tan\frac{\theta}{2}$$

where θ is an acute angle.

9. **Force** The force F (in pounds) on a person's back when he or she bends over at an angle θ is modeled by

$$F = \frac{0.6W \sin(\theta + 90°)}{\sin 12°}$$

where W is the person's weight (in pounds).

(a) Simplify the model.

(b) Use a graphing utility to graph the model, where $W = 185$ and $0° < \theta < 90°$.

(c) At what angle is the force a maximum? At what angle is the force a minimum?

259

10. Hours of Daylight The number of hours of daylight that occur at any location on Earth depends on the time of year and the latitude of the location. The following equations model the numbers of hours of daylight in Seward, Alaska (60° latitude), and New Orleans, Louisiana (30° latitude).

$$D = 12.2 - 6.4 \cos\left[\frac{\pi(t + 0.2)}{182.6}\right] \quad \text{Seward}$$

$$D = 12.2 - 1.9 \cos\left[\frac{\pi(t + 0.2)}{182.6}\right] \quad \text{New Orleans}$$

In these models, D represents the number of hours of daylight and t represents the day, with $t = 0$ corresponding to January 1.

(a) Use a graphing utility to graph both models in the same viewing window. Use a viewing window of $0 \leq t \leq 365$.

(b) Find the days of the year on which both cities receive the same amount of daylight.

(c) Which city has the greater variation in the number of daylight hours? Which constant in each model would you use to determine the difference between the greatest and least numbers of hours of daylight?

(d) Determine the period of each model.

11. Ocean Tide The tide, or depth of the ocean near the shore, changes throughout the day. The water depth d (in feet) of a bay can be modeled by

$$d = 35 - 28 \cos\frac{\pi}{6.2}t$$

where t is the time in hours, with $t = 0$ corresponding to 12:00 A.M.

(a) Algebraically find the times at which the high and low tides occur.

(b) If possible, algebraically find the time(s) at which the water depth is 3.5 feet.

(c) Use a graphing utility to verify your results from parts (a) and (b).

12. Piston Heights The heights h (in inches) of pistons 1 and 2 in an automobile engine can be modeled by

$$h_1 = 3.75 \sin 733t + 7.5$$

and

$$h_2 = 3.75 \sin 733(t + 4\pi/3) + 7.5$$

respectively, where t is measured in seconds.

(a) Use a graphing utility to graph the heights of these pistons in the same viewing window for $0 \leq t \leq 1$.

(b) How often are the pistons at the same height?

13. Index of Refraction The index of refraction n of a transparent material is the ratio of the speed of light in a vacuum to the speed of light in the material. Some common materials and their indices of refraction are air (1.00), water (1.33), and glass (1.50). Triangular prisms are often used to measure the index of refraction based on the formula

$$n = \frac{\sin\left(\dfrac{\theta}{2} + \dfrac{\alpha}{2}\right)}{\sin\dfrac{\theta}{2}}.$$

For the prism shown in the figure, $\alpha = 60°$.

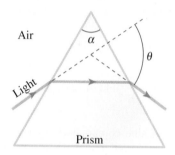

(a) Write the index of refraction as a function of $\cot(\theta/2)$.

(b) Find θ for a prism made of glass.

14. Sum Formulas

(a) Write a sum formula for $\sin(u + v + w)$.

(b) Write a sum formula for $\tan(u + v + w)$.

15. Solving Trigonometric Inequalities Find the solution of each inequality in the interval $[0, 2\pi)$.

(a) $\sin x \geq 0.5$ (b) $\cos x \leq -0.5$

(c) $\tan x < \sin x$ (d) $\cos x \geq \sin x$

16. Sum of Fourth Powers Consider the function $f(x) = \sin^4 x + \cos^4 x$.

(a) Use the power-reducing formulas to write the function in terms of cosine to the first power.

(b) Determine another way of rewriting the function. Use a graphing utility to rule out incorrectly rewritten functions.

(c) Add a trigonometric term to the function so that it becomes a perfect square trinomial. Rewrite the function as a perfect square trinomial minus the term that you added. Use the graphing utility to rule out incorrectly rewritten functions.

(d) Rewrite the result of part (c) in terms of the sine of a double angle. Use the graphing utility to rule out incorrectly rewritten functions.

(e) When you rewrite a trigonometric expression, the result may not be the same as a friend's. Does this mean that one of you is wrong? Explain.

3 Additional Topics in Trigonometry

Work (page 296)

Braking Load
(Exercise 76, page 298)

Navigation (Example 11, page 286)

Engine Design
(Exercise 56, page 277)

Surveying (page 263)

3.1 Law of Sines

You can use the Law of Sines to solve real-life problems involving oblique triangles. For instance, in Exercise 53 on page 269, you will use the Law of Sines to determine the distance from a boat to the shoreline.

■ Use the Law of Sines to solve oblique triangles (AAS or ASA).
■ Use the Law of Sines to solve oblique triangles (SSA).
■ Find the areas of oblique triangles.
■ Use the Law of Sines to model and solve real-life problems.

Introduction

In Chapter 1, you studied techniques for solving right triangles. In this section and the next, you will solve **oblique triangles**—triangles that have no right angles. As standard notation, the angles of a triangle are labeled A, B, and C, and their opposite sides are labeled a, b, and c, as shown below.

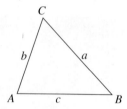

To solve an oblique triangle, you need to know the measure of at least one side and any two other measures of the triangle—either two sides, two angles, or one angle and one side. This breaks down into the following four cases.

1. Two angles and any side (AAS or ASA)
2. Two sides and an angle opposite one of them (SSA)
3. Three sides (SSS)
4. Two sides and their included angle (SAS)

The first two cases can be solved using the **Law of Sines,** whereas the last two cases require the Law of Cosines (see Section 3.2).

Law of Sines

If ABC is a triangle with sides a, b, and c, then

$$\frac{a}{\sin A} = \frac{b}{\sin B} = \frac{c}{\sin C}.$$

A is acute.

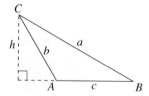

A is obtuse.

The Law of Sines can also be written in the reciprocal form

$$\frac{\sin A}{a} = \frac{\sin B}{b} = \frac{\sin C}{c}.$$

For a proof of the Law of Sines, see Proofs in Mathematics on page 309.

$b = 28$ ft

$102°$

C

a

$29°$

A c B

Figure 3.1

C

b $a = 32$

$30°$ $45°$

A c B

Figure 3.2

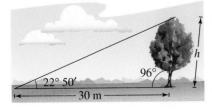

In the 1850s, surveyors used the Law of Sines to calculate the height of Mount Everest. Their calculation was within 30 feet of the currently accepted value.

$22° 50'$

$96°$

h

30 m

Figure 3.3

Daniel Prudek/Shutterstock.com

EXAMPLE 1   **Given Two Angles and One Side — AAS**

For the triangle in Figure 3.1, $C = 102°$, $B = 29°$, and $b = 28$ feet. Find the remaining angle and sides.

Solution   The third angle of the triangle is

$$A = 180° - B - C$$
$$= 180° - 29° - 102°$$
$$= 49°.$$

By the Law of Sines, you have

$$\frac{a}{\sin A} = \frac{b}{\sin B} = \frac{c}{\sin C}.$$

Using $b = 28$ produces

$$a = \frac{b}{\sin B}(\sin A) = \frac{28}{\sin 29°}(\sin 49°) \approx 43.59 \text{ feet}$$

and

$$c = \frac{b}{\sin B}(\sin C) = \frac{28}{\sin 29°}(\sin 102°) \approx 56.49 \text{ feet}.$$

✓ *Checkpoint*))) *Audio-video solution in English & Spanish at LarsonPrecalculus.com.*

For the triangle in Figure 3.2, $A = 30°$, $B = 45°$, and $a = 32$. Find the remaining angle and sides.

EXAMPLE 2   **Given Two Angles and One Side — ASA**

A pole tilts *toward* the sun at an 8° angle from the vertical, and it casts a 22-foot shadow. The angle of elevation from the tip of the shadow to the top of the pole is 43°. How tall is the pole?

Solution   From the figure at the right, note that $A = 43°$ and

$$B = 90° + 8° = 98°.$$

So, the third angle is

$$C = 180° - A - B$$
$$= 180° - 43° - 98°$$
$$= 39°.$$

By the Law of Sines, you have

$$\frac{a}{\sin A} = \frac{c}{\sin C}.$$

Because $c = 22$ feet, the height of the pole is

$$a = \frac{c}{\sin C}(\sin A) = \frac{22}{\sin 39°}(\sin 43°) \approx 23.84 \text{ feet}.$$

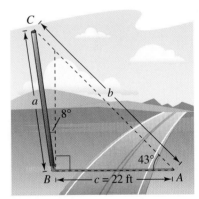

✓ *Checkpoint*))) *Audio-video solution in English & Spanish at LarsonPrecalculus.com.*

Find the height of the tree shown in Figure 3.3.

The Ambiguous Case (SSA)

In Examples 1 and 2, you saw that two angles and one side determine a unique triangle. However, if two sides and one opposite angle are given, then three possible situations can occur: (1) no such triangle exists, (2) one such triangle exists, or (3) two distinct triangles may satisfy the conditions.

The Ambiguous Case (SSA)

Consider a triangle in which you are given a, b, and A.   ($h = b \sin A$)

	A is acute.	A is acute.	A is acute.	A is acute.	A is obtuse.	A is obtuse.
Sketch						
Necessary condition	$a < h$	$a = h$	$a \geq b$	$h < a < b$	$a \leq b$	$a > b$
Triangles possible	None	One	One	Two	None	One

EXAMPLE 3   Single-Solution Case—SSA

For the triangle in Figure 3.4, $a = 22$ inches, $b = 12$ inches, and $A = 42°$. Find the remaining side and angles.

Solution   By the Law of Sines, you have

$$\frac{\sin B}{b} = \frac{\sin A}{a} \qquad \text{Reciprocal form}$$

$$\sin B = b\left(\frac{\sin A}{a}\right) \qquad \text{Multiply each side by } b.$$

$$\sin B = 12\left(\frac{\sin 42°}{22}\right) \qquad \text{Substitute for } A, a, \text{ and } b.$$

$$B \approx 21.41°.$$

Now, you can determine that

$$C \approx 180° - 42° - 21.41° = 116.59°.$$

Then, the remaining side is

$$\frac{c}{\sin C} = \frac{a}{\sin A}$$

$$c = \frac{a}{\sin A}(\sin C)$$

$$= \frac{22}{\sin 42°}(\sin 116.59°)$$

$$\approx 29.40 \text{ inches.}$$

b = 12 in.   *a* = 22 in.   42°

One solution: $a \geq b$

Figure 3.4

✓ **Checkpoint** ◀))) *Audio-video solution in English & Spanish at LarsonPrecalculus.com.*

Given $A = 31°$, $a = 12$, and $b = 5$, find the remaining side and angles of the triangle.

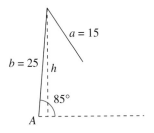

No solution: $a < h$

Figure 3.5

EXAMPLE 4   **No-Solution Case—SSA**

Show that there is no triangle for which $a = 15$, $b = 25$, and $A = 85°$.

Solution   Begin by making the sketch shown in Figure 3.5. From this figure, it appears that no triangle is formed. You can verify this using the Law of Sines.

$$\frac{\sin B}{b} = \frac{\sin A}{a} \qquad \text{Reciprocal form}$$

$$\sin B = b\left(\frac{\sin A}{a}\right) \qquad \text{Multiply each side by } b.$$

$$\sin B = 25\left(\frac{\sin 85°}{15}\right) \approx 1.6603 > 1$$

This contradicts the fact that

$$|\sin B| \leq 1.$$

So, no triangle can be formed having sides $a = 15$ and $b = 25$ and angle $A = 85°$.

✓ **Checkpoint**  *Audio-video solution in English & Spanish at LarsonPrecalculus.com.*

Show that there is no triangle for which $a = 4$, $b = 14$, and $A = 60°$.

EXAMPLE 5   **Two-Solution Case—SSA**

Find two triangles for which $a = 12$ meters, $b = 31$ meters, and $A = 20.5°$.

Solution   By the Law of Sines, you have

$$\frac{\sin B}{b} = \frac{\sin A}{a} \qquad \text{Reciprocal form}$$

$$\sin B = b\left(\frac{\sin A}{a}\right) = 31\left(\frac{\sin 20.5°}{12}\right) \approx 0.9047.$$

There are two angles, $B_1 \approx 64.8°$ and $B_2 \approx 180° - 64.8° = 115.2°$, between 0° and 180° whose sine is 0.9047. For $B_1 \approx 64.8°$, you obtain

$$C \approx 180° - 20.5° - 64.8° = 94.7°$$

$$c = \frac{a}{\sin A}(\sin C) = \frac{12}{\sin 20.5°}(\sin 94.7°) \approx 34.15 \text{ meters.}$$

For $B_2 \approx 115.2°$, you obtain

$$C \approx 180° - 20.5° - 115.2° = 44.3°$$

$$c = \frac{a}{\sin A}(\sin C) = \frac{12}{\sin 20.5°}(\sin 44.3°) \approx 23.93 \text{ meters.}$$

The resulting triangles are shown below.

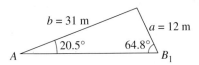

Two solutions: $h < a < b$

✓ **Checkpoint**   *Audio-video solution in English & Spanish at LarsonPrecalculus.com.*

Find two triangles for which $a = 4.5$ feet, $b = 5$ feet, and $A = 58°$.

Area of an Oblique Triangle

The procedure used to prove the Law of Sines leads to a simple formula for the area of an oblique triangle. Referring to the triangles below, note that each triangle has a height of $h = b \sin A$. Consequently, the area of each triangle is

$$\text{Area} = \frac{1}{2}(\text{base})(\text{height})$$

$$= \frac{1}{2}(c)(b \sin A)$$

$$= \frac{1}{2}bc \sin A.$$

By similar arguments, you can develop the formulas

$$\text{Area} = \frac{1}{2}ab \sin C = \frac{1}{2}ac \sin B.$$

• • • • • • • • • • • • • • • • • • • ▷

• • **REMARK**   To see how to obtain the height of the obtuse triangle, notice the use of the reference angle $180° - A$ and the difference formula for sine, as follows.

$h = b \sin(180° - A)$

$= b(\sin 180° \cos A$

$\quad - \cos 180° \sin A)$

$= b[0 \cdot \cos A - (-1) \cdot \sin A]$

$= b \sin A$

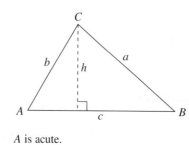

A is acute.

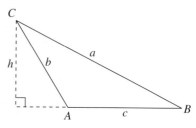

A is obtuse.

Area of an Oblique Triangle

The area of any triangle is one-half the product of the lengths of two sides times the sine of their included angle. That is,

$$\text{Area} = \frac{1}{2}bc \sin A = \frac{1}{2}ab \sin C = \frac{1}{2}ac \sin B.$$

Note that when angle A is 90°, the formula gives the area of a right triangle:

$$\text{Area} = \frac{1}{2}bc(\sin 90°) = \frac{1}{2}bc = \frac{1}{2}(\text{base})(\text{height}). \qquad \text{sin 90° = 1}$$

Similar results are obtained for angles C and B equal to 90°.

EXAMPLE 6   **Finding the Area of a Triangular Lot**

Find the area of a triangular lot having two sides of lengths 90 meters and 52 meters and an included angle of 102°.

Solution   Consider $a = 90$ meters, $b = 52$ meters, and angle $C = 102°$, as shown in Figure 3.6. Then, the area of the triangle is

$$\text{Area} = \frac{1}{2}ab \sin C = \frac{1}{2}(90)(52)(\sin 102°) \approx 2289 \text{ square meters.}$$

✓ **Checkpoint** 🔊)) *Audio-video solution in English & Spanish at LarsonPrecalculus.com.*

Find the area of a triangular lot having two sides of lengths 24 inches and 18 inches and an included angle of 80°.

$b = 52$ m

$102°$

C $a = 90$ m

Figure 3.6

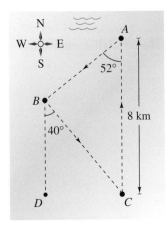

Figure 3.7

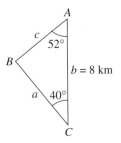

Figure 3.8

Application

EXAMPLE 7   An Application of the Law of Sines

The course for a boat race starts at point A and proceeds in the direction S 52° W to point B, then in the direction S 40° E to point C, and finally back to point A, as shown in Figure 3.7. Point C lies 8 kilometers directly south of point A. Approximate the total distance of the race course.

Solution Because lines BD and AC are parallel, it follows that $\angle BCA \cong \angle CBD$. Consequently, triangle ABC has the measures shown in Figure 3.8. The measure of angle B is $180° - 52° - 40° = 88°$. Using the Law of Sines,

$$\frac{a}{\sin 52°} = \frac{b}{\sin 88°} = \frac{c}{\sin 40°}.$$

Because $b = 8$,

$$a = \frac{8}{\sin 88°}(\sin 52°) \approx 6.31 \quad \text{and} \quad c = \frac{8}{\sin 88°}(\sin 40°) \approx 5.15.$$

The total distance of the course is approximately

Length $\approx 8 + 6.31 + 5.15 = 19.46$ kilometers.

✓ *Checkpoint*   *Audio-video solution in English & Spanish at LarsonPrecalculus.com.*

On a small lake, you swim from point A to point B at a bearing of N 28° E, then to point C at a bearing of N 58° W, and finally back to point A, as shown in the figure below. Point C lies 800 meters directly north of point A. Approximate the total distance that you swim.

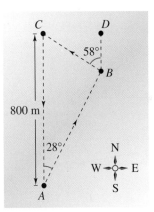

Summarize   (Section 3.1)
1. State the Law of Sines *(page 262)*. For examples of using the Law of Sines to solve oblique triangles (AAS or ASA), see Examples 1 and 2.
2. List the necessary conditions and the numbers of possible triangles for the ambiguous case (SSA) *(page 264)*. For examples of using the Law of Sines to solve oblique triangles (SSA), see Examples 3–5.
3. State the formula for the area of an oblique triangle *(page 266)*. For an example of finding the area of an oblique triangle, see Example 6.
4. Describe how you can use the Law of Sines to model and solve a real-life problem *(page 267, Example 7)*.

3.1 Exercises

See CalcChat.com for tutorial help and worked-out solutions to odd-numbered exercises.

Vocabulary: Fill in the blanks.

1. An _____ triangle is a triangle that has no right angle.

2. For triangle ABC, the Law of Sines is $\dfrac{a}{\sin A} = $ _____ $= \dfrac{c}{\sin C}$.

3. Two _____ and one _____ determine a unique triangle.

4. The area of an oblique triangle is $\frac{1}{2}bc \sin A = \frac{1}{2}ab \sin C = $ _____ .

Skills and Applications

Using the Law of Sines   In Exercises 5–24, use the Law of Sines to solve the triangle. Round your answers to two decimal places.

5.

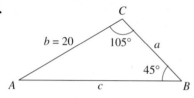

6.

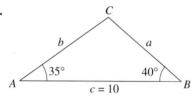

7.

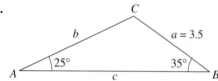

8.

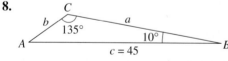

9. $A = 102.4°$, $C = 16.7°$, $a = 21.6$

10. $A = 24.3°$, $C = 54.6°$, $c = 2.68$

11. $A = 83° \, 20'$, $C = 54.6°$, $c = 18.1$

12. $A = 5° \, 40'$, $B = 8° \, 15'$, $b = 4.8$

13. $A = 35°$, $B = 65°$, $c = 10$

14. $A = 120°$, $B = 45°$, $c = 16$

15. $A = 55°$, $B = 42°$, $c = \frac{3}{4}$

16. $B = 28°$, $C = 104°$, $a = 3\frac{5}{8}$

17. $A = 36°$, $a = 8$, $b = 5$

18. $A = 60°$, $a = 9$, $c = 10$

19. $B = 15° \, 30'$, $a = 4.5$, $b = 6.8$

20. $B = 2° \, 45'$, $b = 6.2$, $c = 5.8$

21. $A = 145°$, $a = 14$, $b = 4$

22. $A = 100°$, $a = 125$, $c = 10$

23. $A = 110° \, 15'$, $a = 48$, $b = 16$

24. $C = 95.20°$, $a = 35$, $c = 50$

Using the Law of Sines   In Exercises 25–34, use the Law of Sines to solve (if possible) the triangle. If two solutions exist, find both. Round your answers to two decimal places.

25. $A = 110°$, $a = 125$, $b = 100$

26. $A = 110°$, $a = 125$, $b = 200$

27. $A = 76°$, $a = 18$, $b = 20$

28. $A = 76°$, $a = 34$, $b = 21$

29. $A = 58°$, $a = 11.4$, $b = 12.8$

30. $A = 58°$, $a = 4.5$, $b = 12.8$

31. $A = 120°$, $a = b = 25$

32. $A = 120°$, $a = 25$, $b = 24$

33. $A = 45°$, $a = b = 1$

34. $A = 25° \, 4'$, $a = 9.5$, $b = 22$

Using the Law of Sines   In Exercises 35–38, find values for b such that the triangle has (a) one solution, (b) two solutions, and (c) no solution.

35. $A = 36°$, $a = 5$ 36. $A = 60°$, $a = 10$

37. $A = 10°$, $a = 10.8$

38. $A = 88°$, $a = 315.6$

Finding the Area of a Triangle   In Exercises 39–46, find the area of the triangle having the indicated angle and sides.

39. $C = 120°$, $a = 4$, $b = 6$

40. $B = 130°$, $a = 62$, $c = 20$

41. $A = 150°$, $b = 8$, $c = 10$

42. $C = 170°$, $a = 14$, $b = 24$

43. $A = 43° \, 45'$, $b = 57$, $c = 85$

44. $A = 5° \, 15'$, $b = 4.5$, $c = 22$

45. $B = 72° \, 30'$, $a = 105$, $c = 64$

46. $C = 84° \, 30'$, $a = 16$, $b = 20$

47. Height   Because of prevailing winds, a tree grew so that it was leaning 4° from the vertical. At a point 40 meters from the tree, the angle of elevation to the top of the tree is 30° (see figure). Find the height h of the tree.

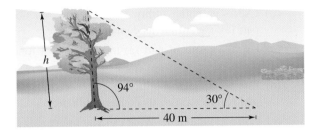

48. Height   A flagpole at a right angle to the horizontal is located on a slope that makes an angle of 12° with the horizontal. The flagpole's shadow is 16 meters long and points directly up the slope. The angle of elevation from the tip of the shadow to the sun is 20°.

(a) Draw a triangle to represent the situation. Show the known quantities on the triangle and use a variable to indicate the height of the flagpole.

(b) Write an equation that can be used to find the height of the flagpole.

(c) Find the height of the flagpole.

49. Angle of Elevation   A 10-meter utility pole casts a 17-meter shadow directly down a slope when the angle of elevation of the sun is 42° (see figure). Find θ, the angle of elevation of the ground.

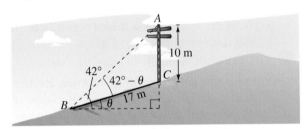

50. Bridge Design   A bridge is to be built across a small lake from a gazebo to a dock (see figure). The bearing from the gazebo to the dock is S 41° W. From a tree 100 meters from the gazebo, the bearings to the gazebo and the dock are S 74° E and S 28° E, respectively. Find the distance from the gazebo to the dock.

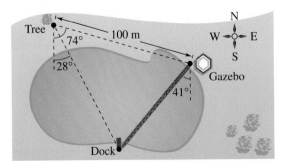

51. Flight Path   A plane flies 500 kilometers with a bearing of 316° from Naples to Elgin (see figure). The plane then flies 720 kilometers from Elgin to Canton (Canton is due west of Naples). Find the bearing of the flight from Elgin to Canton.

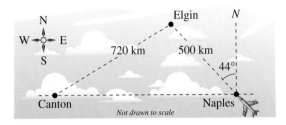

52. Locating a Fire   The bearing from the Pine Knob fire tower to the Colt Station fire tower is N 65° E, and the two towers are 30 kilometers apart. A fire spotted by rangers in each tower has a bearing of N 80° E from Pine Knob and S 70° E from Colt Station (see figure). Find the distance of the fire from each tower.

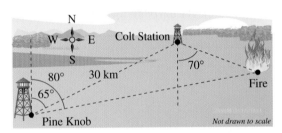

53. Distance   A boat is sailing due east parallel to the shoreline at a speed of 10 miles per hour. At a given time, the bearing to the lighthouse is S 70° E, and 15 minutes later the bearing is S 63° E (see figure). The lighthouse is located at the shoreline. What is the distance from the boat to the shoreline?

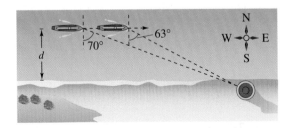

54. Altitude   The angles of elevation to an airplane from two points A and B on level ground are 55° and 72°, respectively. The points A and B are 2.2 miles apart, and the airplane is east of both points in the same vertical plane. Find the altitude of the plane.

55. Distance The angles of elevation θ and ϕ to an airplane from the airport control tower and from an observation post 2 miles away are being continuously monitored (see figure). Write an equation giving the distance d between the plane and observation post in terms of θ and ϕ.

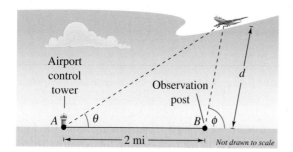

56. The Leaning Tower of Pisa The Leaning Tower of Pisa in Italy leans because it was built on unstable soil—a mixture of clay, sand, and water. The tower is approximately 58.36 meters tall from its foundation (see figure). The top of the tower leans about 5.45 meters off center.

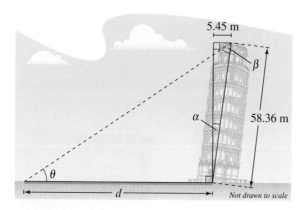

(a) Find the angle of lean α of the tower.

(b) Write β as a function of d and θ, where θ is the angle of elevation to the sun.

(c) Use the Law of Sines to write an equation for the length d of the shadow cast by the tower in terms of θ.

 (d) Use a graphing utility to complete the table.

θ	10°	20°	30°	40°	50°	60°
d						

Exploration

True or False? In Exercises 57–59, determine whether the statement is true or false. Justify your answer.

57. If a triangle contains an obtuse angle, then it must be oblique.

58. Two angles and one side of a triangle do not necessarily determine a unique triangle.

59. If three sides or three angles of an oblique triangle are known, then the triangle can be solved.

 **60. Graphical and Numerical Analysis** In the figure, α and β are positive angles.

(a) Write α as a function of β.

(b) Use a graphing utility to graph the function in part (a). Determine its domain and range.

(c) Use the result of part (a) to write c as a function of β.

(d) Use the graphing utility to graph the function in part (c). Determine its domain and range.

(e) Complete the table. What can you infer?

β	0.4	0.8	1.2	1.6	2.0	2.4	2.8
α							
c							

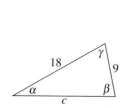

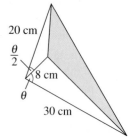

Figure for 60   Figure for 61

61. Graphical Analysis

(a) Write the area A of the shaded region in the figure as a function of θ.

(b) Use a graphing utility to graph the function.

(c) Determine the domain of the function. Explain how decreasing the length of the eight-centimeter line segment would affect the area of the region and the domain of the function.

62. HOW DO YOU SEE IT? In the figure, a triangle is to be formed by drawing a line segment of length a from $(4, 3)$ to the positive x-axis. For what value(s) of a can you form (a) one triangle, (b) two triangles, and (c) no triangles? Explain your reasoning.

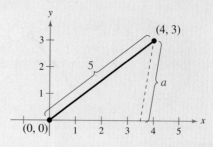

3.2 Law of Cosines

■ Use the Law of Cosines to solve oblique triangles (SSS or SAS).
■ Use the Law of Cosines to model and solve real-life problems.
■ Use Heron's Area Formula to find the area of a triangle.

Introduction

Two cases remain in the list of conditions needed to solve an oblique triangle—SSS and SAS. When you are given three sides (SSS), or two sides and their included angle (SAS), none of the ratios in the Law of Sines would be complete. In such cases, you can use the **Law of Cosines.**

You can use the Law of Cosines to solve real-life problems involving oblique triangles. For instance, in Exercise 56 on page 277, you will use the Law of Cosines to determine the total distance a piston moves in an engine.

Law of Cosines

Standard Form	Alternative Form
$a^2 = b^2 + c^2 - 2bc \cos A$	$\cos A = \dfrac{b^2 + c^2 - a^2}{2bc}$
$b^2 = a^2 + c^2 - 2ac \cos B$	$\cos B = \dfrac{a^2 + c^2 - b^2}{2ac}$
$c^2 = a^2 + b^2 - 2ab \cos C$	$\cos C = \dfrac{a^2 + b^2 - c^2}{2ab}$

For a proof of the Law of Cosines, see Proofs in Mathematics on page 310.

EXAMPLE 1   **Three Sides of a Triangle—SSS**

Find the three angles of the triangle shown below.

B
$a = 8$ ft
$c = 14$ ft
C
$b = 19$ ft
A

Solution   It is a good idea first to find the angle opposite the longest side—side b in this case. Using the alternative form of the Law of Cosines, you find that

$$\cos B = \frac{a^2 + c^2 - b^2}{2ac} = \frac{8^2 + 14^2 - 19^2}{2(8)(14)} \approx -0.45089.$$

Because $\cos B$ is negative, B is an *obtuse* angle given by $B \approx 116.80°$. At this point, it is simpler to use the Law of Sines to determine A.

$$\sin A = a\left(\frac{\sin B}{b}\right) \approx 8\left(\frac{\sin 116.80°}{19}\right) \approx 0.37583$$

Because B is obtuse and a triangle can have at most one obtuse angle, you know that A must be acute. So, $A \approx 22.08°$ and $C \approx 180° - 22.08° - 116.80° = 41.12°$.

✓ **Checkpoint** 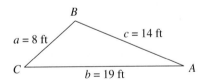   *Audio-video solution in English & Spanish at LarsonPrecalculus.com.*

Find the three angles of the triangle whose sides have lengths $a = 6$, $b = 8$, and $c = 12$. ∎

Do you see why it was wise to find the largest angle *first* in Example 1? Knowing the cosine of an angle, you can determine whether the angle is acute or obtuse. That is,

$$\cos \theta > 0 \quad \text{for} \quad 0° < \theta < 90° \qquad \text{Acute}$$

$$\cos \theta < 0 \quad \text{for} \quad 90° < \theta < 180°. \qquad \text{Obtuse}$$

So, in Example 1, once you found that angle B was obtuse, you knew that angles A and C were both acute. Furthermore, if the largest angle is acute, then the remaining two angles are also acute.

··REMARK When solving an oblique triangle given three sides, you use the alternative form of the Law of Cosines to solve for an angle. When solving an oblique triangle given two sides and their included angle, you use the standard form of the Law of Cosines to solve for an unknown.

EXAMPLE 2 **Two Sides and the Included Angle—SAS**

Find the remaining angles and side of the triangle shown below.

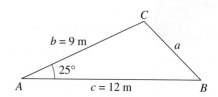

▷ **Solution** Use the Law of Cosines to find the unknown side a in the figure.

$$a^2 = b^2 + c^2 - 2bc \cos A$$

$$a^2 = 9^2 + 12^2 - 2(9)(12) \cos 25°$$

$$a^2 \approx 29.2375$$

$$a \approx 5.4072$$

Because $a \approx 5.4072$ meters, you now know the ratio $(\sin A)/a$, and you can use the reciprocal form of the Law of Sines to solve for B.

$$\frac{\sin B}{b} = \frac{\sin A}{a} \qquad \text{Reciprocal form}$$

$$\sin B = b\left(\frac{\sin A}{a}\right) \qquad \text{Multiply each side by } b.$$

$$\sin B \approx 9\left(\frac{\sin 25°}{5.4072}\right) \qquad \text{Substitute for } A, a, \text{ and } b.$$

$$\sin B \approx 0.7034 \qquad \text{Use a calculator.}$$

There are two angles between $0°$ and $180°$ whose sine is 0.7034, $B_1 \approx 44.7°$ and $B_2 \approx 180° - 44.7° = 135.3°$.

For $B_1 \approx 44.7°$,

$$C_1 \approx 180° - 25° - 44.7° = 110.3°.$$

For $B_2 \approx 135.3°$,

$$C_2 \approx 180° - 25° - 135.3° = 19.7°.$$

Because side c is the longest side of the triangle, C must be the largest angle of the triangle. So, $B \approx 44.7°$ and $C \approx 110.3°$.

✓ **Checkpoint** 🔊))) *Audio-video solution in English & Spanish at LarsonPrecalculus.com.*

Given $A = 80°$, $b = 16$, and $c = 12$, find the remaining angles and side of the triangle.

Applications

EXAMPLE 3   **An Application of the Law of Cosines**

The pitcher's mound on a women's softball field is 43 feet from home plate and the distance between the bases is 60 feet, as shown in Figure 3.9. (The pitcher's mound is *not* halfway between home plate and second base.) How far is the pitcher's mound from first base?

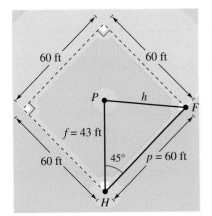

Figure 3.9

Solution   In triangle *HPF*, $H = 45°$ (line *HP* bisects the right angle at *H*), $f = 43$, and $p = 60$. Using the Law of Cosines for this SAS case, you have

$$h^2 = f^2 + p^2 - 2fp \cos H$$
$$= 43^2 + 60^2 - 2(43)(60) \cos 45°$$
$$\approx 1800.3.$$

So, the approximate distance from the pitcher's mound to first base is

$$h \approx \sqrt{1800.3} \approx 42.43 \text{ feet.}$$

✓ *Checkpoint* ◀))) *Audio-video solution in English & Spanish at LarsonPrecalculus.com.*

In a softball game, a batter hits a ball to dead center field, a distance of 240 feet from home plate. The center fielder then throws the ball to third base and gets a runner out. The distance between the bases is 60 feet. How far is the center fielder from third base?

EXAMPLE 4   **An Application of the Law of Cosines**

A ship travels 60 miles due east and then adjusts its course, as shown below. After traveling 80 miles in this new direction, the ship is 139 miles from its point of departure. Describe the bearing from point *B* to point *C*.

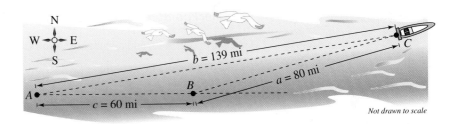

Not drawn to scale

Solution   You have $a = 80$, $b = 139$, and $c = 60$. So, using the alternative form of the Law of Cosines, you have

$$\cos B = \frac{a^2 + c^2 - b^2}{2ac}$$
$$= \frac{80^2 + 60^2 - 139^2}{2(80)(60)}$$
$$\approx -0.97094.$$

So, $B \approx 166.15°$, and thus the bearing measured from due north from point *B* to point *C* is $166.15° - 90° = 76.15°$, or N 76.15° E.

✓ *Checkpoint* ◀))) *Audio-video solution in English & Spanish at LarsonPrecalculus.com.*

A ship travels 40 miles due east and then changes direction, as shown in Figure 3.10. After traveling 30 miles in this new direction, the ship is 56 miles from its point of departure. Describe the bearing from point *B* to point *C*. ▪

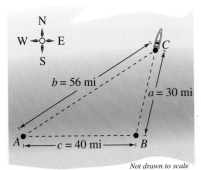

Not drawn to scale

Figure 3.10

Heron's Area Formula

The Law of Cosines can be used to establish the following formula for the area of a triangle. This formula is called **Heron's Area Formula** after the Greek mathematician Heron (ca. 100 B.C.).

Heron's Area Formula

Given any triangle with sides of lengths a, b, and c, the area of the triangle is

$$\text{Area} = \sqrt{s(s-a)(s-b)(s-c)}$$

where

$$s = \frac{a+b+c}{2}.$$

For a proof of Heron's Area Formula, see Proofs in Mathematics on page 311.

EXAMPLE 5   **Using Heron's Area Formula**

Find the area of a triangle having sides of lengths $a = 43$ meters, $b = 53$ meters, and $c = 72$ meters.

Solution   Because $s = (a + b + c)/2 = 168/2 = 84$, Heron's Area Formula yields

$$\begin{aligned}
\text{Area} &= \sqrt{s(s-a)(s-b)(s-c)} \\
&= \sqrt{84(84-43)(84-53)(84-72)} \\
&= \sqrt{84(41)(31)(12)} \\
&\approx 1131.89 \text{ square meters.}
\end{aligned}$$

✓ **Checkpoint**  *Audio-video solution in English & Spanish at LarsonPrecalculus.com.*

Given $a = 5$, $b = 9$, and $c = 8$, use Heron's Area Formula to find the area of the triangle.

You have now studied three different formulas for the area of a triangle.

Standard Formula: $\text{Area} = \dfrac{1}{2}bh$

Oblique Triangle: $\text{Area} = \dfrac{1}{2}bc\sin A = \dfrac{1}{2}ab\sin C = \dfrac{1}{2}ac\sin B$

Heron's Area Formula: $\text{Area} = \sqrt{s(s-a)(s-b)(s-c)}$

Summarize   (Section 3.2)

1. State the Law of Cosines *(page 271)*. For examples of using the Law of Cosines to solve oblique triangles (SSS or SAS), see Examples 1 and 2.

2. Describe real-life problems that can be modeled and solved using the Law of Cosines *(page 273, Examples 3 and 4)*.

3. State Heron's Area Formula *(page 274)*. For an example of using Heron's Area Formula to find the area of a triangle, see Example 5.

3.2 Exercises

See CalcChat.com for tutorial help and worked-out solutions to odd-numbered exercises.

Vocabulary: Fill in the blanks.

1. When you are given three sides of a triangle, you use the Law of _____ to find the three angles of the triangle.

2. When you are given two angles and any side of a triangle, you use the Law of _____ to solve the triangle.

3. The standard form of the Law of Cosines for $\cos B = \dfrac{a^2 + c^2 - b^2}{2ac}$ is _____ .

4. The Law of Cosines can be used to establish a formula for finding the area of a triangle called _____ _____ Formula.

Skills and Applications

Using the Law of Cosines In Exercises 5–24, use the Law of Cosines to solve the triangle. Round your answers to two decimal places.

5.
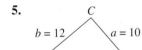

6.
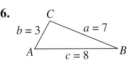

7.
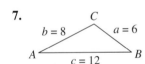

8.
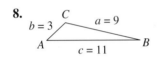

9.
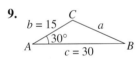

10.
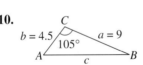

11.
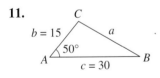

12.
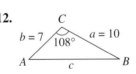

13. $a = 11$, $b = 15$, $c = 21$

14. $a = 55$, $b = 25$, $c = 72$

15. $a = 75.4$, $b = 52$, $c = 52$

16. $a = 1.42$, $b = 0.75$, $c = 1.25$

17. $A = 120°$, $b = 6$, $c = 7$

18. $A = 48°$, $b = 3$, $c = 14$

19. $B = 10° 35'$, $a = 40$, $c = 30$

20. $B = 75° 20'$, $a = 6.2$, $c = 9.5$

21. $B = 125° 40'$, $a = 37$, $c = 37$

22. $C = 15° 15'$, $a = 7.45$, $b = 2.15$

23. $C = 43°$, $a = \frac{4}{9}$, $b = \frac{7}{9}$

24. $C = 101°$, $a = \frac{3}{8}$, $b = \frac{3}{4}$

Finding Measures in a Parallelogram In Exercises 25–30, complete the table by solving the parallelogram shown in the figure. (The lengths of the diagonals are given by c and d.)

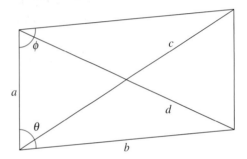

	a	b	c	d	θ	ϕ
25.	5	8			45°	
26.	25	35				120°
27.	10	14	20			
28.	40	60		80		
29.	15		25	20		
30.		25	50	35		

Solving a Triangle In Exercises 31–36, determine whether the Law of Sines or the Law of Cosines is needed to solve the triangle. Then solve (if possible) the triangle. If two solutions exist, find both. Round your answers to two decimal places.

31. $a = 8$, $c = 5$, $B = 40°$

32. $a = 10$, $b = 12$, $C = 70°$

33. $A = 24°$, $a = 4$, $b = 18$

34. $a = 11$, $b = 13$, $c = 7$

35. $A = 42°$, $B = 35°$, $c = 1.2$

36. $a = 160$, $B = 12°$, $C = 7°$

Using Heron's Area Formula **In Exercises 37–44, use Heron's Area Formula to find the area of the triangle.**

37. $a = 8$, $b = 12$, $c = 17$

38. $a = 33$, $b = 36$, $c = 25$

39. $a = 2.5$, $b = 10.2$, $c = 9$

40. $a = 75.4$, $b = 52$, $c = 52$

41. $a = 12.32$, $b = 8.46$, $c = 15.05$

42. $a = 3.05$, $b = 0.75$, $c = 2.45$

43. $a = 1$, $b = \frac{1}{2}$, $c = \frac{3}{4}$

44. $a = \frac{3}{5}$, $b = \frac{5}{8}$, $c = \frac{3}{8}$

45. Navigation A boat race runs along a triangular course marked by buoys A, B, and C. The race starts with the boats headed west for 3700 meters. The other two sides of the course lie to the north of the first side, and their lengths are 1700 meters and 3000 meters. Draw a figure that gives a visual representation of the situation. Then find the bearings for the last two legs of the race.

46. Navigation A plane flies 810 miles from Franklin to Centerville with a bearing of 75°. Then it flies 648 miles from Centerville to Rosemount with a bearing of 32°. Draw a figure that visually represents the situation. Then find the straight-line distance and bearing from Franklin to Rosemount.

47. Surveying To approximate the length of a marsh, a surveyor walks 250 meters from point A to point B, then turns 75° and walks 220 meters to point C (see figure). Approximate the length AC of the marsh.

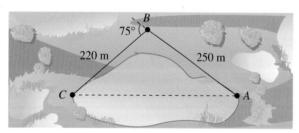

48. Streetlight Design Determine the angle θ in the design of the streetlight shown in the figure.

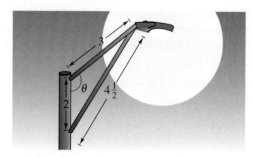

49. Distance Two ships leave a port at 9 A.M. One travels at a bearing of N 53° W at 12 miles per hour, and the other travels at a bearing of S 67° W at 16 miles per hour. Approximate how far apart they are at noon that day.

50. Length A 100-foot vertical tower is to be erected on the side of a hill that makes a 6° angle with the horizontal (see figure). Find the length of each of the two guy wires that will be anchored 75 feet uphill and downhill from the base of the tower.

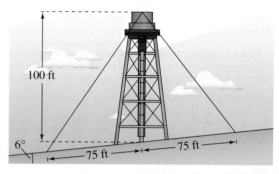

51. Navigation On a map, Minneapolis is 165 millimeters due west of Albany, Phoenix is 216 millimeters from Minneapolis, and Phoenix is 368 millimeters from Albany (see figure).

(a) Find the bearing of Minneapolis from Phoenix.

(b) Find the bearing of Albany from Phoenix.

52. Baseball The baseball player in center field is playing approximately 330 feet from the television camera that is behind home plate. A batter hits a fly ball that goes to the wall 420 feet from the camera (see figure). The camera turns 8° to follow the play. Approximately how far does the center fielder have to run to make the catch?

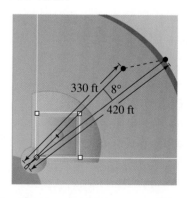

53. Baseball On a baseball diamond with 90-foot sides, the pitcher's mound is 60.5 feet from home plate. How far is it from the pitcher's mound to third base?

54. Surveying A triangular parcel of land has 115 meters of frontage, and the other boundaries have lengths of 76 meters and 92 meters. What angles does the frontage make with the two other boundaries?

55. Surveying A triangular parcel of ground has sides of lengths 725 feet, 650 feet, and 575 feet. Find the measure of the largest angle.

56. Engine Design

An engine has a seven-inch connecting rod fastened to a crank (see figure).

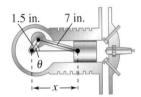

(a) Use the Law of Cosines to write an equation giving the relationship between x and θ.

(b) Write x as a function of θ. (Select the sign that yields positive values of x.)

(c) Use a graphing utility to graph the function in part (b).

(d) Use the graph in part (c) to determine the total distance the piston moves in one cycle.

57. Geometry The lengths of the sides of a triangular parcel of land are approximately 200 feet, 500 feet, and 600 feet. Approximate the area of the parcel.

58. Geometry A parking lot has the shape of a parallelogram (see figure). The lengths of two adjacent sides are 70 meters and 100 meters. The angle between the two sides is 70°. What is the area of the parking lot?

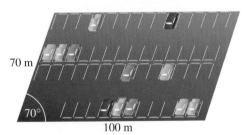

70 m

70°

100 m

59. Geometry You want to buy a triangular lot measuring 510 yards by 840 yards by 1120 yards. The price of the land is $2000 per acre. How much does the land cost? (*Hint:* 1 acre = 4840 square yards)

60. Geometry You want to buy a triangular lot measuring 1350 feet by 1860 feet by 2490 feet. The price of the land is $2200 per acre. How much does the land cost? (*Hint:* 1 acre = 43,560 square feet)

Exploration

True or False? **In Exercises 61 and 62, determine whether the statement is true or false. Justify your answer.**

61. In Heron's Area Formula, s is the average of the lengths of the three sides of the triangle.

62. In addition to SSS and SAS, the Law of Cosines can be used to solve triangles with AAS conditions.

63. Think About It What familiar formula do you obtain when you use the standard form of the Law of Cosines $c^2 = a^2 + b^2 - 2ab \cos C$, and you let $C = 90°$? What is the relationship between the Law of Cosines and this formula?

64. Writing Describe how the Law of Cosines can be used to solve the ambiguous case of the oblique triangle ABC, where $a = 12$ feet, $b = 30$ feet, and $A = 20°$. Is the result the same as when the Law of Sines is used to solve the triangle? Describe the advantages and the disadvantages of each method.

65. Writing In Exercise 64, the Law of Cosines was used to solve a triangle in the two-solution case of SSA. Can the Law of Cosines be used to solve the no-solution and single-solution cases of SSA? Explain.

66. HOW DO YOU SEE IT? Determine whether the Law of Sines or the Law of Cosines is needed to solve the triangle.

(a)
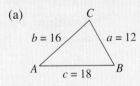

$b = 16$, $a = 12$, $c = 18$

(b)
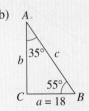

$35°$, $55°$, $a = 18$

(c)
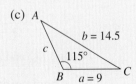

$b = 14.5$, $115°$, $a = 9$

(d)
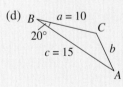

$a = 10$, $20°$, $c = 15$

67. Proof Use the Law of Cosines to prove that

$$\frac{1}{2}bc(1 + \cos A) = \frac{a + b + c}{2} \cdot \frac{-a + b + c}{2}.$$

68. Proof Use the Law of Cosines to prove that

$$\frac{1}{2}bc(1 - \cos A) = \frac{a - b + c}{2} \cdot \frac{a + b - c}{2}.$$

3.3   Vectors in the Plane

You can use vectors to model and solve real-life problems involving magnitude and direction. For instance, in Exercise 102 on page 290, you will use vectors to determine the true direction of a commercial jet.

■ Represent vectors as directed line segments.
■ Write the component forms of vectors.
■ Perform basic vector operations and represent them graphically.
■ Write vectors as linear combinations of unit vectors.
■ Find the direction angles of vectors.
■ Use vectors to model and solve real-life problems.

Introduction

Quantities such as force and velocity involve both *magnitude* and *direction* and cannot be completely characterized by a single real number. To represent such a quantity, you can use a **directed line segment,** as shown in Figure 3.11. The directed line segment $\overrightarrow{PQ}$ has **initial point** P and **terminal point** Q. Its **magnitude** (or length) is denoted by $\|\overrightarrow{PQ}\|$ and can be found using the Distance Formula.

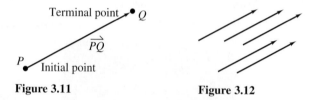

Figure 3.11   **Figure 3.12**

Two directed line segments that have the same magnitude and direction are equivalent. For example, the directed line segments in Figure 3.12 are all equivalent. The set of all directed line segments that are equivalent to the directed line segment $\overrightarrow{PQ}$ is a **vector v in the plane,** written $\mathbf{v} = \overrightarrow{PQ}$. Vectors are denoted by lowercase, boldface letters such as $\mathbf{u}$, $\mathbf{v}$, and $\mathbf{w}$.

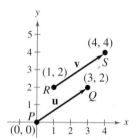

Figure 3.13

| EXAMPLE 1 | **Showing That Two Vectors Are Equivalent** |

Show that $\mathbf{u}$ and $\mathbf{v}$ in Figure 3.13 are equivalent.

Solution   From the Distance Formula, it follows that $\overrightarrow{PQ}$ and $\overrightarrow{RS}$ have the *same magnitude.*

$$\|\overrightarrow{PQ}\| = \sqrt{(3 - 0)^2 + (2 - 0)^2} = \sqrt{13}   \quad \|\overrightarrow{RS}\| = \sqrt{(4 - 1)^2 + (4 - 2)^2} = \sqrt{13}$$

Moreover, both line segments have the *same direction* because they are both directed toward the upper right on lines having a slope of

$$\frac{4 - 2}{4 - 1} = \frac{2 - 0}{3 - 0} = \frac{2}{3}.$$

Because $\overrightarrow{PQ}$ and $\overrightarrow{RS}$ have the same magnitude and direction, $\mathbf{u}$ and $\mathbf{v}$ are equivalent.

✓ *Checkpoint* ◀))) *Audio-video solution in English & Spanish at LarsonPrecalculus.com.*

Show that $\mathbf{u}$ and $\mathbf{v}$ in the figure at the right are equivalent.

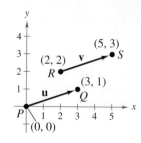

Component Form of a Vector

The directed line segment whose initial point is the origin is often the most convenient representative of a set of equivalent directed line segments. This representative of the vector **v** is in **standard position.**

A vector whose initial point is the origin $(0, 0)$ can be uniquely represented by the coordinates of its terminal point (v_1, v_2). This is the **component form of a vector v,** written as $\mathbf{v} = \langle v_1, v_2 \rangle$. The coordinates v_1 and v_2 are the *components* of **v**. If both the initial point and the terminal point lie at the origin, then **v** is the **zero vector** and is denoted by $\mathbf{0} = \langle 0, 0 \rangle$.

 TECHNOLOGY You can graph vectors with a graphing utility by graphing directed line segments. Consult the user's guide for your graphing utility for specific instructions.

Component Form of a Vector

The component form of the vector with initial point $P(p_1, p_2)$ and terminal point $Q(q_1, q_2)$ is given by

$$\overrightarrow{PQ} = \langle q_1 - p_1, q_2 - p_2 \rangle = \langle v_1, v_2 \rangle = \mathbf{v}.$$

The **magnitude** (or length) of **v** is given by

$$\|\mathbf{v}\| = \sqrt{(q_1 - p_1)^2 + (q_2 - p_2)^2} = \sqrt{v_1^2 + v_2^2}.$$

If $\|\mathbf{v}\| = 1$, then **v** is a **unit vector.** Moreover, $\|\mathbf{v}\| = 0$ if and only if **v** is the zero vector **0**.

Two vectors $\mathbf{u} = \langle u_1, u_2 \rangle$ and $\mathbf{v} = \langle v_1, v_2 \rangle$ are *equal* if and only if $u_1 = v_1$ and $u_2 = v_2$. For instance, in Example 1, the vector **u** from $P(0, 0)$ to $Q(3, 2)$ is $\mathbf{u} = \overrightarrow{PQ} = \langle 3 - 0, 2 - 0 \rangle = \langle 3, 2 \rangle$, and the vector **v** from $R(1, 2)$ to $S(4, 4)$ is $\mathbf{v} = \overrightarrow{RS} = \langle 4 - 1, 4 - 2 \rangle = \langle 3, 2 \rangle$.

EXAMPLE 2   Finding the Component Form of a Vector

Find the component form and magnitude of the vector **v** that has initial point $(4, -7)$ and terminal point $(-1, 5)$.

Algebraic Solution

Let

$$P(4, -7) = (p_1, p_2)$$

and

$$Q(-1, 5) = (q_1, q_2).$$

Then, the components of $\mathbf{v} = \langle v_1, v_2 \rangle$ are

$$v_1 = q_1 - p_1 = -1 - 4 = -5$$

$$v_2 = q_2 - p_2 = 5 - (-7) = 12.$$

So, $\mathbf{v} = \langle -5, 12 \rangle$ and the magnitude of **v** is

$$\|\mathbf{v}\| = \sqrt{(-5)^2 + 12^2}$$

$$= \sqrt{169}$$

$$= 13.$$

Graphical Solution

Use centimeter graph paper to plot the points $P(4, -7)$ and $Q(-1, 5)$. Carefully sketch the vector **v**. Use the sketch to find the components of $\mathbf{v} = \langle v_1, v_2 \rangle$. Then use a centimeter ruler to find the magnitude of **v**. The figure at the right shows that the components of **v** are $v_1 = -5$ and $v_2 = 12$, so $\mathbf{v} = \langle -5, 12 \rangle$. The figure also shows that the magnitude of **v** is $\|\mathbf{v}\| = 13$.

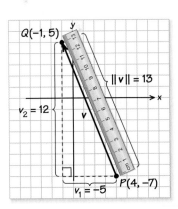

✓ *Checkpoint*))) *Audio-video solution in English & Spanish at LarsonPrecalculus.com.*

Find the component form and magnitude of the vector **v** that has initial point $(-2, 3)$ and terminal point $(-7, 9)$.

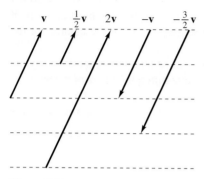

Figure 3.14

Vector Operations

The two basic vector operations are **scalar multiplication** and **vector addition.** In operations with vectors, numbers are usually referred to as **scalars.** In this text, scalars will always be real numbers. Geometrically, the product of a vector **v** and a scalar k is the vector that is $|k|$ times as long as **v**. When k is positive, $k\mathbf{v}$ has the same direction as **v**, and when k is negative, $k\mathbf{v}$ has the direction opposite that of **v**, as shown in Figure 3.14.

To add two vectors **u** and **v** geometrically, first position them (without changing their lengths or directions) so that the initial point of the second vector **v** coincides with the terminal point of the first vector **u**. The sum

u + v

is the vector formed by joining the initial point of the first vector **u** with the terminal point of the second vector **v**, as shown below. This technique is called the **parallelogram law** for vector addition because the vector **u + v**, often called the **resultant** of vector addition, is the diagonal of a parallelogram having adjacent sides **u** and **v**.

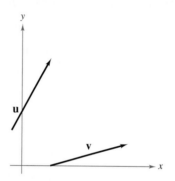
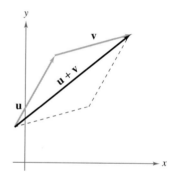

Definitions of Vector Addition and Scalar Multiplication

Let $\mathbf{u} = \langle u_1, u_2 \rangle$ and $\mathbf{v} = \langle v_1, v_2 \rangle$ be vectors and let k be a scalar (a real number). Then the **sum** of **u** and **v** is the vector

$$\mathbf{u} + \mathbf{v} = \langle u_1 + v_1, u_2 + v_2 \rangle \qquad \text{Sum}$$

and the **scalar multiple** of k times **u** is the vector

$$k\mathbf{u} = k\langle u_1, u_2 \rangle = \langle ku_1, ku_2 \rangle. \qquad \text{Scalar multiple}$$

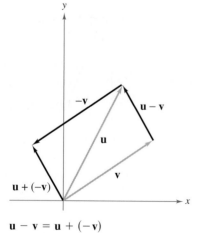

$$\mathbf{u} - \mathbf{v} = \mathbf{u} + (-\mathbf{v})$$

Figure 3.15

The **negative** of $\mathbf{v} = \langle v_1, v_2 \rangle$ is

$$-\mathbf{v} = (-1)\mathbf{v}$$

$$= \langle -v_1, -v_2 \rangle \qquad \text{Negative}$$

and the **difference** of **u** and **v** is

$$\mathbf{u} - \mathbf{v} = \mathbf{u} + (-\mathbf{v}) \qquad \text{Add } (-\mathbf{v}). \text{ See Figure 3.15.}$$

$$= \langle u_1 - v_1, u_2 - v_2 \rangle. \qquad \text{Difference}$$

To represent $\mathbf{u} - \mathbf{v}$ geometrically, you can use directed line segments with the *same* initial point. The difference $\mathbf{u} - \mathbf{v}$ is the vector from the terminal point of **v** to the terminal point of **u**, which is equal to

$$\mathbf{u} + (-\mathbf{v})$$

as shown in Figure 3.15.

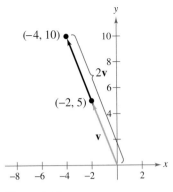

Figure 3.16

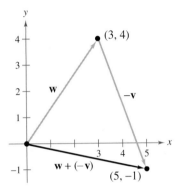

Figure 3.17

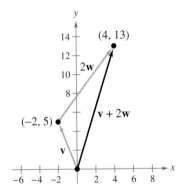

Figure 3.18

The component definitions of vector addition and scalar multiplication are illustrated in Example 3. In this example, notice that each of the vector operations can be interpreted geometrically.

EXAMPLE 3   Vector Operations

Let $\mathbf{v} = \langle -2, 5 \rangle$ and $\mathbf{w} = \langle 3, 4 \rangle$. Find each of the following vectors.

a. $2\mathbf{v}$ **b.** $\mathbf{w} - \mathbf{v}$ **c.** $\mathbf{v} + 2\mathbf{w}$

Solution

a. Because $\mathbf{v} = \langle -2, 5 \rangle$, you have

$$2\mathbf{v} = 2\langle -2, 5 \rangle = \langle 2(-2), 2(5) \rangle = \langle -4, 10 \rangle.$$

A sketch of $2\mathbf{v}$ is shown in Figure 3.16.

b. The difference of $\mathbf{w}$ and $\mathbf{v}$ is

$$\mathbf{w} - \mathbf{v} = \langle 3, 4 \rangle - \langle -2, 5 \rangle$$
$$= \langle 3 - (-2), 4 - 5 \rangle$$
$$= \langle 5, -1 \rangle.$$

A sketch of $\mathbf{w} - \mathbf{v}$ is shown in Figure 3.17. Note that the figure shows the vector difference $\mathbf{w} - \mathbf{v}$ as the sum $\mathbf{w} + (-\mathbf{v})$.

c. The sum of $\mathbf{v}$ and $2\mathbf{w}$ is

$$\mathbf{v} + 2\mathbf{w} = \langle -2, 5 \rangle + 2\langle 3, 4 \rangle$$
$$= \langle -2, 5 \rangle + \langle 2(3), 2(4) \rangle$$
$$= \langle -2, 5 \rangle + \langle 6, 8 \rangle$$
$$= \langle -2 + 6, 5 + 8 \rangle$$
$$= \langle 4, 13 \rangle.$$

A sketch of $\mathbf{v} + 2\mathbf{w}$ is shown in Figure 3.18.

✓ *Checkpoint* ◀))   *Audio-video solution in English & Spanish at LarsonPrecalculus.com.*

Let $\mathbf{u} = \langle 1, 4 \rangle$ and $\mathbf{v} = \langle 3, 2 \rangle$. Find each of the following vectors.

a. $\mathbf{u} + \mathbf{v}$ **b.** $\mathbf{u} - \mathbf{v}$ **c.** $2\mathbf{u} - 3\mathbf{v}$ ■

Vector addition and scalar multiplication share many of the properties of ordinary arithmetic.

••**REMARK** Property 9 can be stated as follows: The magnitude of the vector $c\mathbf{v}$ is the absolute value of c times the magnitude of $\mathbf{v}$.

> **Properties of Vector Addition and Scalar Multiplication**
>
> Let $\mathbf{u}$, $\mathbf{v}$, and $\mathbf{w}$ be vectors and let c and d be scalars. Then the following properties are true.
>
> **1.** $\mathbf{u} + \mathbf{v} = \mathbf{v} + \mathbf{u}$
>
> **2.** $(\mathbf{u} + \mathbf{v}) + \mathbf{w} = \mathbf{u} + (\mathbf{v} + \mathbf{w})$
>
> **3.** $\mathbf{u} + \mathbf{0} = \mathbf{u}$
>
> **4.** $\mathbf{u} + (-\mathbf{u}) = \mathbf{0}$
>
> **5.** $c(d\mathbf{u}) = (cd)\mathbf{u}$
>
> **6.** $(c + d)\mathbf{u} = c\mathbf{u} + d\mathbf{u}$
>
> **7.** $c(\mathbf{u} + \mathbf{v}) = c\mathbf{u} + c\mathbf{v}$
>
> **8.** $1(\mathbf{u}) = \mathbf{u}, \quad 0(\mathbf{u}) = \mathbf{0}$
>
> **9.** $\|c\mathbf{v}\| = |c|\,\|\mathbf{v}\|$

William Rowan Hamilton (1805–1865), an Irish mathematician, did some of the earliest work with vectors. Hamilton spent many years developing a system of vector-like quantities called quaternions. Although Hamilton was convinced of the benefits of quaternions, the operations he defined did not produce good models for physical phenomena. It was not until the latter half of the nineteenth century that the Scottish physicist James Maxwell (1831–1879) restructured Hamilton's quaternions in a form useful for representing physical quantities such as force, velocity, and acceleration.

Unit Vectors

In many applications of vectors, it is useful to find a unit vector that has the same direction as a given nonzero vector $\mathbf{v}$. To do this, you can divide $\mathbf{v}$ by its magnitude to obtain

$$\mathbf{u} = \text{unit vector} = \frac{\mathbf{v}}{\|\mathbf{v}\|} = \left(\frac{1}{\|\mathbf{v}\|}\right)\mathbf{v}. \qquad \text{Unit vector in direction of } \mathbf{v}$$

Note that $\mathbf{u}$ is a scalar multiple of $\mathbf{v}$. The vector $\mathbf{u}$ has a magnitude of 1 and the same direction as $\mathbf{v}$. The vector $\mathbf{u}$ is called a **unit vector in the direction of v.**

EXAMPLE 4   Finding a Unit Vector

Find a unit vector in the direction of $\mathbf{v} = \langle -2, 5 \rangle$ and verify that the result has a magnitude of 1.

Solution   The unit vector in the direction of $\mathbf{v}$ is

$$\frac{\mathbf{v}}{\|\mathbf{v}\|} = \frac{\langle -2, 5 \rangle}{\sqrt{(-2)^2 + 5^2}} = \frac{1}{\sqrt{29}}\langle -2, 5 \rangle = \left\langle \frac{-2}{\sqrt{29}}, \frac{5}{\sqrt{29}} \right\rangle.$$

This vector has a magnitude of 1 because

$$\sqrt{\left(\frac{-2}{\sqrt{29}}\right)^2 + \left(\frac{5}{\sqrt{29}}\right)^2} = \sqrt{\frac{4}{29} + \frac{25}{29}}$$

$$= \sqrt{\frac{29}{29}}$$

$$= 1.$$

✓ **Checkpoint** 🔊))) *Audio-video solution in English & Spanish at LarsonPrecalculus.com.*

Find a unit vector $\mathbf{u}$ in the direction of $\mathbf{v} = \langle 6, -1 \rangle$ and verify that the result has a magnitude of 1. ∎

To find a vector $\mathbf{w}$ with magnitude $\|\mathbf{w}\| = c$ and the same direction as a nonzero vector $\mathbf{v}$, multiply the unit vector $\mathbf{u}$ in the direction of $\mathbf{v}$ by the scalar c to obtain $\mathbf{w} = c\mathbf{u}$.

EXAMPLE 5   Finding a Vector

Find the vector $\mathbf{w}$ with magnitude $\|\mathbf{w}\| = 5$ and the same direction as $\mathbf{v} = \langle -2, 3 \rangle$.

Solution

$$\mathbf{w} = 5\left(\frac{1}{\|\mathbf{v}\|}\mathbf{v}\right)$$

$$= 5\left(\frac{1}{\sqrt{(-2)^2 + 3^2}}\langle -2, 3 \rangle\right)$$

$$= \frac{5}{\sqrt{13}}\langle -2, 3 \rangle$$

$$= \left\langle \frac{-10}{\sqrt{13}}, \frac{15}{\sqrt{13}} \right\rangle$$

✓ **Checkpoint** 🔊))) *Audio-video solution in English & Spanish at LarsonPrecalculus.com.*

Find the vector $\mathbf{w}$ with magnitude $\|\mathbf{w}\| = 6$ and the same direction as $\mathbf{v} = \langle 2, -4 \rangle$. ∎

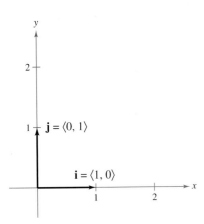

Figure 3.19

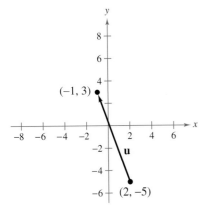

Figure 3.20

The unit vectors $\langle 1, 0 \rangle$ and $\langle 0, 1 \rangle$ are called the **standard unit vectors** and are denoted by

$$\mathbf{i} = \langle 1, 0 \rangle \quad \text{and} \quad \mathbf{j} = \langle 0, 1 \rangle$$

as shown in Figure 3.19. (Note that the lowercase letter $\mathbf{i}$ is written in boldface to distinguish it from the imaginary unit $i = \sqrt{-1}$.) These vectors can be used to represent any vector $\mathbf{v} = \langle v_1, v_2 \rangle$, as follows.

$$
\begin{aligned}
\mathbf{v} &= \langle v_1, v_2 \rangle \\
&= v_1 \langle 1, 0 \rangle + v_2 \langle 0, 1 \rangle \\
&= v_1 \mathbf{i} + v_2 \mathbf{j}
\end{aligned}
$$

The scalars v_1 and v_2 are called the **horizontal** and **vertical components of v,** respectively. The vector sum

$$v_1 \mathbf{i} + v_2 \mathbf{j}$$

is called a **linear combination** of the vectors $\mathbf{i}$ and $\mathbf{j}$. Any vector in the plane can be written as a linear combination of the standard unit vectors $\mathbf{i}$ and $\mathbf{j}$.

EXAMPLE 6   **Writing a Linear Combination of Unit Vectors**

Let $\mathbf{u}$ be the vector with initial point $(2, -5)$ and terminal point $(-1, 3)$. Write $\mathbf{u}$ as a linear combination of the standard unit vectors $\mathbf{i}$ and $\mathbf{j}$.

Solution   Begin by writing the component form of the vector $\mathbf{u}$.

$$
\begin{aligned}
\mathbf{u} &= \langle -1 - 2, 3 - (-5) \rangle \\
&= \langle -3, 8 \rangle \\
&= -3\mathbf{i} + 8\mathbf{j}
\end{aligned}
$$

This result is shown graphically in Figure 3.20.

✓ *Checkpoint* 🔊))) *Audio-video solution in English & Spanish at LarsonPrecalculus.com.*

Let $\mathbf{u}$ be the vector with initial point $(-2, 6)$ and terminal point $(-8, 3)$. Write $\mathbf{u}$ as a linear combination of the standard unit vectors $\mathbf{i}$ and $\mathbf{j}$.

EXAMPLE 7   **Vector Operations**

Let

$$\mathbf{u} = -3\mathbf{i} + 8\mathbf{j} \quad \text{and} \quad \mathbf{v} = 2\mathbf{i} - \mathbf{j}.$$

Find $2\mathbf{u} - 3\mathbf{v}$.

Solution   You could solve this problem by converting $\mathbf{u}$ and $\mathbf{v}$ to component form. This, however, is not necessary. It is just as easy to perform the operations in unit vector form.

$$
\begin{aligned}
2\mathbf{u} - 3\mathbf{v} &= 2(-3\mathbf{i} + 8\mathbf{j}) - 3(2\mathbf{i} - \mathbf{j}) \\
&= -6\mathbf{i} + 16\mathbf{j} - 6\mathbf{i} + 3\mathbf{j} \\
&= -12\mathbf{i} + 19\mathbf{j}
\end{aligned}
$$

✓ *Checkpoint* 🔊))) *Audio-video solution in English & Spanish at LarsonPrecalculus.com.*

Let

$$\mathbf{u} = \mathbf{i} - 2\mathbf{j} \quad \text{and} \quad \mathbf{v} = -3\mathbf{i} + 2\mathbf{j}.$$

Find $5\mathbf{u} - 2\mathbf{v}$.

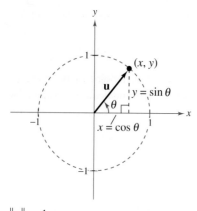

$\|\mathbf{u}\| = 1$
Figure 3.21

Direction Angles

If $\mathbf{u}$ is a *unit vector* such that θ is the angle (measured counterclockwise) from the positive x-axis to $\mathbf{u}$, then the terminal point of $\mathbf{u}$ lies on the unit circle and you have

$$\mathbf{u} = \langle x, y \rangle$$
$$= \langle \cos \theta, \sin \theta \rangle$$
$$= (\cos \theta)\mathbf{i} + (\sin \theta)\mathbf{j}$$

as shown in Figure 3.21. The angle θ is the **direction angle** of the vector $\mathbf{u}$.

Suppose that $\mathbf{u}$ is a unit vector with direction angle θ. If $\mathbf{v} = a\mathbf{i} + b\mathbf{j}$ is any vector that makes an angle θ with the positive x-axis, then it has the same direction as $\mathbf{u}$ and you can write

$$\mathbf{v} = \|\mathbf{v}\|\langle \cos \theta, \sin \theta \rangle$$
$$= \|\mathbf{v}\|(\cos \theta)\mathbf{i} + \|\mathbf{v}\|(\sin \theta)\mathbf{j}.$$

Because $\mathbf{v} = a\mathbf{i} + b\mathbf{j} = \|\mathbf{v}\|(\cos \theta)\mathbf{i} + \|\mathbf{v}\|(\sin \theta)\mathbf{j}$, it follows that the direction angle θ for $\mathbf{v}$ is determined from

$$\tan \theta = \frac{\sin \theta}{\cos \theta} \qquad \text{Quotient identity}$$

$$= \frac{\|\mathbf{v}\| \sin \theta}{\|\mathbf{v}\| \cos \theta} \qquad \text{Multiply numerator and denominator by } \|\mathbf{v}\|.$$

$$= \frac{b}{a}. \qquad \text{Simplify.}$$

EXAMPLE 8   Finding Direction Angles of Vectors

Find the direction angle of each vector.

a. $\mathbf{u} = 3\mathbf{i} + 3\mathbf{j}$ **b.** $\mathbf{v} = 3\mathbf{i} - 4\mathbf{j}$

Solution

a. The direction angle is determined from

$$\tan \theta = \frac{b}{a} = \frac{3}{3} = 1.$$

So, $\theta = 45°$, as shown in Figure 3.22.

b. The direction angle is determined from

$$\tan \theta = \frac{b}{a} = \frac{-4}{3}.$$

Moreover, because $\mathbf{v} = 3\mathbf{i} - 4\mathbf{j}$ lies in Quadrant IV, θ lies in Quadrant IV, and its reference angle is

$$\theta' = \left| \arctan\left(-\frac{4}{3}\right) \right| \approx |-0.9273 \text{ radian}| \approx |-53.13°| = 53.13°.$$

So, it follows that $\theta \approx 360° - 53.13° = 306.87°$, as shown in Figure 3.23.

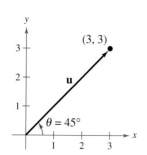

Figure 3.22

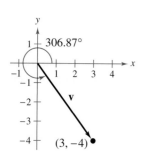

Figure 3.23

✓ *Checkpoint* ◀))) Audio-video solution in English & Spanish at LarsonPrecalculus.com.

Find the direction angle of each vector.

a. $\mathbf{v} = -6\mathbf{i} + 6\mathbf{j}$ **b.** $\mathbf{v} = -7\mathbf{i} - 4\mathbf{j}$

Applications

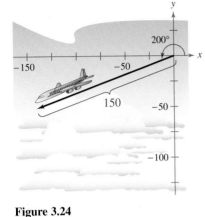

Figure 3.24

EXAMPLE 9 **Finding the Component Form of a Vector**

Find the component form of the vector that represents the velocity of an airplane descending at a speed of 150 miles per hour at an angle 20° below the horizontal, as shown in Figure 3.24.

Solution The velocity vector **v** has a magnitude of 150 and a direction angle of $\theta = 200°$.

$$\mathbf{v} = \|\mathbf{v}\|(\cos \theta)\mathbf{i} + \|\mathbf{v}\|(\sin \theta)\mathbf{j}$$
$$= 150(\cos 200°)\mathbf{i} + 150(\sin 200°)\mathbf{j}$$
$$\approx 150(-0.9397)\mathbf{i} + 150(-0.3420)\mathbf{j}$$
$$\approx -140.96\mathbf{i} - 51.30\mathbf{j}$$
$$= \langle -140.96, -51.30 \rangle$$

You can check that **v** has a magnitude of 150, as follows.

$$\|\mathbf{v}\| \approx \sqrt{(-140.96)^2 + (-51.30)^2} \approx \sqrt{22{,}501.41} \approx 150$$

 Checkpoint Audio-video solution in English & Spanish at LarsonPrecalculus.com.

Find the component form of the vector that represents the velocity of an airplane descending at a speed of 100 miles per hour at an angle 45° below the horizontal $(\theta = 225°)$.

EXAMPLE 10 **Using Vectors to Determine Weight**

A force of 600 pounds is required to pull a boat and trailer up a ramp inclined at 15° from the horizontal. Find the combined weight of the boat and trailer.

Solution Based on Figure 3.25, you can make the following observations.

$\|\overrightarrow{BA}\|$ = force of gravity = combined weight of boat and trailer

$\|\overrightarrow{BC}\|$ = force against ramp

$\|\overrightarrow{AC}\|$ = force required to move boat up ramp = 600 pounds

By construction, triangles BWD and ABC are similar. So, angle ABC is 15°. In triangle ABC, you have

$$\sin 15° = \frac{\|\overrightarrow{AC}\|}{\|\overrightarrow{BA}\|}$$
$$\sin 15° = \frac{600}{\|\overrightarrow{BA}\|}$$
$$\|\overrightarrow{BA}\| = \frac{600}{\sin 15°}$$
$$\|\overrightarrow{BA}\| \approx 2318.$$

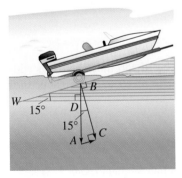

Figure 3.25

So, the combined weight is approximately 2318 pounds. (In Figure 3.25, note that $\overrightarrow{AC}$ is parallel to the ramp.)

 Checkpoint Audio-video solution in English & Spanish at LarsonPrecalculus.com.

A force of 500 pounds is required to pull a boat and trailer up a ramp inclined at 12° from the horizontal. Find the combined weight of the boat and trailer. ∎

•• **REMARK**   Recall from
Section 1.8 that in air navigation,
bearings can be measured in
degrees clockwise from north.

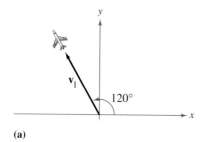

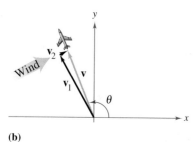

(a)

(b)
Figure 3.26

EXAMPLE 11   **Using Vectors to Find Speed and Direction**

An airplane is traveling at a speed of 500 miles per hour with a bearing of 330° at a fixed altitude with a negligible wind velocity, as shown in Figure 3.26(a). When the airplane reaches a certain point, it encounters a wind with a velocity of 70 miles per hour in the direction N 45° E, as shown in Figure 3.26(b). What are the resultant speed and direction of the airplane?

Solution   Using Figure 3.26, the velocity of the airplane (alone) is

$$\mathbf{v}_1 = 500\langle\cos 120°, \sin 120°\rangle = \langle -250, 250\sqrt{3}\rangle$$

and the velocity of the wind is

$$\mathbf{v}_2 = 70\langle\cos 45°, \sin 45°\rangle = \langle 35\sqrt{2}, 35\sqrt{2}\rangle.$$

So, the velocity of the airplane (in the wind) is

$$\mathbf{v} = \mathbf{v}_1 + \mathbf{v}_2$$
$$= \langle -250 + 35\sqrt{2}, 250\sqrt{3} + 35\sqrt{2}\rangle$$
$$\approx \langle -200.5, 482.5\rangle$$

and the resultant speed of the airplane is

$$\|\mathbf{v}\| \approx \sqrt{(-200.5)^2 + (482.5)^2} \approx 522.5 \text{ miles per hour.}$$

Finally, given that θ is the direction angle of the flight path, you have

$$\tan \theta \approx \frac{482.5}{-200.5} \approx -2.4065$$

which implies that

$$\theta \approx 180° - 67.4° = 112.6°.$$

So, the true direction of the airplane is approximately

$$270° + (180° - 112.6°) = 337.4°.$$

✓ **Checkpoint** ◀))) *Audio-video solution in English & Spanish at LarsonPrecalculus.com.*

Repeat Example 11 for an airplane traveling at a speed of 450 miles per hour with a bearing of 300° that encounters a wind with a velocity of 40 miles per hour in the direction N 30° E.

Airplanes can take advantage of
fast-moving air currents called jet
streams to decrease travel time.

Summarize   **(Section 3.3)**

1. Describe how to represent a vector as a directed line segment *(page 278)*. For an example involving vectors represented as directed line segments, see Example 1.

2. Describe how to write a vector in component form *(page 279)*. For an example of finding the component form of a vector, see Example 2.

3. State the definitions of vector addition and scalar multiplication *(page 280)*. For an example of performing vector operations, see Example 3.

4. Describe how to write a vector as a linear combination of unit vectors *(page 282)*. For examples involving unit vectors, see Examples 4–7.

5. Describe how to find the direction angle of a vector *(page 284)*. For an example of finding the direction angles of vectors, see Example 8.

6. Describe real-life situations that can be modeled and solved using vectors *(pages 285 and 286, Examples 9–11)*.

3.3 Exercises

Vocabulary: Fill in the blanks.

1. A _____ _____ _____ can be used to represent a quantity that involves both magnitude and direction.
2. The directed line segment $\overrightarrow{PQ}$ has _____ point P and _____ point Q.
3. The _____ of the directed line segment $\overrightarrow{PQ}$ is denoted by $\|\overrightarrow{PQ}\|$.
4. The set of all directed line segments that are equivalent to a given directed line segment $\overrightarrow{PQ}$ is a _____ **v** in the plane.
5. In order to show that two vectors are equivalent, you must show that they have the same _____ and the same _____ .
6. The directed line segment whose initial point is the origin is said to be in _____ _____ .
7. A vector that has a magnitude of 1 is called a _____ _____ .
8. The two basic vector operations are scalar _____ and vector _____ .
9. The vector **u** + **v** is called the _____ of vector addition.
10. The vector sum $v_1\mathbf{i} + v_2\mathbf{j}$ is called a _____ _____ of the vectors **i** and **j**, and the scalars v_1 and v_2 are called the _____ and _____ components of **v**, respectively.

Skills and Applications

Showing That Two Vectors Are Equivalent In Exercises 11 and 12, show that u and v are equivalent.

11.
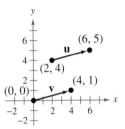

12.
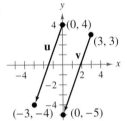

17.
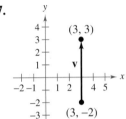

18.
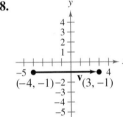

Initial Point	Terminal Point
19. $(-3, -5)$	$(5, 1)$
20. $(-2, 7)$	$(5, -17)$
21. $(1, 3)$	$(-8, -9)$
22. $(1, 11)$	$(9, 3)$
23. $(-1, 5)$	$(15, 12)$
24. $(-3, 11)$	$(9, 40)$

Finding the Component Form of a Vector In Exercises 13–24, find the component form and magnitude of the vector v.

13.
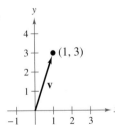

14.
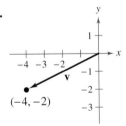

15.
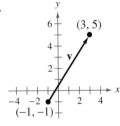

16.
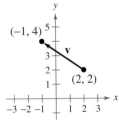

Sketching the Graph of a Vector In Exercises 25–30, use the figure to sketch a graph of the specified vector. To print an enlarged copy of the graph, go to _MathGraphs.com_.

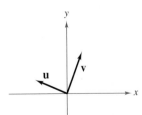

25. $-\mathbf{v}$

26. $5\mathbf{v}$

27. $\mathbf{u} + \mathbf{v}$

28. $\mathbf{u} + 2\mathbf{v}$

29. $\mathbf{u} - \mathbf{v}$

30. $\mathbf{v} - \frac{1}{2}\mathbf{u}$

Vector Operations   In Exercises 31–38, find (a) $\mathbf{u} + \mathbf{v}$, (b) $\mathbf{u} - \mathbf{v}$, and (c) $2\mathbf{u} - 3\mathbf{v}$. Then sketch each resultant vector.

31. $\mathbf{u} = \langle 2, 1 \rangle$, $\mathbf{v} = \langle 1, 3 \rangle$ **32.** $\mathbf{u} = \langle 2, 3 \rangle$, $\mathbf{v} = \langle 4, 0 \rangle$

33. $\mathbf{u} = \langle -5, 3 \rangle$, $\mathbf{v} = \langle 0, 0 \rangle$ **34.** $\mathbf{u} = \langle 0, 0 \rangle$, $\mathbf{v} = \langle 2, 1 \rangle$

35. $\mathbf{u} = \mathbf{i} + \mathbf{j}$, $\mathbf{v} = 2\mathbf{i} - 3\mathbf{j}$

36. $\mathbf{u} = -2\mathbf{i} + \mathbf{j}$, $\mathbf{v} = 3\mathbf{j}$

37. $\mathbf{u} = 2\mathbf{i}$, $\mathbf{v} = \mathbf{j}$ **38.** $\mathbf{u} = 2\mathbf{j}$, $\mathbf{v} = 3\mathbf{i}$

Finding a Unit Vector   In Exercises 39–48, find a unit vector in the direction of the given vector. Verify that the result has a magnitude of 1.

39. $\mathbf{u} = \langle 3, 0 \rangle$ **40.** $\mathbf{u} = \langle 0, -2 \rangle$

41. $\mathbf{v} = \langle -2, 2 \rangle$ **42.** $\mathbf{v} = \langle 5, -12 \rangle$

43. $\mathbf{v} = \mathbf{i} + \mathbf{j}$ **44.** $\mathbf{v} = 6\mathbf{i} - 2\mathbf{j}$

45. $\mathbf{w} = 4\mathbf{j}$ **46.** $\mathbf{w} = -6\mathbf{i}$

47. $\mathbf{w} = \mathbf{i} - 2\mathbf{j}$ **48.** $\mathbf{w} = 7\mathbf{j} - 3\mathbf{i}$

Finding a Vector   In Exercises 49–52, find the vector w with the given magnitude and the same direction as v.

	Magnitude	Direction
49.	$\|\mathbf{w}\| = 10$	$\mathbf{v} = \langle -3, 4 \rangle$
50.	$\|\mathbf{w}\| = 3$	$\mathbf{v} = \langle -12, -5 \rangle$
51.	$\|\mathbf{w}\| = 9$	$\mathbf{v} = \langle 2, 5 \rangle$
52.	$\|\mathbf{w}\| = 8$	$\mathbf{v} = \langle 3, 3 \rangle$

Writing a Linear Combination of Unit Vectors   In Exercises 53–56, the initial and terminal points of a vector are given. Write the vector as a linear combination of the standard unit vectors i and j.

	Initial Point	Terminal Point
53.	$(-2, 1)$	$(3, -2)$
54.	$(0, -2)$	$(3, 6)$
55.	$(-6, 4)$	$(0, 1)$
56.	$(-1, -5)$	$(2, 3)$

Vector Operations   In Exercises 57–62, find the component form of v and sketch the specified vector operations geometrically, where $\mathbf{u} = 2\mathbf{i} - \mathbf{j}$ and $\mathbf{w} = \mathbf{i} + 2\mathbf{j}$.

57. $\mathbf{v} = \frac{3}{2}\mathbf{u}$ **58.** $\mathbf{v} = \frac{3}{4}\mathbf{w}$

59. $\mathbf{v} = \mathbf{u} + 2\mathbf{w}$ **60.** $\mathbf{v} = -\mathbf{u} + \mathbf{w}$

61. $\mathbf{v} = \frac{1}{2}(3\mathbf{u} + \mathbf{w})$ **62.** $\mathbf{v} = \mathbf{u} - 2\mathbf{w}$

Finding the Direction Angle of a Vector   In Exercises 63–66, find the magnitude and direction angle of the vector v.

63. $\mathbf{v} = 6\mathbf{i} - 6\mathbf{j}$ **64.** $\mathbf{v} = -5\mathbf{i} + 4\mathbf{j}$

65. $\mathbf{v} = 3(\cos 60°\mathbf{i} + \sin 60°\mathbf{j})$

66. $\mathbf{v} = 8(\cos 135°\mathbf{i} + \sin 135°\mathbf{j})$

Finding the Component Form of a Vector   In Exercises 67–74, find the component form of v given its magnitude and the angle it makes with the positive x-axis. Sketch v.

	Magnitude	Angle
67.	$\|\mathbf{v}\| = 3$	$\theta = 0°$
68.	$\|\mathbf{v}\| = 1$	$\theta = 45°$
69.	$\|\mathbf{v}\| = \frac{7}{2}$	$\theta = 150°$
70.	$\|\mathbf{v}\| = \frac{3}{4}$	$\theta = 150°$
71.	$\|\mathbf{v}\| = 2\sqrt{3}$	$\theta = 45°$
72.	$\|\mathbf{v}\| = 4\sqrt{3}$	$\theta = 90°$
73.	$\|\mathbf{v}\| = 3$	v in the direction $3\mathbf{i} + 4\mathbf{j}$
74.	$\|\mathbf{v}\| = 2$	v in the direction $\mathbf{i} + 3\mathbf{j}$

Finding the Component Form of a Vector   In Exercises 75–78, find the component form of the sum of u and v with direction angles $\theta_\mathbf{u}$ and $\theta_\mathbf{v}$.

	Magnitude	Angle
75.	$\|\mathbf{u}\| = 5$	$\theta_\mathbf{u} = 0°$
	$\|\mathbf{v}\| = 5$	$\theta_\mathbf{v} = 90°$
76.	$\|\mathbf{u}\| = 4$	$\theta_\mathbf{u} = 60°$
	$\|\mathbf{v}\| = 4$	$\theta_\mathbf{v} = 90°$
77.	$\|\mathbf{u}\| = 20$	$\theta_\mathbf{u} = 45°$
	$\|\mathbf{v}\| = 50$	$\theta_\mathbf{v} = 180°$
78.	$\|\mathbf{u}\| = 50$	$\theta_\mathbf{u} = 30°$
	$\|\mathbf{v}\| = 30$	$\theta_\mathbf{v} = 110°$

Using the Law of Cosines   In Exercises 79 and 80, use the Law of Cosines to find the angle α between the vectors. (Assume $0° \le \alpha \le 180°$.)

79. $\mathbf{v} = \mathbf{i} + \mathbf{j}$, $\mathbf{w} = 2\mathbf{i} - 2\mathbf{j}$

80. $\mathbf{v} = \mathbf{i} + 2\mathbf{j}$, $\mathbf{w} = 2\mathbf{i} - \mathbf{j}$

Resultant Force   In Exercises 81 and 82, find the angle between the forces given the magnitude of their resultant. (*Hint:* Write force 1 as a vector in the direction of the positive x-axis and force 2 as a vector at an angle θ with the positive x-axis.)

	Force 1	Force 2	Resultant Force
81.	45 pounds	60 pounds	90 pounds
82.	3000 pounds	1000 pounds	3750 pounds

83. Velocity   A gun with a muzzle velocity of 1200 feet per second is fired at an angle of 6° above the horizontal. Find the vertical and horizontal components of the velocity.

84. Velocity   Pitcher Joel Zumaya was recorded throwing a pitch at a velocity of 104 miles per hour. Assuming he threw the pitch at an angle of 3.5° below the horizontal, find the vertical and horizontal components of the velocity. (*Source: Damon Lichtenwalner, Baseball Info Solutions*)

85. Resultant Force   Forces with magnitudes of 125 newtons and 300 newtons act on a hook (see figure). The angle between the two forces is 45°. Find the direction and magnitude of the resultant of these forces.

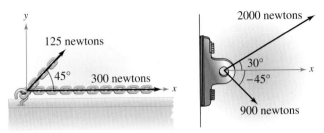

Figure for 85   Figure for 86

86. Resultant Force   Forces with magnitudes of 2000 newtons and 900 newtons act on a machine part at angles of 30° and −45°, respectively, with the x-axis (see figure). Find the direction and magnitude of the resultant of these forces.

87. Resultant Force   Three forces with magnitudes of 75 pounds, 100 pounds, and 125 pounds act on an object at angles of 30°, 45°, and 120°, respectively, with the positive x-axis. Find the direction and magnitude of the resultant of these forces.

88. Resultant Force   Three forces with magnitudes of 70 pounds, 40 pounds, and 60 pounds act on an object at angles of −30°, 45°, and 135°, respectively, with the positive x-axis. Find the direction and magnitude of the resultant of these forces.

89. Cable Tension   The cranes shown in the figure are lifting an object that weighs 20,240 pounds. Find the tension in the cable of each crane.

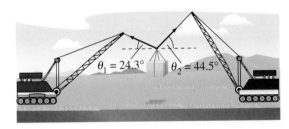

90. Cable Tension   Repeat Exercise 89 for $\theta_1 = 35.6°$ and $\theta_2 = 40.4°$.

Cable Tension   In Exercises 91 and 92, use the figure to determine the tension in each cable supporting the load.

91.
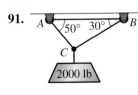

92.
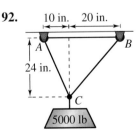

93. Tow Line Tension   A loaded barge is being towed by two tugboats, and the magnitude of the resultant is 6000 pounds directed along the axis of the barge (see figure). Find the tension in the tow lines when they each make an 18° angle with the axis of the barge.

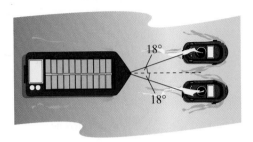

94. Rope Tension   To carry a 100-pound cylindrical weight, two people lift on the ends of short ropes that are tied to an eyelet on the top center of the cylinder. Each rope makes a 20° angle with the vertical. Draw a figure that gives a visual representation of the situation. Then find the tension in the ropes.

Inclined Ramp   In Exercises 95–98, a force of F pounds is required to pull an object weighing W pounds up a ramp inclined at θ degrees from the horizontal.

95. Find F when $W = 100$ pounds and $\theta = 12°$.

96. Find W when $F = 600$ pounds and $\theta = 14°$.

97. Find θ when $F = 5000$ pounds and $W = 15{,}000$ pounds.

98. Find F when $W = 5000$ pounds and $\theta = 26°$.

99. Work   A heavy object is pulled 30 feet across a floor, using a force of 100 pounds. The force is exerted at an angle of 50° above the horizontal (see figure). Find the work done. (Use the formula for work, $W = FD$, where F is the component of the force in the direction of motion and D is the distance.)

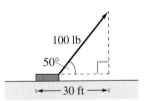

100. Rope Tension   A tetherball weighing 1 pound is pulled outward from the pole by a horizontal force **u** until the rope makes a 45° angle with the pole (see figure). Determine the resulting tension in the rope and the magnitude of **u**.

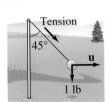

101. Navigation An airplane is flying in the direction of 148° with an airspeed of 875 kilometers per hour. Because of the wind, its groundspeed and direction are 800 kilometers per hour and 140°, respectively (see figure). Find the direction and speed of the wind.

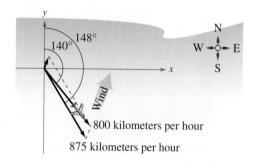

102. Navigation

A commercial jet is flying from Miami to Seattle. The jet's velocity with respect to the air is 580 miles per hour, and its bearing is 332°. The wind, at the altitude of the plane, is blowing from the southwest with a velocity of 60 miles per hour.

(a) Draw a figure that gives a visual representation of the situation.

(b) Write the velocity of the wind as a vector in component form.

(c) Write the velocity of the jet relative to the air in component form.

(d) What is the speed of the jet with respect to the ground?

(e) What is the true direction of the jet?

Exploration

True or False? In Exercises 103–106, determine whether the statement is true or false. Justify your answer.

103. If **u** and **v** have the same magnitude and direction, then **u** and **v** are equivalent.

104. If **u** is a unit vector in the direction of **v**, then $\mathbf{v} = \|\mathbf{v}\|\mathbf{u}$.

105. If $\mathbf{v} = a\mathbf{i} + b\mathbf{j} = \mathbf{0}$, then $a = -b$.

106. If $\mathbf{u} = a\mathbf{i} + b\mathbf{j}$ is a unit vector, then $a^2 + b^2 = 1$.

107. Proof Prove that $(\cos\theta)\mathbf{i} + (\sin\theta)\mathbf{j}$ is a unit vector for any value of θ.

108. Technology Write a program for your graphing utility that graphs two vectors and their difference given the vectors in component form.

Finding the Difference of Two Vectors In Exercises 109 and 110, use the program in Exercise 108 to find the difference of the vectors shown in the figure.

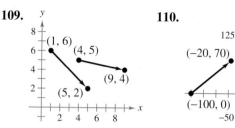

111. Graphical Reasoning Consider two forces

$$\mathbf{F}_1 = \langle 10, 0 \rangle \text{ and } \mathbf{F}_2 = 5\langle \cos\theta, \sin\theta \rangle.$$

(a) Find $\|\mathbf{F}_1 + \mathbf{F}_2\|$ as a function of θ.

(b) Use a graphing utility to graph the function in part (a) for $0 \le \theta < 2\pi$.

(c) Use the graph in part (b) to determine the range of the function. What is its maximum, and for what value of θ does it occur? What is its minimum, and for what value of θ does it occur?

(d) Explain why the magnitude of the resultant is never 0.

112. HOW DO YOU SEE IT? Use the figure to determine whether each statement is true or false. Justify your answer.

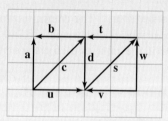

(a) $\mathbf{a} = -\mathbf{d}$ (b) $\mathbf{c} = \mathbf{s}$

(c) $\mathbf{a} + \mathbf{u} = \mathbf{c}$ (d) $\mathbf{v} + \mathbf{w} = -\mathbf{s}$

(e) $\mathbf{a} + \mathbf{w} = -2\mathbf{d}$ (f) $\mathbf{a} + \mathbf{d} = \mathbf{0}$

(g) $\mathbf{u} - \mathbf{v} = -2(\mathbf{b} + \mathbf{t})$ (h) $\mathbf{t} - \mathbf{w} = \mathbf{b} - \mathbf{a}$

113. Writing Give geometric descriptions of the operations of addition of vectors and multiplication of a vector by a scalar.

114. Writing Identify the quantity as a scalar or as a vector. Explain your reasoning.

(a) The muzzle velocity of a bullet

(b) The price of a company's stock

(c) The air temperature in a room

(d) The weight of an automobile

3.4 Vectors and Dot Products

- Find the dot product of two vectors and use the properties of the dot product.
- Find the angle between two vectors and determine whether two vectors are orthogonal.
- Write a vector as the sum of two vector components.
- Use vectors to find the work done by a force.

The Dot Product of Two Vectors

So far you have studied two vector operations—vector addition and multiplication by a scalar—each of which yields another vector. In this section, you will study a third vector operation, the **dot product.** This operation yields a scalar, rather than a vector.

You can use the dot product of two vectors to solve real-life problems involving two vector quantities. For instance, in Exercise 76 on page 298, you will use the dot product to find the force necessary to keep a sport utility vehicle from rolling down a hill.

Definition of the Dot Product

The **dot product** of $\mathbf{u} = \langle u_1, u_2 \rangle$ and $\mathbf{v} = \langle v_1, v_2 \rangle$ is

$$\mathbf{u} \cdot \mathbf{v} = u_1 v_1 + u_2 v_2.$$

Properties of the Dot Product

Let $\mathbf{u}$, $\mathbf{v}$, and $\mathbf{w}$ be vectors in the plane or in space and let c be a scalar.

1. $\mathbf{u} \cdot \mathbf{v} = \mathbf{v} \cdot \mathbf{u}$

2. $\mathbf{0} \cdot \mathbf{v} = 0$

3. $\mathbf{u} \cdot (\mathbf{v} + \mathbf{w}) = \mathbf{u} \cdot \mathbf{v} + \mathbf{u} \cdot \mathbf{w}$

4. $\mathbf{v} \cdot \mathbf{v} = \|\mathbf{v}\|^2$

5. $c(\mathbf{u} \cdot \mathbf{v}) = c\mathbf{u} \cdot \mathbf{v} = \mathbf{u} \cdot c\mathbf{v}$

For proofs of the properties of the dot product, see Proofs in Mathematics on page 312.

REMARK In Example 1, be sure you see that the dot product of two vectors is a scalar (a real number), not a vector. Moreover, notice that the dot product can be positive, zero, or negative.

EXAMPLE 1   Finding Dot Products

a. $\langle 4, 5 \rangle \cdot \langle 2, 3 \rangle = 4(2) + 5(3)$

$\qquad\qquad\qquad = 8 + 15$

$\qquad\qquad\qquad = 23$

b. $\langle 2, -1 \rangle \cdot \langle 1, 2 \rangle = 2(1) + (-1)(2)$

$\qquad\qquad\qquad\quad = 2 - 2$

$\qquad\qquad\qquad\quad = 0$

c. $\langle 0, 3 \rangle \cdot \langle 4, -2 \rangle = 0(4) + 3(-2)$

$\qquad\qquad\qquad\quad = 0 - 6$

$\qquad\qquad\qquad\quad = -6$

✓ *Checkpoint* 🔊)) *Audio-video solution in English & Spanish at LarsonPrecalculus.com.*

Find the dot product of $\mathbf{u} = \langle 3, 4 \rangle$ and $\mathbf{v} = \langle 2, -3 \rangle$.

EXAMPLE 2 **Using Properties of Dot Products**

Let $\mathbf{u} = \langle -1, 3 \rangle$, $\mathbf{v} = \langle 2, -4 \rangle$, and $\mathbf{w} = \langle 1, -2 \rangle$. Use the vectors and the properties of the dot product to find each quantity.

a. $(\mathbf{u} \cdot \mathbf{v})\mathbf{w}$ b. $\mathbf{u} \cdot 2\mathbf{v}$ c. $\|\mathbf{u}\|$

Solution   Begin by finding the dot product of $\mathbf{u}$ and $\mathbf{v}$ and the dot product of $\mathbf{u}$ and $\mathbf{u}$.

$$\mathbf{u} \cdot \mathbf{v} = \langle -1, 3 \rangle \cdot \langle 2, -4 \rangle = -1(2) + 3(-4) = -14$$

$$\mathbf{u} \cdot \mathbf{u} = \langle -1, 3 \rangle \cdot \langle -1, 3 \rangle = -1(-1) + 3(3) = 10$$

a. $(\mathbf{u} \cdot \mathbf{v})\mathbf{w} = -14\langle 1, -2 \rangle = \langle -14, 28 \rangle$

b. $\mathbf{u} \cdot 2\mathbf{v} = 2(\mathbf{u} \cdot \mathbf{v}) = 2(-14) = -28$

c. Because $\|\mathbf{u}\|^2 = \mathbf{u} \cdot \mathbf{u} = 10$, it follows that $\|\mathbf{u}\| = \sqrt{\mathbf{u} \cdot \mathbf{u}} = \sqrt{10}$.

In Example 2, notice that the product in part (a) is a vector, whereas the product in part (b) is a scalar. Can you see why?

✓ **Checkpoint** 🔊 Audio-video solution in English & Spanish at LarsonPrecalculus.com.

Let $\mathbf{u} = \langle 3, 4 \rangle$ and $\mathbf{v} = \langle -2, 6 \rangle$. Use the vectors and the properties of the dot product to find each quantity.

a. $(\mathbf{u} \cdot \mathbf{v})\mathbf{v}$ b. $\mathbf{u} \cdot (\mathbf{u} + \mathbf{v})$ c. $\|\mathbf{v}\|$

The Angle Between Two Vectors

The **angle between two nonzero vectors** is the angle θ, $0 \le \theta \le \pi$, between their respective standard position vectors, as shown in Figure 3.27. This angle can be found using the dot product.

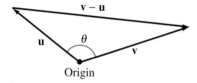

Figure 3.27

> **Angle Between Two Vectors**
>
> If θ is the angle between two nonzero vectors $\mathbf{u}$ and $\mathbf{v}$, then
>
> $$\cos \theta = \frac{\mathbf{u} \cdot \mathbf{v}}{\|\mathbf{u}\| \, \|\mathbf{v}\|}.$$

For a proof of the angle between two vectors, see Proofs in Mathematics on page 312.

EXAMPLE 3 **Finding the Angle Between Two Vectors**

Find the angle θ between $\mathbf{u} = \langle 4, 3 \rangle$ and $\mathbf{v} = \langle 3, 5 \rangle$ (see Figure 3.28).

Solution

$$\cos \theta = \frac{\mathbf{u} \cdot \mathbf{v}}{\|\mathbf{u}\| \, \|\mathbf{v}\|} = \frac{\langle 4, 3 \rangle \cdot \langle 3, 5 \rangle}{\|\langle 4, 3 \rangle\| \, \|\langle 3, 5 \rangle\|} = \frac{4(3) + 3(5)}{\sqrt{4^2 + 3^2} \, \sqrt{3^2 + 5^2}} = \frac{27}{5\sqrt{34}}$$

This implies that the angle between the two vectors is

$$\theta = \cos^{-1} \frac{27}{5\sqrt{34}} \approx 0.3869 \text{ radian.} \qquad \text{Use a calculator.}$$

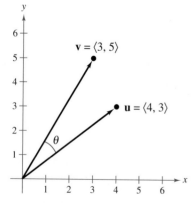

Figure 3.28

✓ **Checkpoint** 🔊 Audio-video solution in English & Spanish at LarsonPrecalculus.com.

Find the angle θ between $\mathbf{u} = \langle 2, 1 \rangle$ and $\mathbf{v} = \langle 1, 3 \rangle$.

Rewriting the expression for the angle between two vectors in the form

$$\mathbf{u} \cdot \mathbf{v} = \|\mathbf{u}\| \, \|\mathbf{v}\| \cos \theta \qquad \text{Alternative form of dot product}$$

produces an alternative way to calculate the dot product. From this form, you can see that because $\|\mathbf{u}\|$ and $\|\mathbf{v}\|$ are always positive, $\mathbf{u} \cdot \mathbf{v}$ and $\cos \theta$ will always have the same sign. The five possible orientations of two vectors are shown below.

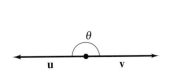

$\theta = \pi$

$\cos \theta = -1$
Opposite direction

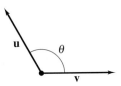

$\dfrac{\pi}{2} < \theta < \pi$

$-1 < \cos \theta < 0$
Obtuse angle

$\theta = \dfrac{\pi}{2}$

$\cos \theta = 0$
90° angle

$0 < \theta < \dfrac{\pi}{2}$

$0 < \cos \theta < 1$
Acute angle

$\theta = 0$

$\cos \theta = 1$
Same direction

Definition of Orthogonal Vectors

The vectors $\mathbf{u}$ and $\mathbf{v}$ are **orthogonal** if and only if $\mathbf{u} \cdot \mathbf{v} = 0$.

The terms *orthogonal* and *perpendicular* mean essentially the same thing—meeting at right angles. Note that the zero vector is orthogonal to every vector $\mathbf{u}$, because $\mathbf{0} \cdot \mathbf{u} = 0$.

▷ **TECHNOLOGY** The graphing utility program, Finding the Angle Between Two Vectors, found on the website for this text at *LarsonPrecalculus.com*, graphs two vectors $\mathbf{u} = \langle a, b \rangle$ and $\mathbf{v} = \langle c, d \rangle$ in standard position and finds the measure of the angle between them. Use the program to verify the solutions for Examples 3 and 4.

EXAMPLE 4   **Determining Orthogonal Vectors**

Are the vectors $\mathbf{u} = \langle 2, -3 \rangle$ and $\mathbf{v} = \langle 6, 4 \rangle$ orthogonal?

Solution   Find the dot product of the two vectors.

$$\mathbf{u} \cdot \mathbf{v} = \langle 2, -3 \rangle \cdot \langle 6, 4 \rangle$$
$$= 2(6) + (-3)(4)$$
$$= 0$$

Because the dot product is 0, the two vectors are orthogonal (see below).

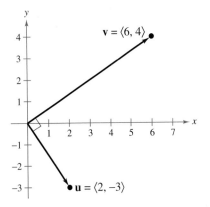

✓ Checkpoint *Audio-video solution in English & Spanish at LarsonPrecalculus.com.*

Are the vectors $\mathbf{u} = \langle 6, 10 \rangle$ and $\mathbf{v} = \left\langle -\frac{1}{3}, \frac{1}{5} \right\rangle$ orthogonal?

Figure 3.29

Finding Vector Components

You have already seen applications in which two vectors are added to produce a resultant vector. Many applications in physics and engineering pose the reverse problem—decomposing a given vector into the sum of two **vector components.**

Consider a boat on an inclined ramp, as shown in Figure 3.29. The force **F** due to gravity pulls the boat *down* the ramp and *against* the ramp. These two orthogonal forces, $\mathbf{w}_1$ and $\mathbf{w}_2$, are vector components of **F**. That is,

$$\mathbf{F} = \mathbf{w}_1 + \mathbf{w}_2. \qquad \text{Vector components of } \mathbf{F}$$

The negative of component $\mathbf{w}_1$ represents the force needed to keep the boat from rolling down the ramp, whereas $\mathbf{w}_2$ represents the force that the tires must withstand against the ramp. A procedure for finding $\mathbf{w}_1$ and $\mathbf{w}_2$ is shown below.

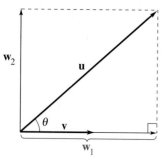

θ is acute.

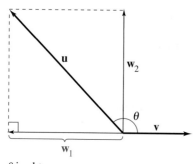

θ is obtuse.

Figure 3.30

Definition of Vector Components

Let **u** and **v** be nonzero vectors such that

$$\mathbf{u} = \mathbf{w}_1 + \mathbf{w}_2$$

where $\mathbf{w}_1$ and $\mathbf{w}_2$ are orthogonal and $\mathbf{w}_1$ is parallel to (or a scalar multiple of) **v**, as shown in Figure 3.30. The vectors $\mathbf{w}_1$ and $\mathbf{w}_2$ are called **vector components** of **u**. The vector $\mathbf{w}_1$ is the **projection** of **u** onto **v** and is denoted by

$$\mathbf{w}_1 = \text{proj}_{\mathbf{v}}\mathbf{u}.$$

The vector $\mathbf{w}_2$ is given by

$$\mathbf{w}_2 = \mathbf{u} - \mathbf{w}_1.$$

From the definition of vector components, you can see that it is easy to find the component $\mathbf{w}_2$ once you have found the projection of **u** onto **v**. To find the projection, you can use the dot product, as follows.

$$\mathbf{u} = \mathbf{w}_1 + \mathbf{w}_2$$

$$\mathbf{u} = c\mathbf{v} + \mathbf{w}_2 \qquad \mathbf{w}_1 \text{ is a scalar multiple of } \mathbf{v}.$$

$$\mathbf{u} \cdot \mathbf{v} = (c\mathbf{v} + \mathbf{w}_2) \cdot \mathbf{v} \qquad \text{Take dot product of each side with } \mathbf{v}.$$

$$\mathbf{u} \cdot \mathbf{v} = c\mathbf{v} \cdot \mathbf{v} + \mathbf{w}_2 \cdot \mathbf{v}$$

$$\mathbf{u} \cdot \mathbf{v} = c\|\mathbf{v}\|^2 + 0 \qquad \mathbf{w}_2 \text{ and } \mathbf{v} \text{ are orthogonal.}$$

So,

$$c = \frac{\mathbf{u} \cdot \mathbf{v}}{\|\mathbf{v}\|^2}$$

and

$$\mathbf{w}_1 = \text{proj}_{\mathbf{v}}\mathbf{u} = c\mathbf{v} = \left(\frac{\mathbf{u} \cdot \mathbf{v}}{\|\mathbf{v}\|^2}\right)\mathbf{v}.$$

Projection of u onto v

Let **u** and **v** be nonzero vectors. The projection of **u** onto **v** is

$$\text{proj}_{\mathbf{v}}\mathbf{u} = \left(\frac{\mathbf{u} \cdot \mathbf{v}}{\|\mathbf{v}\|^2}\right)\mathbf{v}.$$

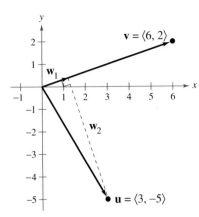

Figure 3.31

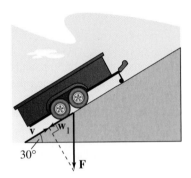

Figure 3.32

EXAMPLE 5 **Decomposing a Vector into Components**

Find the projection of $\mathbf{u} = \langle 3, -5 \rangle$ onto $\mathbf{v} = \langle 6, 2 \rangle$. Then write $\mathbf{u}$ as the sum of two orthogonal vectors, one of which is $\text{proj}_{\mathbf{v}}\mathbf{u}$.

Solution The projection of $\mathbf{u}$ onto $\mathbf{v}$ is

$$\mathbf{w}_1 = \text{proj}_{\mathbf{v}}\mathbf{u} = \left(\frac{\mathbf{u} \cdot \mathbf{v}}{\|\mathbf{v}\|^2} \right)\mathbf{v} = \left(\frac{8}{40} \right)\langle 6, 2 \rangle = \left\langle \frac{6}{5}, \frac{2}{5} \right\rangle$$

as shown in Figure 3.31. The other component, $\mathbf{w}_2$, is

$$\mathbf{w}_2 = \mathbf{u} - \mathbf{w}_1 = \langle 3, -5 \rangle - \left\langle \frac{6}{5}, \frac{2}{5} \right\rangle = \left\langle \frac{9}{5}, -\frac{27}{5} \right\rangle.$$

So,

$$\mathbf{u} = \mathbf{w}_1 + \mathbf{w}_2 = \left\langle \frac{6}{5}, \frac{2}{5} \right\rangle + \left\langle \frac{9}{5}, -\frac{27}{5} \right\rangle = \langle 3, -5 \rangle.$$

✓ **Checkpoint** 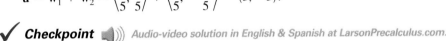 Audio-video solution in English & Spanish at LarsonPrecalculus.com.

Find the projection of $\mathbf{u} = \langle 3, 4 \rangle$ onto $\mathbf{v} = \langle 8, 2 \rangle$. Then write $\mathbf{u}$ as the sum of two orthogonal vectors, one of which is $\text{proj}_{\mathbf{v}}\mathbf{u}$.

EXAMPLE 6 **Finding a Force**

A 200-pound cart sits on a ramp inclined at $30°$, as shown in Figure 3.32. What force is required to keep the cart from rolling down the ramp?

Solution Because the force due to gravity is vertical and downward, you can represent the gravitational force by the vector

$$\mathbf{F} = -200\mathbf{j}. \qquad \text{Force due to gravity}$$

To find the force required to keep the cart from rolling down the ramp, project $\mathbf{F}$ onto a unit vector $\mathbf{v}$ in the direction of the ramp, as follows.

$$\mathbf{v} = (\cos 30°)\mathbf{i} + (\sin 30°)\mathbf{j}$$
$$= \frac{\sqrt{3}}{2}\mathbf{i} + \frac{1}{2}\mathbf{j} \qquad \text{Unit vector along ramp}$$

So, the projection of $\mathbf{F}$ onto $\mathbf{v}$ is

$$\mathbf{w}_1 = \text{proj}_{\mathbf{v}}\mathbf{F}$$
$$= \left(\frac{\mathbf{F} \cdot \mathbf{v}}{\|\mathbf{v}\|^2} \right)\mathbf{v}$$
$$= (\mathbf{F} \cdot \mathbf{v})\mathbf{v}$$
$$= (-200)\left(\frac{1}{2} \right)\mathbf{v}$$
$$= -100\left(\frac{\sqrt{3}}{2}\mathbf{i} + \frac{1}{2}\mathbf{j} \right).$$

The magnitude of this force is 100. So, a force of 100 pounds is required to keep the cart from rolling down the ramp.

✓ **Checkpoint**  Audio-video solution in English & Spanish at LarsonPrecalculus.com.

Rework Example 6 for a 150-pound cart sitting on a ramp inclined at $15°$. ▪

Force acts along the line of motion.
Figure 3.33

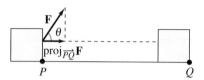

Force acts at angle θ with the line of motion.
Figure 3.34

Figure 3.35

Work is done only when an object is moved. It does not matter how much force is applied—if an object does not move, then no work has been done.

Work

The work W done by a *constant* force $\mathbf{F}$ acting along the line of motion of an object is given by

$$W = (\text{magnitude of force})(\text{distance})$$

$$= \|\mathbf{F}\| \, \|\overrightarrow{PQ}\|$$

as shown in Figure 3.33 When the constant force $\mathbf{F}$ is not directed along the line of motion, as shown in Figure 3.34, the work W done by the force is given by

$$W = \|\text{proj}_{\overrightarrow{PQ}} \mathbf{F}\| \, \|\overrightarrow{PQ}\| \qquad \text{Projection form for work}$$

$$= (\cos \theta)\|\mathbf{F}\| \, \|\overrightarrow{PQ}\| \qquad \|\text{proj}_{\overrightarrow{PQ}} \mathbf{F}\| = (\cos \theta)\|\mathbf{F}\|$$

$$= \mathbf{F} \cdot \overrightarrow{PQ}. \qquad \text{Dot product form for work}$$

This notion of work is summarized in the following definition.

Definition of Work

The **work** W done by a constant force $\mathbf{F}$ as its point of application moves along the vector $\overrightarrow{PQ}$ is given by either of the following.

1. $W = \|\text{proj}_{\overrightarrow{PQ}}\mathbf{F}\| \, \|\overrightarrow{PQ}\|$ Projection form

2. $W = \mathbf{F} \cdot \overrightarrow{PQ}$ Dot product form

---

EXAMPLE 7 **Finding Work**

To close a sliding barn door, a person pulls on a rope with a constant force of 50 pounds at a constant angle of 60°, as shown in Figure 3.35. Find the work done in moving the barn door 12 feet to its closed position.

Solution Using a projection, you can calculate the work as follows.

$$W = \|\text{proj}_{\overrightarrow{PQ}} \mathbf{F}\| \, \|\overrightarrow{PQ}\| = (\cos 60°)\|\mathbf{F}\| \, \|\overrightarrow{PQ}\| = \frac{1}{2}(50)(12) = 300 \text{ foot-pounds}$$

So, the work done is 300 foot-pounds. You can verify this result by finding the vectors $\mathbf{F}$ and $\overrightarrow{PQ}$ and calculating their dot product.

✓ **Checkpoint**  Audio-video solution in English & Spanish at LarsonPrecalculus.com.

A wagon is pulled by exerting a force of 35 pounds on a handle that makes a 30° angle with the horizontal. Find the work done in pulling the wagon 40 feet. ∎

Summarize (Section 3.4)

1. State the definition of the dot product *(page 291)*. For examples of finding dot products and using the properties of the dot product, see Examples 1 and 2.

2. Describe how to find the angle between two vectors *(page 292)*. For examples involving the angle between two vectors, see Examples 3 and 4.

3. Describe how to decompose a vector into components *(page 294)*. For examples involving vector components, see Examples 5 and 6.

4. State the definition of work *(page 296)*. For an example of finding the work done by a constant force, see Example 7.

3.4 Exercises

See **CalcChat.com** for tutorial help and worked-out solutions to odd-numbered exercises.

Vocabulary: Fill in the blanks.

1. The _____ _____ of two vectors yields a scalar, rather than a vector.
2. The dot product of $\mathbf{u} = \langle u_1, u_2 \rangle$ and $\mathbf{v} = \langle v_1, v_2 \rangle$ is $\mathbf{u} \cdot \mathbf{v} = $ _____ .
3. If θ is the angle between two nonzero vectors $\mathbf{u}$ and $\mathbf{v}$, then $\cos \theta = $ _____ .
4. The vectors $\mathbf{u}$ and $\mathbf{v}$ are _____ when $\mathbf{u} \cdot \mathbf{v} = 0$.
5. The projection of $\mathbf{u}$ onto $\mathbf{v}$ is given by $\text{proj}_{\mathbf{v}}\mathbf{u} = $ _____ .
6. The work W done by a constant force $\mathbf{F}$ as its point of application moves along the vector $\overrightarrow{PQ}$ is given by $W = $ _____ or $W = $ _____ .

Skills and Applications

Finding a Dot Product In Exercises 7–14, find $\mathbf{u} \cdot \mathbf{v}$.

7. $\mathbf{u} = \langle 7, 1 \rangle$
   $\mathbf{v} = \langle -3, 2 \rangle$
8. $\mathbf{u} = \langle 6, 10 \rangle$
   $\mathbf{v} = \langle -2, 3 \rangle$
9. $\mathbf{u} = \langle -4, 1 \rangle$
   $\mathbf{v} = \langle 2, -3 \rangle$
10. $\mathbf{u} = \langle -2, 5 \rangle$
    $\mathbf{v} = \langle -1, -8 \rangle$
11. $\mathbf{u} = 4\mathbf{i} - 2\mathbf{j}$
    $\mathbf{v} = \mathbf{i} - \mathbf{j}$
12. $\mathbf{u} = 3\mathbf{i} + 4\mathbf{j}$
    $\mathbf{v} = 7\mathbf{i} - 2\mathbf{j}$
13. $\mathbf{u} = 3\mathbf{i} + 2\mathbf{j}$
    $\mathbf{v} = -2\mathbf{i} - 3\mathbf{j}$
14. $\mathbf{u} = \mathbf{i} - 2\mathbf{j}$
    $\mathbf{v} = -2\mathbf{i} + \mathbf{j}$

Using Properties of Dot Products In Exercises 15–24, use the vectors $\mathbf{u} = \langle 3, 3 \rangle$, $\mathbf{v} = \langle -4, 2 \rangle$, and $\mathbf{w} = \langle 3, -1 \rangle$ to find the indicated quantity. State whether the result is a vector or a scalar.

15. $\mathbf{u} \cdot \mathbf{u}$
16. $3\mathbf{u} \cdot \mathbf{v}$
17. $(\mathbf{u} \cdot \mathbf{v})\mathbf{v}$
18. $(\mathbf{v} \cdot \mathbf{u})\mathbf{w}$
19. $(3\mathbf{w} \cdot \mathbf{v})\mathbf{u}$
20. $(\mathbf{u} \cdot 2\mathbf{v})\mathbf{w}$
21. $\|\mathbf{w}\| - 1$
22. $2 - \|\mathbf{u}\|$
23. $(\mathbf{u} \cdot \mathbf{v}) - (\mathbf{u} \cdot \mathbf{w})$
24. $(\mathbf{v} \cdot \mathbf{u}) - (\mathbf{w} \cdot \mathbf{v})$

Finding the Magnitude of a Vector In Exercises 25–30, use the dot product to find the magnitude of $\mathbf{u}$.

25. $\mathbf{u} = \langle -8, 15 \rangle$
26. $\mathbf{u} = \langle 4, -6 \rangle$
27. $\mathbf{u} = 20\mathbf{i} + 25\mathbf{j}$
28. $\mathbf{u} = 12\mathbf{i} - 16\mathbf{j}$
29. $\mathbf{u} = 6\mathbf{j}$
30. $\mathbf{u} = -21\mathbf{i}$

Finding the Angle Between Two Vectors In Exercises 31–40, find the angle θ between the vectors.

31. $\mathbf{u} = \langle 1, 0 \rangle$
    $\mathbf{v} = \langle 0, -2 \rangle$
32. $\mathbf{u} = \langle 3, 2 \rangle$
    $\mathbf{v} = \langle 4, 0 \rangle$
33. $\mathbf{u} = 3\mathbf{i} + 4\mathbf{j}$
    $\mathbf{v} = -2\mathbf{j}$
34. $\mathbf{u} = 2\mathbf{i} - 3\mathbf{j}$
    $\mathbf{v} = \mathbf{i} - 2\mathbf{j}$
35. $\mathbf{u} = 2\mathbf{i} - \mathbf{j}$
    $\mathbf{v} = 6\mathbf{i} + 4\mathbf{j}$
36. $\mathbf{u} = -6\mathbf{i} - 3\mathbf{j}$
    $\mathbf{v} = -8\mathbf{i} + 4\mathbf{j}$

37. $\mathbf{u} = 5\mathbf{i} + 5\mathbf{j}$
    $\mathbf{v} = -6\mathbf{i} + 6\mathbf{j}$
38. $\mathbf{u} = 2\mathbf{i} - 3\mathbf{j}$
    $\mathbf{v} = 4\mathbf{i} + 3\mathbf{j}$
39. $\mathbf{u} = \cos\left(\dfrac{\pi}{3}\right)\mathbf{i} + \sin\left(\dfrac{\pi}{3}\right)\mathbf{j}$
    $\mathbf{v} = \cos\left(\dfrac{3\pi}{4}\right)\mathbf{i} + \sin\left(\dfrac{3\pi}{4}\right)\mathbf{j}$
40. $\mathbf{u} = \cos\left(\dfrac{\pi}{4}\right)\mathbf{i} + \sin\left(\dfrac{\pi}{4}\right)\mathbf{j}$
    $\mathbf{v} = \cos\left(\dfrac{\pi}{2}\right)\mathbf{i} + \sin\left(\dfrac{\pi}{2}\right)\mathbf{j}$

Finding the Angle Between Two Vectors In Exercises 41–44, graph the vectors and find the degree measure of the angle θ between the vectors.

41. $\mathbf{u} = 3\mathbf{i} + 4\mathbf{j}$
    $\mathbf{v} = -7\mathbf{i} + 5\mathbf{j}$
42. $\mathbf{u} = 6\mathbf{i} + 3\mathbf{j}$
    $\mathbf{v} = -4\mathbf{i} + 4\mathbf{j}$
43. $\mathbf{u} = 5\mathbf{i} + 5\mathbf{j}$
    $\mathbf{v} = -8\mathbf{i} + 8\mathbf{j}$
44. $\mathbf{u} = 2\mathbf{i} - 3\mathbf{j}$
    $\mathbf{v} = 8\mathbf{i} + 3\mathbf{j}$

Finding the Angles in a Triangle In Exercises 45–48, use vectors to find the interior angles of the triangle with the given vertices.

45. $(1, 2), (3, 4), (2, 5)$
46. $(-3, -4), (1, 7), (8, 2)$
47. $(-3, 0), (2, 2), (0, 6)$
48. $(-3, 5), (-1, 9), (7, 9)$

Using the Angle Between Two Vectors In Exercises 49–52, find $\mathbf{u} \cdot \mathbf{v}$, where θ is the angle between $\mathbf{u}$ and $\mathbf{v}$.

49. $\|\mathbf{u}\| = 4, \|\mathbf{v}\| = 10, \theta = \dfrac{2\pi}{3}$
50. $\|\mathbf{u}\| = 100, \|\mathbf{v}\| = 250, \theta = \dfrac{\pi}{6}$
51. $\|\mathbf{u}\| = 9, \|\mathbf{v}\| = 36, \theta = \dfrac{3\pi}{4}$
52. $\|\mathbf{u}\| = 4, \|\mathbf{v}\| = 12, \theta = \dfrac{\pi}{3}$

Determining Orthogonal Vectors In Exercises 53–58, determine whether **u** and **v** are orthogonal.

53. $\mathbf{u} = \langle -12, 30 \rangle$
    $\mathbf{v} = \langle \frac{1}{2}, -\frac{5}{4} \rangle$

54. $\mathbf{u} = \langle 3, 15 \rangle$
    $\mathbf{v} = \langle -1, 5 \rangle$

55. $\mathbf{u} = \frac{1}{4}(3\mathbf{i} - \mathbf{j})$
    $\mathbf{v} = 5\mathbf{i} + 6\mathbf{j}$

56. $\mathbf{u} = \mathbf{i}$
    $\mathbf{v} = -2\mathbf{i} + 2\mathbf{j}$

57. $\mathbf{u} = 2\mathbf{i} - 2\mathbf{j}$
    $\mathbf{v} = -\mathbf{i} - \mathbf{j}$

58. $\mathbf{u} = \langle \cos\theta, \sin\theta \rangle$
    $\mathbf{v} = \langle \sin\theta, -\cos\theta \rangle$

Decomposing a Vector into Components In Exercises 59–62, find the projection of **u** onto **v**. Then write **u** as the sum of two orthogonal vectors, one of which is proj$_v$**u**.

59. $\mathbf{u} = \langle 2, 2 \rangle$
    $\mathbf{v} = \langle 6, 1 \rangle$

60. $\mathbf{u} = \langle 4, 2 \rangle$
    $\mathbf{v} = \langle 1, -2 \rangle$

61. $\mathbf{u} = \langle 0, 3 \rangle$
    $\mathbf{v} = \langle 2, 15 \rangle$

62. $\mathbf{u} = \langle -3, -2 \rangle$
    $\mathbf{v} = \langle -4, -1 \rangle$

Finding the Projection of u onto v In Exercises 63–66, use the graph to find the projection of **u** onto **v**. (The coordinates of the terminal points of the vectors in standard position are given.) Use the formula for the projection of **u** onto **v** to verify your result.

63.

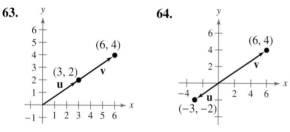

64.

65.

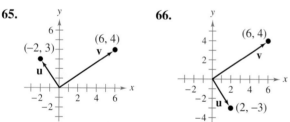

66.

Finding Orthogonal Vectors In Exercises 67–70, find two vectors in opposite directions that are orthogonal to the vector **u**. (There are many correct answers.)

67. $\mathbf{u} = \langle 3, 5 \rangle$

68. $\mathbf{u} = \langle -8, 3 \rangle$

69. $\mathbf{u} = \frac{1}{2}\mathbf{i} - \frac{2}{3}\mathbf{j}$

70. $\mathbf{u} = -\frac{5}{2}\mathbf{i} - 3\mathbf{j}$

Work In Exercises 71 and 72, find the work done in moving a particle from P to Q when the magnitude and direction of the force are given by **v**.

71. $P(0, 0)$, $Q(4, 7)$, $\mathbf{v} = \langle 1, 4 \rangle$

72. $P(1, 3)$, $Q(-3, 5)$, $\mathbf{v} = -2\mathbf{i} + 3\mathbf{j}$

73. **Business** The vector $\mathbf{u} = \langle 1225, 2445 \rangle$ gives the numbers of hours worked by employees of a temp agency at two pay levels. The vector $\mathbf{v} = \langle 12.20, 8.50 \rangle$ gives the hourly wage (in dollars) paid at each level, respectively.

    (a) Find the dot product $\mathbf{u} \cdot \mathbf{v}$ and explain its meaning in the context of the problem.

    (b) Identify the vector operation used to increase wages by 2%.

74. **Revenue** The vector $\mathbf{u} = \langle 3140, 2750 \rangle$ gives the numbers of hamburgers and hot dogs, respectively, sold at a fast-food stand in one month. The vector $\mathbf{v} = \langle 2.25, 1.75 \rangle$ gives the prices (in dollars) of the food items.

    (a) Find the dot product $\mathbf{u} \cdot \mathbf{v}$ and interpret the result in the context of the problem.

    (b) Identify the vector operation used to increase the prices by 2.5%.

75. **Braking Load** A truck with a gross weight of 30,000 pounds is parked on a slope of $d°$ (see figure). Assume that the only force to overcome is the force of gravity.

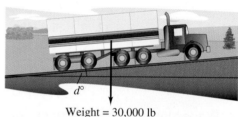

Weight = 30,000 lb

    (a) Find the force required to keep the truck from rolling down the hill in terms of the slope d.

    (b) Use a graphing utility to complete the table.

d	0°	1°	2°	3°	4°	5°
Force						

d	6°	7°	8°	9°	10°
Force					

    (c) Find the force perpendicular to the hill when $d = 5°$.

76. **Braking Load**
A sport utility vehicle with a gross weight of 5400 pounds is parked on a slope of 10°. Assume that the only force to overcome is the force of gravity. Find the force required to keep the vehicle from rolling down the hill. Find the force perpendicular to the hill.

77. Work Determine the work done by a person lifting a 245-newton bag of sugar 3 meters.

78. Work Determine the work done by a crane lifting a 2400-pound car 5 feet.

79. Work A force of 45 pounds, exerted at an angle of 30° with the horizontal, is required to slide a table across a floor. Determine the work done in sliding the table 20 feet.

80. Work A force of 50 pounds, exerted at an angle of 25° with the horizontal, is required to slide a desk across a floor. Determine the work done in sliding the desk 15 feet.

81. Work A tractor pulls a log 800 meters, and the tension in the cable connecting the tractor and log is approximately 15,691 newtons. The direction of the force is 35° above the horizontal. Approximate the work done in pulling the log.

82. Work One of the events in a strength competition is to pull a cement block 100 feet. One competitor pulls the block by exerting a force of 250 pounds on a rope attached to the block at an angle of 30° with the horizontal (see figure). Find the work done in pulling the block.

30°
──100 ft──
Not drawn to scale

83. Work A toy wagon is pulled by exerting a force of 25 pounds on a handle that makes a 20° angle with the horizontal (see figure). Find the work done in pulling the wagon 50 feet.

20°

84. Work A ski patroller pulls a rescue toboggan across a flat snow surface by exerting a force of 35 pounds on a handle that makes an angle of 22° with the horizontal. Find the work done in pulling the toboggan 200 feet.

22°

85. Programming Given vectors **u** and **v** in component form, write a program for your graphing utility in which the output is (a) $\|\mathbf{u}\|$, (b) $\|\mathbf{v}\|$, and (c) the angle between **u** and **v**.

86. Programming Use the program you wrote in Exercise 85 to find the angle between the given vectors.
(a) $\mathbf{u} = \langle 8, -4 \rangle$ and $\mathbf{v} = \langle 2, 5 \rangle$
(b) $\mathbf{u} = \langle 2, -6 \rangle$ and $\mathbf{v} = \langle 4, 1 \rangle$

87. Programming Given vectors **u** and **v** in component form, write a program for your graphing utility in which the output is the component form of the projection of **u** onto **v**.

88. Programming Use the program you wrote in Exercise 87 to find the projection of **u** onto **v** for the given vectors.
(a) $\mathbf{u} = \langle 5, 6 \rangle$ and $\mathbf{v} = \langle -1, 3 \rangle$
(b) $\mathbf{u} = \langle 3, -2 \rangle$ and $\mathbf{v} = \langle -2, 1 \rangle$

Exploration

True or False? **In Exercises 89 and 90, determine whether the statement is true or false. Justify your answer.**

89. The work W done by a constant force **F** acting along the line of motion of an object is represented by a vector.

90. A sliding door moves along the line of vector $\overrightarrow{PQ}$. If a force is applied to the door along a vector that is orthogonal to $\overrightarrow{PQ}$, then no work is done.

91. Proof Use vectors to prove that the diagonals of a rhombus are perpendicular.

92. HOW DO YOU SEE IT? What is known about θ, the angle between two nonzero vectors **u** and **v**, under each condition (see figure)?

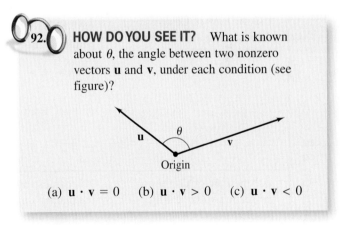
θ
u
v
Origin

(a) $\mathbf{u} \cdot \mathbf{v} = 0$ (b) $\mathbf{u} \cdot \mathbf{v} > 0$ (c) $\mathbf{u} \cdot \mathbf{v} < 0$

93. Think About It What can be said about the vectors **u** and **v** under each condition?
(a) The projection of **u** onto **v** equals **u**.
(b) The projection of **u** onto **v** equals **0**.

94. Proof Prove the following.
$$\|\mathbf{u} - \mathbf{v}\|^2 = \|\mathbf{u}\|^2 + \|\mathbf{v}\|^2 - 2\mathbf{u} \cdot \mathbf{v}$$

Chapter Summary

What Did You Learn?	Explanation/Examples	Review Exercises
Section 3.1 Use the Law of Sines to solve oblique triangles (AAS or ASA) *(p. 262)*.	**Law of Sines** If ABC is a triangle with sides a, b, and c, then $$\frac{a}{\sin A} = \frac{b}{\sin B} = \frac{c}{\sin C}.$$ A is acute. A is obtuse.	1–12
Use the Law of Sines to solve oblique triangles (SSA) *(p. 264)*.	If two sides and one opposite angle are given, then three possible situations can occur: (1) no such triangle exists (see Example 4), (2) one such triangle exists (see Example 3), or (3) two distinct triangles may satisfy the conditions (see Example 5).	1–12, 31–34
Find the areas of oblique triangles *(p. 266)*.	The area of any triangle is one-half the product of the lengths of two sides times the sine of their included angle. That is, $$\text{Area} = \frac{1}{2}bc \sin A = \frac{1}{2}ab \sin C = \frac{1}{2}ac \sin B.$$	13–16
Use the Law of Sines to model and solve real-life problems *(p. 267)*.	You can use the Law of Sines to approximate the total distance of a boat race course. (See Example 7.)	17–20
Section 3.2 Use the Law of Cosines to solve oblique triangles (SSS or SAS) *(p. 271)*.	**Law of Cosines** **Standard Form** $\qquad\qquad$ **Alternative Form** $a^2 = b^2 + c^2 - 2bc \cos A \qquad \cos A = \dfrac{b^2 + c^2 - a^2}{2bc}$ $b^2 = a^2 + c^2 - 2ac \cos B \qquad \cos B = \dfrac{a^2 + c^2 - b^2}{2ac}$ $c^2 = a^2 + b^2 - 2ab \cos C \qquad \cos C = \dfrac{a^2 + b^2 - c^2}{2ab}$	21–34
Use the Law of Cosines to model and solve real-life problems *(p. 273)*.	You can use the Law of Cosines to find the distance between the pitcher's mound and first base on a women's softball field. (See Example 3.)	35–38
Use Heron's Area Formula to find the area of a triangle *(p. 274)*.	**Heron's Area Formula:** Given any triangle with sides of lengths a, b, and c, the area of the triangle is $$\text{Area} = \sqrt{s(s-a)(s-b)(s-c)}$$ where $$s = \frac{a+b+c}{2}.$$	39–42

What Did You Learn?	**Explanation/Examples**	**Review Exercises**
Represent vectors as directed line segments *(p. 278)*.	Terminal point • Q $\overrightarrow{PQ}$ P • Initial point	43, 44
Write the component forms of vectors *(p. 279)*.	The component form of the vector with initial point $P(p_1, p_2)$ and terminal point $Q(q_1, q_2)$ is given by $\overrightarrow{PQ} = \langle q_1 - p_1, q_2 - p_2 \rangle = \langle v_1, v_2 \rangle = \mathbf{v}.$	45–50
Perform basic vector operations and represent them graphically *(p. 280)*.	Let $\mathbf{u} = \langle u_1, u_2 \rangle$ and $\mathbf{v} = \langle v_1, v_2 \rangle$ be vectors and let k be a scalar (a real number). $\mathbf{u} + \mathbf{v} = \langle u_1 + v_1, u_2 + v_2 \rangle$ $k\mathbf{u} = \langle ku_1, ku_2 \rangle$ $-\mathbf{v} = \langle -v_1, -v_2 \rangle$ $\mathbf{u} - \mathbf{v} = \langle u_1 - v_1, u_2 - v_2 \rangle$	51–64
Write vectors as linear combinations of unit vectors *(p. 282)*.	The vector sum $\mathbf{v} = \langle v_1, v_2 \rangle = v_1 \langle 1, 0 \rangle + v_2 \langle 0, 1 \rangle = v_1 \mathbf{i} + v_2 \mathbf{j}$ is a linear combination of the vectors $\mathbf{i}$ and $\mathbf{j}$.	65–68
Find the direction angles of vectors *(p. 284)*.	If $\mathbf{u} = 2\mathbf{i} + 2\mathbf{j}$, then the direction angle is determined from $\tan \theta = \dfrac{2}{2} = 1.$ So, $\theta = 45°$.	69–74
Use vectors to model and solve real-life problems *(p. 285)*.	You can use vectors to find the resultant speed and direction of an airplane. (See Example 11.)	75–78
Find the dot product of two vectors and use the properties of the dot product *(p. 291)*.	The dot product of $\mathbf{u} = \langle u_1, u_2 \rangle$ and $\mathbf{v} = \langle v_1, v_2 \rangle$ is $\mathbf{u} \cdot \mathbf{v} = u_1 v_1 + u_2 v_2.$	79–90
Find the angle between two vectors and determine whether two vectors are orthogonal *(p. 292)*.	If θ is the angle between two nonzero vectors $\mathbf{u}$ and $\mathbf{v}$, then $\cos \theta = \dfrac{\mathbf{u} \cdot \mathbf{v}}{\|\mathbf{u}\| \, \|\mathbf{v}\|}.$ Vectors $\mathbf{u}$ and $\mathbf{v}$ are orthogonal if and only if $\mathbf{u} \cdot \mathbf{v} = 0$.	91–98
Write a vector as the sum of two vector components *(p. 294)*.	Many applications in physics and engineering require the decomposition of a given vector into the sum of two vector components. (See Example 6.)	99–102
Use vectors to find the work done by a force *(p. 296)*.	The work W done by a constant force $\mathbf{F}$ as its point of application moves along the vector $\overrightarrow{PQ}$ is given by either of the following. **1.** $W = \|\mathrm{proj}_{\overrightarrow{PQ}} \mathbf{F}\| \, \|\overrightarrow{PQ}\|$ **2.** $W = \mathbf{F} \cdot \overrightarrow{PQ}$	103–106

Section 3.3 *(rows for Represent vectors through Use vectors to model real-life problems)*

Section 3.4 *(rows for Find the dot product through Use vectors to find the work done by a force)*

Review Exercises   See CalcChat.com for tutorial help and worked-out solutions to odd-numbered exercises.

3.1 **Using the Law of Sines**   In Exercises 1–12, use the Law of Sines to solve (if possible) the triangle. If two solutions exist, find both. Round your answers to two decimal places.

1.
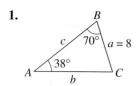

2.
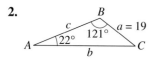

3. $B = 72°$, $C = 82°$, $b = 54$

4. $B = 10°$, $C = 20°$, $c = 33$

5. $A = 16°$, $B = 98°$, $c = 8.4$

6. $A = 95°$, $B = 45°$, $c = 104.8$

7. $A = 24°$, $C = 48°$, $b = 27.5$

8. $B = 64°$, $C = 36°$, $a = 367$

9. $B = 150°$, $b = 30$, $c = 10$

10. $B = 150°$, $a = 10$, $b = 3$

11. $A = 75°$, $a = 51.2$, $b = 33.7$

12. $B = 25°$, $a = 6.2$, $b = 4$

Finding the Area of a Triangle   In Exercises 13–16, find the area of the triangle having the indicated angle and sides.

13. $A = 33°$, $b = 7$, $c = 10$

14. $B = 80°$, $a = 4$, $c = 8$

15. $C = 119°$, $a = 18$, $b = 6$

16. $A = 11°$, $b = 22$, $c = 21$

17. **Height**   From a certain distance, the angle of elevation to the top of a building is 17°. At a point 50 meters closer to the building, the angle of elevation is 31°. Approximate the height of the building.

18. **Geometry**   Find the length of the side w of the parallelogram.

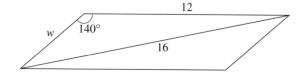

19. **Height**   A tree stands on a hillside of slope 28° from the horizontal. From a point 75 feet down the hill, the angle of elevation to the top of the tree is 45° (see figure). Find the height of the tree.

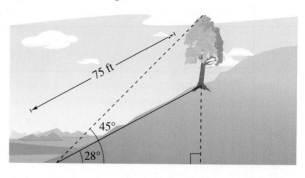

20. **River Width**   A surveyor finds that a tree on the opposite bank of a river flowing due east has a bearing of N 22° 30′ E from a certain point and a bearing of N 15° W from a point 400 feet downstream. Find the width of the river.

3.2 **Using the Law of Cosines**   In Exercises 21–30, use the Law of Cosines to solve the triangle. Round your answers to two decimal places.

21.
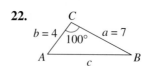

22.

23. $a = 6$, $b = 9$, $c = 14$

24. $a = 75$, $b = 50$, $c = 110$

25. $a = 2.5$, $b = 5.0$, $c = 4.5$

26. $a = 16.4$, $b = 8.8$, $c = 12.2$

27. $B = 108°$, $a = 11$, $c = 11$

28. $B = 150°$, $a = 10$, $c = 20$

29. $C = 43°$, $a = 22.5$, $b = 31.4$

30. $A = 62°$, $b = 11.34$, $c = 19.52$

Solving a Triangle In Exercises 31–34, determine whether the Law of Sines or the Law of Cosines is needed to solve the triangle. Then solve (if possible) the triangle. If two solutions exist, find both. Round your answers to two decimal places.

31. $b = 9$, $c = 13$, $C = 64°$

32. $a = 4$, $c = 5$, $B = 52°$

33. $a = 13$, $b = 15$, $c = 24$

34. $A = 44°$, $B = 31°$, $c = 2.8$

35. Geometry The lengths of the diagonals of a parallelogram are 10 feet and 16 feet. Find the lengths of the sides of the parallelogram when the diagonals intersect at an angle of 28°.

36. Geometry The lengths of the diagonals of a parallelogram are 30 meters and 40 meters. Find the lengths of the sides of the parallelogram when the diagonals intersect at an angle of 34°.

37. Surveying To approximate the length of a marsh, a surveyor walks 425 meters from point A to point B. Then the surveyor turns 65° and walks 300 meters to point C (see figure). Find the length AC of the marsh.

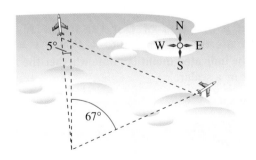

38. Navigation Two planes leave an airport at approximately the same time. One is flying 425 miles per hour at a bearing of 355°, and the other is flying 530 miles per hour at a bearing of 67° (see figure). Determine the distance between the planes after they have flown for 2 hours.

Using Heron's Area Formula In Exercises 39–42, use Heron's Area Formula to find the area of the triangle.

39. $a = 3$, $b = 6$, $c = 8$

40. $a = 15$, $b = 8$, $c = 10$

41. $a = 12.3$, $b = 15.8$, $c = 3.7$

42. $a = \frac{4}{5}$, $b = \frac{3}{4}$, $c = \frac{5}{8}$

3.3 Showing That Two Vectors Are Equivalent In Exercises 43 and 44, show that u and v are equivalent.

43.
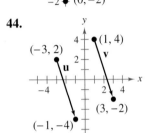

44.
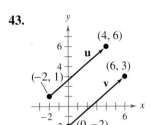

Finding the Component Form of a Vector In Exercises 45–50, find the component form of the vector v satisfying the given conditions.

45.
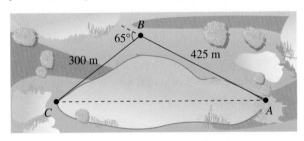

46.
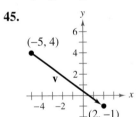

47. Initial point: $(0, 10)$; terminal point: $(7, 3)$

48. Initial point: $(1, 5)$; terminal point: $(15, 9)$

49. $\|\mathbf{v}\| = 8$, $\theta = 120°$

50. $\|\mathbf{v}\| = \frac{1}{2}$, $\theta = 225°$

Vector Operations   **In Exercises 51–58, find (a) u + v, (b) u − v, (c) 4u, and (d) 3v + 5u. Then sketch each resultant vector.**

51. $\mathbf{u} = \langle -1, -3 \rangle$, $\mathbf{v} = \langle -3, 6 \rangle$

52. $\mathbf{u} = \langle 4, 5 \rangle$, $\mathbf{v} = \langle 0, -1 \rangle$

53. $\mathbf{u} = \langle -5, 2 \rangle$, $\mathbf{v} = \langle 4, 4 \rangle$

54. $\mathbf{u} = \langle 1, -8 \rangle$, $\mathbf{v} = \langle 3, -2 \rangle$

55. $\mathbf{u} = 2\mathbf{i} - \mathbf{j}$, $\mathbf{v} = 5\mathbf{i} + 3\mathbf{j}$

56. $\mathbf{u} = -7\mathbf{i} - 3\mathbf{j}$, $\mathbf{v} = 4\mathbf{i} - \mathbf{j}$

57. $\mathbf{u} = 4\mathbf{i}$, $\mathbf{v} = -\mathbf{i} + 6\mathbf{j}$

58. $\mathbf{u} = -6\mathbf{j}$, $\mathbf{v} = \mathbf{i} + \mathbf{j}$

Vector Operations   **In Exercises 59–64, find the component form of w and sketch the specified vector operations geometrically, where $\mathbf{u} = 6\mathbf{i} - 5\mathbf{j}$ and $\mathbf{v} = 10\mathbf{i} + 3\mathbf{j}$.**

59. $\mathbf{w} = 3\mathbf{v}$

60. $\mathbf{w} = \frac{1}{2}\mathbf{v}$

61. $\mathbf{w} = 2\mathbf{u} + \mathbf{v}$

62. $\mathbf{w} = 4\mathbf{u} - 5\mathbf{v}$

63. $\mathbf{w} = 5\mathbf{u} - 4\mathbf{v}$

64. $\mathbf{w} = -3\mathbf{u} + 2\mathbf{v}$

Writing a Linear Combination of Unit Vectors   **In Exercises 65–68, the initial and terminal points of a vector are given. Write the vector as a linear combination of the standard unit vectors i and j.**

Initial Point	Terminal Point
65. $(2, 3)$	$(1, 8)$
66. $(4, -2)$	$(-2, -10)$
67. $(3, 4)$	$(9, 8)$
68. $(-2, 7)$	$(5, -9)$

Finding the Direction Angle of a Vector   **In Exercises 69–74, find the magnitude and direction angle of the vector v.**

69. $\mathbf{v} = 7(\cos 60°\mathbf{i} + \sin 60°\mathbf{j})$

70. $\mathbf{v} = 3(\cos 150°\mathbf{i} + \sin 150°\mathbf{j})$

71. $\mathbf{v} = 5\mathbf{i} + 4\mathbf{j}$

72. $\mathbf{v} = -4\mathbf{i} + 7\mathbf{j}$

73. $\mathbf{v} = -3\mathbf{i} - 3\mathbf{j}$

74. $\mathbf{v} = 8\mathbf{i} - \mathbf{j}$

75. Navigation   An airplane has an airspeed of 430 miles per hour at a bearing of 135°. The wind velocity is 35 miles per hour in the direction of N 30° E. Find the resultant speed and direction of the airplane.

76. Navigation   An airplane has an airspeed of 724 kilometers per hour at a bearing of 30°. The wind velocity is 32 kilometers per hour from the west. Find the resultant speed and direction of the airplane.

77. Cable Tension   In a manufacturing process, an electric hoist lifts 200-pound ingots. Find the tension in the support cables (see figure).

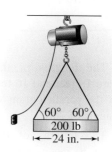

78. Rope Tension   A 180-pound weight is supported by two ropes, as shown in the figure. Find the tension in each rope.

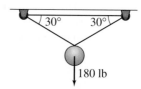

3.4 **Finding a Dot Product**   **In Exercises 79–82, find the dot product of u and v.**

79. $\mathbf{u} = \langle 6, 7 \rangle$
  $\mathbf{v} = \langle -3, 9 \rangle$

80. $\mathbf{u} = \langle -7, 12 \rangle$
  $\mathbf{v} = \langle -4, -14 \rangle$

81. $\mathbf{u} = 3\mathbf{i} + 7\mathbf{j}$
  $\mathbf{v} = 11\mathbf{i} - 5\mathbf{j}$

82. $\mathbf{u} = -7\mathbf{i} + 2\mathbf{j}$
  $\mathbf{v} = 16\mathbf{i} - 12\mathbf{j}$

Using Properties of Dot Products   **In Exercises 83–90, use the vectors $\mathbf{u} = \langle -4, 2 \rangle$ and $\mathbf{v} = \langle 5, 1 \rangle$ to find the indicated quantity. State whether the result is a vector or a scalar.**

83. $2\mathbf{u} \cdot \mathbf{u}$

84. $3\mathbf{u} \cdot \mathbf{v}$

85. $4 - \|\mathbf{u}\|$

86. $\|\mathbf{v}\|^2$

87. $\mathbf{u}(\mathbf{u} \cdot \mathbf{v})$

88. $(\mathbf{u} \cdot \mathbf{v})\mathbf{v}$

89. $(\mathbf{u} \cdot \mathbf{u}) - (\mathbf{u} \cdot \mathbf{v})$

90. $(\mathbf{v} \cdot \mathbf{v}) - (\mathbf{v} \cdot \mathbf{u})$

Finding the Angle Between Two Vectors In Exercises 91–94, find the angle θ between the vectors.

91. $\mathbf{u} = \cos\dfrac{7\pi}{4}\mathbf{i} + \sin\dfrac{7\pi}{4}\mathbf{j}$

$\mathbf{v} = \cos\dfrac{5\pi}{6}\mathbf{i} + \sin\dfrac{5\pi}{6}\mathbf{j}$

92. $\mathbf{u} = \cos 45°\mathbf{i} + \sin 45°\mathbf{j}$

$\mathbf{v} = \cos 300°\mathbf{i} + \sin 300°\mathbf{j}$

93. $\mathbf{u} = \langle 2\sqrt{2}, -4 \rangle$

$\mathbf{v} = \langle -\sqrt{2}, 1 \rangle$

94. $\mathbf{u} = \langle 3, \sqrt{3} \rangle$

$\mathbf{v} = \langle 4, 3\sqrt{3} \rangle$

Determining Orthogonal Vectors In Exercises 95–98, determine whether u and v are orthogonal.

95. $\mathbf{u} = \langle -3, 8 \rangle$

$\mathbf{v} = \langle 8, 3 \rangle$

96. $\mathbf{u} = \left\langle \dfrac{1}{4}, -\dfrac{1}{2} \right\rangle$

$\mathbf{v} = \langle -2, 4 \rangle$

97. $\mathbf{u} = -\mathbf{i}$

$\mathbf{v} = \mathbf{i} + 2\mathbf{j}$

98. $\mathbf{u} = -2\mathbf{i} + \mathbf{j}$

$\mathbf{v} = 3\mathbf{i} + 6\mathbf{j}$

Decomposing a Vector into Components In Exercises 99–102, find the projection of u onto v. Then write u as the sum of two orthogonal vectors, one of which is $\text{proj}_\mathbf{v}\,\mathbf{u}$.

99. $\mathbf{u} = \langle -4, 3 \rangle$, $\mathbf{v} = \langle -8, -2 \rangle$

100. $\mathbf{u} = \langle 5, 6 \rangle$, $\mathbf{v} = \langle 10, 0 \rangle$

101. $\mathbf{u} = \langle 2, 7 \rangle$, $\mathbf{v} = \langle 1, -1 \rangle$

102. $\mathbf{u} = \langle -3, 5 \rangle$, $\mathbf{v} = \langle -5, 2 \rangle$

Work In Exercises 103 and 104, find the work done in moving a particle from P to Q when the magnitude and direction of the force are given by v.

103. $P(5, 3)$, $Q(8, 9)$, $\mathbf{v} = \langle 2, 7 \rangle$

104. $P(-2, -9)$, $Q(-12, 8)$, $\mathbf{v} = 3\mathbf{i} - 6\mathbf{j}$

105. Work Determine the work done (in foot-pounds) by a crane lifting an 18,000-pound truck 48 inches.

106. Work A mover exerts a horizontal force of 25 pounds on a crate as it is pushed up a ramp that is 12 feet long and inclined at an angle of 20° above the horizontal. Find the work done in pushing the crate.

Exploration

True or False? In Exercises 107 and 108, determine whether the statement is true or false. Justify your answer.

107. The Law of Sines is true when one of the angles in the triangle is a right angle.

108. When the Law of Sines is used, the solution is always unique.

109. Law of Sines State the Law of Sines from memory.

110. Law of Cosines State the Law of Cosines from memory.

111. Reasoning What characterizes a vector in the plane?

112. Think About It Which vectors in the figure appear to be equivalent?

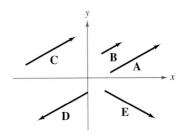

113. Think About It The vectors $\mathbf{u}$ and $\mathbf{v}$ have the same magnitudes in the two figures. In which figure will the magnitude of the sum be greater? Give a reason for your answer.

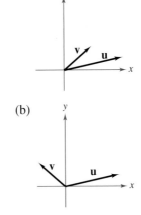

114. Geometry Describe geometrically the scalar multiple $k\mathbf{u}$ of the vector $\mathbf{u}$, for $k > 0$ and $k < 0$.

115. Geometry Describe geometrically the sum of the vectors $\mathbf{u}$ and $\mathbf{v}$.

Chapter Test

See **CalcChat.com** for tutorial help and worked-out solutions to odd-numbered exercises.

Take this test as you would take a test in class. When you are finished, check your work against the answers given in the back of the book.

In Exercises 1–6, determine whether the Law of Sines or the Law of Cosines is needed to solve the triangle. Then solve (if possible) the triangle. If two solutions exist, find both. Round your answers to two decimal places.

1. $A = 24°$, $B = 68°$, $a = 12.2$
2. $B = 110°$, $C = 28°$, $a = 15.6$
3. $A = 24°$, $a = 11.2$, $b = 13.4$
4. $a = 4.0$, $b = 7.3$, $c = 12.4$
5. $B = 100°$, $a = 15$, $b = 23$
6. $C = 121°$, $a = 34$, $b = 55$

7. A triangular parcel of land has borders of lengths 60 meters, 70 meters, and 82 meters. Find the area of the parcel of land.

8. An airplane flies 370 miles from point A to point B with a bearing of 24°. It then flies 240 miles from point B to point C with a bearing of 37° (see figure). Find the distance and bearing from point A to point C.

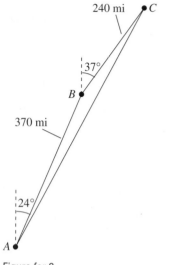

Figure for 8

In Exercises 9 and 10, find the component form of the vector v satisfying the given conditions.

9. Initial point of **v**: $(-3, 7)$
   Terminal point of **v**: $(11, -16)$
10. Magnitude of **v**: $\|\mathbf{v}\| = 12$
    Direction of **v**: $\mathbf{u} = \langle 3, -5 \rangle$

In Exercises 11–14, u $= \langle 2, 7 \rangle$ and v $= \langle -6, 5 \rangle$. Find the resultant vector and sketch its graph.

11. $\mathbf{u} + \mathbf{v}$
12. $\mathbf{u} - \mathbf{v}$
13. $5\mathbf{u} - 3\mathbf{v}$
14. $4\mathbf{u} + 2\mathbf{v}$

15. Find a unit vector in the direction of $\mathbf{u} = \langle 24, -7 \rangle$.
16. Forces with magnitudes of 250 pounds and 130 pounds act on an object at angles of 45° and $-60°$, respectively, with the positive x-axis. Find the direction and magnitude of the resultant of these forces.
17. Find the angle between the vectors $\mathbf{u} = \langle -1, 5 \rangle$ and $\mathbf{v} = \langle 3, -2 \rangle$.
18. Are the vectors $\mathbf{u} = \langle 6, -10 \rangle$ and $\mathbf{v} = \langle 5, 3 \rangle$ orthogonal?
19. Find the projection of $\mathbf{u} = \langle 6, 7 \rangle$ onto $\mathbf{v} = \langle -5, -1 \rangle$. Then write $\mathbf{u}$ as the sum of two orthogonal vectors, one of which is $\text{proj}_\mathbf{v}\mathbf{u}$.
20. A 500-pound motorcycle is headed up a hill inclined at 12°. What force is required to keep the motorcycle from rolling down the hill when stopped at a red light?

Cumulative Test for Chapters 1–3

See **CalcChat.com** for tutorial help and worked-out solutions to odd-numbered exercises.

Take this test as you would take a test in class. When you are finished, check your work against the answers given in the back of the book.

1. Consider the angle $\theta = -120°$.
   (a) Sketch the angle in standard position.
   (b) Determine a coterminal angle in the interval $[0°, 360°)$.
   (c) Rewrite the angle in radian measure as a multiple of π.
   (d) Find the reference angle θ'.
   (e) Find the exact values of the six trigonometric functions of θ.

2. Convert the angle $\theta = -1.45$ radians to degrees. Round the answer to one decimal place.

3. Find $\cos \theta$ when $\tan \theta = -\frac{21}{20}$ and $\sin \theta < 0$.

In Exercises 4–6, sketch the graph of the function. (Include two full periods.)

4. $f(x) = 3 - 2 \sin \pi x$

5. $g(x) = \frac{1}{2} \tan\left(x - \frac{\pi}{2}\right)$

6. $h(x) = -\sec(x + \pi)$

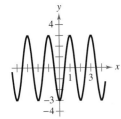

Figure for 7

7. Find a, b, and c such that the graph of the function $h(x) = a \cos(bx + c)$ matches the graph shown in the figure.

8. Sketch the graph of the function $f(x) = \frac{1}{2}x \sin x$ on the interval $-3\pi \le x \le 3\pi$.

In Exercises 9 and 10, find the exact value of the expression without using a calculator.

9. $\tan(\arctan 4.9)$

10. $\tan\left(\arcsin \frac{3}{5}\right)$

11. Write an algebraic expression equivalent to $\sin(\arccos 2x)$.

12. Use the fundamental identities to simplify: $\cos\left(\frac{\pi}{2} - x\right) \csc x$.

13. Subtract and simplify: $\dfrac{\sin \theta - 1}{\cos \theta} - \dfrac{\cos \theta}{\sin \theta - 1}$.

In Exercises 14–16, verify the identity.

14. $\cot^2 \alpha(\sec^2 \alpha - 1) = 1$

15. $\sin(x + y) \sin(x - y) = \sin^2 x - \sin^2 y$

16. $\sin^2 x \cos^2 x = \frac{1}{8}(1 - \cos 4x)$

In Exercises 17 and 18, find all solutions of the equation in the interval $[0, 2\pi)$.

17. $2 \cos^2 \beta - \cos \beta = 0$

18. $3 \tan \theta - \cot \theta = 0$

19. Use the Quadratic Formula to solve the equation in the interval $[0, 2\pi)$: $\sin^2 x + 2 \sin x + 1 = 0$.

20. Given that $\sin u = \frac{12}{13}$, $\cos v = \frac{3}{5}$, and angles u and v are both in Quadrant I, find $\tan(u - v)$.

21. Given that $\tan \theta = \frac{1}{2}$, find the exact value of $\tan(2\theta)$.

22. Given that $\tan \theta = \frac{4}{3}$, find the exact value of $\sin \frac{\theta}{2}$.

23. Write the product $5 \sin \frac{3\pi}{4} \cdot \cos \frac{7\pi}{4}$ as a sum or difference.

24. Write $\cos 9x - \cos 7x$ as a product.

In Exercises 25–30, determine whether the Law of Sines or the Law of Cosines is needed to solve the triangle at the left, then solve the triangle. Round your answers to two decimal places.

25. $A = 30°$, $a = 9$, $b = 8$

26. $A = 30°$, $b = 8$, $c = 10$

27. $A = 30°$, $C = 90°$, $b = 10$

28. $a = 4.7$, $b = 8.1$, $c = 10.3$

29. $A = 45°$, $B = 26°$, $c = 20$

30. $a = 1.2$, $b = 10$, $C = 80°$

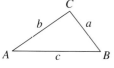

Figure for 25–30

Figure for 40

31. Two sides of a triangle have lengths 7 inches and 12 inches. Their included angle measures 99°. Find the area of the triangle.

32. Find the area of a triangle with sides of lengths 30 meters, 41 meters, and 45 meters.

33. Write the vector

$$\mathbf{u} = \langle 7, 8 \rangle$$

as a linear combination of the standard unit vectors $\mathbf{i}$ and $\mathbf{j}$.

34. Find a unit vector in the direction of $\mathbf{v} = \mathbf{i} + \mathbf{j}$.

35. Find $\mathbf{u} \cdot \mathbf{v}$ for $\mathbf{u} = 3\mathbf{i} + 4\mathbf{j}$ and $\mathbf{v} = \mathbf{i} - 2\mathbf{j}$.

36. Find the projection of $\mathbf{u} = \langle 8, -2 \rangle$ onto $\mathbf{v} = \langle 1, 5 \rangle$. Then write $\mathbf{u}$ as the sum of two orthogonal vectors, one of which is $\text{proj}_{\mathbf{v}}\mathbf{u}$.

37. A ceiling fan with 21-inch blades makes 63 revolutions per minute. Find the angular speed of the fan in radians per minute. Find the linear speed of the tips of the blades in inches per minute.

38. Find the area of the sector of a circle with a radius of 12 yards and a central angle of 105°.

39. From a point 200 feet from a flagpole, the angles of elevation to the bottom and top of the flag are 16° 45′ and 18°, respectively. Approximate the height of the flag to the nearest foot.

40. To determine the angle of elevation of a star in the sky, you get the star and the top of the backboard of a basketball hoop that is 5 feet higher than your eyes in your line of vision (see figure). Your horizontal distance from the backboard is 12 feet. What is the angle of elevation of the star?

41. Write a model for a particle in simple harmonic motion with a displacement of 4 inches and a period of 8 seconds.

42. An airplane has an airspeed of 500 kilometers per hour at a bearing of 30°. The wind velocity is 50 kilometers per hour in the direction of N 60° E. Find the resultant speed and direction of the airplane.

43. A force of 85 pounds, exerted at an angle of 60° with the horizontal, is required to slide an object across a floor. Determine the work done in sliding the object 10 feet.

Proofs in Mathematics ▪ ▪ ▪ ▪ ▪ ▪ ▪ ▪ ▪ ▪ ▪ ▪ ▪ ▪ ▪

LAW OF TANGENTS

Besides the Law of Sines and the Law of Cosines, there is also a Law of Tangents, which was developed by Francois Viète (1540–1603). The Law of Tangents follows from the Law of Sines and the sum-to-product formulas for sine and is defined as follows.

$$\frac{a + b}{a - b} = \frac{\tan[(A + B)/2]}{\tan[(A - B)/2]}$$

The Law of Tangents can be used to solve a triangle when two sides and the included angle are given (SAS). Before calculators were invented, the Law of Tangents was used to solve the SAS case instead of the Law of Cosines because computation with a table of tangent values was easier.

Law of Sines   *(p. 262)*

If ABC is a triangle with sides a, b, and c, then

$$\frac{a}{\sin A} = \frac{b}{\sin B} = \frac{c}{\sin C}.$$

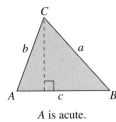
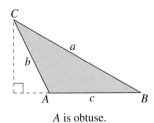

A is acute. A is obtuse.

Proof

Let h be the altitude of either triangle in the figure above. Then you have

$$\sin A = \frac{h}{b} \quad \text{or} \quad h = b \sin A$$

$$\sin B = \frac{h}{a} \quad \text{or} \quad h = a \sin B.$$

Equating these two values of h, you have

$$a \sin B = b \sin A \quad \text{or} \quad \frac{a}{\sin A} = \frac{b}{\sin B}.$$

Note that $\sin A \neq 0$ and $\sin B \neq 0$ because no angle of a triangle can have a measure of $0°$ or $180°$. In a similar manner, construct an altitude h from vertex B to side AC (extended in the obtuse triangle), as shown at the left. Then you have

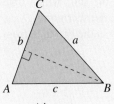

A is acute.

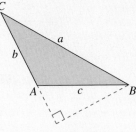

A is obtuse.

$$\sin A = \frac{h}{c} \quad \text{or} \quad h = c \sin A$$

$$\sin C = \frac{h}{a} \quad \text{or} \quad h = a \sin C.$$

Equating these two values of h, you have

$$a \sin C = c \sin A \quad \text{or} \quad \frac{a}{\sin A} = \frac{c}{\sin C}.$$

By the Transitive Property of Equality, you know that

$$\frac{a}{\sin A} = \frac{b}{\sin B} = \frac{c}{\sin C}.$$

So, the Law of Sines is established. ▪

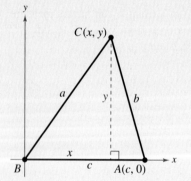

Law of Cosines   *(p. 271)*

Standard Form	**Alternative Form**
$a^2 = b^2 + c^2 - 2bc \cos A$	$\cos A = \dfrac{b^2 + c^2 - a^2}{2bc}$
$b^2 = a^2 + c^2 - 2ac \cos B$	$\cos B = \dfrac{a^2 + c^2 - b^2}{2ac}$
$c^2 = a^2 + b^2 - 2ab \cos C$	$\cos C = \dfrac{a^2 + b^2 - c^2}{2ab}$

Proof

To prove the first formula, consider the top triangle at the left, which has three acute angles. Note that vertex B has coordinates $(c, 0)$. Furthermore, C has coordinates (x, y), where $x = b \cos A$ and $y = b \sin A$. Because a is the distance from vertex C to vertex B, it follows that

$$a = \sqrt{(x - c)^2 + (y - 0)^2} \qquad \text{Distance Formula}$$
$$a^2 = (x - c)^2 + (y - 0)^2 \qquad \text{Square each side.}$$
$$a^2 = (b \cos A - c)^2 + (b \sin A)^2 \qquad \text{Substitute for } x \text{ and } y.$$
$$a^2 = b^2 \cos^2 A - 2bc \cos A + c^2 + b^2 \sin^2 A \qquad \text{Expand.}$$
$$a^2 = b^2(\sin^2 A + \cos^2 A) + c^2 - 2bc \cos A \qquad \text{Factor out } b^2.$$
$$a^2 = b^2 + c^2 - 2bc \cos A. \qquad \sin^2 A + \cos^2 A = 1$$

To prove the second formula, consider the bottom triangle at the left, which also has three acute angles. Note that vertex A has coordinates $(c, 0)$. Furthermore, C has coordinates (x, y), where $x = a \cos B$ and $y = a \sin B$. Because b is the distance from vertex C to vertex A, it follows that

$$b = \sqrt{(x - c)^2 + (y - 0)^2} \qquad \text{Distance Formula}$$
$$b^2 = (x - c)^2 + (y - 0)^2 \qquad \text{Square each side.}$$
$$b^2 = (a \cos B - c)^2 + (a \sin B)^2 \qquad \text{Substitute for } x \text{ and } y.$$
$$b^2 = a^2 \cos^2 B - 2ac \cos B + c^2 + a^2 \sin^2 B \qquad \text{Expand.}$$
$$b^2 = a^2(\sin^2 B + \cos^2 B) + c^2 - 2ac \cos B \qquad \text{Factor out } a^2.$$
$$b^2 = a^2 + c^2 - 2ac \cos B. \qquad \sin^2 B + \cos^2 B = 1$$

A similar argument is used to establish the third formula.

Heron's Area Formula   *(p. 274)*

Given any triangle with sides of lengths a, b, and c, the area of the triangle is

$$\text{Area} = \sqrt{s(s - a)(s - b)(s - c)}$$

where $s = \dfrac{a + b + c}{2}$.

Proof

From Section 3.1, you know that

$$\text{Area} = \frac{1}{2}bc \sin A \qquad \text{Formula for the area of an oblique triangle}$$

$$(\text{Area})^2 = \frac{1}{4}b^2c^2 \sin^2 A \qquad \text{Square each side.}$$

$$\text{Area} = \sqrt{\frac{1}{4}b^2c^2 \sin^2 A} \qquad \text{Take the square root of each side.}$$

$$= \sqrt{\frac{1}{4}b^2c^2(1 - \cos^2 A)} \qquad \text{Pythagorean identity}$$

$$= \sqrt{\left[\frac{1}{2}bc(1 + \cos A)\right]\left[\frac{1}{2}bc(1 - \cos A)\right]}. \qquad \text{Factor.}$$

Using the Law of Cosines, you can show that

$$\frac{1}{2}bc(1 + \cos A) = \frac{a + b + c}{2} \cdot \frac{-a + b + c}{2}$$

and

$$\frac{1}{2}bc(1 - \cos A) = \frac{a - b + c}{2} \cdot \frac{a + b - c}{2}.$$

Letting $s = (a + b + c)/2$, these two equations can be rewritten as

$$\frac{1}{2}bc(1 + \cos A) = s(s - a)$$

and

$$\frac{1}{2}bc(1 - \cos A) = (s - b)(s - c).$$

By substituting into the last formula for area, you can conclude that

$$\text{Area} = \sqrt{s(s - a)(s - b)(s - c)}.$$

Properties of the Dot Product *(p. 291)*

Let **u**, **v**, and **w** be vectors in the plane or in space and let c be a scalar.

1. $\mathbf{u} \cdot \mathbf{v} = \mathbf{v} \cdot \mathbf{u}$ 2. $\mathbf{0} \cdot \mathbf{v} = 0$

3. $\mathbf{u} \cdot (\mathbf{v} + \mathbf{w}) = \mathbf{u} \cdot \mathbf{v} + \mathbf{u} \cdot \mathbf{w}$ 4. $\mathbf{v} \cdot \mathbf{v} = \|\mathbf{v}\|^2$

5. $c(\mathbf{u} \cdot \mathbf{v}) = c\mathbf{u} \cdot \mathbf{v} = \mathbf{u} \cdot c\mathbf{v}$

Proof

Let $\mathbf{u} = \langle u_1, u_2 \rangle$, $\mathbf{v} = \langle v_1, v_2 \rangle$, $\mathbf{w} = \langle w_1, w_2 \rangle$, $\mathbf{0} = \langle 0, 0 \rangle$, and let c be a scalar.

1. $\mathbf{u} \cdot \mathbf{v} = u_1 v_1 + u_2 v_2 = v_1 u_1 + v_2 u_2 = \mathbf{v} \cdot \mathbf{u}$

2. $\mathbf{0} \cdot \mathbf{v} = 0 \cdot v_1 + 0 \cdot v_2 = 0$

3. $\mathbf{u} \cdot (\mathbf{v} + \mathbf{w}) = \mathbf{u} \cdot \langle v_1 + w_1, v_2 + w_2 \rangle$

$$= u_1(v_1 + w_1) + u_2(v_2 + w_2)$$

$$= u_1 v_1 + u_1 w_1 + u_2 v_2 + u_2 w_2$$

$$= (u_1 v_1 + u_2 v_2) + (u_1 w_1 + u_2 w_2) = \mathbf{u} \cdot \mathbf{v} + \mathbf{u} \cdot \mathbf{w}$$

4. $\mathbf{v} \cdot \mathbf{v} = v_1^2 + v_2^2 = \left(\sqrt{v_1^2 + v_2^2} \right)^2 = \|\mathbf{v}\|^2$

5. $c(\mathbf{u} \cdot \mathbf{v}) = c(\langle u_1, u_2 \rangle \cdot \langle v_1, v_2 \rangle)$

$$= c(u_1 v_1 + u_2 v_2)$$

$$= (cu_1)v_1 + (cu_2)v_2$$

$$= \langle cu_1, cu_2 \rangle \cdot \langle v_1, v_2 \rangle$$

$$= c\mathbf{u} \cdot \mathbf{v}$$

Angle Between Two Vectors *(p. 292)*

If θ is the angle between two nonzero vectors **u** and **v**, then $\cos \theta = \dfrac{\mathbf{u} \cdot \mathbf{v}}{\|\mathbf{u}\| \, \|\mathbf{v}\|}$.

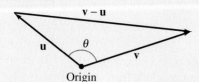

Proof

Consider the triangle determined by vectors **u**, **v**, and $\mathbf{v} - \mathbf{u}$, as shown at the left. By the Law of Cosines, you can write

$$\|\mathbf{v} - \mathbf{u}\|^2 = \|\mathbf{u}\|^2 + \|\mathbf{v}\|^2 - 2\|\mathbf{u}\| \, \|\mathbf{v}\| \cos \theta$$

$$(\mathbf{v} - \mathbf{u}) \cdot (\mathbf{v} - \mathbf{u}) = \|\mathbf{u}\|^2 + \|\mathbf{v}\|^2 - 2\|\mathbf{u}\| \, \|\mathbf{v}\| \cos \theta$$

$$(\mathbf{v} - \mathbf{u}) \cdot \mathbf{v} - (\mathbf{v} - \mathbf{u}) \cdot \mathbf{u} = \|\mathbf{u}\|^2 + \|\mathbf{v}\|^2 - 2\|\mathbf{u}\| \, \|\mathbf{v}\| \cos \theta$$

$$\mathbf{v} \cdot \mathbf{v} - \mathbf{u} \cdot \mathbf{v} - \mathbf{v} \cdot \mathbf{u} + \mathbf{u} \cdot \mathbf{u} = \|\mathbf{u}\|^2 + \|\mathbf{v}\|^2 - 2\|\mathbf{u}\| \, \|\mathbf{v}\| \cos \theta$$

$$\|\mathbf{v}\|^2 - 2\mathbf{u} \cdot \mathbf{v} + \|\mathbf{u}\|^2 = \|\mathbf{u}\|^2 + \|\mathbf{v}\|^2 - 2\|\mathbf{u}\| \, \|\mathbf{v}\| \cos \theta$$

$$\cos \theta = \frac{\mathbf{u} \cdot \mathbf{v}}{\|\mathbf{u}\| \, \|\mathbf{v}\|}.$$

P.S. Problem Solving ■ ■ ■ ■ ■ ■ ■ ■ ■ ■ ■ ■ ■

1. **Distance** In the figure, a beam of light is directed at the blue mirror, reflected to the red mirror, and then reflected back to the blue mirror. Find PT, the distance that the light travels from the red mirror back to the blue mirror.

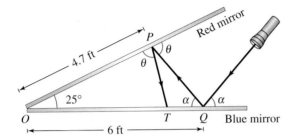

2. **Correcting a Course** A triathlete sets a course to swim S 25° E from a point on shore to a buoy $\frac{3}{4}$ mile away. After swimming 300 yards through a strong current, the triathlete is off course at a bearing of S 35° E. Find the bearing and distance the triathlete needs to swim to correct her course.

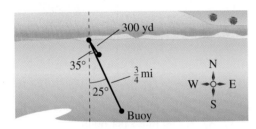

3. **Locating Lost Hikers** A group of hikers is lost in a national park. Two ranger stations have received an emergency SOS signal from the hikers. Station B is 75 miles due east of station A. The bearing from station A to the signal is S 60° E and the bearing from station B to the signal is S 75° W.

   (a) Draw a diagram that gives a visual representation of the problem.

   (b) Find the distance from each station to the SOS signal.

   (c) A rescue party is in the park 20 miles from station A at a bearing of S 80° E. Find the distance and the bearing the rescue party must travel to reach the lost hikers.

4. **Seeding a Courtyard** You are seeding a triangular courtyard. One side of the courtyard is 52 feet long and another side is 46 feet long. The angle opposite the 52-foot side is 65°.

   (a) Draw a diagram that gives a visual representation of the situation.

   (b) How long is the third side of the courtyard?

   (c) One bag of grass seed covers an area of 50 square feet. How many bags of grass seed will you need to cover the courtyard?

5. **Finding Magnitudes** For each pair of vectors, find the following.

   (i) $\|\mathbf{u}\|$

   (ii) $\|\mathbf{v}\|$

   (iii) $\|\mathbf{u} + \mathbf{v}\|$

   (iv) $\left\| \dfrac{\mathbf{u}}{\|\mathbf{u}\|} \right\|$

   (v) $\left\| \dfrac{\mathbf{v}}{\|\mathbf{v}\|} \right\|$

   (vi) $\left\| \dfrac{\mathbf{u} + \mathbf{v}}{\|\mathbf{u} + \mathbf{v}\|} \right\|$

   (a) $\mathbf{u} = \langle 1, -1 \rangle$
   $\mathbf{v} = \langle -1, 2 \rangle$

   (b) $\mathbf{u} = \langle 0, 1 \rangle$
   $\mathbf{v} = \langle 3, -3 \rangle$

   (c) $\mathbf{u} = \left\langle 1, \frac{1}{2} \right\rangle$
   $\mathbf{v} = \langle 2, 3 \rangle$

   (d) $\mathbf{u} = \langle 2, -4 \rangle$
   $\mathbf{v} = \langle 5, 5 \rangle$

6. **Writing a Vector in Terms of Other Vectors** Write the vector $\mathbf{w}$ in terms of $\mathbf{u}$ and $\mathbf{v}$, given that the terminal point of $\mathbf{w}$ bisects the line segment (see figure).

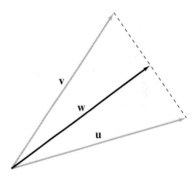

7. **Proof** Prove that if $\mathbf{u}$ is orthogonal to $\mathbf{v}$ and $\mathbf{w}$, then $\mathbf{u}$ is orthogonal to

$$c\mathbf{v} + d\mathbf{w}$$

for any scalars c and d.

8. **Comparing Work** Two forces of the same magnitude $\mathbf{F}_1$ and $\mathbf{F}_2$ act at angles θ_1 and θ_2, respectively. Use a diagram to compare the work done by $\mathbf{F}_1$ with the work done by $\mathbf{F}_2$ in moving along the vector PQ when

   (a) $\theta_1 = -\theta_2$

   (b) $\theta_1 = 60°$ and $\theta_2 = 30°$.

9. Skydiving A skydiver is falling at a constant downward velocity of 120 miles per hour. In the figure, vector **u** represents the skydiver's velocity. A steady breeze pushes the skydiver to the east at 40 miles per hour. Vector **v** represents the wind velocity.

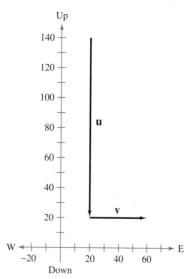

(a) Write the vectors **u** and **v** in component form.

(b) Let

$$s = u + v.$$

Use the figure to sketch **s**. To print an enlarged copy of the graph, go to *MathGraphs.com*.

(c) Find the magnitude of **s**. What information does the magnitude give you about the skydiver's fall?

(d) If there were no wind, then the skydiver would fall in a path perpendicular to the ground. At what angle to the ground is the path of the skydiver when affected by the 40-mile-per-hour wind from due west?

(e) The skydiver is blown to the west at 30 miles per hour. Draw a new figure that gives a visual representation of the problem and find the skydiver's new velocity.

10. Speed and Velocity of an Airplane Four basic forces are in action during flight: weight, lift, thrust, and drag. To fly through the air, an object must overcome its own *weight*. To do this, it must create an upward force called *lift*. To generate lift, a forward motion called *thrust* is needed. The thrust must be great enough to overcome air resistance, which is called *drag*.

For a commercial jet aircraft, a quick climb is important to maximize efficiency because the performance of an aircraft at high altitudes is enhanced. In addition, it is necessary to clear obstacles such as buildings and mountains and to reduce noise in residential areas. In the diagram, the angle θ is called the climb angle. The velocity of the plane can be represented by a vector **v** with a vertical component $\|\mathbf{v}\| \sin \theta$ (called climb speed) and a horizontal component $\|\mathbf{v}\| \cos \theta$, where $\|\mathbf{v}\|$ is the speed of the plane.

When taking off, a pilot must decide how much of the thrust to apply to each component. The more the thrust is applied to the horizontal component, the faster the airplane gains speed. The more the thrust is applied to the vertical component, the quicker the airplane climbs.

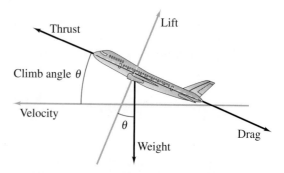

(a) Complete the table for an airplane that has a speed of $\|\mathbf{v}\| = 100$ miles per hour.

θ	0.5°	1.0°	1.5°	2.0°	2.5°	3.0°
$\|\mathbf{v}\| \sin \theta$						
$\|\mathbf{v}\| \cos \theta$						

(b) Does an airplane's speed equal the sum of the vertical and horizontal components of its velocity? If not, how could you find the speed of an airplane whose velocity components were known?

(c) Use the result of part (b) to find the speed of an airplane with the given velocity components.

(i) $\|\mathbf{v}\| \sin \theta = 5.235$ miles per hour
$\|\mathbf{v}\| \cos \theta = 149.909$ miles per hour

(ii) $\|\mathbf{v}\| \sin \theta = 10.463$ miles per hour
$\|\mathbf{v}\| \cos \theta = 149.634$ miles per hour

6.7 Polar Coordinates

You can use polar coordinates in mathematical modeling. For instance, in Exercise 127 on page 472, you will use polar coordinates to model the path of a passenger car on a Ferris wheel.

- ■ Plot points in the polar coordinate system.
- ■ Convert points from rectangular to polar form and vice versa.
- ■ Convert equations from rectangular to polar form and vice versa.

Introduction

So far, you have been representing graphs of equations as collections of points (x, y) in the rectangular coordinate system, where x and y represent the directed distances from the coordinate axes to the point (x, y). In this section, you will study a different system called the **polar coordinate system.**

To form the polar coordinate system in the plane, fix a point O, called the **pole** (or **origin**), and construct from O an initial ray called the **polar axis,** as shown below. Then each point P in the plane can be assigned **polar coordinates** (r, θ) as follows.

1. $r = directed\ distance$ from O to P

2. $\theta = directed\ angle$, counterclockwise from polar axis to segment $\overline{OP}$

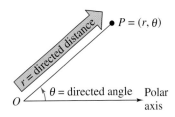

EXAMPLE 1 Plotting Points in the Polar Coordinate System

a. The point $(r, \theta) = (2, \pi/3)$ lies two units from the pole on the terminal side of the angle $\theta = \pi/3$, as shown in Figure 6.45.

b. The point $(r, \theta) = (3, -\pi/6)$ lies three units from the pole on the terminal side of the angle $\theta = -\pi/6$, as shown in Figure 6.46.

c. The point $(r, \theta) = (3, 11\pi/6)$ coincides with the point $(3, -\pi/6)$, as shown in Figure 6.47.

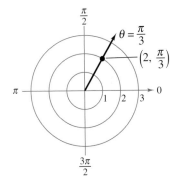

Figure 6.45

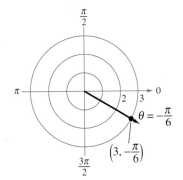

Figure 6.46

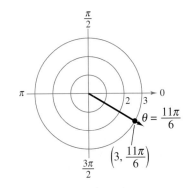

Figure 6.47

✓ **Checkpoint** ◀)) Audio-video solution in English & Spanish at LarsonPrecalculus.com.

Plot each point given in polar coordinates.

a. $(3, \pi/4)$ **b.** $(2, -\pi/3)$ **c.** $(2, 5\pi/3)$

Pavzyuk Svitlana/Shutterstock.com

In rectangular coordinates, each point (x, y) has a unique representation. This is not true for polar coordinates. For instance, the coordinates

$$(r, \theta) \quad \text{and} \quad (r, \theta + 2\pi)$$

represent the same point, as illustrated in Example 1. Another way to obtain multiple representations of a point is to use negative values for r. Because r is a *directed distance,* the coordinates

$$(r, \theta) \quad \text{and} \quad (-r, \theta + \pi)$$

represent the same point. In general, the point (r, θ) can be represented as

$$(r, \theta) = (r, \theta \pm 2n\pi) \quad \text{or} \quad (r, \theta) = (-r, \theta \pm (2n + 1)\pi)$$

where n is any integer. Moreover, the pole is represented by $(0, \theta)$, where θ is any angle.

EXAMPLE 2   Multiple Representations of Points

Plot the point

$$\left(3, -\frac{3\pi}{4}\right)$$

and find three additional polar representations of this point, using

$$-2\pi < \theta < 2\pi.$$

Solution   The point is shown below. Three other representations are as follows.

$$\left(3, -\frac{3\pi}{4} + 2\pi\right) = \left(3, \frac{5\pi}{4}\right) \qquad\qquad \text{Add } 2\pi \text{ to } \theta.$$

$$\left(-3, -\frac{3\pi}{4} - \pi\right) = \left(-3, -\frac{7\pi}{4}\right) \qquad\qquad \text{Replace } r \text{ by } -r; \text{ subtract } \pi \text{ from } \theta.$$

$$\left(-3, -\frac{3\pi}{4} + \pi\right) = \left(-3, \frac{\pi}{4}\right) \qquad\qquad \text{Replace } r \text{ by } -r; \text{ add } \pi \text{ to } \theta.$$

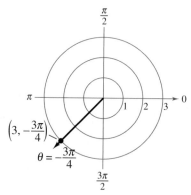

$$\left(3, -\tfrac{3\pi}{4}\right) = \left(3, \tfrac{5\pi}{4}\right) = \left(-3, -\tfrac{7\pi}{4}\right) = \left(-3, \tfrac{\pi}{4}\right) = \dots$$

✓ *Checkpoint*))) Audio-video solution in English & Spanish at LarsonPrecalculus.com.

Plot each point and find three additional polar representations of the point, using $-2\pi < \theta < 2\pi$.

a. $\left(3, \dfrac{4\pi}{3}\right)$ **b.** $\left(2, -\dfrac{5\pi}{6}\right)$ **c.** $\left(-1, \dfrac{3\pi}{4}\right)$ ■

Coordinate Conversion

To establish the relationship between polar and rectangular coordinates, let the polar axis coincide with the positive x-axis and the pole with the origin, as shown in Figure 6.48. Because (x, y) lies on a circle of radius r, it follows that $r^2 = x^2 + y^2$. Moreover, for $r > 0$, the definitions of the trigonometric functions imply that

$$\tan \theta = \frac{y}{x}, \quad \cos \theta = \frac{x}{r}, \quad \text{and} \quad \sin \theta = \frac{y}{r}.$$

You can show that the same relationships hold for $r < 0$.

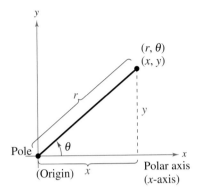

Figure 6.48

Coordinate Conversion

The polar coordinates (r, θ) are related to the rectangular coordinates (x, y) as follows.

Polar-to-Rectangular	Rectangular-to-Polar
$x = r \cos \theta$	$\tan \theta = \dfrac{y}{x}$
$y = r \sin \theta$	$r^2 = x^2 + y^2$

EXAMPLE 3   Polar-to-Rectangular Conversion

Convert $\left(\sqrt{3}, \dfrac{\pi}{6} \right)$ to rectangular coordinates.

Solution   For the point $(r, \theta) = \left(\sqrt{3}, \pi/6 \right)$, you have the following.

$$x = r \cos \theta = \sqrt{3} \cos \frac{\pi}{6} = \sqrt{3}\left(\frac{\sqrt{3}}{2} \right) = \frac{3}{2}$$

$$y = r \sin \theta = \sqrt{3} \sin \frac{\pi}{6} = \sqrt{3}\left(\frac{1}{2} \right) = \frac{\sqrt{3}}{2}$$

The rectangular coordinates are $(x, y) = \left(\dfrac{3}{2}, \dfrac{\sqrt{3}}{2} \right)$. (See Figure 6.49.)

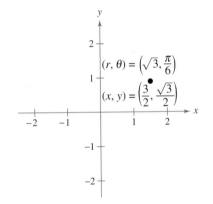

✓ **Checkpoint** 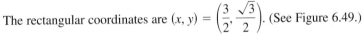 Audio-video solution in English & Spanish at LarsonPrecalculus.com.

Convert $(2, \pi)$ to rectangular coordinates.

Figure 6.49

EXAMPLE 4   Rectangular-to-Polar Conversion

Convert $(-1, 1)$ to polar coordinates.

Solution   For the second-quadrant point $(x, y) = (-1, 1)$, you have

$$\tan \theta = \frac{y}{x} = \frac{1}{-1} = -1 \implies \theta = \pi + \arctan(-1) = \frac{3\pi}{4}.$$

Because θ lies in the same quadrant as (x, y), use positive r.

$$r = \sqrt{x^2 + y^2} = \sqrt{(-1)^2 + (1)^2} = \sqrt{2}$$

So, *one* set of polar coordinates is $(r, \theta) = \left(\sqrt{2}, 3\pi/4 \right)$, as shown in Figure 6.50.

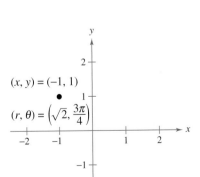

Figure 6.50

✓ **Checkpoint**  Audio-video solution in English & Spanish at LarsonPrecalculus.com.

Convert $(0, 2)$ to polar coordinates.

Equation Conversion

To convert a rectangular equation to polar form, replace x by $r \cos \theta$ and y by $r \sin \theta$. For instance, the rectangular equation $y = x^2$ can be written in polar form as follows.

$$y = x^2 \qquad \text{Rectangular equation}$$

$$r \sin \theta = (r \cos \theta)^2 \qquad \text{Polar equation}$$

$$r = \sec \theta \tan \theta \qquad \text{Simplest form}$$

Converting a polar equation to rectangular form requires considerable ingenuity. Example 5 demonstrates several polar-to-rectangular conversions that enable you to sketch the graphs of some polar equations.

EXAMPLE 5   Converting Polar Equations to Rectangular Form

Convert each polar equation to a rectangular equation.

a. $r = 2$ **b.** $\theta = \pi/3$ **c.** $r = \sec \theta$

Solution

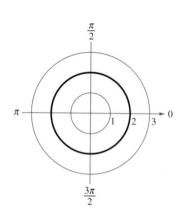

Figure 6.51

a. The graph of the polar equation $r = 2$ consists of all points that are two units from the pole. In other words, this graph is a circle centered at the origin with a radius of 2, as shown in Figure 6.51. You can confirm this by converting to rectangular form, using the relationship $r^2 = x^2 + y^2$.

$$\underbrace{r = 2}_{\text{Polar equation}} \implies r^2 = 2^2 \implies \underbrace{x^2 + y^2 = 2^2}_{\text{Rectangular equation}}$$

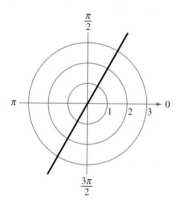

Figure 6.52

b. The graph of the polar equation $\theta = \pi/3$ consists of all points on the line that makes an angle of $\pi/3$ with the positive polar axis, as shown in Figure 6.52. To convert to rectangular form, make use of the relationship $\tan \theta = y/x$.

$$\underbrace{\theta = \pi/3}_{\text{Polar equation}} \implies \tan \theta = \sqrt{3} \implies \underbrace{y = \sqrt{3}x}_{\text{Rectangular equation}}$$

c. The graph of the polar equation $r = \sec \theta$ is not evident by simple inspection, so convert to rectangular form by using the relationship $r \cos \theta = x$.

$$\underbrace{r = \sec \theta}_{\text{Polar equation}} \implies r \cos \theta = 1 \implies \underbrace{x = 1}_{\text{Rectangular equation}}$$

Now you see that the graph is a vertical line, as shown in Figure 6.53.

✓ *Checkpoint* ◀))) Audio-video solution in English & Spanish at LarsonPrecalculus.com.

Convert $r = 6 \sin \theta$ to a rectangular equation. ▪

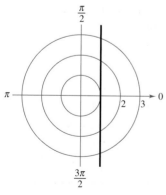

Figure 6.53

Summarize **(Section 6.7)**

1. Describe how to plot the point (r, θ) in the polar coordinate system *(page 467)*. For examples of plotting points in the polar coordinate system, see Examples 1 and 2.

2. Describe how to convert points from rectangular to polar form and vice versa *(page 469)*. For examples of converting between forms, see Examples 3 and 4.

3. Describe how to convert equations from rectangular to polar form and vice versa *(page 470)*. For an example of converting polar equations to rectangular form, see Example 5.

6.7 Exercises

Vocabulary: Fill in the blanks.

1. The origin of the polar coordinate system is called the _____.
2. For the point (r, θ), r is the _____ _____ from O to P and θ is the _____ _____ , counterclockwise from the polar axis to the line segment $\overline{OP}$.
3. To plot the point (r, θ), use the _____ coordinate system.
4. The polar coordinates (r, θ) are related to the rectangular coordinates (x, y) as follows:

   $x =$ _____ $y =$ _____ $\tan \theta =$ _____ $r^2 =$ _____

Skills and Applications

Plotting Points in the Polar Coordinate System   In Exercises 5–18, plot the point given in polar coordinates and find two additional polar representations of the point, using $-2\pi < \theta < 2\pi$.

5. $(2, 5\pi/6)$
6. $(3, 5\pi/4)$
7. $(4, -\pi/3)$
8. $(-1, -3\pi/4)$
9. $(2, 3\pi)$
10. $(4, 5\pi/2)$
11. $(-2, 2\pi/3)$
12. $(-3, 11\pi/6)$
13. $(0, -7\pi/6)$
14. $(0, -7\pi/2)$
15. $\left(\sqrt{2}, 2.36\right)$
16. $(2\sqrt{2}, 4.71)$
17. $(-3, -1.57)$
18. $(-5, -2.36)$

Polar-to-Rectangular Conversion   In Exercises 19–34, a point in polar coordinates is given. Convert the point to rectangular coordinates.

19. $(0, \pi)$
20. $(0, -\pi)$
21. $(3, \pi/2)$
22. $(3, 3\pi/2)$
23. $(2, 3\pi/4)$
24. $(1, 5\pi/4)$
25. $(-1, 5\pi/4)$
26. $(-2, 7\pi/4)$
27. $(-2, 7\pi/6)$
28. $(-3, 5\pi/6)$
29. $(-3, -\pi/3)$
30. $(-2, -4\pi/3)$
31. $(2, 2.74)$
32. $(1.5, 3.67)$
33. $(-2.5, 1.1)$
34. $(-2, 5.76)$

Using a Graphing Utility to Find Rectangular Coordinates   In Exercises 35–42, use a graphing utility to find the rectangular coordinates of the point given in polar coordinates. Round your results to two decimal places.

35. $(2, 2\pi/9)$
36. $(4, 11\pi/9)$
37. $(-4.5, 1.3)$
38. $(8.25, 3.5)$
39. $(2.5, 1.58)$
40. $(5.4, 2.85)$
41. $(-4.1, -0.5)$
42. $(8.2, -3.2)$

Rectangular-to-Polar Conversion   In Exercises 43–60, a point in rectangular coordinates is given. Convert the point to polar coordinates.

43. $(1, 1)$
44. $(2, 2)$
45. $(-3, -3)$
46. $(-4, -4)$
47. $(-6, 0)$
48. $(3, 0)$
49. $(0, -5)$
50. $(0, 5)$
51. $(-3, 4)$
52. $(-4, -3)$
53. $\left(-\sqrt{3}, -\sqrt{3}\right)$
54. $\left(-\sqrt{3}, \sqrt{3}\right)$
55. $\left(\sqrt{3}, -1\right)$
56. $\left(-1, \sqrt{3}\right)$
57. $(6, 9)$
58. $(6, 2)$
59. $(5, 12)$
60. $(7, 15)$

Using a Graphing Utility to Find Polar Coordinates   In Exercises 61–70, use a graphing utility to find one set of polar coordinates of the point given in rectangular coordinates.

61. $(3, -2)$
62. $(-4, -2)$
63. $(-5, 2)$
64. $(7, -2)$
65. $\left(\sqrt{3}, 2\right)$
66. $\left(5, -\sqrt{2}\right)$
67. $\left(\frac{5}{2}, \frac{4}{3}\right)$
68. $\left(\frac{9}{5}, \frac{11}{2}\right)$
69. $\left(\frac{7}{4}, \frac{3}{2}\right)$
70. $\left(-\frac{7}{9}, -\frac{3}{4}\right)$

Converting a Rectangular Equation to Polar Form   In Exercises 71–90, convert the rectangular equation to polar form. Assume $a > 0$.

71. $x^2 + y^2 = 9$
72. $x^2 + y^2 = 16$
73. $y = x$
74. $y = -x$
75. $x = 10$
76. $x = a$
77. $y = 1$
78. $y = -2$
79. $3x - y + 2 = 0$
80. $3x + 5y - 2 = 0$
81. $xy = 16$
82. $2xy = 1$
83. $x^2 + y^2 = a^2$
84. $x^2 + y^2 = 9a^2$
85. $x^2 + y^2 - 2ax = 0$
86. $x^2 + y^2 - 2ay = 0$
87. $(x^2 + y^2)^2 = x^2 - y^2$
88. $(x^2 + y^2)^2 = 9(x^2 - y^2)$
89. $y^3 = x^2$
90. $y^2 = x^3$

Converting a Polar Equation to Rectangular Form   In Exercises 91–116, convert the polar equation to rectangular form.

91. $r = 4 \sin \theta$ 92. $r = 2 \cos \theta$

93. $r = -2 \cos \theta$ 94. $r = -5 \sin \theta$

95. $\theta = 2\pi/3$ 96. $\theta = 5\pi/3$

97. $\theta = 11\pi/6$ 98. $\theta = 5\pi/6$

99. $r = 4$ 100. $r = 10$

101. $r = 4 \csc \theta$ 102. $r = 2 \csc \theta$

103. $r = -3 \sec \theta$ 104. $r = -\sec \theta$

105. $r^2 = \cos \theta$ 106. $r^2 = 2 \sin \theta$

107. $r^2 = \sin 2\theta$ 108. $r^2 = \cos 2\theta$

109. $r = 2 \sin 3\theta$ 110. $r = 3 \cos 2\theta$

111. $r = \dfrac{2}{1 + \sin \theta}$ 112. $r = \dfrac{1}{1 - \cos \theta}$

113. $r = \dfrac{6}{2 - 3 \sin \theta}$ 114. $r = \dfrac{5}{1 - 4 \cos \theta}$

115. $r = \dfrac{6}{2 \cos \theta - 3 \sin \theta}$ 116. $r = \dfrac{5}{\sin \theta - 4 \cos \theta}$

Converting a Polar Equation to Rectangular Form   In Exercises 117–126, convert the polar equation to rectangular form. Then sketch its graph.

117. $r = 6$ 118. $r = 8$

119. $\theta = \pi/6$ 120. $\theta = 3\pi/4$

121. $r = 2 \sin \theta$ 122. $r = 4 \cos \theta$

123. $r = -6 \cos \theta$ 124. $r = -3 \sin \theta$

125. $r = 3 \sec \theta$ 126. $r = 2 \csc \theta$

127. Ferris Wheel • • • • • • • • • • • • • • • • • •

The center of a Ferris wheel lies at the pole of the polar coordinate system, where the distances are in feet. Passengers enter a car at $(30, -\pi/2)$. It takes 45 seconds for the wheel to complete one clockwise revolution.

(a) Write a polar equation that models the possible positions of a passenger car.

(b) Passengers enter a car. Find and interpret their coordinates after 15 seconds of rotation.

(c) Convert the point in part (b) to rectangular coordinates. Interpret the coordinates.

128. Ferris Wheel   Repeat Exercise 127 when the distance from a passenger car to the center is 35 feet and it takes 60 seconds to complete one clockwise revolution.

Exploration

True or False?   In Exercises 129–132, determine whether the statement is true or false. Justify your answer.

129. If $\theta_1 = \theta_2 + 2\pi n$ for some integer n, then (r, θ_1) and (r, θ_2) represent the same point in the polar coordinate system.

130. If (r_1, θ_1) and (r_2, θ_2) represent the same point in the polar coordinate system, then $\theta_1 = \theta_2 + 2\pi n$ for some integer n.

131. If $|r_1| = |r_2|$, then (r_1, θ) and (r_2, θ) represent the same point in the polar coordinate system.

132. If (r_1, θ_1) and (r_2, θ_2) represent the same point in the polar coordinate system, then $|r_1| = |r_2|$.

133. **Converting a Polar Equation to Rectangular Form**   Convert the polar equation

$$r = 2(h \cos \theta + k \sin \theta)$$

to rectangular form and verify that it is the equation of a circle. Find the radius of the circle and the rectangular coordinates of the center of the circle.

134. **Converting a Polar Equation to Rectangular Form**   Convert the polar equation $r = \cos \theta + 3 \sin \theta$ to rectangular form and identify the graph.

135. **Think About It**

(a) Show that the distance between the points (r_1, θ_1) and (r_2, θ_2) is $\sqrt{r_1^2 + r_2^2 - 2r_1 r_2 \cos(\theta_1 - \theta_2)}$.

(b) Simplify the Distance Formula for $\theta_1 = \theta_2$. Is the simplification what you expected? Explain.

(c) Simplify the Distance Formula for $\theta_1 - \theta_2 = 90°$. Is the simplification what you expected? Explain.

136. **HOW DO YOU SEE IT?**   Use the polar coordinate system shown below.

(a) Identify the polar coordinates of the points.

(b) Which points lie on the graph of $r = 3$?

(c) Which points lie on the graph of $\theta = \pi/4$?

6.8 Graphs of Polar Equations

■ Graph polar equations by point plotting.
■ Use symmetry, zeros, and maximum r-values to sketch graphs of polar equations.
■ Recognize special polar graphs.

Introduction

In previous chapters, you learned how to sketch graphs in rectangular coordinate systems. You began with the basic point-plotting method. Then you used sketching aids such as symmetry, intercepts, asymptotes, periods, and shifts to further investigate the natures of graphs. This section approaches curve sketching in the polar coordinate system similarly, beginning with a demonstration of point plotting.

You can use graphs of polar equations in mathematical modeling. For instance, in Exercise 69 on page 480, you will graph the pickup pattern of a microphone in a polar coordinate system.

EXAMPLE 1 **Graphing a Polar Equation by Point Plotting**

Sketch the graph of the polar equation $r = 4 \sin \theta$.

Solution The sine function is periodic, so you can get a full range of r-values by considering values of θ in the interval $0 \le \theta \le 2\pi$, as shown in the following table.

θ	0	$\dfrac{\pi}{6}$	$\dfrac{\pi}{3}$	$\dfrac{\pi}{2}$	$\dfrac{2\pi}{3}$	$\dfrac{5\pi}{6}$	π	$\dfrac{7\pi}{6}$	$\dfrac{3\pi}{2}$	$\dfrac{11\pi}{6}$	2π
r	0	2	$2\sqrt{3}$	4	$2\sqrt{3}$	2	0	-2	-4	-2	0

By plotting these points, as shown in Figure 6.54, it appears that the graph is a circle of radius 2 whose center is at the point $(x, y) = (0, 2)$.

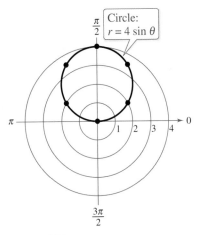

Figure 6.54

✓ **Checkpoint** ◀))) *Audio-video solution in English & Spanish at LarsonPrecalculus.com.*

Sketch the graph of the polar equation $r = 6 \cos \theta$.

You can confirm the graph in Figure 6.54 by converting the polar equation to rectangular form and then sketching the graph of the rectangular equation. You can also use a graphing utility set to *polar* mode and graph the polar equation or set the graphing utility to *parametric* mode and graph a parametric representation.

Symmetry, Zeros, and Maximum *r*-Values

In Figure 6.54 on the preceding page, note that as θ increases from 0 to 2π the graph is traced out twice. Moreover, note that the graph is *symmetric with respect to the line* $\theta = \pi/2$. Had you known about this symmetry and retracing ahead of time, you could have used fewer points. The three important types of symmetry to consider in polar curve sketching are shown below.

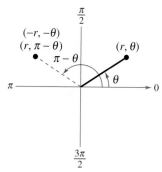

Symmetry with Respect to the Line $\theta = \dfrac{\pi}{2}$

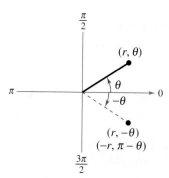

Symmetry with Respect to the Polar Axis

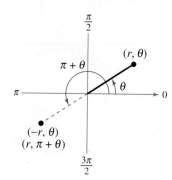

Symmetry with Respect to the Pole

Tests for Symmetry in Polar Coordinates

The graph of a polar equation is symmetric with respect to the following when the given substitution yields an equivalent equation.

1. **The line $\theta = \pi/2$:**   Replace (r, θ) by $(r, \pi - \theta)$ or $(-r, -\theta)$.

2. **The polar axis:**   Replace (r, θ) by $(r, -\theta)$ or $(-r, \pi - \theta)$.

3. **The pole:**   Replace (r, θ) by $(r, \pi + \theta)$ or $(-r, \theta)$.

EXAMPLE 2   **Using Symmetry to Sketch a Polar Graph**

Use symmetry to sketch the graph of $r = 3 + 2 \cos \theta$.

Solution   Replacing (r, θ) by $(r, -\theta)$ produces

$$r = 3 + 2\cos(-\theta) = 3 + 2\cos\theta. \qquad \cos(-\theta) = \cos\theta$$

So, you can conclude that the curve is symmetric with respect to the polar axis. Plotting the points in the table and using polar axis symmetry, you obtain the graph shown in Figure 6.55. This graph is called a **limaçon.**

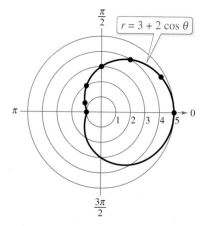

Figure 6.55

θ	0	$\dfrac{\pi}{6}$	$\dfrac{\pi}{3}$	$\dfrac{\pi}{2}$	$\dfrac{2\pi}{3}$	$\dfrac{5\pi}{6}$	π
r	5	$3 + \sqrt{3}$	4	3	2	$3 - \sqrt{3}$	1

✓ *Checkpoint* ◀))) *Audio-video solution in English & Spanish at LarsonPrecalculus.com.*

Use symmetry to sketch the graph of $r = 3 + 2 \sin \theta$.

Note in Example 2 that $\cos(-\theta) = \cos\theta$. This is because the cosine function is *even*. Recall from Section 1.2 that the cosine function is even and the sine function is odd. That is, $\sin(-\theta) = -\sin\theta$.

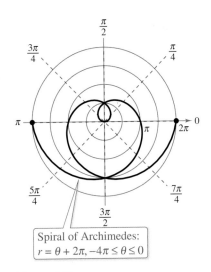

Spiral of Archimedes:
$r = \theta + 2\pi, -4\pi \le \theta \le 0$

Figure 6.56

The three tests for symmetry in polar coordinates listed on page 474 are sufficient to guarantee symmetry, but they are not necessary. For instance, Figure 6.56 shows the graph of

$$r = \theta + 2\pi \qquad \text{Spiral of Archimedes}$$

to be symmetric with respect to the line $\theta = \pi/2$, and yet the tests on page 474 fail to indicate symmetry because neither of the following replacements yields an equivalent equation.

Original Equation	**Replacement**	**New Equation**
$r = \theta + 2\pi$	(r, θ) by $(-r, -\theta)$	$-r = -\theta + 2\pi$
$r = \theta + 2\pi$	(r, θ) by $(r, \pi - \theta)$	$r = -\theta + 3\pi$

The equations discussed in Examples 1 and 2 are of the form

$$r = 4 \sin \theta = f(\sin \theta) \quad \text{and} \quad r = 3 + 2 \cos \theta = g(\cos \theta).$$

The graph of the first equation is symmetric with respect to the line $\theta = \pi/2$, and the graph of the second equation is symmetric with respect to the polar axis. This observation can be generalized to yield the following tests.

Quick Tests for Symmetry in Polar Coordinates

1. The graph of $r = f(\sin \theta)$ is symmetric with respect to the line $\theta = \dfrac{\pi}{2}$.

2. The graph of $r = g(\cos \theta)$ is symmetric with respect to the polar axis.

Two additional aids to sketching graphs of polar equations involve knowing the θ-values for which $|r|$ is maximum and knowing the θ-values for which $r = 0$. For instance, in Example 1, the maximum value of $|r|$ for $r = 4 \sin \theta$ is $|r| = 4$, and this occurs when $\theta = \pi/2$, as shown in Figure 6.54. Moreover, $r = 0$ when $\theta = 0$.

EXAMPLE 3   **Sketching a Polar Graph**

Sketch the graph of $r = 1 - 2 \cos \theta$.

Solution   From the equation $r = 1 - 2 \cos \theta$, you can obtain the following.

Symmetry: With respect to the polar axis

Maximum value of $|r|$: $r = 3$ when $\theta = \pi$

Zero of r: $r = 0$ when $\theta = \pi/3$

The table shows several θ-values in the interval $[0, \pi]$. By plotting the corresponding points, you can sketch the graph shown in Figure 6.57.

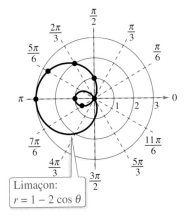

Limaçon:
$r = 1 - 2 \cos \theta$

Figure 6.57

θ	0	$\dfrac{\pi}{6}$	$\dfrac{\pi}{3}$	$\dfrac{\pi}{2}$	$\dfrac{2\pi}{3}$	$\dfrac{5\pi}{6}$	π
r	-1	$1 - \sqrt{3}$	0	1	2	$1 + \sqrt{3}$	3

Note how the negative r-values determine the *inner loop* of the graph in Figure 6.57. This graph, like the one in Figure 6.55, is a limaçon.

✓ **Checkpoint** ◀))) *Audio-video solution in English & Spanish at LarsonPrecalculus.com.*

Sketch the graph of $r = 1 + 2 \sin \theta$. ∎

Some curves reach their zeros and maximum r-values at more than one point, as shown in Example 4.

> **EXAMPLE 4** **Sketching a Polar Graph**

Sketch the graph of

$$r = 2 \cos 3\theta.$$

Solution

Symmetry: With respect to the polar axis

Maximum value of $|r|$: $|r| = 2$ when $3\theta = 0,\ \pi,\ 2\pi,\ 3\pi$ or $\theta = 0,\ \dfrac{\pi}{3},\ \dfrac{2\pi}{3},\ \pi$

Zeros of r: $r = 0$ when $3\theta = \dfrac{\pi}{2},\ \dfrac{3\pi}{2},\ \dfrac{5\pi}{2}$ or $\theta = \dfrac{\pi}{6},\ \dfrac{\pi}{2},\ \dfrac{5\pi}{6}$

θ	0	$\dfrac{\pi}{12}$	$\dfrac{\pi}{6}$	$\dfrac{\pi}{4}$	$\dfrac{\pi}{3}$	$\dfrac{5\pi}{12}$	$\dfrac{\pi}{2}$
r	2	$\sqrt{2}$	0	$-\sqrt{2}$	-2	$-\sqrt{2}$	0

By plotting these points and using the specified symmetry, zeros, and maximum values, you can obtain the graph, as shown below. This graph is called a **rose curve,** and each loop on the graph is called a *petal*. Note how the entire curve is generated as θ increases from 0 to π.

$0 \le \theta \le \dfrac{\pi}{6}$

$0 \le \theta \le \dfrac{\pi}{3}$

$0 \le \theta \le \dfrac{\pi}{2}$

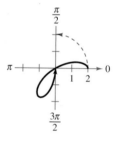

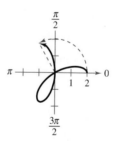

$0 \le \theta \le \dfrac{2\pi}{3}$

$0 \le \theta \le \dfrac{5\pi}{6}$

$0 \le \theta \le \pi$

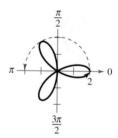

▷ **TECHNOLOGY** Use a graphing utility in *polar* mode to verify the graph of $r = 2 \cos 3\theta$ shown in Example 4.

✓ *Checkpoint* ◀))) Audio-video solution in English & Spanish at LarsonPrecalculus.com.

Sketch the graph of

$$r = 2 \sin 3\theta.$$

Special Polar Graphs

Several important types of graphs have equations that are simpler in polar form than in rectangular form. For example, the circle

$$r = 4 \sin \theta$$

in Example 1 has the more complicated rectangular equation

$$x^2 + (y - 2)^2 = 4.$$

Several other types of graphs that have simple polar equations are shown below.

Limaçons

$$r = a \pm b \cos \theta, \ r = a \pm b \sin \theta \quad (a > 0, b > 0)$$

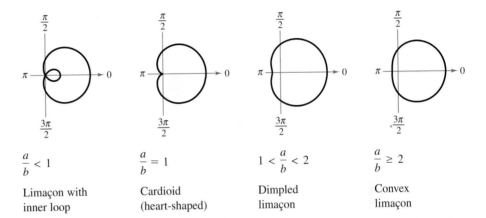

$\dfrac{a}{b} < 1$	$\dfrac{a}{b} = 1$	$1 < \dfrac{a}{b} < 2$	$\dfrac{a}{b} \geq 2$
Limaçon with inner loop	Cardioid (heart-shaped)	Dimpled limaçon	Convex limaçon

Rose Curves

n petals when n is odd, $2n$ petals when n is even $\quad (n \geq 2)$

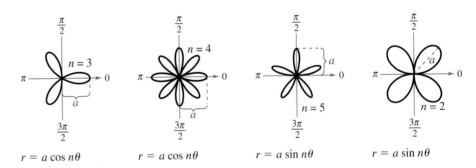

$r = a \cos n\theta$	$r = a \cos n\theta$	$r = a \sin n\theta$	$r = a \sin n\theta$

Circles and Lemniscates

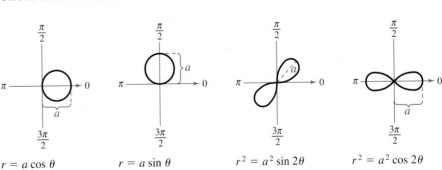

$r = a \cos \theta$	$r = a \sin \theta$	$r^2 = a^2 \sin 2\theta$	$r^2 = a^2 \cos 2\theta$
Circle	Circle	Lemniscate	Lemniscate

θ	r
0	3
$\dfrac{\pi}{6}$	$\dfrac{3}{2}$
$\dfrac{\pi}{4}$	0
$\dfrac{\pi}{3}$	$-\dfrac{3}{2}$

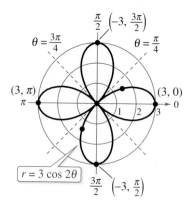

Figure 6.58

θ	$r = \pm 3\sqrt{\sin 2\theta}$
0	0
$\dfrac{\pi}{12}$	$\pm\dfrac{3}{\sqrt{2}}$
$\dfrac{\pi}{4}$	± 3
$\dfrac{5\pi}{12}$	$\pm\dfrac{3}{\sqrt{2}}$
$\dfrac{\pi}{2}$	0

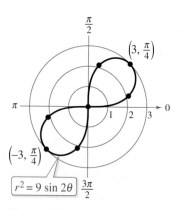

Figure 6.59

EXAMPLE 5　Sketching a Rose Curve

Sketch the graph of $r = 3 \cos 2\theta$.

Solution

Type of curve:　　　Rose curve with $2n = 4$ petals

Symmetry:　　　With respect to the polar axis, the line $\theta = \pi/2$, and the pole

Maximum value of $|r|$:　$|r| = 3$ when $\theta = 0, \pi/2, \pi, 3\pi/2$

Zeros of r:　　　$r = 0$ when $\theta = \pi/4, 3\pi/4$

Using this information together with the additional points shown in the table at the left, you obtain the graph shown in Figure 6.58.

✓ **Checkpoint** 🔊))) *Audio-video solution in English & Spanish at LarsonPrecalculus.com.*

Sketch the graph of $r = 3 \cos 3\theta$.

EXAMPLE 6　Sketching a Lemniscate

Sketch the graph of $r^2 = 9 \sin 2\theta$.

Solution

Type of curve:　　Lemniscate

Symmetry:　　With respect to the pole

Maximum value of $|r|$:　$|r| = 3$ when $\theta = \pi/4$

Zeros of r:　　$r = 0$ when $\theta = 0, \pi/2$

When $\sin 2\theta < 0$, this equation has no solution points. So, you restrict the values of θ to those for which $\sin 2\theta \geq 0$.

$$0 \leq \theta \leq \frac{\pi}{2} \quad \text{or} \quad \pi \leq \theta \leq \frac{3\pi}{2}$$

Using symmetry, you need to consider only the first of these two intervals. By finding a few additional points (see table at the left), you obtain the graph shown in Figure 6.59.

✓ **Checkpoint** 🔊))) *Audio-video solution in English & Spanish at LarsonPrecalculus.com.*

Sketch the graph of $r^2 = 4 \cos 2\theta$.　　■

Summarize　(Section 6.8)

1. Describe how to graph polar equations by point plotting *(page 473)*. For an example of graphing a polar equation by point plotting, see Example 1.

2. State the tests for symmetry in polar coordinates *(page 474)*. For an example of using symmetry to sketch the graph of a polar equation, see Example 2.

3. Describe how to use zeros and maximum r-values to sketch graphs of polar equations *(pages 475 and 476)*. For examples of using zeros and maximum r-values to sketch graphs of polar equations, see Examples 3 and 4.

4. State and give an example of a special polar graph covered in this lesson *(page 477)*. For examples of sketching special polar graphs, see Examples 5 and 6.

6.8 Exercises

See **CalcChat.com** for tutorial help and worked-out solutions to odd-numbered exercises.

Vocabulary: Fill in the blanks.

1. The graph of $r = f(\sin \theta)$ is symmetric with respect to the line _____.
2. The graph of $r = g(\cos \theta)$ is symmetric with respect to the _____ _____.
3. The equation $r = 2 + \cos \theta$ represents a _____ _____.
4. The equation $r = 2 \cos \theta$ represents a _____.
5. The equation $r^2 = 4 \sin 2\theta$ represents a _____.
6. The equation $r = 1 + \sin \theta$ represents a _____.

Skills and Applications

Identifying Types of Polar Graphs In Exercises 7–12, identify the type of polar graph.

7.

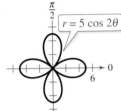

$r = 5 \cos 2\theta$

8.

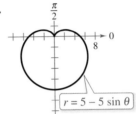

$r = 5 - 5 \sin \theta$

9.

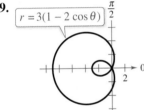

$r = 3(1 - 2 \cos \theta)$

10.

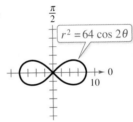

$r^2 = 64 \cos 2\theta$

11.

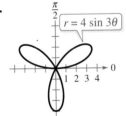

$r = 4 \sin 3\theta$

12.

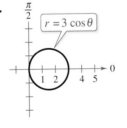

$r = 3 \cos \theta$

Testing for Symmetry In Exercises 13–18, test for symmetry with respect to the line $\theta = \pi/2$, the polar axis, and the pole.

13. $r = 5 + 4 \cos \theta$
14. $r = 9 \cos 3\theta$
15. $r = \dfrac{2}{1 + \sin \theta}$
16. $r = \dfrac{3}{2 + \cos \theta}$
17. $r^2 = 36 \cos 2\theta$
18. $r^2 = 25 \sin 2\theta$

Finding the Maximum Value of $|r|$ and Zeros of r In Exercises 19–22, find the maximum value of $|r|$ and any zeros of r.

19. $r = 10 - 10 \sin \theta$
20. $r = 6 + 12 \cos \theta$
21. $r = 4 \cos 3\theta$
22. $r = 3 \sin 2\theta$

Sketching the Graph of a Polar Equation In Exercises 23–48, sketch the graph of the polar equation using symmetry, zeros, maximum r-values, and any other additional points.

23. $r = 4$
24. $r = -7$
25. $r = \pi/3$
26. $r = -3\pi/4$
27. $r = \sin \theta$
28. $r = 4 \cos \theta$
29. $r = 3(1 - \cos \theta)$
30. $r = 4(1 - \sin \theta)$
31. $r = 4(1 + \sin \theta)$
32. $r = 2(1 + \cos \theta)$
33. $r = 4 + 3 \cos \theta$
34. $r = 4 - 3 \sin \theta$
35. $r = 1 - 2 \sin \theta$
36. $r = 2 - 4 \cos \theta$
37. $r = 3 - 4 \cos \theta$
38. $r = 3 + 6 \sin \theta$
39. $r = 5 \sin 2\theta$
40. $r = 2 \cos 2\theta$
41. $r = 6 \cos 3\theta$
42. $r = 3 \sin 3\theta$
43. $r = 2 \sec \theta$
44. $r = 5 \csc \theta$
45. $r = \dfrac{3}{\sin \theta - 2 \cos \theta}$
46. $r = \dfrac{6}{2 \sin \theta - 3 \cos \theta}$
47. $r^2 = 9 \cos 2\theta$
48. $r^2 = 4 \sin \theta$

Graphing a Polar Equation In Exercises 49–58, use a graphing utility to graph the polar equation. Describe your viewing window.

49. $r = 9/4$
50. $r = -5/2$
51. $r = 5\pi/8$
52. $r = -\pi/10$
53. $r = 8 \cos \theta$
54. $r = \cos 2\theta$
55. $r = 3(2 - \sin \theta)$
56. $r = 2 \cos(3\theta - 2)$
57. $r = 8 \sin \theta \cos^2 \theta$
58. $r = 2 \csc \theta + 5$

Finding an Interval In Exercises 59–64, use a graphing utility to graph the polar equation. Find an interval for θ for which the graph is traced *only once*.

59. $r = 3 - 8 \cos \theta$
60. $r = 5 + 4 \cos \theta$
61. $r = 2 \cos\left(\dfrac{3\theta}{2}\right)$
62. $r = 3 \sin\left(\dfrac{5\theta}{2}\right)$
63. $r^2 = 16 \sin 2\theta$
64. $r^2 = 1/\theta$

Asymptote of a Graph of a Polar Equation In Exercises 65–68, use a graphing utility to graph the polar equation and show that the indicated line is an asymptote of the graph.

Name of Graph	Polar Equation	Asymptote
65. Conchoid	$r = 2 - \sec \theta$	$x = -1$
66. Conchoid	$r = 2 + \csc \theta$	$y = 1$
67. Hyperbolic spiral	$r = \dfrac{3}{\theta}$	$y = 3$
68. Strophoid	$r = 2 \cos 2\theta \sec \theta$	$x = -2$

69. Microphone

The sound pickup pattern of a microphone is modeled by the polar equation

$r = 5 + 5 \cos \theta$

where $|r|$ measures how sensitive the microphone is to sounds coming from the angle θ.

(a) Sketch the graph of the model and identify the type of polar graph.

(b) At what angle is the microphone most sensitive to sound?

Exploration

70. **Area** The area of the lemniscate $r^2 = a^2 \cos 2\theta$ is a^2.

(a) Sketch the graph of $r^2 = 16 \cos 2\theta$.

(b) Find the area of one loop of the graph from part (a).

True or False? In Exercises 71 and 72, determine whether the statement is true or false. Justify your answer.

71. The graph of $r = 10 \sin 5\theta$ is a rose curve with five petals.

72. A rose curve will always have symmetry with respect to the line $\theta = \pi/2$.

73. **Sketching the Graph of a Polar Equation** Sketch the graph of $r = 6 \cos \theta$ over each interval. Describe the part of the graph obtained in each case.

(a) $0 \le \theta \le \dfrac{\pi}{2}$ (b) $\dfrac{\pi}{2} \le \theta \le \pi$

(c) $-\dfrac{\pi}{2} \le \theta \le \dfrac{\pi}{2}$ (d) $\dfrac{\pi}{4} \le \theta \le \dfrac{3\pi}{4}$

74. **Graphical Reasoning** Use a graphing utility to graph the polar equation $r = 6[1 + \cos(\theta - \phi)]$ for (a) $\phi = 0$, (b) $\phi = \pi/4$, and (c) $\phi = \pi/2$. Use the graphs to describe the effect of the angle ϕ. Write the equation as a function of $\sin \theta$ for part (c).

75. **Rotation Through an Angle** The graph of $r = f(\theta)$ is rotated about the pole through an angle ϕ. Show that the equation of the rotated graph is $r = f(\theta - \phi)$.

76. **Rotation Through an Angle** Consider the graph of $r = f(\sin \theta)$.

(a) Show that when the graph is rotated counterclockwise $\pi/2$ radians about the pole, the equation of the rotated graph is $r = f(-\cos \theta)$.

(b) Show that when the graph is rotated counterclockwise π radians about the pole, the equation of the rotated graph is $r = f(-\sin \theta)$.

(c) Show that when the graph is rotated counterclockwise $3\pi/2$ radians about the pole, the equation of the rotated graph is $r = f(\cos \theta)$.

Rotation Through an Angle In Exercises 77 and 78, use the results of Exercises 75 and 76.

77. Write an equation for the limaçon $r = 2 - \sin \theta$ after it has been rotated through the given angle.

(a) $\dfrac{\pi}{4}$ (b) $\dfrac{\pi}{2}$ (c) π (d) $\dfrac{3\pi}{2}$

78. Write an equation for the rose curve $r = 2 \sin 2\theta$ after it has been rotated through the given angle.

(a) $\dfrac{\pi}{6}$ (b) $\dfrac{\pi}{2}$ (c) $\dfrac{2\pi}{3}$ (d) π

79. **Graphing a Polar Equation** Consider the equation $r = 3 \sin k\theta$.

(a) Use a graphing utility to graph the equation for $k = 1.5$. Find the interval for θ over which the graph is traced only once.

(b) Use the graphing utility to graph the equation for $k = 2.5$. Find the interval for θ over which the graph is traced only once.

(c) Is it possible to find an interval for θ over which the graph is traced only once for any rational number k? Explain.

80. **HOW DO YOU SEE IT?** Determine which graph matches each polar equation.

(a) $r = 5 \sin \theta$

(b) $r = 2 + 5 \sin \theta$

(c) $r = 5 \cos 2\theta$

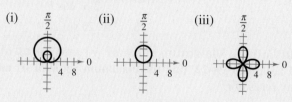

6.9   Polar Equations of Conics

■ Define conics in terms of eccentricity, and write and graph equations of conics in polar form.
■ Use equations of conics in polar form to model real-life problems.

Alternative Definition and Polar Equations of Conics

In Sections 6.3 and 6.4, you learned that the rectangular equations of ellipses and hyperbolas take simple forms when the origin lies at their *centers*. As it happens, there are many important applications of conics in which it is more convenient to use one of the *foci* as the origin. In this section, you will learn that polar equations of conics take simple forms when one of the foci lies at the pole.

To begin, consider the following alternative definition of a conic that uses the concept of eccentricity.

You can model the orbits of planets and satellites with polar equations. For instance, in Exercise 64 on page 486, you will use a polar equation to model the orbit of a satellite.

Alternative Definition of a Conic

The locus of a point in the plane that moves such that its distance from a fixed point (focus) is in a constant ratio to its distance from a fixed line (directrix) is a **conic**. The constant ratio is the eccentricity of the conic and is denoted by e. Moreover, the conic is an **ellipse** when $0 < e < 1$, a **parabola** when $e = 1$, and a **hyperbola** when $e > 1$. (See the figures below.)

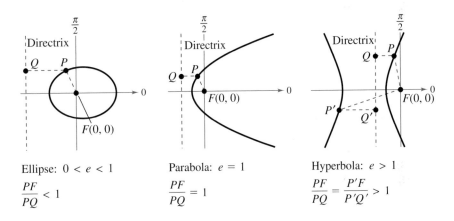

Ellipse: $0 < e < 1$
$$\frac{PF}{PQ} < 1$$

Parabola: $e = 1$
$$\frac{PF}{PQ} = 1$$

Hyperbola: $e > 1$
$$\frac{PF}{PQ} = \frac{P'F}{P'Q'} > 1$$

In the figures, note that for each type of conic, the focus is at the pole. The benefit of locating a focus of a conic at the pole is that the equation of the conic takes on a simpler form. For a proof of the polar equations of conics, see Proofs in Mathematics on page 498.

Polar Equations of Conics

The graph of a polar equation of the form

1. $r = \dfrac{ep}{1 \pm e \cos \theta}$ or   **2.** $r = \dfrac{ep}{1 \pm e \sin \theta}$

is a conic, where $e > 0$ is the eccentricity and $|p|$ is the distance between the focus (pole) and the directrix.

An equation of the form

$$r = \frac{ep}{1 \pm e \cos \theta}$$ Vertical directrix

corresponds to a conic with a vertical directrix and symmetry with respect to the polar axis. An equation of the form

$$r = \frac{ep}{1 \pm e \sin \theta}$$ Horizontal directrix

corresponds to a conic with a horizontal directrix and symmetry with respect to the line $\theta = \pi/2$. Moreover, the converse is also true—that is, any conic with a focus at the pole and having a horizontal or vertical directrix can be represented by one of these equations.

EXAMPLE 1   Identifying a Conic from Its Equation

Identify the type of conic represented by the equation $r = \dfrac{15}{3 - 2 \cos \theta}$.

Algebraic Solution

To identify the type of conic, rewrite the equation in the form

$$r = \frac{ep}{1 \pm e \cos \theta}.$$

$$r = \frac{15}{3 - 2 \cos \theta}$$ Write original equation.

$$= \frac{5}{1 - (2/3) \cos \theta}$$ Divide numerator and denominator by 3.

Because $e = \frac{2}{3} < 1$, you can conclude that the graph is an ellipse.

Graphical Solution

Use a graphing utility in *polar* mode and be sure to use a square setting, as shown below.

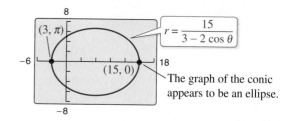

The graph of the conic appears to be an ellipse.

✓ *Checkpoint* ◀))) Audio-video solution in English & Spanish at LarsonPrecalculus.com.

Identify the type of conic represented by the equation $r = \dfrac{8}{2 - 3 \sin \theta}$.   ▪

For the ellipse in Example 1, the major axis is horizontal and the vertices lie at $(15, 0)$ and $(3, \pi)$. So, the length of the *major* axis is $2a = 18$. To find the length of the *minor* axis, you can use the equations $e = c/a$ and $b^2 = a^2 - c^2$ to conclude that

$$b^2 = a^2 - c^2$$

$$= a^2 - (ea)^2$$

$$= a^2(1 - e^2).$$ Ellipse

Because $e = \frac{2}{3}$, you have

$$b^2 = 9^2 \left[1 - \left(\frac{2}{3} \right)^2 \right] = 45$$

which implies that $b = \sqrt{45} = 3\sqrt{5}$. So, the length of the minor axis is $2b = 6\sqrt{5}$. A similar analysis for hyperbolas yields

$$b^2 = c^2 - a^2$$

$$= (ea)^2 - a^2$$

$$= a^2(e^2 - 1).$$ Hyperbola

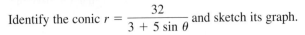

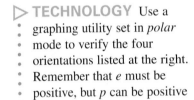

Figure 6.60

EXAMPLE 2 **Sketching a Conic from Its Polar Equation**

Identify the conic $r = \dfrac{32}{3 + 5 \sin \theta}$ and sketch its graph.

Solution Dividing the numerator and denominator by 3, you have

$$r = \frac{32/3}{1 + (5/3) \sin \theta}.$$

Because $e = \frac{5}{3} > 1$, the graph is a hyperbola. The transverse axis of the hyperbola lies on the line $\theta = \pi/2$, and the vertices occur at $(4, \pi/2)$ and $(-16, 3\pi/2)$. Because the length of the transverse axis is 12, you can see that $a = 6$. To find b, write

$$b^2 = a^2(e^2 - 1) = 6^2\left[\left(\tfrac{5}{3}\right)^2 - 1\right] = 64.$$

So, $b = 8$. You can use a and b to determine that the asymptotes of the hyperbola are $y = 10 \pm \frac{3}{4}x$. The graph is shown in Figure 6.60.

✓ **Checkpoint** ◀))) *Audio-video solution in English & Spanish at LarsonPrecalculus.com.*

Identify the conic $r = \dfrac{3}{2 - 4 \sin \theta}$ and sketch its graph. ■

In the next example, you are asked to find a polar equation of a specified conic. To do this, let p be the distance between the pole and the directrix.

1. Horizontal directrix above the pole: $r = \dfrac{ep}{1 + e \sin \theta}$

2. Horizontal directrix below the pole: $r = \dfrac{ep}{1 - e \sin \theta}$

3. Vertical directrix to the right of the pole: $r = \dfrac{ep}{1 + e \cos \theta}$

4. Vertical directrix to the left of the pole: $r = \dfrac{ep}{1 - e \cos \theta}$

EXAMPLE 3 **Finding the Polar Equation of a Conic**

Find a polar equation of the parabola whose focus is the pole and whose directrix is the line $y = 3$.

Solution Because the directrix is horizontal and above the pole, use an equation of the form

$$r = \frac{ep}{1 + e \sin \theta}.$$

Moreover, because the eccentricity of a parabola is $e = 1$ and the distance between the pole and the directrix is $p = 3$, you have the equation

$$r = \frac{3}{1 + \sin \theta}.$$

The parabola is shown in Figure 6.61.

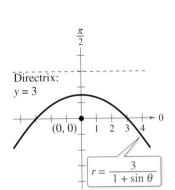

Figure 6.61

✓ **Checkpoint** ◀))) *Audio-video solution in English & Spanish at LarsonPrecalculus.com.*

Find a polar equation of the parabola whose focus is the pole and whose directrix is the line $x = -2$. ■

Application

Kepler's Laws (listed below), named after the German astronomer Johannes Kepler (1571–1630), can be used to describe the orbits of the planets about the sun.

1. Each planet moves in an elliptical orbit with the sun at one focus.

2. A ray from the sun to a planet sweeps out equal areas in equal times.

3. The square of the period (the time it takes for a planet to orbit the sun) is proportional to the cube of the mean distance between the planet and the sun.

Although Kepler stated these laws on the basis of observation, they were later validated by Isaac Newton (1642–1727). In fact, Newton showed that these laws apply to the orbits of all heavenly bodies, including comets and satellites. This is illustrated in the next example, which involves the comet named after the English mathematician and physicist Edmund Halley (1656–1742).

If you use Earth as a reference with a period of 1 year and a distance of 1 astronomical unit (about 93 million miles), then the proportionality constant in Kepler's third law is 1. For example, because Mars has a mean distance to the sun of $d \approx 1.524$ astronomical units, its period P is given by $d^3 = P^2$. So, the period of Mars is $P \approx 1.88$ years.

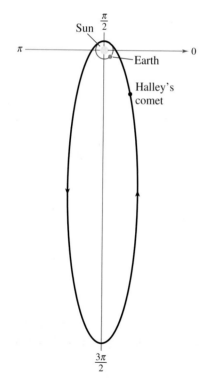

Figure 6.62

EXAMPLE 4 Halley's Comet

Halley's comet has an elliptical orbit with an eccentricity of $e \approx 0.967$. The length of the major axis of the orbit is approximately 35.88 astronomical units. Find a polar equation for the orbit. How close does Halley's comet come to the sun?

Solution Using a vertical major axis, as shown in Figure 6.62, choose an equation of the form $r = ep/(1 + e \sin \theta)$. Because the vertices of the ellipse occur when $\theta = \pi/2$ and $\theta = 3\pi/2$, you can determine the length of the major axis to be the sum of the r-values of the vertices. That is,

$$2a = \frac{0.967p}{1 + 0.967} + \frac{0.967p}{1 - 0.967} \approx 29.79p \approx 35.88.$$

So, $p \approx 1.204$ and $ep \approx (0.967)(1.204) \approx 1.164$. Using this value of ep in the equation, you have

$$r = \frac{1.164}{1 + 0.967 \sin \theta}$$

where r is measured in astronomical units. To find the closest point to the sun (the focus), substitute $\theta = \pi/2$ into this equation to obtain

$$r = \frac{1.164}{1 + 0.967 \sin(\pi/2)} \approx 0.59 \text{ astronomical unit} \approx 55,000,000 \text{ miles}.$$

✓ **Checkpoint**  Audio-video solution in English & Spanish at LarsonPrecalculus.com.

Encke's comet has an elliptical orbit with an eccentricity of $e \approx 0.848$. The length of the major axis of the orbit is approximately 4.429 astronomical units. Find a polar equation for the orbit. How close does Encke's comet come to the sun?

Summarize (Section 6.9)

1. State the definition of a conic in terms of eccentricity *(page 481)*. For examples of identifying and sketching conics in polar form, see Examples 1 and 2.

2. Describe a real-life problem that can be modeled by an equation of a conic in polar form *(page 484, Example 4)*.

6.9 Exercises

See CalcChat.com for tutorial help and worked-out solutions to odd-numbered exercises.

Vocabulary

In Exercises 1–3, fill in the blanks.

1. The locus of a point in the plane that moves such that its distance from a fixed point (focus) is in a constant ratio to its distance from a fixed line (directrix) is a _____.

2. The constant ratio is the _____ of the conic and is denoted by _____.

3. An equation of the form $r = \dfrac{ep}{1 + e \cos \theta}$ has a _____ directrix to the _____ of the pole.

4. Match the conic with its eccentricity.

    (a) $0 < e < 1$ (b) $e = 1$ (c) $e > 1$

    (i) parabola   (ii) hyperbola   (iii) ellipse

Skills and Applications

Identifying a Conic In Exercises 5–8, write the polar equation of the conic for $e = 1$, $e = 0.5$, and $e = 1.5$. Identify the conic for each equation. Verify your answers with a graphing utility.

5. $r = \dfrac{2e}{1 + e \cos \theta}$

6. $r = \dfrac{2e}{1 - e \cos \theta}$

7. $r = \dfrac{2e}{1 - e \sin \theta}$

8. $r = \dfrac{2e}{1 + e \sin \theta}$

Matching In Exercises 9–14, match the polar equation with its graph. [The graphs are labeled (a), (b), (c), (d), (e), and (f).]

(a)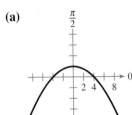

(b)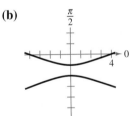

(c)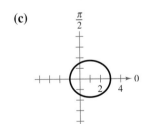

(d)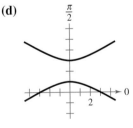

(e)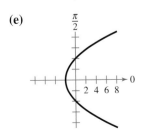

(f)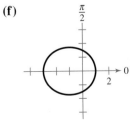

9. $r = \dfrac{4}{1 - \cos \theta}$

10. $r = \dfrac{3}{2 - \cos \theta}$

11. $r = \dfrac{3}{1 + 2 \sin \theta}$

12. $r = \dfrac{3}{2 + \cos \theta}$

13. $r = \dfrac{4}{1 + \sin \theta}$

14. $r = \dfrac{4}{1 - 3 \sin \theta}$

Sketching a Conic In Exercises 15–26, identify the conic and sketch its graph.

15. $r = \dfrac{3}{1 - \cos \theta}$

16. $r = \dfrac{7}{1 + \sin \theta}$

17. $r = \dfrac{5}{1 + \sin \theta}$

18. $r = \dfrac{6}{1 + \cos \theta}$

19. $r = \dfrac{2}{2 - \cos \theta}$

20. $r = \dfrac{4}{4 + \sin \theta}$

21. $r = \dfrac{6}{2 + \sin \theta}$

22. $r = \dfrac{9}{3 - 2 \cos \theta}$

23. $r = \dfrac{3}{2 + 4 \sin \theta}$

24. $r = \dfrac{5}{-1 + 2 \cos \theta}$

25. $r = \dfrac{3}{2 - 6 \cos \theta}$

26. $r = \dfrac{3}{2 + 6 \sin \theta}$

Graphing a Polar Equation In Exercises 27–34, use a graphing utility to graph the polar equation. Identify the graph.

27. $r = \dfrac{-1}{1 - \sin \theta}$

28. $r = \dfrac{-5}{2 + 4 \sin \theta}$

29. $r = \dfrac{3}{-4 + 2 \cos \theta}$

30. $r = \dfrac{4}{1 - 2 \cos \theta}$

31. $r = \dfrac{4}{3 - \cos \theta}$

32. $r = \dfrac{2}{2 + 3 \sin \theta}$

33. $r = \dfrac{14}{14 + 17 \sin \theta}$

34. $r = \dfrac{12}{2 - \cos \theta}$

Graphing a Rotated Conic
In Exercises 35–38, use a graphing utility to graph the rotated conic.

35. $r = \dfrac{3}{1 - \cos(\theta - \pi/4)}$ (See Exercise 15.)

36. $r = \dfrac{4}{4 + \sin(\theta - \pi/3)}$ (See Exercise 20.)

37. $r = \dfrac{6}{2 + \sin(\theta + \pi/6)}$ (See Exercise 21.)

38. $r = \dfrac{5}{-1 + 2\cos(\theta + 2\pi/3)}$ (See Exercise 24.)

Finding the Polar Equation of a Conic
In Exercises 39–54, find a polar equation of the conic with its focus at the pole.

Conic	Eccentricity	Directrix
39. Parabola	$e = 1$	$x = -1$
40. Parabola	$e = 1$	$y = -4$
41. Ellipse	$e = \frac{1}{2}$	$y = 1$
42. Ellipse	$e = \frac{3}{4}$	$y = -2$
43. Hyperbola	$e = 2$	$x = 1$
44. Hyperbola	$e = \frac{3}{2}$	$x = -1$

Conic	Vertex or Vertices
45. Parabola	$(1, -\pi/2)$
46. Parabola	$(8, 0)$
47. Parabola	$(5, \pi)$
48. Parabola	$(10, \pi/2)$
49. Ellipse	$(2, 0), (10, \pi)$
50. Ellipse	$(2, \pi/2), (4, 3\pi/2)$
51. Ellipse	$(20, 0), (4, \pi)$
52. Hyperbola	$(2, 0), (8, 0)$
53. Hyperbola	$(1, 3\pi/2), (9, 3\pi/2)$
54. Hyperbola	$(4, \pi/2), (1, \pi/2)$

55. **Planetary Motion** The planets travel in elliptical orbits with the sun at one focus. Assume that the focus is at the pole, the major axis lies on the polar axis, and the length of the major axis is $2a$ (see figure). Show that the polar equation of the orbit is $r = a(1 - e^2)/(1 - e\cos\theta)$, where e is the eccentricity.

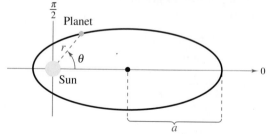

Cristi Matei/Shutterstock.com

56. **Planetary Motion** Use the result of Exercise 55 to show that the minimum distance (*perihelion*) from the sun to the planet is

$r = a(1 - e)$

and the maximum distance (*aphelion*) is

$r = a(1 + e)$.

Planetary Motion In Exercises 57–62, use the results of Exercises 55 and 56 to find (a) the polar equation of the planet's orbit and (b) the perihelion and aphelion.

57. Earth $a \approx 9.2956 \times 10^7$ miles, $e \approx 0.0167$

58. Saturn $a \approx 1.4267 \times 10^9$ kilometers, $e \approx 0.0542$

59. Venus $a \approx 1.0821 \times 10^8$ kilometers, $e \approx 0.0068$

60. Mercury $a \approx 3.5983 \times 10^7$ miles, $e \approx 0.2056$

61. Mars $a \approx 1.4163 \times 10^8$ miles, $e \approx 0.0934$

62. Jupiter $a \approx 7.7841 \times 10^8$ kilometers, $e \approx 0.0484$

63. **Astronomy** The comet Hale-Bopp has an elliptical orbit with an eccentricity of $e \approx 0.995$. The length of the major axis of the orbit is approximately 500 astronomical units. Find a polar equation for the orbit. How close does the comet come to the sun?

64. **Satellite Orbit**
A satellite in a 100-mile-high circular orbit around Earth has a velocity of approximately 17,500 miles per hour. If this velocity is multiplied by $\sqrt{2}$, then the satellite will have the minimum velocity necessary to escape Earth's gravity and will follow a parabolic path with the center of Earth as the focus (see figure).

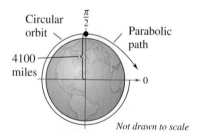

Not drawn to scale

(a) Find a polar equation of the parabolic path of the satellite (assume the radius of Earth is 4000 miles).

(b) Use a graphing utility to graph the equation you found in part (a).

(c) Find the distance between the surface of the Earth and the satellite when $\theta = 30°$.

(d) Find the distance between the surface of Earth and the satellite when $\theta = 60°$.

Exploration

True or False?   **In Exercises 65–68, determine whether the statement is true or false. Justify your answer.**

65. For values of $e > 1$ and $0 \le \theta \le 2\pi$, the graphs of the following equations are the same.

$$r = \frac{ex}{1 - e \cos \theta} \quad \text{and} \quad r = \frac{e(-x)}{1 + e \cos \theta}$$

66. The graph of $r = 4/(-3 - 3 \sin \theta)$ has a horizontal directrix above the pole.

67. The conic represented by the following equation is an ellipse.

$$r^2 = \frac{16}{9 - 4 \cos\left(\theta + \dfrac{\pi}{4}\right)}$$

68. The conic represented by the following equation is a parabola.

$$r = \frac{6}{3 - 2 \cos \theta}$$

69. Writing   Explain how the graph of each conic differs from the graph of $r = \dfrac{5}{1 + \sin \theta}$. (See Exercise 17.)

(a) $r = \dfrac{5}{1 - \cos \theta}$ (b) $r = \dfrac{5}{1 - \sin \theta}$

(c) $r = \dfrac{5}{1 + \cos \theta}$ (d) $r = \dfrac{5}{1 - \sin[\theta - (\pi/4)]}$

70.   HOW DO YOU SEE IT?   The graph of

$$r = \frac{e}{1 - e \sin \theta}$$

is shown for different values of e. Determine which graph matches each value of e.

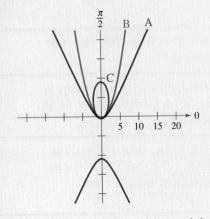

(a) $e = 0.9$ (b) $e = 1.0$ (c) $e = 1.1$

71. Verifying a Polar Equation   Show that the polar equation of the ellipse

$$\frac{x^2}{a^2} + \frac{y^2}{b^2} = 1 \quad \text{is} \quad r^2 = \frac{b^2}{1 - e^2 \cos^2 \theta}.$$

72. Verifying a Polar Equation   Show that the polar equation of the hyperbola

$$\frac{x^2}{a^2} - \frac{y^2}{b^2} = 1 \quad \text{is} \quad r^2 = \frac{-b^2}{1 - e^2 \cos^2 \theta}.$$

Writing a Polar Equation   **In Exercises 73–78, use the results of Exercises 71 and 72 to write the polar form of the equation of the conic.**

73. $\dfrac{x^2}{169} + \dfrac{y^2}{144} = 1$

74. $\dfrac{x^2}{25} + \dfrac{y^2}{16} = 1$

75. $\dfrac{x^2}{9} - \dfrac{y^2}{16} = 1$

76. $\dfrac{x^2}{36} - \dfrac{y^2}{4} = 1$

77. Hyperbola
One focus: $(5, 0)$
Vertices: $(4, 0), (4, \pi)$

78. Ellipse
One focus: $(4, 0)$
Vertices: $(5, 0), (5, \pi)$

79. Reasoning   Consider the polar equation

$$r = \frac{4}{1 - 0.4 \cos \theta}.$$

(a) Identify the conic without graphing the equation.

(b) Without graphing the following polar equations, describe how each differs from the given polar equation.

$$r_1 = \frac{4}{1 + 0.4 \cos \theta}$$

$$r_2 = \frac{4}{1 - 0.4 \sin \theta}$$

(c) Use a graphing utility to verify your results in part (b).

80. Reasoning   The equation

$$r = \frac{ep}{1 \pm e \sin \theta}$$

is the equation of an ellipse with $e < 1$. What happens to the lengths of both the major axis and the minor axis when the value of e remains fixed and the value of p changes? Use an example to explain your reasoning.

Chapter Summary

	What Did You Learn?	Explanation/Examples	Review Exercises
Section 6.1	Find the inclination of a line (*p. 414*).	If a nonvertical line has inclination θ and slope m, then $m = \tan\theta$.	1–4
	Find the angle between two lines (*p. 415*).	If two nonperpendicular lines have slopes m_1 and m_2, then the tangent of the angle between the lines is $\tan\theta = \lvert (m_2 - m_1)/(1 + m_1 m_2) \rvert$.	5–8
	Find the distance between a point and a line (*p. 416*).	The distance between the point (x_1, y_1) and the line $Ax + By + C = 0$ is $d = \lvert Ax_1 + By_1 + C \rvert / \sqrt{A^2 + B^2}$.	9, 10
Section 6.2	Recognize a conic as the intersection of a plane and a double-napped cone (*p. 421*).	In the formation of the four basic conics, the intersecting plane does not pass through the vertex of the cone. (See Figure 6.7.)	11, 12
	Write equations of parabolas in standard form (*p. 422*).	**Horizontal Axis** $(y - k)^2 = 4p(x - h),\ p \neq 0$ **Vertical Axis** $(x - h)^2 = 4p(y - k),\ p \neq 0$	13–16
	Use the reflective property of parabolas to solve real-life problems (*p. 424*).	The tangent line to a parabola at a point P makes equal angles with (1) the line passing through P and the focus and (2) the axis of the parabola.	17–20
Section 6.3	Write equations of ellipses in standard form and graph ellipses (*p. 431*).	**Horizontal Major Axis** $\dfrac{(x - h)^2}{a^2} + \dfrac{(y - k)^2}{b^2} = 1$ **Vertical Major Axis** $\dfrac{(x - h)^2}{b^2} + \dfrac{(y - k)^2}{a^2} = 1$	21–24, 27–30
	Use properties of ellipses to model and solve real-life problems (*p. 434*).	You can use the properties of ellipses to find distances from Earth's center to the moon's center in the moon's orbit. (See Example 4.)	25, 26
	Find eccentricities (*p. 435*).	The eccentricity e of an ellipse is given by $e = c/a$.	27–30
Section 6.4	Write equations of hyperbolas in standard form (*p. 440*), and find asymptotes of and graph hyperbolas (*p. 441*)	**Horizontal Transverse Axis** $\dfrac{(x - h)^2}{a^2} - \dfrac{(y - k)^2}{b^2} = 1$ **Vertical Transverse Axis** $\dfrac{(y - k)^2}{a^2} - \dfrac{(x - h)^2}{b^2} = 1$	31–38
	Use properties of hyperbolas to solve real-life problems (*p. 444*).	You can use the properties of hyperbolas in radar and other detection systems. (See Example 5.)	39, 40
	Classify conics from their general equations (*p. 445*).	The graph of $Ax^2 + Cy^2 + Dx + Ey + F = 0$ is, except in degenerate cases, a circle ($A = C$), a parabola ($AC = 0$), an ellipse ($AC > 0$), or a hyperbola ($AC < 0$).	41–44
Section 6.5	Rotate the coordinate axes to eliminate the xy-term in equations of conics (*p. 449*).	The equation $Ax^2 + Bxy + Cy^2 + Dx + Ey + F = 0$ can be rewritten as $A'(x')^2 + C'(y')^2 + D'x' + E'y' + F' = 0$ by rotating the coordinate axes through an angle θ, where $\cot 2\theta = (A - C)/B$.	45–48
	Use the discriminant to classify conics (*p. 453*).	The graph of $Ax^2 + Bxy + Cy^2 + Dx + Ey + F = 0$ is, except in degenerate cases, an ellipse or a circle $\left(B^2 - 4AC < 0\right)$, a parabola $\left(B^2 - 4AC = 0\right)$, or a hyperbola $\left(B^2 - 4AC > 0\right)$.	49–52

What Did You Learn?	**Explanation/Examples**	**Review Exercises**
Section 6.6 Evaluate sets of parametric equations for given values of the parameter *(p. 457)*.	If f and g are continuous functions of t on an interval I, then the set of ordered pairs $(f(t), g(t))$ is a plane curve C. The equations $x = f(t)$ and $y = g(t)$ are parametric equations for C, and t is the parameter.	53, 54
Sketch curves that are represented by sets of parametric equations *(p. 458)*.	Sketching a curve represented by parametric equations requires plotting points in the xy-plane. Each set of coordinates (x, y) is determined from a value chosen for t.	55–60
Rewrite sets of parametric equations as single rectangular equations by eliminating the parameter *(p. 459)*.	To eliminate the parameter in a pair of parametric equations, solve for t in one equation and substitute the value of t into the other equation. The result is the corresponding rectangular equation.	55–60
Find sets of parametric equations for graphs *(p. 461)*.	When finding a set of parametric equations for a given graph, remember that the parametric equations are not unique.	61–66
Section 6.7 Plot points in the polar coordinate system *(p. 467)*.	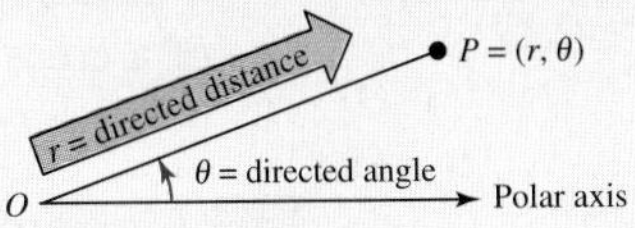	67–70
Convert points *(p. 469)* and equations *(p. 470)* from rectangular to polar form and vice versa.	**Polar Coordinates (r, θ) and Rectangular Coordinates (x, y)** Polar-to-Rectangular: $x = r\cos\theta$, $y = r\sin\theta$ Rectangular-to-Polar: $\tan\theta = y/x$, $r^2 = x^2 + y^2$	71–90
Section 6.8 Graph polar equations by point plotting *(p. 473)*.	Graphing a polar equation by point plotting is similar to graphing a rectangular equation.	91–100
Use symmetry, zeros, and maximum r-values to sketch graphs of polar equations *(p. 474)*.	The graph of a polar equation is symmetric with respect to the following when the given substitution yields an equivalent equation. **1.** Line $\theta = \pi/2$: Replace (r, θ) by $(r, \pi - \theta)$ or $(-r, -\theta)$. **2.** Polar axis: Replace (r, θ) by $(r, -\theta)$ or $(-r, \pi - \theta)$. **3.** Pole: Replace (r, θ) by $(r, \pi + \theta)$ or $(-r, \theta)$. Other aids to graphing polar equations are the θ-values for which $|r|$ is maximum and the θ-values for which $r = 0$.	91–100
Recognize special polar graphs *(p. 477)*.	Several types of graphs, such as limaçons, rose curves, circles, and lemniscates, have equations that are simpler in polar form than in rectangular form. (See page 477.)	101–104
Section 6.9 Define conics in terms of eccentricity, and write and graph equations of conics in polar form *(p. 481)*.	The eccentricity of a conic is denoted by e. **ellipse:** $0 < e < 1$ **parabola:** $e = 1$ **hyperbola:** $e > 1$ The graph of a polar equation of the form (1) $r = (ep)/(1 \pm e\cos\theta)$ or (2) $r = (ep)/(1 \pm e\sin\theta)$ is a conic, where $e > 0$ is the eccentricity and $|p|$ is the distance between the focus (pole) and the directrix.	105–112
Use equations of conics in polar form to model real-life problems *(p. 484)*.	You can use the equation of a conic in polar form to model the orbit of Halley's comet. (See Example 4.)	113, 114

Review Exercises   See CalcChat.com for tutorial help and worked-out solutions to odd-numbered exercises.

6.1 **Finding the Inclination of a Line**   **In Exercises 1–4, find the inclination θ (in radians and degrees) of the line with the given characteristics.**

1. Passes through the points $(-1, 2)$ and $(2, 5)$
2. Passes through the points $(3, 4)$ and $(-2, 7)$
3. Equation: $y = 2x + 4$
4. Equation: $x - 5y = 7$

Finding the Angle Between Two Lines   **In Exercises 5–8, find the angle θ (in radians and degrees) between the lines.**

5. $4x + y = 2$
   $-5x + y = -1$

6. $-5x + 3y = 3$
   $-2x + 3y = 1$

7. $2x - 7y = 8$
   $\frac{2}{5}x + y = 0$

8. $0.02x + 0.07y = 0.18$
   $0.09x - 0.04y = 0.17$

Finding the Distance Between a Point and a Line   **In Exercises 9 and 10, find the distance between the point and the line.**

Point	Line
9. $(5, 3)$	$x - y = 10$
10. $(0, 4)$	$x + 2y = 2$

6.2 **Forming a Conic Section**   **In Exercises 11 and 12, state what type of conic is formed by the intersection of the plane and the double-napped cone.**

11. 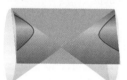   12.

Finding the Standard Equation of a Parabola   **In Exercises 13–16, find the standard form of the equation of the parabola with the given characteristics. Then sketch the parabola.**

13. Vertex: $(0, 0)$
    Focus: $(4, 0)$

14. Vertex: $(2, 0)$
    Focus: $(0, 0)$

15. Vertex: $(0, 2)$
    Directrix: $x = -3$

16. Vertex: $(-3, -3)$
    Directrix: $y = 0$

Finding the Tangent Line at a Point on a Parabola   **In Exercises 17 and 18, find the equation of the tangent line to the parabola at the given point.**

17. $y = 2x^2$, $(-1, 2)$
18. $x^2 = -2y$, $(-4, -8)$

19. **Architecture**   A parabolic archway is 10 meters high at the vertex. At a height of 8 meters, the width of the archway is 6 meters (see figure). How wide is the archway at ground level?

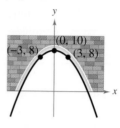

Figure for 19   Figure for 20

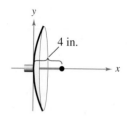

20. **Parabolic Microphone**   The receiver of a parabolic microphone is at the focus of its parabolic reflector, 4 inches from the vertex (see figure). Write an equation of a cross section of the reflector with its focus on the positive x-axis and its vertex at the origin.

6.3 **Finding the Standard Equation of an Ellipse**   **In Exercises 21–24, find the standard form of the equation of the ellipse with the given characteristics.**

21. Vertices: $(-2, 0)$, $(8, 0)$; foci: $(0, 0)$, $(6, 0)$
22. Vertices: $(4, 3)$, $(4, 7)$; foci: $(4, 4)$, $(4, 6)$
23. Vertices: $(0, 1)$, $(4, 1)$; endpoints of the minor axis: $(2, 0)$, $(2, 2)$
24. Vertices: $(-4, -1)$, $(-4, 11)$; endpoints of the minor axis: $(-6, 5)$, $(-2, 5)$

25. **Architecture**   A contractor plans to construct a semielliptical arch 10 feet wide and 4 feet high. Where should the foci be placed in order to sketch the arch?

26. **Wading Pool**   You are building a wading pool that is in the shape of an ellipse. Your plans give an equation for the elliptical shape of the pool measured in feet as

$$\frac{x^2}{324} + \frac{y^2}{196} = 1.$$

Find the longest distance across the pool, the shortest distance, and the distance between the foci.

Sketching an Ellipse   **In Exercises 27–30, find the center, vertices, foci, and eccentricity of the ellipse. Then sketch the ellipse.**

27. $\dfrac{(x + 1)^2}{25} + \dfrac{(y - 2)^2}{49} = 1$

28. $\dfrac{(x - 5)^2}{1} + \dfrac{(y + 3)^2}{36} = 1$

29. $16x^2 + 9y^2 - 32x + 72y + 16 = 0$

30. $4x^2 + 25y^2 + 16x - 150y + 141 = 0$

6.4 Finding the Standard Equation of a Hyperbola In Exercises 31–34, find the standard form of the equation of the hyperbola with the given characteristics.

31. Vertices: $(0, \pm 1)$; foci: $(0, \pm 2)$

32. Vertices: $(3, 3), (-3, 3)$; foci: $(4, 3), (-4, 3)$

33. Foci: $(\pm 5, 0)$; asymptotes: $y = \pm \frac{3}{4} x$

34. Foci: $(0, \pm 13)$; asymptotes: $y = \pm \frac{5}{12} x$

Sketching a Hyperbola In Exercises 35–38, find the center, vertices, foci, and the equations of the asymptotes of the hyperbola. Then sketch the hyperbola using the asymptotes as an aid.

35. $\dfrac{(x-5)^2}{36} - \dfrac{(y+3)^2}{16} = 1$ **36.** $\dfrac{(y-1)^2}{4} - x^2 = 1$

37. $9x^2 - 16y^2 - 18x - 32y - 151 = 0$

38. $-4x^2 + 25y^2 - 8x + 150y + 121 = 0$

39. Navigation Radio transmitting station A is located 200 miles east of transmitting station B. A ship is in an area to the north and 40 miles west of station A. Synchronized radio pulses transmitted at 186,000 miles per second by the two stations are received 0.0005 second sooner from station A than from station B. How far north is the ship?

40. Locating an Explosion Two of your friends live 4 miles apart and on the same "east-west" street, and you live halfway between them. You are having a three-way phone conversation when you hear an explosion. Six seconds later, your friend to the east hears the explosion, and your friend to the west hears it 8 seconds after you do. Find equations of two hyperbolas that would locate the explosion. (Assume that the coordinate system is measured in feet and that sound travels at 1100 feet per second.)

Classifying a Conic from a General Equation In Exercises 41–44, classify the graph of the equation as a circle, a parabola, an ellipse, or a hyperbola.

41. $5x^2 - 2y^2 + 10x - 4y + 17 = 0$

42. $-4y^2 + 5x + 3y + 7 = 0$

43. $3x^2 + 2y^2 - 12x + 12y + 29 = 0$

44. $4x^2 + 4y^2 - 4x + 8y - 11 = 0$

6.5 Rotation of Axes In Exercises 45–48, rotate the axes to eliminate the xy-term in the equation. Then write the equation in standard form. Sketch the graph of the resulting equation, showing both sets of axes.

45. $xy + 3 = 0$

46. $x^2 - 4xy + y^2 + 9 = 0$

47. $5x^2 - 2xy + 5y^2 - 12 = 0$

48. $4x^2 + 8xy + 4y^2 + 7\sqrt{2}x + 9\sqrt{2}y = 0$

Rotation and Graphing Utilities In Exercises 49–52, (a) use the discriminant to classify the graph of the equation, (b) use the Quadratic Formula to solve for y, and (c) use a graphing utility to graph the equation.

49. $16x^2 - 24xy + 9y^2 - 30x - 40y = 0$

50. $13x^2 - 8xy + 7y^2 - 45 = 0$

51. $x^2 + y^2 + 2xy + 2\sqrt{2}x - 2\sqrt{2}y + 2 = 0$

52. $x^2 - 10xy + y^2 + 1 = 0$

6.6 Sketching a Curve In Exercises 53 and 54, (a) create a table of x- and y-values for the parametric equations using $t = -2, -1, 0, 1,$ and $2,$ and (b) plot the points (x, y) generated in part (a) and sketch a graph of the parametric equations.

53. $x = 3t - 2$ and $y = 7 - 4t$

54. $x = \dfrac{1}{4}t$ and $y = \dfrac{6}{t+3}$

Sketching a Curve In Exercises 55–60, (a) sketch the curve represented by the parametric equations (indicate the orientation of the curve) and (b) eliminate the parameter and write the resulting rectangular equation whose graph represents the curve. Adjust the domain of the rectangular equation, if necessary. (c) Verify your result with a graphing utility.

55. $x = 2t$ **56.** $x = 1 + 4t$

  $y = 4t$ $y = 2 - 3t$

57. $x = t^2$ **58.** $x = t + 4$

  $y = \sqrt{t}$ $y = t^2$

59. $x = 3\cos\theta$ **60.** $x = 3 + 3\cos\theta$

  $y = 3\sin\theta$ $y = 2 + 5\sin\theta$

Finding Parametric Equations for a Graph In Exercises 61–64, find a set of parametric equations to represent the graph of the rectangular equation using (a) $t = x$, (b) $t = x + 1$, and (c) $t = 3 - x$.

61. $y = 2x + 3$

62. $y = 4 - 3x$

63. $y = x^2 + 3$

64. $y = 2 - x^2$

65. $y = 2x^2 + 2$

66. $y = 1 - 4x^2$

6.7 Plotting Points in the Polar Coordinate System In Exercises 67–70, plot the point given in polar coordinates and find two additional polar representations of the point, using $-2\pi < \theta < 2\pi$.

67. $\left(2, \dfrac{\pi}{4}\right)$ **68.** $\left(-5, -\dfrac{\pi}{3}\right)$

69. $(-7, 4.19)$ **70.** $\left(\sqrt{3}, 2.62\right)$

Polar-to-Rectangular Conversion In Exercises 71–74, a point in polar coordinates is given. Convert the point to rectangular coordinates.

71. $\left(-1, \dfrac{\pi}{3}\right)$ **72.** $\left(2, \dfrac{5\pi}{4}\right)$

73. $\left(3, \dfrac{3\pi}{4}\right)$ **74.** $\left(0, \dfrac{\pi}{2}\right)$

Rectangular-to-Polar Conversion In Exercises 75–78, a point in rectangular coordinates is given. Convert the point to polar coordinates.

75. $(0, 1)$ **76.** $\left(-\sqrt{5}, \sqrt{5}\right)$

77. $(4, 6)$ **78.** $(3, -4)$

Converting a Rectangular Equation to Polar Form In Exercises 79–84, convert the rectangular equation to polar form.

79. $x^2 + y^2 = 81$ **80.** $x^2 + y^2 = 48$

81. $x^2 + y^2 - 6y = 0$ **82.** $x^2 + y^2 - 4x = 0$

83. $xy = 5$ **84.** $xy = -2$

Converting a Polar Equation to Rectangular Form In Exercises 85–90, convert the polar equation to rectangular form.

85. $r = 5$ **86.** $r = 12$

87. $r = 3 \cos \theta$ **88.** $r = 8 \sin \theta$

89. $r^2 = \sin \theta$ **90.** $r^2 = 4 \cos 2\theta$

6.8 **Sketching the Graph of a Polar Equation** In Exercises 91–100, sketch the graph of the polar equation using symmetry, zeros, maximum r-values, and any other additional points.

91. $r = 6$ **92.** $r = 11$

93. $r = 4 \sin 2\theta$ **94.** $r = \cos 5\theta$

95. $r = -2(1 + \cos \theta)$ **96.** $r = 1 - 4 \cos \theta$

97. $r = 2 + 6 \sin \theta$ **98.** $r = 5 - 5 \cos \theta$

99. $r = -3 \cos 2\theta$ **100.** $r^2 = \cos 2\theta$

Identifying Types of Polar Graphs In Exercises 101–104, identify the type of polar graph and use a graphing utility to graph the equation.

101. $r = 3(2 - \cos \theta)$ **102.** $r = 5(1 - 2 \cos \theta)$

103. $r = 8 \cos 3\theta$ **104.** $r^2 = 2 \sin 2\theta$

6.9 **Sketching a Conic** In Exercises 105–108, identify the conic and sketch its graph.

105. $r = \dfrac{1}{1 + 2 \sin \theta}$ **106.** $r = \dfrac{6}{1 + \sin \theta}$

107. $r = \dfrac{4}{5 - 3 \cos \theta}$ **108.** $r = \dfrac{16}{4 + 5 \cos \theta}$

Finding the Polar Equation of a Conic In Exercises 109–112, find a polar equation of the conic with its focus at the pole.

109. Parabola   Vertex: $(2, \pi)$

110. Parabola   Vertex: $(2, \pi/2)$

111. Ellipse   Vertices: $(5, 0), (1, \pi)$

112. Hyperbola   Vertices: $(1, 0), (7, 0)$

113. Explorer 18 On November 27, 1963, the United States launched Explorer 18. Its low and high points above the surface of Earth were 110 miles and 122,800 miles, respectively. The center of Earth was at one focus of the orbit (see figure). Find the polar equation of the orbit and find the distance between the surface of Earth (assume Earth has a radius of 4000 miles) and the satellite when $\theta = \pi/3$.

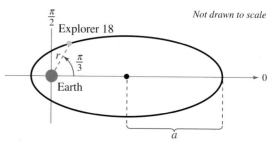

Not drawn to scale

114. Asteroid An asteroid takes a parabolic path with Earth as its focus. It is about 6,000,000 miles from Earth at its closest approach. Write the polar equation of the path of the asteroid with its vertex at $\theta = \pi/2$. Find the distance between the asteroid and Earth when $\theta = -\pi/3$.

Exploration

True or False? In Exercises 115–117, determine whether the statement is true or false. Justify your answer.

115. The graph of $\frac{1}{4}x^2 - y^4 = 1$ is a hyperbola.

116. Only one set of parametric equations can represent the line $y = 3 - 2x$.

117. There is a unique polar coordinate representation of each point in the plane.

118. Think About It Consider an ellipse with the major axis horizontal and 10 units in length. The number b in the standard form of the equation of the ellipse must be less than what real number? Explain the change in the shape of the ellipse as b approaches this number.

119. Think About It What is the relationship between the graphs of the rectangular and polar equations?

(a) $x^2 + y^2 = 25, \quad r = 5$

(b) $x - y = 0, \quad \theta = \dfrac{\pi}{4}$

Chapter Test

See **CalcChat.com** for tutorial help and worked-out solutions to odd-numbered exercises.

Take this test as you would take a test in class. When you are finished, check your work against the answers given in the back of the book.

1. Find the inclination of the line $2x - 5y + 5 = 0$.

2. Find the angle between the lines $3x + 2y = 4$ and $4x - y = -6$.

3. Find the distance between the point $(7, 5)$ and the line $y = 5 - x$.

In Exercises 4–7, identify the conic and write the equation in standard form. Find the center, vertices, foci, and the equations of the asymptotes (if applicable). Then sketch the conic.

4. $y^2 - 2x + 2 = 0$

5. $x^2 - 4y^2 - 4x = 0$

6. $9x^2 + 16y^2 + 54x - 32y - 47 = 0$

7. $2x^2 + 2y^2 - 8x - 4y + 9 = 0$

8. Find the standard form of the equation of the parabola with vertex $(2, -3)$ and a vertical axis that passes through the point $(4, 0)$.

9. Find the standard form of the equation of the hyperbola with foci $(0, \pm 2)$ and asymptotes $y = \pm\frac{1}{9}x$.

10. (a) Determine the number of degrees the axes must be rotated to eliminate the xy-term of the conic $x^2 + 6xy + y^2 - 6 = 0$.

    (b) Sketch the conic from part (a) and use a graphing utility to confirm your result.

11. Sketch the curve represented by the parametric equations $x = 2 + 3\cos\theta$ and $y = 2\sin\theta$. Eliminate the parameter and write the resulting rectangular equation.

12. Find a set of parametric equations to represent the graph of the rectangular equation $y = 3 - x^2$ using (a) $t = x$ and (b) $t = x + 2$.

13. Convert the polar coordinates $\left(-2, \frac{5\pi}{6}\right)$ to rectangular form.

14. Convert the rectangular coordinates $(2, -2)$ to polar form and find two additional polar representations of the point, using $-2\pi < \theta < 2\pi$.

15. Convert the rectangular equation $x^2 + y^2 - 3x = 0$ to polar form.

In Exercises 16–19, sketch the graph of the polar equation. Identify the type of graph.

16. $r = \dfrac{4}{1 + \cos\theta}$

17. $r = \dfrac{4}{2 + \sin\theta}$

18. $r = 2 + 3\sin\theta$

19. $r = 2\sin 4\theta$

20. Find a polar equation of the ellipse with focus at the pole, eccentricity $e = \frac{1}{4}$, and directrix $y = 4$.

21. A straight road rises with an inclination of 0.15 radian from the horizontal. Find the slope of the road and the change in elevation over a one-mile stretch of the road.

22. A baseball is hit at a point 3 feet above the ground toward the left field fence. The fence is 10 feet high and 375 feet from home plate. The path of the baseball can be modeled by the parametric equations $x = (115\cos\theta)t$ and $y = 3 + (115\sin\theta)t - 16t^2$. Will the baseball go over the fence when it is hit at an angle of $\theta = 30°$? Will the baseball go over the fence when $\theta = 35°$?

Cumulative Test for Chapters 4–6

See CalcChat.com for tutorial help and worked-out solutions to odd-numbered exercises.

Take this test as you would take a test in class. When you are finished, check your work against the answers given in the back of the book.

1. Write the complex number $6 - \sqrt{-49}$ in standard form.

In Exercises 2–4, perform the operation and write the result in standard form.

2. $6i - \left(2 + \sqrt{-81}\right)$

3. $(5i - 2)^2$

4. $\left(\sqrt{3} + i\right)\left(\sqrt{3} - i\right)$

5. Write the quotient in standard form: $\dfrac{8i}{10 + 2i}$.

In Exercises 6 and 7, find all the zeros of the function.

6. $f(x) = x^3 + 2x^2 + 4x + 8$

7. $f(x) = x^4 + 4x^3 - 21x^2$

8. Find a polynomial function with real coefficients that has -6, -3, and $4 + \sqrt{5}i$ as its zeros. (There are many correct answers.)

9. Write the complex number $z = -2 + 2i$ in trigonometric form.

10. Find the product of $\left[4(\cos 30° + i \sin 30°)\right]$ and $\left[6(\cos 120° + i \sin 120°)\right]$. Write the answer in standard form.

In Exercises 11 and 12, use DeMoivre's Theorem to find the indicated power of the complex number. Write the result in standard form.

11. $\left[2\left(\cos \dfrac{2\pi}{3} + i \sin \dfrac{2\pi}{3}\right)\right]^4$

12. $\left(-\sqrt{3} - i\right)^6$

13. Find the three cube roots of 1.

14. Find all solutions of the equation $x^4 - 81i = 0$.

In Exercises 15 and 16, use the graph of f to describe the transformation that yields the graph of g.

15. $f(x) = \left(\frac{2}{5}\right)^x$, $g(x) = -\left(\frac{2}{5}\right)^{-x+3}$

16. $f(x) = 2.2^x$, $g(x) = -2.2^x + 4$

In Exercises 17–20, use a calculator to evaluate the expression. Round your result to three decimal places.

17. $\log 98$

18. $\log \frac{6}{7}$

19. $\ln \sqrt{31}$

20. $\ln\left(\sqrt{30} - 4\right)$

In Exercises 21–23, evaluate the logarithm using the change-of-base formula. Round your result to three decimal places.

21. $\log_5 4.3$

22. $\log_3 0.149$

23. $\log_{1/2} 17$

24. Use the properties of logarithms to expand $\ln\left(\dfrac{x^2 - 16}{x^4}\right)$, where $x > 4$.

25. Condense $2 \ln x - \frac{1}{2} \ln(x + 5)$ to the logarithm of a single quantity.

In Exercises 26–29, solve the equation algebraically. Approximate the result to three decimal places.

26. $6e^{2x} = 72$

27. $4^{x-5} + 21 = 30$

28. $\log_2 x + \log_2 5 = 6$

29. $\ln 4x - \ln 2 = 8$

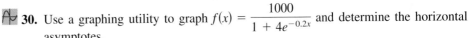 **30.** Use a graphing utility to graph $f(x) = \dfrac{1000}{1 + 4e^{-0.2x}}$ and determine the horizontal asymptotes.

31. The number N of bacteria in a culture is given by the model $N = 175e^{kt}$, where t is the time in hours. If $N = 420$ when $t = 8$, then estimate the time required for the population to double in size.

32. The populations P of Texas (in millions) from 2001 through 2010 can be approximated by the model $P = 20.871e^{0.0188t}$, where t represents the year, with $t = 1$ corresponding to 2001. According to this model, when will the population reach 30 million? *(Source: U.S. Census Bureau)*

33. Find the angle between the lines $2x + y = 3$ and $x - 3y = -6$.

34. Find the distance between the point $(6, -3)$ and the line $y = 2x - 4$.

In Exercises 35–38, identify the conic and write the equation in standard form. Find the center, vertices, foci, and the equations of the asymptotes (if applicable). Then sketch the conic.

35. $9x^2 + 4y^2 - 36x + 8y + 4 = 0$

36. $4x^2 - y^2 - 4 = 0$

37. $x^2 + y^2 + 2x - 6y - 12 = 0$

38. $y^2 + 2x + 2 = 0$

39. Find the standard form of the equation of the circle with center $(2, -4)$ that passes through the point $(0, 4)$.

40. Find the standard form of the equation of the hyperbola with foci $(0, \pm 5)$ and asymptotes $y = \pm \frac{4}{3}x$.

41. (a) Determine the number of degrees the axes must be rotated to eliminate the xy-term of the conic $x^2 + xy + y^2 + 2x - 3y - 30 = 0$.

    (b) Sketch the conic from part (a) and use a graphing utility to confirm your result.

42. Sketch the curve represented by the parametric equations $x = 3 + 4\cos\theta$ and $y = \sin\theta$. Eliminate the parameter and write the resulting rectangular equation.

43. Find a set of parametric equations to represent the graph of the rectangular equation $y = 1 - x$ using (a) $t = x$ and (b) $t = 2 - x$.

44. Plot the point $(-2, -3\pi/4)$ and find three additional polar representations of the point, using $-2\pi < \theta < 2\pi$.

45. Convert the rectangular equation $x^2 + y^2 - 16y = 0$ to polar form.

46. Convert the polar equation $r = \dfrac{2}{4 - 5\cos\theta}$ to rectangular form.

In Exercises 47 and 48, identify the conic and sketch its graph.

47. $r = \dfrac{4}{2 + \cos\theta}$

48. $r = \dfrac{8}{1 + \sin\theta}$

49. Match each polar equation with its graph at the left.

    (a) $r = 2 + 3\sin\theta$

    (b) $r = 3\sin\theta$

    (c) $r = 3\sin 2\theta$

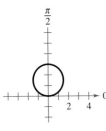

(i)

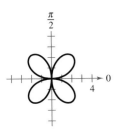

(ii)

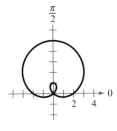

(iii)

Figure for 49

Proofs in Mathematics ■ ■ ■ ■ ■ ■ ■ ■ ■ ■ ■ ■ ■

Proof

If $m = 0$, then the line is horizontal and $\theta = 0$. So, the result is true for horizontal lines because $m = 0 = \tan 0$.

If the line has a positive slope, then it will intersect the x-axis. Label this point $(x_1, 0)$, as shown in the figure. If (x_2, y_2) is a second point on the line, then the slope is

$$m = \frac{y_2 - 0}{x_2 - x_1} = \frac{y_2}{x_2 - x_1} = \tan\theta.$$

The case in which the line has a negative slope can be proved in a similar manner. ■

Proof

For simplicity, assume that the given line is neither horizontal nor vertical (see figure). By writing the equation $Ax + By + C = 0$ in slope-intercept form

$$y = -\frac{A}{B}x - \frac{C}{B}$$

you can see that the line has a slope of $m = -A/B$. So, the slope of the line passing through (x_1, y_1) and perpendicular to the given line is B/A, and its equation is $y - y_1 = (B/A)(x - x_1)$. These two lines intersect at the point (x_2, y_2), where

$$x_2 = \frac{B(Bx_1 - Ay_1) - AC}{A^2 + B^2} \quad \text{and} \quad y_2 = \frac{A(-Bx_1 + Ay_1) - BC}{A^2 + B^2}.$$

Finally, the distance between (x_1, y_1) and (x_2, y_2) is

$$d = \sqrt{(x_2 - x_1)^2 + (y_2 - y_1)^2}$$

$$= \sqrt{\left(\frac{B^2x_1 - ABy_1 - AC}{A^2 + B^2} - x_1\right)^2 + \left(\frac{-ABx_1 + A^2y_1 - BC}{A^2 + B^2} - y_1\right)^2}$$

$$= \sqrt{\frac{A^2(Ax_1 + By_1 + C)^2 + B^2(Ax_1 + By_1 + C)^2}{(A^2 + B^2)^2}}$$

$$= \frac{|Ax_1 + By_1 + C|}{\sqrt{A^2 + B^2}}.$$

■

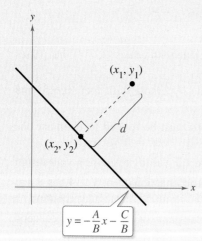

PARABOLIC PATHS

There are many natural occurrences of parabolas in real life. For instance, the famous astronomer Galileo discovered in the 17th century that an object that is projected upward and obliquely to the pull of gravity travels in a parabolic path. Examples of this are the center of gravity of a jumping dolphin and the path of water molecules in a drinking fountain.

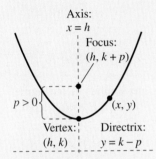

Axis:
$x = h$

Focus:
$(h, k + p)$

$p > 0$

(x, y)

Vertex:
(h, k)

Directrix:
$y = k - p$

Parabola with vertical axis

Directrix:
$x = h - p$

$p > 0$

(x, y)

Axis:
$y = k$

Focus:
$(h + p, k)$

Vertex: (h, k)

Parabola with horizontal axis

Standard Equation of a Parabola   *(p. 422)*

The standard form of the equation of a parabola with vertex at (h, k) is as follows.

$$(x - h)^2 = 4p(y - k), \quad p \neq 0 \qquad \text{Vertical axis; directrix: } y = k - p$$

$$(y - k)^2 = 4p(x - h), \quad p \neq 0 \qquad \text{Horizontal axis; directrix: } x = h - p$$

The focus lies on the axis p units (*directed distance*) from the vertex. If the vertex is at the origin, then the equation takes one of the following forms.

$$x^2 = 4py \qquad\qquad\qquad\qquad \text{Vertical axis}$$

$$y^2 = 4px \qquad\qquad\qquad\qquad \text{Horizontal axis}$$

Proof

For the case in which the directrix is parallel to the x-axis and the focus lies above the vertex, as shown in the top figure, if (x, y) is any point on the parabola, then, by definition, it is equidistant from the focus

$$(h, k + p)$$

and the directrix

$$y = k - p.$$

So, you have

$$\sqrt{(x - h)^2 + [y - (k + p)]^2} = y - (k - p)$$

$$(x - h)^2 + [y - (k + p)]^2 = [y - (k - p)]^2$$

$$(x - h)^2 + y^2 - 2y(k + p) + (k + p)^2 = y^2 - 2y(k - p) + (k - p)^2$$

$$(x - h)^2 + y^2 - 2ky - 2py + k^2 + 2pk + p^2 = y^2 - 2ky + 2py + k^2 - 2pk + p^2$$

$$(x - h)^2 - 2py + 2pk = 2py - 2pk$$

$$(x - h)^2 = 4p(y - k).$$

For the case in which the directrix is parallel to the y-axis and the focus lies to the right of the vertex, as shown in the bottom figure, if (x, y) is any point on the parabola, then, by definition, it is equidistant from the focus

$$(h + p, k)$$

and the directrix

$$x = h - p.$$

So, you have

$$\sqrt{[x - (h + p)]^2 + (y - k)^2} = x - (h - p)$$

$$[x - (h + p)]^2 + (y - k)^2 = [x - (h - p)]^2$$

$$x^2 - 2x(h + p) + (h + p)^2 + (y - k)^2 = x^2 - 2x(h - p) + (h - p)^2$$

$$x^2 - 2hx - 2px + h^2 + 2ph + p^2 + (y - k)^2 = x^2 - 2hx + 2px + h^2 - 2ph + p^2$$

$$-2px + 2ph + (y - k)^2 = 2px - 2ph$$

$$(y - k)^2 = 4p(x - h).$$

Note that if a parabola is centered at the origin, then the two equations above would simplify to $x^2 = 4py$ and $y^2 = 4px$, respectively. ■

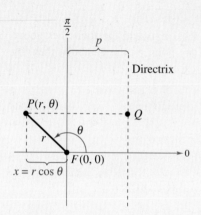

Polar Equations of Conics   (p. 481)

The graph of a polar equation of the form

1. $r = \dfrac{ep}{1 \pm e \cos \theta}$

or

2. $r = \dfrac{ep}{1 \pm e \sin \theta}$

is a conic, where $e > 0$ is the eccentricity and $|p|$ is the distance between the focus (pole) and the directrix.

Proof

A proof for

$$r = \frac{ep}{1 + e \cos \theta}$$

with $p > 0$ is shown here. The proofs of the other cases are similar. In the figure, consider a vertical directrix, p units to the right of the focus $F(0, 0)$. If $P(r, \theta)$ is a point on the graph of

$$r = \frac{ep}{1 + e \cos \theta}$$

then the distance between P and the directrix is

$$\begin{aligned}
PQ &= |p - x| \\
&= |p - r \cos \theta| \\
&= \left| p - \left(\frac{ep}{1 + e \cos \theta} \right) \cos \theta \right| \\
&= \left| p \left(1 - \frac{e \cos \theta}{1 + e \cos \theta} \right) \right| \\
&= \left| \frac{p}{1 + e \cos \theta} \right| \\
&= \left| \frac{r}{e} \right|.
\end{aligned}$$

Moreover, because the distance between P and the pole is simply $PF = |r|$, the ratio of PF to PQ is

$$\begin{aligned}
\frac{PF}{PQ} &= \frac{|r|}{\left| \dfrac{r}{e} \right|} \\
&= |e| \\
&= e
\end{aligned}$$

and, by definition, the graph of the equation must be a conic. ■

498

P.S. Problem Solving ▪ ▪ ▪ ▪ ▪ ▪ ▪ ▪ ▪ ▪ ▪ ▪ ▪ ▪ 🔳

1. **Mountain Climbing** Several mountain climbers are located in a mountain pass between two peaks. The angles of elevation to the two peaks are 0.84 radian and 1.10 radians. A range finder shows that the distances to the peaks are 3250 feet and 6700 feet, respectively (see figure).

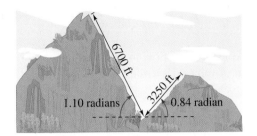

(a) Find the angle between the two lines.

(b) Approximate the amount of vertical climb that is necessary to reach the summit of each peak.

2. **Finding the Equation of a Parabola** Find the general equation of a parabola that has the x-axis as the axis of symmetry and the focus at the origin.

3. **Area** Find the area of the square inscribed in the ellipse, as shown below.

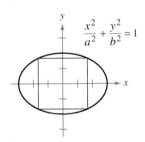

$$\frac{x^2}{a^2} + \frac{y^2}{b^2} = 1$$

4. **Involute** The *involute* of a circle is described by the endpoint P of a string that is held taut as it is unwound from a spool (see figure). The spool does not rotate. Show that

$$x = r(\cos\theta + \theta\sin\theta)$$

and

$$y = r(\sin\theta - \theta\cos\theta)$$

is a parametric representation of the involute of a circle.

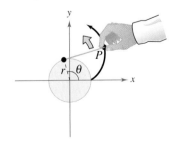

5. **Tour Boat** A tour boat travels between two islands that are 12 miles apart (see figure). There is enough fuel for a 20-mile trip.

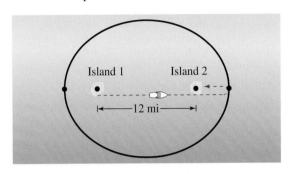

(a) Explain why the region in which the boat can travel is bounded by an ellipse.

(b) Let $(0, 0)$ represent the center of the ellipse. Find the coordinates of each island.

(c) The boat travels from Island 1, past Island 2 to a vertex of the ellipse, and then to Island 2. How many miles does the boat travel? Use your answer to find the coordinates of the vertex.

(d) Use the results from parts (b) and (c) to write an equation of the ellipse that bounds the region in which the boat can travel.

6. **Finding the Equation of a Hyperbola** Find an equation of the hyperbola such that for any point on the hyperbola, the absolute value of the difference of its distances from the points $(2, 2)$ and $(10, 2)$ is 6.

7. **Proof** Prove that the graph of the equation

$$Ax^2 + Cy^2 + Dx + Ey + F = 0$$

is one of the following (except in degenerate cases).

Conic	Condition
(a) Circle	$A = C$
(b) Parabola	$A = 0$ or $C = 0$ (but not both)
(c) Ellipse	$AC > 0$
(d) Hyperbola	$AC < 0$

8. **Proof** Prove that

$$c^2 = a^2 + b^2$$

for the equation of the hyperbola

$$\frac{x^2}{a^2} - \frac{y^2}{b^2} = 1$$

where the distance from the center of the hyperbola $(0, 0)$ to a focus is c.

9. Projectile Motion The following sets of parametric equations model projectile motion.

$$x = (v_0 \cos \theta)t \qquad\qquad x = (v_0 \cos \theta)t$$

$$y = (v_0 \sin \theta)t \qquad\qquad y = h + (v_0 \sin \theta)t - 16t^2$$

(a) Under what circumstances would you use each model?

(b) Eliminate the parameter for each set of equations.

(c) In which case is the path of the moving object not affected by a change in the velocity v? Explain.

10. Orientation of an Ellipse As t increases, the ellipse given by the parametric equations

$$x = \cos t$$

and

$$y = 2 \sin t$$

is traced out *counterclockwise*. Find a parametric representation for which the same ellipse is traced out *clockwise*.

11. Rose Curves The rose curves described in this chapter are of the form

$$r = a \cos n\theta$$

or

$$r = a \sin n\theta$$

where n is a positive integer that is greater than or equal to 2. Use a graphing utility to graph $r = a \cos n\theta$ and $r = a \sin n\theta$ for some noninteger values of n. Describe the graphs.

12. Strophoid The curve given by the parametric equations

$$x = \frac{1 - t^2}{1 + t^2}$$

and

$$y = \frac{t(1 - t^2)}{1 + t^2}$$

is called a **strophoid.**

(a) Find a rectangular equation of the strophoid.

(b) Find a polar equation of the strophoid.

(c) Use a graphing utility to graph the strophoid.

13. Hypocycloid A **hypocycloid** has the parametric equations

$$x = (a - b) \cos t + b \cos\left(\frac{a - b}{b}t\right)$$

and

$$y = (a - b) \sin t - b \sin\left(\frac{a - b}{b}t\right).$$

Use a graphing utility to graph the hypocycloid for each pair of values. Describe each graph.

(a) $a = 2, b = 1$

(b) $a = 3, b = 1$

(c) $a = 4, b = 1$

(d) $a = 10, b = 1$

(e) $a = 3, b = 2$

(f) $a = 4, b = 3$

14. Butterfly Curve The graph of the polar equation

$$r = e^{\cos \theta} - 2 \cos 4\theta + \sin^5\left(\frac{\theta}{12}\right)$$

is called the *butterfly curve,* as shown in the figure.

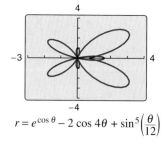

$$r = e^{\cos \theta} - 2 \cos 4\theta + \sin^5\left(\frac{\theta}{12}\right)$$

(a) The graph shown was produced using $0 \le \theta \le 2\pi$. Does this show the entire graph? Explain your reasoning.

(b) Approximate the maximum r-value of the graph. Does this value change when you use $0 \le \theta \le 4\pi$ instead of $0 \le \theta \le 2\pi$? Explain.

15. Writing Use a graphing utility to graph the polar equation

$$r = \cos 5\theta + n \cos \theta$$

for the integers $n = -5$ to $n = 5$ using $0 \le \theta \le \pi$. As you graph these equations, you should see the graph's shape change from a heart to a bell. Write a short paragraph explaining what values of n produce the heart portion of the curve and what values of n produce the bell portion.

Answers to Odd-Numbered Exercises and Tests

Chapter P

Section P.1 *(page 12)*

1. irrational   **3.** absolute value   **5.** terms
7. (a) 5, 1, 2   (b) 0, 5, 1, 2   (c) $-9, 5, 0, 1, -4, 2, -11$
(d) $-\frac{7}{2}, \frac{2}{3}, -9, 5, 0, 1, -4, 2, -11$ (e) $\sqrt{2}$
9. (a) 1   (b) 1   (c) $-13, 1, -6$
(d) $2.01, -13, 1, -6, 0.666\ldots$ (e) $0.010110111\ldots$
11. (a)   (b)

(c)   (d)

13.   **15.**

$-4 > -8$ $\frac{5}{6} > \frac{2}{3}$

17. (a) $x \leq 5$ denotes the set of all real numbers less than or equal to 5.
(b)   (c) Unbounded

19. (a) $[4, \infty)$ denotes the set of all real numbers greater than or equal to 4.
(b)   (c) Unbounded

21. (a) $-2 < x < 2$ denotes the set of all real numbers greater than -2 and less than 2.
(b)   (c) Bounded

23. (a) $[-5, 2)$ denotes the set of all real numbers greater than or equal to -5 and less than 2.
(b)   (c) Bounded

Inequality	Interval
25. $y \geq 0$	$[0, \infty)$
27. $10 \leq t \leq 22$	$[10, 22]$
29. $W > 65$	$(65, \infty)$

31. 10   **33.** 5   **35.** -1 **37.** -1 **39.** -1
41. $|-4| = |4|$ **43.** $-|-6| < |-6|$ **45.** 51   **47.** $\frac{5}{2}$
49. $\frac{128}{75}$ **51.** $|x - 5| \leq 3$ **53.** $|y - a| \leq 2$
55. $1880.1 billion; $412.7 billion
57. $2524.0 billion; $458.5 billion
59. $7x$ and 4 are the terms; 7 is the coefficient.
61. $4x^3, 0.5x$, and -5 are the terms; 4 and 0.5 are the coefficients.
63. (a) -10 (b) -6 **65.** (a) -10 (b) 0
67. Multiplicative Inverse Property   **69.** Distributive Property
71. Associative and Commutative Properties of Multiplication
73. $\frac{3}{8}$ **75.** $\frac{5x}{12}$
77. (a) Negative   (b) Negative   (c) Positive   (d) Positive
79. False. Zero is nonnegative, but not positive.

81. (a)

n	0.0001	0.01	1	100	10,000
$\frac{5}{n}$	50,000	500	5	0.05	0.0005

(b) (i) The value of $5/n$ approaches infinity as n approaches 0.
(ii) The value of $5/n$ approaches 0 as n increases without bound.

Section P.2 *(page 24)*

1. equation   **3.** extraneous   **5.** 4   **7.** -9 **9.** 1
11. No solution   **13.** $-\frac{96}{23}$ **15.** 4   **17.** 3
19. No solution. The variable is divided out.
21. No solution. The solution is extraneous.
23. 5   **25.** $0, -\frac{1}{2}$ **27.** $4, -2$ **29.** -5 **31.** $2, -6$
33. $-\frac{20}{3}, -4$ **35.** ± 7 **37.** $\pm 3\sqrt{3} = \pm 5.20$
39. 8, 16   **41.** $\frac{1 \pm 3\sqrt{2}}{2} \approx 2.62, -1.62$ **43.** $4, -8$
45. $-3 \pm \sqrt{7}$ **47.** $1 \pm \frac{\sqrt{6}}{3}$ **49.** $\frac{-5 \pm \sqrt{89}}{4}$
51. $\frac{1}{2}, -1$ **53.** $1 \pm \sqrt{3}$ **55.** $\frac{3}{4} \pm \frac{\sqrt{41}}{4}$
57. $\frac{2}{3} \pm \frac{\sqrt{7}}{3}$ **59.** $-\frac{5}{3}$ **61.** $2 \pm \frac{\sqrt{6}}{2}$ **63.** $6 \pm \sqrt{11}$
65. $0.976, -0.643$ **67.** $1.687, -0.488$ **69.** $1 \pm \sqrt{2}$
71. $6, -12$ **73.** $\frac{1}{2} \pm \sqrt{3}$ **75.** $-\frac{1}{2}$ **77.** $0, \pm \frac{\sqrt{21}}{3}$
79. $-3, 0$ **81.** 48   **83.** -16 **85.** $2, -5$ **87.** 9
89. 9   **91.** $\pm \sqrt{14}$ **93.** $8, -3$ **95.** $-6, -3, 3$
97. 65.8 in.   **99.** False. See Example 14 on page 23.
101. True. There is no value that satisfies this equation.
103. Equivalent equations have the same solution set, and one is derived from the other by steps for generating equivalent equations. $2x = 5, 2x + 3 = 8$

Section P.3 *(page 36)*

1. Cartesian   **3.** Midpoint Formula   **5.** graph
7. y-axis
9. A: $(2, 6)$, B: $(-6, -2)$, C: $(4, -4)$, D: $(-3, 2)$
11.

13. $(-3, 4)$ **15.** Quadrant IV   **17.** Quadrant III

19.

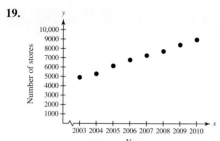
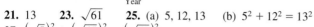

21. 13   **23.** $\sqrt{61}$ **25.** (a) 5, 12, 13   (b) $5^2 + 12^2 = 13^2$
27. $\left(\sqrt{5}\right)^2 + \left(\sqrt{45}\right)^2 = \left(\sqrt{50}\right)^2$
29. Distances between the points: $\sqrt{29}, \sqrt{58}, \sqrt{29}$
31. (a)   **33.** (a)

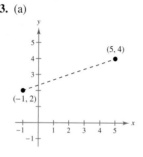

   (b) 8   (c) (6, 1)   (b) $2\sqrt{10}$ (c) (2, 3)
35. $30\sqrt{41} \approx 192$ km   **37.** $27,343.5 million
39. (a) Yes   (b) Yes   **41.** (a) Yes   (b) No
43. (a) No   (b) Yes
45.

x	-1	0	1	2	$\frac{5}{2}$
y	7	5	3	1	0
(x, y)	$(-1, 7)$	$(0, 5)$	$(1, 3)$	$(2, 1)$	$\left(\frac{5}{2}, 0\right)$

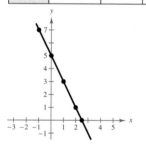

47.

x	-1	0	1	2	3
y	4	0	-2	-2	0
(x, y)	$(-1, 4)$	$(0, 0)$	$(1, -2)$	$(2, -2)$	$(3, 0)$

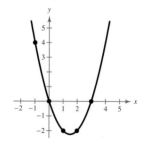

49. x-intercepts: $(\pm 2, 0)$ **51.** x-intercept: $\left(\frac{6}{5}, 0\right)$
   y-intercept: $(0, 16)$ y-intercept: $(0, -6)$
53. x-intercept: $(-4, 0)$ **55.** x-intercept: $\left(\frac{7}{3}, 0\right)$
   y-intercept: $(0, 2)$ y-intercept: $(0, 7)$
57. x-intercepts: $(0, 0), (2, 0)$ **59.** x-intercept: $(6, 0)$
   y-intercept: $(0, 0)$ y-intercepts: $\left(0, \pm\sqrt{6}\right)$
61.   **63.**

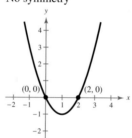
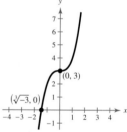

65. y-axis symmetry   **67.** Origin symmetry
69. Origin symmetry   **71.** x-axis symmetry
73. x-intercept: $\left(\frac{1}{3}, 0\right)$
   y-intercept: $(0, 1)$
   No symmetry

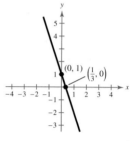

75. x-intercepts: $(0, 0), (2, 0)$ **77.** x-intercept: $\left(\sqrt[3]{-3}, 0\right)$
   y-intercept: $(0, 0)$ y-intercept: $(0, 3)$
   No symmetry   No symmetry

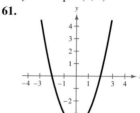

79. x-intercept: $(3, 0)$ **81.** x-intercept: $(6, 0)$
   y-intercept: None y-intercept: $(0, 6)$
   No symmetry   No symmetry

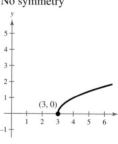
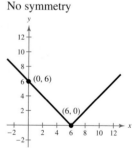

83. $x^2 + y^2 = 16$ **85.** $(x + 1)^2 + (y - 2)^2 = 5$
87. $(x - 3)^2 + (y - 4)^2 = 25$

89. Center: $(0, 0)$; Radius: 5 **91.** Center: $\left(\frac{1}{2}, \frac{1}{2}\right)$; Radius: $\frac{3}{2}$

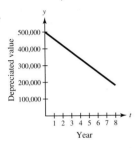

11.

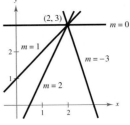

13. $\frac{3}{2}$

93.

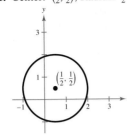

15. $m = 5$
y-intercept: $(0, 3)$

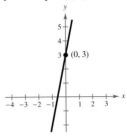

17. $m = -\frac{1}{2}$
y-intercept: $(0, 4)$

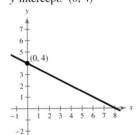

95. (a)

x	5	10	20	30	40
y	414.8	103.7	25.93	11.52	6.48

x	50	60	70	80	90	100
y	4.15	2.88	2.12	1.62	1.28	1.04

19. $m = 0$
y-intercept: $(0, 3)$

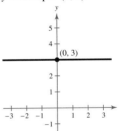

21. m is undefined.
There is no y-intercept.

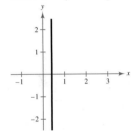

(b)

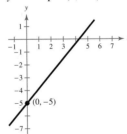

When $x = 85.5$, the resistance is 1.4 ohms.
(c) 1.42 ohms
(d) As the diameter of the copper wire increases, the resistance decreases.

23. $m = \frac{7}{6}$
y-intercept: $(0, -5)$

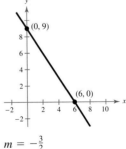

97. False. The Midpoint Formula would be used 15 times.
99. False. A graph is symmetric with respect to the x-axis if, whenever (x, y) is on the graph, $(x, -y)$ is also on the graph.
101. Point on x-axis: $y = 0$; Point on y-axis: $x = 0$
103. Use the Midpoint Formula to prove that the diagonals of the parallelogram bisect each other.
$$\left(\frac{b + a}{2}, \frac{c + 0}{2}\right) = \left(\frac{a + b}{2}, \frac{c}{2}\right)$$
$$\left(\frac{a + b + 0}{2}, \frac{c + 0}{2}\right) = \left(\frac{a + b}{2}, \frac{c}{2}\right)$$
105. $(2x_m - x_1, 2y_m - y_1)$ (a) $(7, 0)$ (b) $(9, -3)$

25.  $m = -\frac{3}{2}$ **27.** 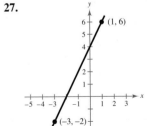 $m = 2$

Section P.4   (page 49)

1. linear   **3.** point-slope   **5.** perpendicular
7. linear extrapolation   **9.** (a) L_2 (b) L_3 (c) L_1

CHAPTER P

29.

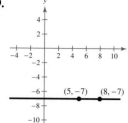

$m = 0$

31.

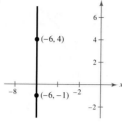

m is undefined.

33.

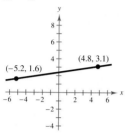

$m = 0.15$

35. $(0, 1), (3, 1), (-1, 1)$ **37.** $(-8, 0), (-8, 2), (-8, 3)$

39. $(-4, 6), (-3, 8), (-2, 10)$

41. $(-3, -5), (1, -7), (5, -9)$

43. $y = 3x - 2$

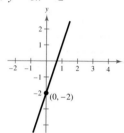

45. $y = -2x$

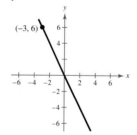

47. $y = -\frac{1}{3}x + \frac{4}{3}$

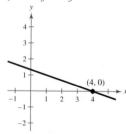

49. $y = -\frac{1}{2}x - 2$

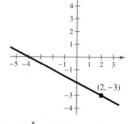

51. $x = 6$

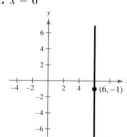

53. $y = \frac{5}{2}$

55. $y = -\frac{3}{5}x + 2$

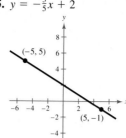

57. $x = -8$

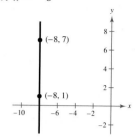

59. $y = -\frac{1}{2}x + \frac{3}{2}$

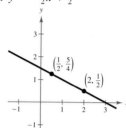

61. $y = 0.4x + 0.2$

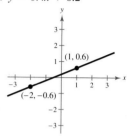

63. $y = -1$

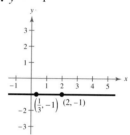

65. Parallel   **67.** Neither   **69.** Perpendicular

71. Parallel   **73.** (a) $y = 2x - 3$ (b) $y = -\frac{1}{2}x + 2$

75. (a) $y = -\frac{3}{4}x + \frac{3}{8}$ (b) $y = \frac{4}{3}x + \frac{127}{72}$

77. (a) $y = 0$ (b) $x = -1$

79. (a) $y = x + 4.3$ (b) $y = -x + 9.3$

81. $3x + 2y - 6 = 0$ **83.** $12x + 3y + 2 = 0$

85. $x + y - 3 = 0$

87. (a) Sales increasing 135 units/yr

(b) No change in sales

(c) Sales decreasing 40 units/yr

89. 12 ft   **91.** $V(t) = -125t + 4165, \ 13 \leq t \leq 18$

93. C-intercept: fixed initial cost; Slope: cost of producing an additional laptop bag

95. $V(t) = -175t + 875, \ 0 \leq t \leq 5$

97. $F = 1.8C + 32$ or $C = \frac{5}{9}F - \frac{160}{9}$

99. (a) $C = 21t + 42,000$ (b) $R = 45t$

(c) $P = 24t - 42,000$ (d) 1750 h

101. False. The slope with the greatest magnitude corresponds to the steepest line.

103. Find the slopes of the lines containing each two points and use the relationship $m_1 = -\dfrac{1}{m_2}$.

105. No. The slope cannot be determined without knowing the scale on the y-axis. The slopes could be the same.

107. No. The slopes of two perpendicular lines have opposite signs (assume that neither line is vertical or horizontal).

109. The line $y = 4x$ rises most quickly, and the line $y = -4x$ falls most quickly. The greater the magnitude of the slope (the absolute value of the slope), the faster the line rises or falls.

111. $3x - 2y - 1 = 0$ **113.** $80x + 12y + 139 = 0$

Section P.5   (page 62)

1. domain; range; function   **3.** implied domain

5. Yes, each input value has exactly one output value.

7. No, the input values 7 and 10 each have two different output values.

9. (a) Function

(b) Not a function, because the element 1 in A corresponds to two elements, -2 and 1, in B.

(c) Function

(d) Not a function, because not every element in A is matched with an element in B.

11. Not a function   **13.** Function   **15.** Function

17. Function   **19.** Function

21. (a) -1 (b) -9 (c) $2x - 5$

23. (a) 15   (b) $4t^2 - 19t + 27$ (c) $4t^2 - 3t - 10$

25. (a) 1   (b) 2.5   (c) $3 - 2|x|$

27. (a) $-\dfrac{1}{9}$ (b) Undefined   (c) $\dfrac{1}{y^2 + 6y}$

29. (a) 1   (b) -1 (c) $\dfrac{|x - 1|}{x - 1}$

31. (a) -1 (b) 2   (c) 6

33.

x	1	2	3	4	5
$f(x)$	8	5	0	1	2

35.

t	-5	-4	-3	-2	-1
$h(t)$	1	$\frac{1}{2}$	0	$\frac{1}{2}$	1

37. 5   **39.** $\frac{4}{3}$ **41.** ± 3 **43.** $0, \pm 1$ **45.** $-1, 2$

47. $0, \pm 2$ **49.** All real numbers x

51. All real numbers t except $t = 0$

53. All real numbers y such that $y \geq 10$

55. All real numbers x except $x = 0, -2$

57. All real numbers s such that $s \geq 1$ except $s = 4$

59. All real numbers x such that $x > 0$

61. (a) The maximum volume is 1024 cubic centimeters.

(b) Yes, V is a function of x.

(graph with axis labeled Volume vertically and Height horizontally, points plotted)

(c) $V = x(24 - 2x)^2, \quad 0 < x < 12$

63. $A = \dfrac{P^2}{16}$ **65.** Yes, the ball will be at a height of 6 feet.

67. $A = \dfrac{x^2}{2(x - 2)}, \quad x > 2$

69. 2004: 45.58%

2005: 50.15%

2006: 54.72%

2007: 59.29%

2008: 64.40%

2009: 67.75%

2010: 71.10%

71. (a) $C = 12.30x + 98,000$ (b) $R = 17.98x$

(c) $P = 5.68x - 98,000$

73. (a)

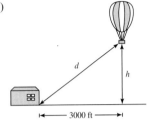

(b) $h = \sqrt{d^2 - 3000^2}, \quad d \geq 3000$

75. (a) $R = \dfrac{240n - n^2}{20}, \quad n \geq 80$

(b)

n	90	100	110	120	130	140	150
$R(n)$	\$675	\$700	\$715	\$720	\$715	\$700	\$675

The revenue is maximum when 120 people take the trip.

77. $3 + h, \ h \neq 0$ **79.** $3x^2 + 3xh + h^2 + 3, \ h \neq 0$

81. $-\dfrac{x + 3}{9x^2}, \ x \neq 3$ **83.** $\dfrac{\sqrt{5x} - 5}{x - 5}$

85. $g(x) = cx^2; c = -2$ **87.** $r(x) = \dfrac{c}{x}; c = 32$

89. False. A function is a special type of relation.

91. False. The range is $[-1, \infty)$.

93. Domain of $f(x)$: all real numbers $x \geq 1$

Domain of $g(x)$: all real numbers $x > 1$

Notice that the domain of $f(x)$ includes $x = 1$ and the domain of $g(x)$ does not because you cannot divide by 0.

95. No; x is the independent variable, f is the name of the function.

97. (a) Yes. The amount you pay in sales tax will increase as the price of the item purchased increases.

(b) No. The length of time that you study will not necessarily determine how well you do on an exam.

Section P.6   (page 74)

1. Vertical Line Test   **3.** decreasing

5. average rate of change; secant

7. Domain: $(-\infty, \infty)$; Range: $[-4, \infty)$

(a) 0   (b) -1 (c) 0   (d) -2

9. Domain: $(-\infty, \infty)$; Range: $(-2, \infty)$

(a) 0   (b) 1   (c) 2   (d) 3

11. Function   **13.** Not a function   **15.** $-\frac{5}{2}, 6$ **17.** 0

19. $0, \pm \sqrt{2}$ **21.** $\pm 3, 4$ **23.** $\frac{1}{2}$

25.

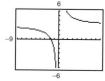

$-\dfrac{5}{3}$

27.

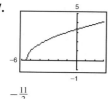

$-\dfrac{11}{2}$

29.

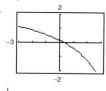

$\dfrac{1}{3}$

31. Increasing on $(-\infty, \infty)$

33. Increasing on $(-\infty, 0)$ and $(2, \infty)$; Decreasing on $(0, 2)$

35. Increasing on $(1, \infty)$; Decreasing on $(-\infty, -1)$
Constant on $(-1, 1)$

37. Increasing on $(-\infty, 0)$ and $(2, \infty)$; Constant on $(0, 2)$

39.

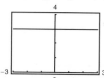

Constant on $(-\infty, \infty)$

41.

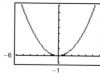

Decreasing on $(-\infty, 0)$
Increasing on $(0, \infty)$

43.

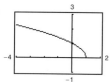

Decreasing on $(-\infty, 1)$

45.

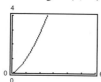

Increasing on $(0, \infty)$

47.

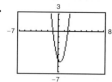

Relative minimum:
$\left(\dfrac{1}{3}, -\dfrac{16}{3}\right)$

49.

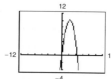

Relative maximum:
$(2.25, 10.125)$

51.

Relative maximum:
$(-0.15, 1.08)$
Relative minimum:
$(2.15, -5.08)$

53.

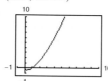

Relative minimum:
$(0.33, -0.38)$

55.

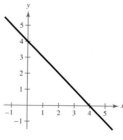

$(-\infty, 4]$

57.

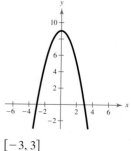

$[-3, 3]$

59.

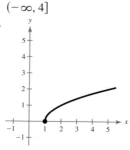

$[1, \infty)$

61. The average rate of change from $x_1 = 0$ to $x_2 = 3$ is -2.

63. The average rate of change from $x_1 = 1$ to $x_2 = 3$ is 0.

65. (a)

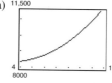

(b) 484.75 million; The amount
the U.S. Department of
Energy spent for research
and development increased
by about \$484.75 million
each year from 2005 to
2010.

67. (a) $s = -16t^2 + 64t + 6$

(b)

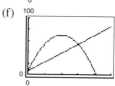

(c) Average rate of change $= 16$

(d) The slope of the secant line
is positive.

(e) Secant line: $16t + 6$

(f)

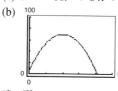

69. (a) $s = -16t^2 + 120t$

(b)

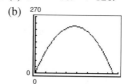

(c) Average rate of
change $= -8$

(d) The slope of the secant line is negative.

(e) Secant line: $-8t + 240$

(f)

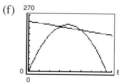

71. Even; y-axis symmetry   **73.** Odd; origin symmetry

75. Neither; no symmetry

77.

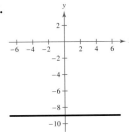

Even

79.

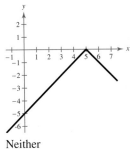

Neither

81.

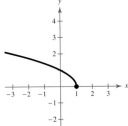

Neither

83. $h = 3 - 4x + x^2$ **85.** $L = 2 - \sqrt[3]{2y}$

87. (a)

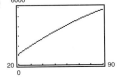

(b) 30 W

89. (a) Ten thousands   (b) Ten millions   (c) Percents

91. (a)

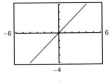

(b)

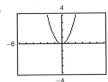

(c)

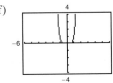

(d)

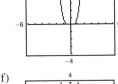

(e)

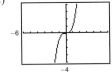

(f)

All the graphs pass through the origin. The graphs of the odd powers of x are symmetric with respect to the origin, and the graphs of the even powers are symmetric with respect to the y-axis. As the powers increase, the graphs become flatter in the interval $-1 < x < 1$.

93. False. The function $f(x) = \sqrt{x^2 + 1}$ has a domain of all real numbers.

95. (a) $\left(\frac{5}{3}, -7\right)$ (b) $\left(\frac{5}{3}, 7\right)$

97.

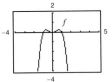

Even   Neither

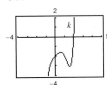

Odd   Even

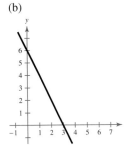

Neither   Odd

Equations of odd functions contain only odd powers of x. Equations of even functions contain only even powers of x. Odd functions have all variables raised to odd powers and even functions have all variables raised to even powers. A function that has variables raised to even and odd powers is neither odd nor even.

Section P.7   *(page 83)*

1. g   **2.** i   **3.** h   **4.** a   **5.** b   **6.** e   **7.** f

8. c   **9.** d

11. (a) $f(x) = -2x + 6$ **13.** (a) $f(x) = -1$

(b)   (b)

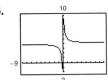   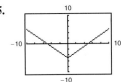

15.

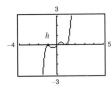

17.

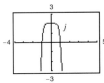

19.

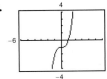

21.

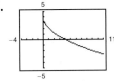

23.

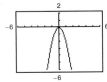

25.

27. (a) 2   (b) 2   (c) −4   (d) 3
29. (a) 8   (b) 2   (c) 6   (d) 13
31.

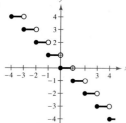

33.

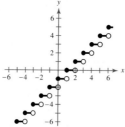

35.

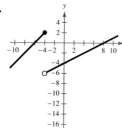

37.

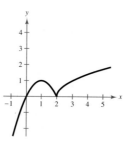

39.

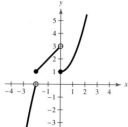

41. (a)

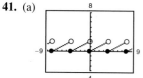

(b) Domain: $(-\infty, \infty)$
    Range: $[0, 2)$

43. (a) $W(30) = 420$; $W(40) = 560$;
    $W(45) = 665$; $W(50) = 770$

(b) $W(h) = \begin{cases} 14h, & 0 < h \le 45 \\ 21(h - 45) + 630, & h > 45 \end{cases}$

45.

Interval	Input Pipe	Drain Pipe 1	Drain Pipe 2
$[0, 5]$	Open	Closed	Closed
$[5, 10]$	Open	Open	Closed
$[10, 20]$	Closed	Closed	Closed
$[20, 30]$	Closed	Closed	Open
$[30, 40]$	Open	Open	Open
$[40, 45]$	Open	Closed	Open
$[45, 50]$	Open	Open	Open
$[50, 60]$	Open	Open	Closed

47. $f(t) = \begin{cases} t, & 0 \le t \le 2 \\ 2t - 2, & 2 < t \le 8 \\ \frac{1}{2}t + 10, & 8 < t \le 9 \end{cases}$

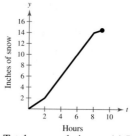

Total accumulation = 14.5 in.

49. False. A piecewise-defined function is a function that is defined by two or more equations over a specified domain. That domain may or may not include *x*- and *y*-intercepts.

Section P.8   *(page 90)*

1. rigid   **3.** vertical stretch; vertical shrink
5. (a)   (b)

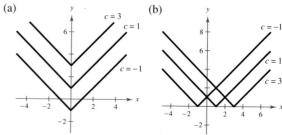

7. (a)   (b)

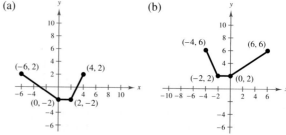

9. (a)   (b)

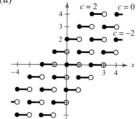

(c)   (d)

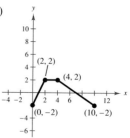

(e)

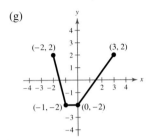

(f)

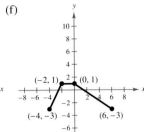

(g)

11. (a) $y = x^2 - 1$ (b) $y = 1 - (x + 1)^2$
13. (a) $y = -|x + 3|$ (b) $y = |x - 2| - 4$
15. Horizontal shift of $y = x^3$; $y = (x - 2)^3$
17. Reflection in the x-axis of $y = x^2$; $y = -x^2$
19. Reflection in the x-axis and vertical shift of $y = \sqrt{x}$;
    $y = 1 - \sqrt{x}$
21. (a) $f(x) = x^2$
    (b) Reflection in the x-axis and vertical shift 12 units up
    (c)   (d) $g(x) = 12 - f(x)$

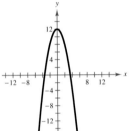

23. (a) $f(x) = x^3$ (b) Vertical shift seven units up
    (c)   (d) $g(x) = f(x) + 7$

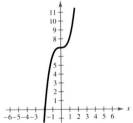

25. (a) $f(x) = x^2$
    (b) Vertical shrink of two-thirds and vertical shift four units up
    (c)   (d) $g(x) = \frac{2}{3}f(x) + 4$

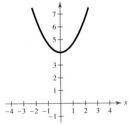

27. (a) $f(x) = x^2$
    (b) Reflection in the x-axis, horizontal shift five units to the
        left, and vertical shift two units up
    (c)   (d) $g(x) = 2 - f(x + 5)$

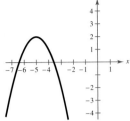

29. (a) $f(x) = \sqrt{x}$ (b) Horizontal shrink of one-third
    (c)   (d) $g(x) = f(3x)$

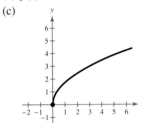

31. (a) $f(x) = x^3$
    (b) Vertical shift two units up and horizontal shift one unit to
        the right
    (c)   (d) $g(x) = f(x - 1) + 2$

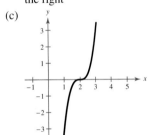

33. (a) $f(x) = x^3$
    (b) Vertical stretch of three and horizontal shift two units to
        the right
    (c)   (d) $g(x) = 3f(x - 2)$

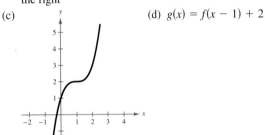

35. (a) $f(x) = |x|$
    (b) Reflection in the x-axis and vertical shift two units down
    (c)   (d) $g(x) = -f(x) - 2$

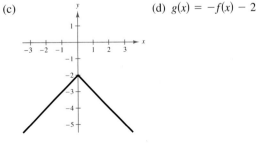

CHAPTER P

37. (a) $f(x) = |x|$

(b) Reflection in the x-axis, horizontal shift four units to the left, and vertical shift eight units up

(c)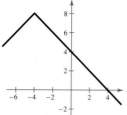

(d) $g(x) = -f(x + 4) + 8$

39. (a) $f(x) = |x|$

(b) Reflection in the x-axis, vertical stretch of two, horizontal shift one unit to the right, and vertical shift four units down

(c)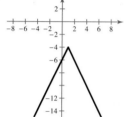

(d) $g(x) = -2f(x - 1) - 4$

41. (a) $f(x) = [\![x]\!]$

(b) Reflection in the x-axis and vertical shift three units up

(c)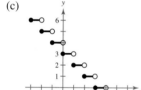

(d) $g(x) = 3 - f(x)$

43. (a) $f(x) = \sqrt{x}$ (b) Horizontal shift nine units to the right

(c)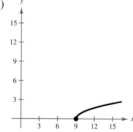

(d) $g(x) = f(x - 9)$

45. (a) $f(x) = \sqrt{x}$

(b) Reflection in the y-axis, horizontal shift seven units to the right, and vertical shift two units down

(c)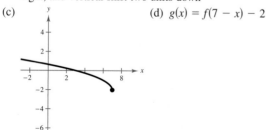

(d) $g(x) = f(7 - x) - 2$

47. $g(x) = (x - 3)^2 - 7$ **49.** $g(x) = (x - 13)^3$

51. $g(x) = -|x| + 12$ **53.** $g(x) = -\sqrt{-x + 6}$

55. (a) $y = -3x^2$ (b) $y = 4x^2 + 3$

57. (a) $y = -\frac{1}{2}|x|$ (b) $y = 3|x| - 3$

59. Vertical stretch of $y = x^3$; $y = 2x^3$

61. Reflection in the x-axis and vertical shrink of $y = x^2$; $y = -\frac{1}{2}x^2$

63. Reflection in the y-axis and vertical shrink of $y = \sqrt{x}$; $y = \frac{1}{2}\sqrt{-x}$

65. $y = -(x - 2)^3 + 2$ **67.** $y = -\sqrt{x} - 3$

69. (a)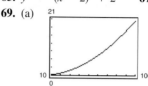

(b) $H\left(\dfrac{x}{1.6}\right) = 0.00078x^2 + 0.003x - 0.029,\ 16 \le x \le 160$;

Horizontal stretch

71. False. The graph of $y = f(-x)$ is a reflection of the graph of $f(x)$ in the y-axis.

73. True. $|-x| = |x|$

75. $(-2, 0), (-1, 1), (0, 2)$

77. (a)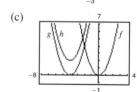

g is a right shift of four units. h is a right shift of four units and a shift of three units up.

(b)

g is a left shift of one unit. h is a left shift of one unit and a shift of two units down.

(c)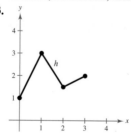

g is a left shift of four units. h is a left shift of four units and a shift of two units up.

79. (a) $g(t) = \frac{3}{4}f(t)$ (b) $g(t) = f(t) + 10,000$

(c) $g(t) = f(t - 2)$

Section P.9   (page 99)

1. addition; subtraction; multiplication; division

3.

5. (a) $2x$ (b) 4   (c) $x^2 - 4$

(d) $\dfrac{x+2}{x-2}$; all real numbers x except $x = 2$

7. (a) $x^2 + 4x - 5$ (b) $x^2 - 4x + 5$ (c) $4x^3 - 5x^2$

(d) $\dfrac{x^2}{4x-5}$; all real numbers x except $x = \dfrac{5}{4}$

9. (a) $x^2 + 6 + \sqrt{1-x}$ (b) $x^2 + 6 - \sqrt{1-x}$

(c) $(x^2 + 6)\sqrt{1-x}$

(d) $\dfrac{(x^2+6)\sqrt{1-x}}{1-x}$; all real numbers x such that $x < 1$

11. (a) $\dfrac{x+1}{x^2}$ (b) $\dfrac{x-1}{x^2}$ (c) $\dfrac{1}{x^3}$

(d) x; all real numbers x except $x = 0$

13. 3   **15.** 5   **17.** $9t^2 - 3t + 5$ **19.** 74

21. 26   **23.** $\frac{3}{5}$

25.

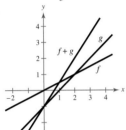

27.

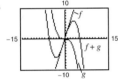

$f(x), g(x)$

29.

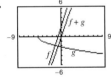

$f(x), f(x)$

31. (a) $(x-1)^2$ (b) $x^2 - 1$ (c) $x - 2$

33. (a) x (b) x (c) $x^9 + 3x^6 + 3x^3 + 2$

35. (a) $\sqrt{x^2 + 4}$ (b) $x + 4$

Domains of f and $g \circ f$: all real numbers x such that $x \geq -4$

Domains of g and $f \circ g$: all real numbers x

37. (a) $x + 1$ (b) $\sqrt{x^2 + 1}$

Domains of f and $g \circ f$: all real numbers x

Domains of g and $f \circ g$: all real numbers x such that $x \geq 0$

39. (a) $|x + 6|$ (b) $|x| + 6$

Domains of $f, g, f \circ g$, and $g \circ f$: all real numbers x

41. (a) $\dfrac{1}{x+3}$ (b) $\dfrac{1}{x} + 3$

Domains of f and $g \circ f$: all real numbers x except $x = 0$

Domain of g: all real numbers x

Domain of $f \circ g$: all real numbers x except $x = -3$

43. (a) 3   (b) 0   **45.** (a) 0   (b) 4

47. $f(x) = x^2$, $g(x) = 2x + 1$

49. $f(x) = \sqrt[3]{x}$, $g(x) = x^2 - 4$

51. $f(x) = \dfrac{1}{x}$, $g(x) = x + 2$ **53.** $f(x) = \dfrac{x+3}{4+x}$, $g(x) = -x^2$

55. (a) $T = \frac{3}{4}x + \frac{1}{15}x^2$

(b)

```
Distance traveled (in feet)
300
250          T
200             B
150
100
50              R
     10 20 30 40 50 60
     Speed (in miles per hour)
```

(c) The braking function $B(x)$. As x increases, $B(x)$ increases at a faster rate than $R(x)$.

57. (a) $p(t) = d(t) + c(t)$

(b) $p(13)$ is the number of dogs and cats in the year 2013.

(c) $h(t) = \dfrac{d(t) + c(t)}{n(t)}$;

The function $h(t)$ represents the number of dogs and cats per capita.

59. (a) $r(x) = \dfrac{x}{2}$

(b) $A(r) = \pi r^2$

(c) $(A \circ r)(x) = \pi\left(\dfrac{x}{2}\right)^2$;

$(A \circ r)(x)$ represents the area of the circular base of the tank on the square foundation with side length x.

61. $g(f(x))$ represents 3 percent of an amount over \$500,000.

63. (a) $O(M(Y)) = 2\left(6 + \frac{1}{2}Y\right) = 12 + Y$

(b) Middle child is 8 years old; youngest child is 4 years old.

65. False. $(f \circ g)(x) = 6x + 1$ and $(g \circ f)(x) = 6x + 6$

67–69. Proofs

Section P.10   *(page 108)*

1. inverse   **3.** range; domain   **5.** one-to-one

7. $f^{-1}(x) = \dfrac{1}{6}x$ **9.** $f^{-1}(x) = \dfrac{x-1}{3}$ **11.** $f^{-1}(x) = x^3$

13. $f(g(x)) = f\left(-\dfrac{2x+6}{7}\right) = -\dfrac{7}{2}\left(-\dfrac{2x+6}{7}\right) - 3 = x$

$g(f(x)) = g\left(-\dfrac{7}{2}x - 3\right) = -\dfrac{2\left(-\frac{7}{2}x - 3\right) + 6}{7} = x$

15. $f(g(x)) = f(\sqrt[3]{x-5}) = (\sqrt[3]{x-5})^3 + 5 = x$

$g(f(x)) = g(x^3 + 5) = \sqrt[3]{(x^3 + 5) - 5} = x$

17.

```
       y
       3
       2
       1
   -3 -1   1 2 3  x
      -1
      -2
      -3
```

19.

```
       y
       4
       3
       2
       1
  -1    1 2 3 4  x
      -1
```

21. (a) $f(g(x)) = f\left(\dfrac{x}{2}\right) = 2\left(\dfrac{x}{2}\right) = x$

$g(f(x)) = g(2x) = \dfrac{(2x)}{2} = x$

(b)

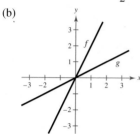

23. (a) $f(g(x)) = f\left(\dfrac{x-1}{7}\right) = 7\left(\dfrac{x-1}{7}\right) + 1 = x$

$g(f(x)) = g(7x+1) = \dfrac{(7x+1)-1}{7} = x$

(b)

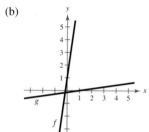

25. (a) $f(g(x)) = f\left(\sqrt[3]{8x}\right) = \dfrac{\left(\sqrt[3]{8x}\right)^3}{8} = x$

$g(f(x)) = g\left(\dfrac{x^3}{8}\right) = \sqrt[3]{8\left(\dfrac{x^3}{8}\right)} = x$

(b)

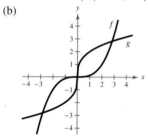

27. (a) $f(g(x)) = f(x^2 + 4), \; x \geq 0$

$= \sqrt{(x^2+4)-4} = x$

$g(f(x)) = g\left(\sqrt{x-4}\right)$

$= \left(\sqrt{x-4}\right)^2 + 4 = x$

(b)

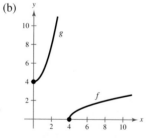

29. (a) $f(g(x)) = f\left(\sqrt{9-x}\right), \; x \leq 9$

$= 9 - \left(\sqrt{9-x}\right)^2 = x$

$g(f(x)) = g(9-x^2), \; x \geq 0$

$= \sqrt{9-(9-x^2)} = x$

(b)

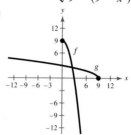

31. (a) $f(g(x)) = f\left(-\dfrac{5x+1}{x-1}\right) = \dfrac{-\left(\dfrac{5x+1}{x-1}\right)-1}{-\left(\dfrac{5x+1}{x-1}\right)+5}$

$= \dfrac{-5x-1-x+1}{-5x-1+5x-5} = x$

$g(f(x)) = g\left(\dfrac{x-1}{x+5}\right) = \dfrac{-5\left(\dfrac{x-1}{x+5}\right)-1}{\dfrac{x-1}{x+5}-1}$

$= \dfrac{-5x+5-x-5}{x-1-x-5} = x$

(b)

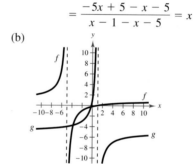

33. No   **35.**

x	-2	0	2	4	6	8
$f^{-1}(x)$	-2	-1	0	1	2	3

37. Yes   **39.** No

41.

The function has an inverse function.

43.

The function does not have an inverse function.

45. (a) $f^{-1}(x) = \dfrac{x+3}{2}$

(b)

(c) The graph of f^{-1} is the reflection of the graph of f in the line $y = x$.

(d) The domains and ranges of f and f^{-1} are all real numbers.

47. (a) $f^{-1}(x) = \sqrt[5]{x} + 2$
(b)

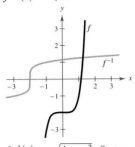

(c) The graph of f^{-1} is the reflection of the graph of f in the line $y = x$.
(d) The domains and ranges of f and f^{-1} are all real numbers.

49. (a) $f^{-1}(x) = \sqrt{4 - x^2}$, $0 \le x \le 2$
(b)

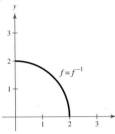

(c) The graph of f^{-1} is the same as the graph of f.
(d) The domains and ranges of f and f^{-1} are all real numbers x such that $0 \le x \le 2$.

51. (a) $f^{-1}(x) = \dfrac{4}{x}$
(b)

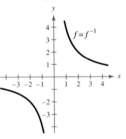

(c) The graph of f^{-1} is the same as the graph of f.
(d) The domains and ranges of f and f^{-1} are all real numbers x except $x = 0$.

53. (a) $f^{-1}(x) = \dfrac{2x + 1}{x - 1}$
(b)

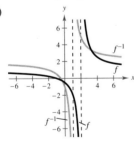

(c) The graph of f^{-1} is the reflection of the graph of f in the line $y = x$.
(d) The domain of f and the range of f^{-1} are all real numbers x except $x = 2$. The domain of f^{-1} and the range of f are all real numbers x except $x = 1$.

55. (a) $f^{-1}(x) = x^3 + 1$
(b)

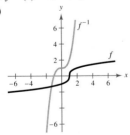

(c) The graph of f^{-1} is the reflection of the graph of f in the line $y = x$.
(d) The domains and ranges of f and f^{-1} are all real numbers.

57. No inverse function
59. $g^{-1}(x) = 8x$
61. No inverse function
63. $f^{-1}(x) = \sqrt{x} - 3$
65. No inverse function
67. No inverse function
69. $f^{-1}(x) = \dfrac{x^2 - 3}{2}$, $x \ge 0$
71. $f^{-1}(x) = \dfrac{5x - 4}{6 - 4x}$

73. $f^{-1}(x) = \sqrt{x} + 2$
The domain of f and the range of f^{-1} are all real numbers x such that $x \ge 2$. The domain of f^{-1} and the range of f are all real numbers x such that $x \ge 0$.

75. $f^{-1}(x) = x - 2$
The domain of f and the range of f^{-1} are all real numbers x such that $x \ge -2$. The domain of f^{-1} and the range of f are all real numbers x such that $x \ge 0$.

77. $f^{-1}(x) = \sqrt{x} - 6$
The domain of f and the range of f^{-1} are all real numbers x such that $x \ge -6$. The domain of f^{-1} and the range of f are all real numbers x such that $x \ge 0$.

79. $f^{-1}(x) = \dfrac{\sqrt{-2(x - 5)}}{2}$
The domain of f and the range of f^{-1} are all real numbers x such that $x \ge 0$. The domain of f^{-1} and the range of f are all real numbers x such that $x \le 5$.

81. $f^{-1}(x) = x + 3$
The domain of f and the range of f^{-1} are all real numbers x such that $x \ge 4$. The domain of f^{-1} and the range of f are all real numbers x such that $x \ge 1$.

83. 32   **85.** 600   **87.** $2\sqrt[3]{x + 3}$ **89.** $\dfrac{x + 1}{2}$

91. $\dfrac{x + 1}{2}$

93. (a) $y = \dfrac{x - 10}{0.75}$
  $x = $ hourly wage; $y = $ number of units produced
(b) 19 units

95. False. $f(x) = x^2$ has no inverse function.

97.

x	1	3	4	6
y	1	2	6	7

x	1	2	6	7
$f^{-1}(x)$	1	3	4	6

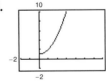

99. Proof   **101.** $k = \frac{1}{4}$
103.

There is an inverse function $f^{-1}(x) = \sqrt{x} - 1$ because the domain of f is equal to the range of f^{-1} and the range of f is equal to the domain of f^{-1}.

105. This situation could be represented by a one-to-one function if the runner does not stop to rest. The inverse function would represent the time in hours for a given number of miles completed.

Review Exercises   (page 114)

1. (a) 11   (b) 11   (c) 11   (d) 11, $-\frac{8}{9}, \frac{5}{2}, 0.4$ (e) $\sqrt{6}$
3.

$\frac{5}{4} > \frac{7}{8}$
5. 122   **7.** $|x - 7| \ge 4$ **9.** (a) -1 (b) -3

11. Associative Property of Addition

13. Additive Identity Property

15. Commutative Property of Addition   **17.** -11 **19.** $\frac{1}{12}$

21. -144 **23.** 20 **25.** -30 **27.** $-\frac{3}{2}, -1$

29. $-4 \pm 3\sqrt{2}$ **31.** $0, \frac{12}{5}$ **33.** 5

35. $-124, 126$ **37.** $-5, 15$

39. (a)   (b) 5   (c) $\left(3, \frac{5}{2}\right)$

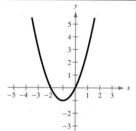

41.

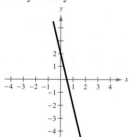

Actual temperature (in °F)

43.

x	1	2	3	4	5
y	-4	-2	0	2	4

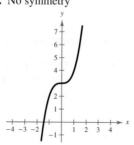

45.

x	-3	-2	-1	0	1
y	3	0	-1	0	3

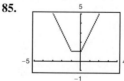

47. x-intercepts: $(1, 0), (5, 0)$

 y-intercept: $(0, 5)$

49. No symmetry   **51.** No symmetry

53. $(x - 2)^2 + (y + 3)^2 = 13$

55. Center: $(0, 0)$;

 Radius: 3

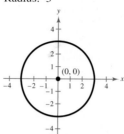

57. Slope: -2

 y-intercept: -7

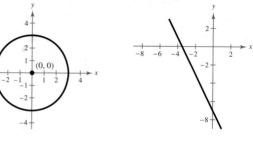

59. Slope: 0

 y-intercept: 6

61.

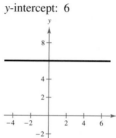

63. $y = -\frac{1}{2}x + 2$

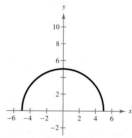

65. $y = \frac{2}{7}x + \frac{2}{7}$ **67.** (a) $y = \frac{5}{4}x - \frac{23}{4}$ (b) $y = -\frac{4}{5}x + \frac{2}{5}$

69. $W = 0.75x + 12.25$ **71.** No   **73.** Yes

75. (a) 16   (b) $(t + 1)^{4/3}$ (c) 81   (d) $x^{4/3}$

77. All real numbers x such that $-5 \le x \le 5$

79. (a) 16 ft/sec   (b) 1.5 sec   **81.** Function   **83.** $-\frac{1}{2}$

85.

 Increasing on $(0, \infty)$

 Decreasing on $(-\infty, -1)$

 Constant on $(-1, 0)$

87.

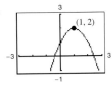

89. 4   **91.** Neither even nor odd; no symmetry

93. Odd; origin symmetry

95. (a) $f(x) = -3x$

(b)

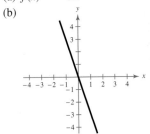

97.

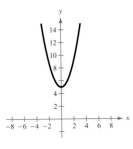

99.

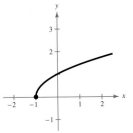

101.

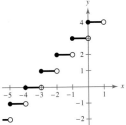

103. (a) $f(x) = x^2$

(b) Reflection in the x-axis, horizontal shift two units to the left, and vertical shift three units up

(c)   (d) $h(x) = -f(x+2) + 3$

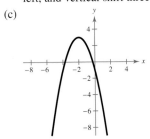

105. (a) $f(x) = x^3$

(b) Reflection in the x-axis and vertical shrink

(c)   (d) $h(x) = -\frac{1}{3}f(x)$

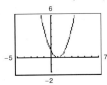

107. (a) $f(x) = \sqrt{x}$

(b) Reflection in the x-axis and vertical shift four units up

(c)   (d) $h(x) = -f(x) + 4$

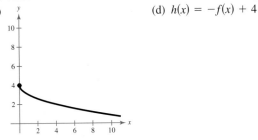

109. (a) $f(x) = |x|$

(b) Horizontal shift three units to the left and vertical shift five units down

(c)   (d) $h(x) = f(x+3) - 5$

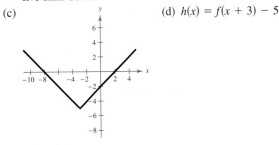

111. (a) $f(x) = [\![x]\!]$

(b) Reflection in the x-axis and vertical shift six units up

(c)   (d) $h(x) = -f(x) + 6$

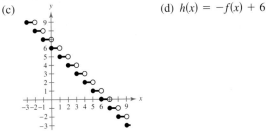

113. (a) $x^2 + 2x + 2$ (b) $x^2 - 2x + 4$

(c) $2x^3 - x^2 + 6x - 3$

(d) $\dfrac{x^2 + 3}{2x - 1}$; all real numbers x except $x = \dfrac{1}{2}$

115. (a) $x - \frac{8}{3}$ (b) $x - 8$

Domains of f, g, $f \circ g$, and $g \circ f$: all real numbers x

117. $N(T(t)) = 100t^2 + 275$

The composition function $N(T(t))$ represents the number of bacteria in the food as a function of time.

119. $f^{-1}(x) = 5x + 4$

$$f(f^{-1}(x)) = \frac{5x + 4 - 4}{5} = x$$

$$f^{-1}(f(x)) = 5\left(\frac{x-4}{5}\right) + 4 = x$$

121.

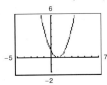

The function does not have an inverse function.

CHAPTER P

123. (a) $f^{-1}(x) = 2x + 6$

(b)

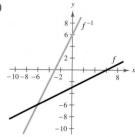

(c) The graphs are reflections of each other in the line $y = x$.

(d) Both f and f^{-1} have domains and ranges that are all real numbers.

125. $x > 4; f^{-1}(x) = \sqrt{\dfrac{x}{2}} + 4, x \neq 0$

127. False. The graph is reflected in the x-axis, shifted 9 units to the left, and then shifted 13 units down.

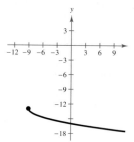

Chapter Test   *(page 117)*

1. $-\dfrac{10}{3} > -|-4|$ **2.** 9.15

3. Additive Identity Property   **4.** $\dfrac{128}{11}$ **5.** $-4, 5$

6. No solution   **7.** $\pm\sqrt{2}$

8.

Distance: $\sqrt{89}$;

Midpoint: $\left(2, \dfrac{5}{2}\right)$

9. No symmetry

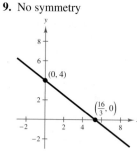

10. y-axis symmetry

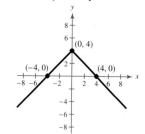

11. Origin symmetry

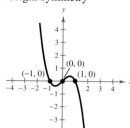

12. Center: $(3, 0)$; Radius: 3

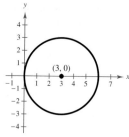

13. $y = -2x + 1$

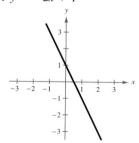

14. $y = -1.7x + 5.9$

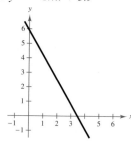

15. (a) $y = -\dfrac{5}{2}x + 4$ (b) $y = \dfrac{2}{5}x + 4$

16. (a) -9 (b) 1   (c) $|x - 4| - 15$

17. (a) 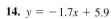   (b) All real numbers x

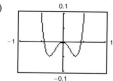

(c) Increasing on $(-0.31, 0)$, $(0.31, \infty)$

Decreasing on $(-\infty, -0.31)$, $(0, 0.31)$

(d) Even

18. (a)

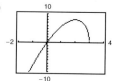

(b) All real numbers x such that $x \leq 3$

(c) Increasing on $(-\infty, 2)$

Decreasing on $(2, 3)$

(d) Neither

19. (a)

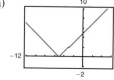

(b) All real numbers x

(c) Increasing on $(-5, \infty)$

Decreasing on $(-\infty, -5)$

(d) Neither

20. (a) $f(x) = [\![x]\!]$ (b) Vertical stretch of $y = [\![x]\!]$

(c)

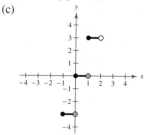

21. (a) $f(x) = \sqrt{x}$

(b) Reflection in the x-axis, vertical shift eight units up, and horizontal shift five units to the left

(c)

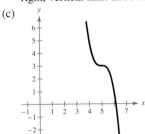

22. (a) $f(x) = x^3$

(b) Reflection in the y-axis, horizontal shift five units to the right, vertical shift three units up, and vertical stretch

(c)

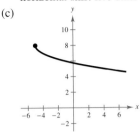

23. (a) $2x^2 - 4x - 2$ (b) $4x^2 + 4x - 12$

(c) $-3x^4 - 12x^3 + 22x^2 + 28x - 35$

(d) $\dfrac{3x^2 - 7}{-x^2 - 4x + 5}, \ x \neq 1, -5$

(e) $3x^4 + 24x^3 + 18x^2 - 120x + 68$

(f) $-9x^4 + 30x^2 - 16$

24. (a) $\dfrac{1 + 2x^{3/2}}{x}, \ x > 0$ (b) $\dfrac{1 - 2x^{3/2}}{x}, \ x > 0$

(c) $\dfrac{2\sqrt{x}}{x}, \ x > 0$ (d) $\dfrac{1}{2x^{3/2}}, \ x > 0$

(e) $\dfrac{\sqrt{x}}{2x}, \ x > 0$ (f) $\dfrac{2\sqrt{x}}{x}, \ x > 0$

25. $f^{-1}(x) = \sqrt[3]{x - 8}$ **26.** No inverse

27. $f^{-1}(x) = \left(\dfrac{x}{3}\right)^{2/3}, \ x \geq 0$

Problem Solving   *(page 119)*

1. (a) $W_1 = 2000 + 0.07S$ (b) $W_2 = 2300 + 0.05S$

(c)

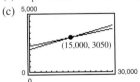

Both jobs pay the same monthly salary when sales equal $15,000.

(d) No. Job 1 would pay $3400 and job 2 would pay $3300.

3. (a) The function will be even.   (b) The function will be odd.

(c) The function will be neither even nor odd.

5. $f(x) = a_{2n}x^{2n} + a_{2n-2}x^{2n-2} + \cdots + a_2 x^2 + a_0$

$f(-x) = a_{2n}(-x)^{2n} + a_{2n-2}(-x)^{2n-2}$
$+ \cdots + a_2(-x)^2 + a_0$

$= f(x)$

7. (a) $81\frac{2}{3}$ h

(b) $25\frac{5}{7}$ mi/h

(c) $y = \dfrac{-180}{7}x + 3400$

Domain: $0 \leq x \leq \dfrac{1190}{9}$

Range: $0 \leq y \leq 3400$

(d)

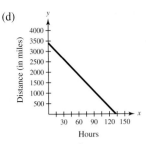

9. (a) $(f \circ g)(x) = 4x + 24$ (b) $(f \circ g)^{-1}(x) = \frac{1}{4}x - 6$

(c) $f^{-1}(x) = \frac{1}{4}x; \ g^{-1}(x) = x - 6$

(d) $(g^{-1} \circ f^{-1})(x) = \frac{1}{4}x - 6$; They are the same.

(e) $(f \circ g)(x) = 8x^3 + 1; \ (f \circ g)^{-1}(x) = \frac{1}{2}\sqrt[3]{x - 1};$
$f^{-1}(x) = \sqrt[3]{x - 1}; \ g^{-1}(x) = \frac{1}{2}x;$
$(g^{-1} \circ f^{-1})(x) = \frac{1}{2}\sqrt[3]{x - 1}$

(f) Answers will vary.

(g) $(f \circ g)^{-1}(x) = (g^{-1} \circ f^{-1})(x)$

11. (a)   (b)   (c)   (d)   (e)   (f)

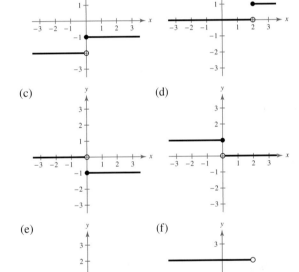

13. Proof

15. (a)

x	-4	-2	0	4
$f(f^{-1}(x))$	-4	-2	0	4

(b)

x	-3	-2	0	1
$(f + f^{-1})(x)$	5	1	-3	-5

(c)

x	-3	-2	0	1
$(f \cdot f^{-1})(x)$	4	0	2	6

(d)

x	-4	-3	0	4
$|f^{-1}(x)|$	2	1	1	3

Chapter 1

Section 1.1   *(page 129)*

1. coterminal   **3.** complementary; supplementary
5. linear; angular   **7.** 1 rad   **9.** -3 rad
11. (a) Quadrant I   (b) Quadrant III
13. (a)   (b)

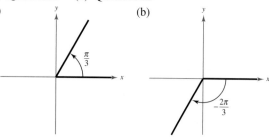

15. *Sample answers:* (a) $\dfrac{13\pi}{6}, -\dfrac{11\pi}{6}$ (b) $\dfrac{19\pi}{6}, -\dfrac{5\pi}{6}$

17. (a) Complement: $\dfrac{\pi}{6}$; Supplement: $\dfrac{2\pi}{3}$

   (b) Complement: $\dfrac{\pi}{4}$; Supplement: $\dfrac{3\pi}{4}$

19. (a) Complement: $\dfrac{\pi}{2} - 1 \approx 0.57$;

   Supplement: $\pi - 1 \approx 2.14$
   (b) Complement: none; Supplement: $\pi - 2 \approx 1.14$

21. $210°$ **23.** $-60°$ **25.** (a) Quadrant II   (b) Quadrant I
27. (a)   (b)

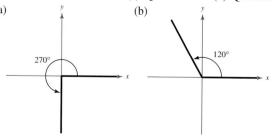

29. *Sample answers:* (a) $405°, -315°$ (b) $324°, -396°$
31. (a) Complement: $72°$; Supplement: $162°$
   (b) Complement: $5°$; Supplement: $95°$
33. (a) Complement: none; Supplement: $30°$
   (b) Complement: $11°$; Supplement: $101°$
35. (a) $\dfrac{2\pi}{3}$ (b) $-\dfrac{\pi}{9}$ **37.** (a) $270°$ (b) $210°$
39. 0.785   **41.** 0.009   **43.** $81.818°$ **45.** $-756.000°$
47. (a) $54.75°$ (b) $-128.5°$
49. (a) $240°36'$ (b) $-145°48'$
51. 10π in. ≈ 31.42 in.   **53.** $\dfrac{15}{8}$ rad   **55.** 4 rad
57. 18π mm$^2 \approx 56.55$ mm^2 **59.** 591.3 mi   **61.** $23.87°$
63. (a) $10{,}000\pi$ rad/min $\approx 31{,}415.93$ rad/min
   (b) 9490.23 ft/min
65. (a) $[400\pi, 1000\pi]$ rad/min   (b) $[2400\pi, 6000\pi]$ cm/min
67. (a) 35.70 mi/h   (b) 739.50 revolutions/min
69.

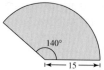

$A = 87.5\pi$ m$^2 \approx 274.89$ m^2

71. False. A measurement of 4π radians corresponds to two complete revolutions from the initial side to the terminal side of an angle.
73. False. The terminal side of the angle lies on the x-axis.
75. The speed increases. The linear velocity is proportional to the radius.
77. The arc length is increasing. If θ is constant, the length of the arc is proportional to the radius $(s = r\theta)$.

Section 1.2   *(page 137)*

1. unit circle   **3.** period
5. $\sin t = \dfrac{5}{13}$ $\csc t = \dfrac{13}{5}$
   $\cos t = \dfrac{12}{13}$ $\sec t = \dfrac{13}{12}$
   $\tan t = \dfrac{5}{12}$ $\cot t = \dfrac{12}{5}$

7. $\sin t = -\dfrac{3}{5}$ $\csc t = -\dfrac{5}{3}$
   $\cos t = -\dfrac{4}{5}$ $\sec t = -\dfrac{5}{4}$
   $\tan t = \dfrac{3}{4}$ $\cot t = \dfrac{4}{3}$

9. $(0, 1)$ **11.** $\left(-\dfrac{\sqrt{3}}{2}, \dfrac{1}{2}\right)$

13. $\sin \dfrac{\pi}{4} = \dfrac{\sqrt{2}}{2}$

   $\cos \dfrac{\pi}{4} = \dfrac{\sqrt{2}}{2}$

   $\tan \dfrac{\pi}{4} = 1$

15. $\sin\left(-\dfrac{\pi}{6}\right) = -\dfrac{1}{2}$

   $\cos\left(-\dfrac{\pi}{6}\right) = \dfrac{\sqrt{3}}{2}$

   $\tan\left(-\dfrac{\pi}{6}\right) = -\dfrac{\sqrt{3}}{3}$

17. $\sin\left(-\dfrac{7\pi}{4}\right) = \dfrac{\sqrt{2}}{2}$

   $\cos\left(-\dfrac{7\pi}{4}\right) = \dfrac{\sqrt{2}}{2}$

   $\tan\left(-\dfrac{7\pi}{4}\right) = 1$

19. $\sin \dfrac{11\pi}{6} = -\dfrac{1}{2}$

   $\cos \dfrac{11\pi}{6} = \dfrac{\sqrt{3}}{2}$

   $\tan \dfrac{11\pi}{6} = -\dfrac{\sqrt{3}}{3}$

21. $\sin\left(-\dfrac{3\pi}{2}\right) = 1$

   $\cos\left(-\dfrac{3\pi}{2}\right) = 0$

   $\tan\left(-\dfrac{3\pi}{2}\right)$ is undefined.

23. $\sin \dfrac{2\pi}{3} = \dfrac{\sqrt{3}}{2}$ $\csc \dfrac{2\pi}{3} = \dfrac{2\sqrt{3}}{3}$

   $\cos \dfrac{2\pi}{3} = -\dfrac{1}{2}$ $\sec \dfrac{2\pi}{3} = -2$

   $\tan \dfrac{2\pi}{3} = -\sqrt{3}$ $\cot \dfrac{2\pi}{3} = -\dfrac{\sqrt{3}}{3}$

25. $\sin \dfrac{4\pi}{3} = -\dfrac{\sqrt{3}}{2}$ $\csc \dfrac{4\pi}{3} = -\dfrac{2\sqrt{3}}{3}$

   $\cos \dfrac{4\pi}{3} = -\dfrac{1}{2}$ $\sec \dfrac{4\pi}{3} = -2$

   $\tan \dfrac{4\pi}{3} = \sqrt{3}$ $\cot \dfrac{4\pi}{3} = \dfrac{\sqrt{3}}{3}$

27. $\sin\left(-\dfrac{5\pi}{3}\right) = \dfrac{\sqrt{3}}{2}$ $\csc\left(-\dfrac{5\pi}{3}\right) = \dfrac{2\sqrt{3}}{3}$

   $\cos\left(-\dfrac{5\pi}{3}\right) = \dfrac{1}{2}$ $\sec\left(-\dfrac{5\pi}{3}\right) = 2$

   $\tan\left(-\dfrac{5\pi}{3}\right) = \sqrt{3}$ $\cot\left(-\dfrac{5\pi}{3}\right) = \dfrac{\sqrt{3}}{3}$

29. $\sin\left(-\dfrac{\pi}{2}\right) = -1$ $\csc\left(-\dfrac{\pi}{2}\right) = -1$

   $\cos\left(-\dfrac{\pi}{2}\right) = 0$ $\sec\left(-\dfrac{\pi}{2}\right)$ is undefined.

   $\tan\left(-\dfrac{\pi}{2}\right)$ is undefined. $\cot\left(-\dfrac{\pi}{2}\right) = 0$

31. $\sin 4\pi = \sin 0 = 0$ **33.** $\cos \dfrac{7\pi}{3} = \cos \dfrac{\pi}{3} = \dfrac{1}{2}$

35. $\sin \dfrac{19\pi}{6} = \sin \dfrac{7\pi}{6} = -\dfrac{1}{2}$ **37.** (a) $-\dfrac{1}{2}$ (b) -2

39. (a) $-\dfrac{1}{5}$ (b) -5 **41.** (a) $\dfrac{4}{5}$ (b) $-\dfrac{4}{5}$ **43.** 1.7321

45. 1.3940   **47.** -4.4014

49. (a) 0.25 ft   (b) 0.02 ft   (c) -0.25 ft

51. False. $\sin(-t) = -\sin(t)$ means that the function is odd, not that the sine of a negative angle is a negative number.

53. True. The tangent function has a period of π.

55. (a) y-axis symmetry   (b) $\sin t_1 = \sin(\pi - t_1)$
   (c) $\cos(\pi - t_1) = -\cos t_1$

57. Answers will vary.

59. (a)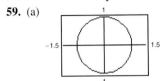

Circle of radius 1 centered at $(0, 0)$

(b) The t-values represent the central angle in radians. The x- and y-values represent the location in the coordinate plane.

(c) $-1 \le x \le 1, -1 \le y \le 1$

61. It is an odd function.

Section 1.3   *(page 146)*

1. (a) v   (b) iv   (c) vi   (d) iii   (e) i   (f) ii

3. complementary

5. $\sin \theta = \dfrac{3}{5}$ $\csc \theta = \dfrac{5}{3}$ **7.** $\sin \theta = \dfrac{9}{41}$ $\csc \theta = \dfrac{41}{9}$
$\cos \theta = \dfrac{4}{5}$ $\sec \theta = \dfrac{5}{4}$ $\cos \theta = \dfrac{40}{41}$ $\sec \theta = \dfrac{41}{40}$
$\tan \theta = \dfrac{3}{4}$ $\cot \theta = \dfrac{4}{3}$ $\tan \theta = \dfrac{9}{40}$ $\cot \theta = \dfrac{40}{9}$

9. $\sin \theta = \dfrac{8}{17}$ $\csc \theta = \dfrac{17}{8}$
$\cos \theta = \dfrac{15}{17}$ $\sec \theta = \dfrac{17}{15}$
$\tan \theta = \dfrac{8}{15}$ $\cot \theta = \dfrac{15}{8}$
The triangles are similar, and corresponding sides are proportional.

11. $\sin \theta = \dfrac{1}{3}$ $\csc \theta = 3$
$\cos \theta = \dfrac{2\sqrt{2}}{3}$ $\sec \theta = \dfrac{3\sqrt{2}}{4}$
$\tan \theta = \dfrac{\sqrt{2}}{4}$ $\cot \theta = 2\sqrt{2}$
The triangles are similar, and corresponding sides are proportional.

13.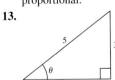

$\sin \theta = \dfrac{3}{5}$ $\csc \theta = \dfrac{5}{3}$
$\cos \theta = \dfrac{4}{5}$ $\sec \theta = \dfrac{5}{4}$
            $\cot \theta = \dfrac{4}{3}$

15.

$\sin \theta = \dfrac{\sqrt{5}}{3}$ $\csc \theta = \dfrac{3\sqrt{5}}{5}$
$\cos \theta = \dfrac{2}{3}$
$\tan \theta = \dfrac{\sqrt{5}}{2}$ $\cot \theta = \dfrac{2\sqrt{5}}{5}$

17.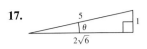

$\csc \theta = 5$
$\cos \theta = \dfrac{2\sqrt{6}}{5}$ $\sec \theta = \dfrac{5\sqrt{6}}{12}$
$\tan \theta = \dfrac{\sqrt{6}}{12}$ $\cot \theta = 2\sqrt{6}$

19.

$\sin \theta = \dfrac{\sqrt{10}}{10}$ $\csc \theta = \sqrt{10}$
$\cos \theta = \dfrac{3\sqrt{10}}{10}$ $\sec \theta = \dfrac{\sqrt{10}}{3}$
$\tan \theta = \dfrac{1}{3}$

21. $\dfrac{\pi}{6}; \dfrac{1}{2}$ **23.** $45°; \sqrt{2}$ **25.** $60°; \dfrac{\pi}{3}$ **27.** $30°; 2$

29. $45°; \dfrac{\pi}{4}$ **31.** (a) 0.1736   (b) 0.1736

33. (a) 0.2815   (b) 3.5523   **35.** (a) 0.9964   (b) 1.0036

37. (a) 5.0273   (b) 0.1989   **39.** (a) 1.8527   (b) 0.9817

41. (a) $\dfrac{1}{2}$ (b) $\dfrac{\sqrt{3}}{2}$ (c) $\sqrt{3}$ (d) $\dfrac{\sqrt{3}}{3}$

43. (a) $\dfrac{2\sqrt{2}}{3}$ (b) $2\sqrt{2}$ (c) 3   (d) 3

45. (a) $\dfrac{1}{5}$ (b) $\sqrt{26}$ (c) $\dfrac{1}{5}$ (d) $\dfrac{5\sqrt{26}}{26}$

47–55. Answers will vary.   **57.** (a) $30° = \dfrac{\pi}{6}$ (b) $30° = \dfrac{\pi}{6}$

59. (a) $60° = \dfrac{\pi}{3}$ (b) $45° = \dfrac{\pi}{4}$

61. (a) $60° = \dfrac{\pi}{3}$ (b) $45° = \dfrac{\pi}{4}$ **63.** $x = 9, y = 9\sqrt{3}$

65. $x = \dfrac{32\sqrt{3}}{3}, r = \dfrac{64\sqrt{3}}{3}$ **67.** 443.2 m; 323.3 m

69. $30° = \dfrac{\pi}{6}$ **71.** (a) 219.9 ft   (b) 160.9 ft

73. $(x_1, y_1) = (28\sqrt{3}, 28)$
     $(x_2, y_2) = (28, 28\sqrt{3})$

75. $\sin 20° \approx 0.34, \cos 20° \approx 0.94, \tan 20° \approx 0.36,$
     $\csc 20° \approx 2.92, \sec 20° \approx 1.06, \cot 20° \approx 2.75$

77. (a) 519.33 ft   (b) 1174.17 ft   (c) 173.11 ft/min

79. True, $\csc x = \dfrac{1}{\sin x}$.   **81.** False, $\dfrac{\sqrt{2}}{2} + \dfrac{\sqrt{2}}{2} \neq 1$.

83. False, $1.7321 \neq 0.0349$.

85. Yes, $\tan \theta$ is equal to opp/adj. You can find the value of the hypotenuse by the Pythagorean Theorem. Then you can find $\sec \theta$, which is equal to hyp/adj.

87. (a)

θ	0°	18°	36°	54°	72°	90°
$\sin \theta$	0	0.3090	0.5878	0.8090	0.9511	1
$\cos \theta$	1	0.9511	0.8090	0.5878	0.3090	0

(b) Increasing function   (c) Decreasing function

(d) As the angle increases, the length of the side opposite the angle increases relative to the length of the hypotenuse and the length of the side adjacent to the angle decreases relative to the length of the hypotenuse. Thus, the sine increases and the cosine decreases.

CHAPTER 1

Section 1.4 *(page 156)*

1. $\dfrac{y}{r}$ **3.** $\dfrac{y}{x}$ **5.** $\cos \theta$ **7.** zero; defined

9. (a) $\sin \theta = \dfrac{3}{5}$ $\csc \theta = \dfrac{5}{3}$

$\cos \theta = \dfrac{4}{5}$ $\sec \theta = \dfrac{5}{4}$

$\tan \theta = \dfrac{3}{4}$ $\cot \theta = \dfrac{4}{3}$

(b) $\sin \theta = \dfrac{15}{17}$ $\csc \theta = \dfrac{17}{15}$

$\cos \theta = -\dfrac{8}{17}$ $\sec \theta = -\dfrac{17}{8}$

$\tan \theta = -\dfrac{15}{8}$ $\cot \theta = -\dfrac{8}{15}$

11. (a) $\sin \theta = -\dfrac{1}{2}$ $\csc \theta = -2$

$\cos \theta = -\dfrac{\sqrt{3}}{2}$ $\sec \theta = -\dfrac{2\sqrt{3}}{3}$

$\tan \theta = \dfrac{\sqrt{3}}{3}$ $\cot \theta = \sqrt{3}$

(b) $\sin \theta = -\dfrac{\sqrt{17}}{17}$ $\csc \theta = -\sqrt{17}$

$\cos \theta = \dfrac{4\sqrt{17}}{17}$ $\sec \theta = \dfrac{\sqrt{17}}{4}$

$\tan \theta = -\dfrac{1}{4}$ $\cot \theta = -4$

13. $\sin \theta = \dfrac{12}{13}$ $\csc \theta = \dfrac{13}{12}$

$\cos \theta = \dfrac{5}{13}$ $\sec \theta = \dfrac{13}{5}$

$\tan \theta = \dfrac{12}{5}$ $\cot \theta = \dfrac{5}{12}$

15. $\sin \theta = -\dfrac{2\sqrt{29}}{29}$ $\csc \theta = -\dfrac{\sqrt{29}}{2}$

$\cos \theta = -\dfrac{5\sqrt{29}}{29}$ $\sec \theta = -\dfrac{\sqrt{29}}{5}$

$\tan \theta = \dfrac{2}{5}$ $\cot \theta = \dfrac{5}{2}$

17. $\sin \theta = \dfrac{4}{5}$ $\csc \theta = \dfrac{5}{4}$

$\cos \theta = -\dfrac{3}{5}$ $\sec \theta = -\dfrac{5}{3}$

$\tan \theta = -\dfrac{4}{3}$ $\cot \theta = -\dfrac{3}{4}$

19. Quadrant I   **21.** Quadrant II

23. $\sin \theta = \dfrac{15}{17}$ **25.** $\sin \theta = \dfrac{3}{5}$

$\cos \theta = -\dfrac{8}{17}$ $\cos \theta = -\dfrac{4}{5}$

$\tan \theta = -\dfrac{15}{8}$ $\tan \theta = -\dfrac{3}{4}$

$\csc \theta = \dfrac{17}{15}$ $\csc \theta = \dfrac{5}{3}$

$\sec \theta = -\dfrac{17}{8}$ $\sec \theta = -\dfrac{5}{4}$

$\cot \theta = -\dfrac{8}{15}$ $\cot \theta = -\dfrac{4}{3}$

27. $\sin \theta = -\dfrac{\sqrt{10}}{10}$ $\csc \theta = -\sqrt{10}$

$\cos \theta = \dfrac{3\sqrt{10}}{10}$ $\sec \theta = \dfrac{\sqrt{10}}{3}$

$\tan \theta = -\dfrac{1}{3}$ $\cot \theta = -3$

29. $\sin \theta = -\dfrac{\sqrt{3}}{2}$ $\csc \theta = -\dfrac{2\sqrt{3}}{3}$

$\cos \theta = -\dfrac{1}{2}$ $\sec \theta = -2$

$\tan \theta = \sqrt{3}$ $\cot \theta = \dfrac{\sqrt{3}}{3}$

31. $\sin \theta = 0$ $\csc \theta$ is undefined.

$\cos \theta = -1$ $\sec \theta = -1$

$\tan \theta = 0$ $\cot \theta$ is undefined.

33. $\sin \theta = \dfrac{\sqrt{2}}{2}$ $\csc \theta = \sqrt{2}$

$\cos \theta = -\dfrac{\sqrt{2}}{2}$ $\sec \theta = -\sqrt{2}$

$\tan \theta = -1$ $\cot \theta = -1$

35. $\sin \theta = -\dfrac{2\sqrt{5}}{5}$ $\csc \theta = -\dfrac{\sqrt{5}}{2}$

$\cos \theta = -\dfrac{\sqrt{5}}{5}$ $\sec \theta = -\sqrt{5}$

$\tan \theta = 2$ $\cot \theta = \dfrac{1}{2}$

37. 0   **39.** Undefined   **41.** 1   **43.** Undefined

45. $\theta' = 20°$ **47.** $\theta' = 55°$

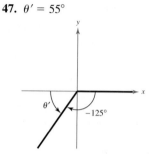

49. $\theta' = \dfrac{\pi}{3}$ **51.** $\theta' = 2\pi - 4.8$

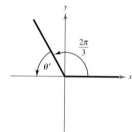

53. $\sin 225° = -\dfrac{\sqrt{2}}{2}$ **55.** $\sin 750° = \dfrac{1}{2}$

$\cos 225° = -\dfrac{\sqrt{2}}{2}$ $\cos 750° = \dfrac{\sqrt{3}}{2}$

$\tan 225° = 1$ $\tan 750° = \dfrac{\sqrt{3}}{3}$

57. $\sin(-840°) = -\dfrac{\sqrt{3}}{2}$ **59.** $\sin \dfrac{2\pi}{3} = \dfrac{\sqrt{3}}{2}$

$\cos(-840°) = -\dfrac{1}{2}$ $\cos \dfrac{2\pi}{3} = -\dfrac{1}{2}$

$\tan(-840°) = \sqrt{3}$ $\tan \dfrac{2\pi}{3} = -\sqrt{3}$

61. $\sin \dfrac{5\pi}{4} = -\dfrac{\sqrt{2}}{2}$ **63.** $\sin\left(-\dfrac{\pi}{6}\right) = -\dfrac{1}{2}$

$\cos \dfrac{5\pi}{4} = -\dfrac{\sqrt{2}}{2}$ $\cos\left(-\dfrac{\pi}{6}\right) = \dfrac{\sqrt{3}}{2}$

$\tan \dfrac{5\pi}{4} = 1$ $\tan\left(-\dfrac{\pi}{6}\right) = -\dfrac{\sqrt{3}}{3}$

65. $\sin \dfrac{9\pi}{4} = \dfrac{\sqrt{2}}{2}$ **67.** $\sin\left(-\dfrac{3\pi}{2}\right) = 1$

$\cos \dfrac{9\pi}{4} = \dfrac{\sqrt{2}}{2}$ $\cos\left(-\dfrac{3\pi}{2}\right) = 0$

$\tan \dfrac{9\pi}{4} = 1$ $\tan\left(-\dfrac{3\pi}{2}\right)$ is undefined.

69. $\dfrac{4}{5}$ **71.** $-\dfrac{\sqrt{13}}{2}$ **73.** $\dfrac{8}{5}$ **75.** 0.1736

77. -0.3420 **79.** -1.4826 **81.** 3.2361   **83.** 4.6373

85. 0.3640   **87.** -0.6052 **89.** -0.4142

91. (a) $30° = \dfrac{\pi}{6}$, $150° = \dfrac{5\pi}{6}$ (b) $210° = \dfrac{7\pi}{6}$, $330° = \dfrac{11\pi}{6}$

93. (a) $60° = \dfrac{\pi}{3}$, $120° = \dfrac{2\pi}{3}$ (b) $135° = \dfrac{3\pi}{4}$, $315° = \dfrac{7\pi}{4}$

95. (a) $45° = \dfrac{\pi}{4}$, $225° = \dfrac{5\pi}{4}$ (b) $150° = \dfrac{5\pi}{6}$, $330° = \dfrac{11\pi}{6}$

97. (a) 12 mi   (b) 6 mi   (c) 6.9 mi

99. (a) $N = 22.099 \sin(0.522t - 2.219) + 55.008$
      $F = 36.641 \sin(0.502t - 1.831) + 25.610$
  (b) February: $N = 34.6°$, $F = -1.4°$
      March: $N = 41.6°$, $F = 13.9°$
      May: $N = 63.4°$, $F = 48.6°$
      June: $N = 72.5°$, $F = 59.5°$
      August: $N = 75.5°$, $F = 55.6°$
      September: $N = 68.6°$, $F = 41.7°$
      November: $N = 46.8°$, $F = 6.5°$
  (c) Answers will vary.

101. (a) 270.63 ft   (b) 307.75 ft   (c) 270.63 ft

103. False. In each of the four quadrants, the signs of the secant function and the cosine function are the same because these functions are reciprocals of each other.

105. (a) $\sin t = y$ (b) $r = 1$ because it is a unit circle.
      $\cos t = x$
  (c) $\sin \theta = y$ (d) $\sin t = \sin \theta$ and $\cos t = \cos \theta$
      $\cos \theta = x$

Section 1.5 (page 166)

1. cycle   **3.** phase shift   **5.** Period: $\dfrac{2\pi}{5}$; Amplitude: 2

7. Period: 4π; Amplitude: $\dfrac{3}{4}$ **9.** Period: 6; Amplitude: $\dfrac{1}{2}$

11. Period: 2π; Amplitude: 4   **13.** Period: $\dfrac{\pi}{5}$; Amplitude: 3

15. Period: $\dfrac{5\pi}{2}$; Amplitude: $\dfrac{5}{3}$ **17.** Period: 1; Amplitude: $\dfrac{1}{4}$

19. g is a shift of f π units to the right.

21. g is a reflection of f in the x-axis.

23. The period of f is twice the period of g.

25. g is a shift of f three units up.

27. The graph of g has twice the amplitude of the graph of f.

29. The graph of g is a horizontal shift of the graph of f π units to the right.

31.
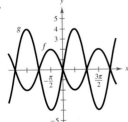

33.
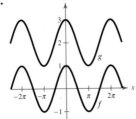

35.
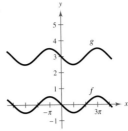

37.
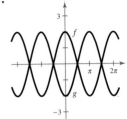

39.
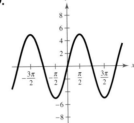

41.
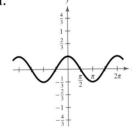

43.
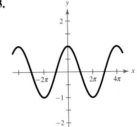

45.
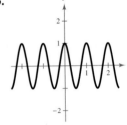

47.
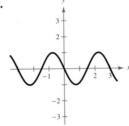

49.
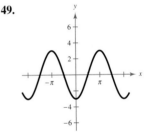

51.
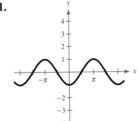

53.
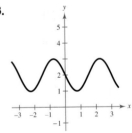

55.
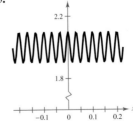

57.
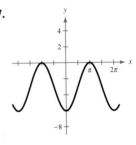

59.

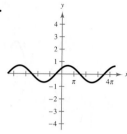

61. (a) $g(x)$ is obtained by a horizontal shrink of four, and one cycle of $g(x)$ corresponds to the interval $[\pi/4, 3\pi/4]$.

(b)   (c) $g(x) = f(4x - \pi)$

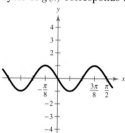

63. (a) One cycle of $g(x)$ corresponds to the interval $[\pi, 3\pi]$, and $g(x)$ is obtained by shifting $f(x)$ up two units.

(b)   (c) $g(x) = f(x - \pi) + 2$

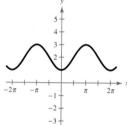

65. (a) One cycle of $g(x)$ is $[\pi/4, 3\pi/4]$. $g(x)$ is also shifted down three units and has an amplitude of two.

(b)   (c) $g(x) = 2f(4x - \pi) - 3$

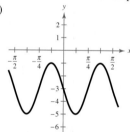

67. 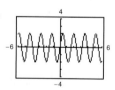   **69.**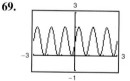

71. 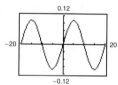   **73.** $a = 2, d = 1$

75. $a = -4, d = 4$ **77.** $a = -3, b = 2, c = 0$

79. $a = 2, b = 1, c = -\dfrac{\pi}{4}$

81. 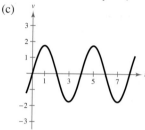   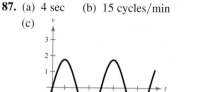

83. $y = 1 + 2\sin(2x - \pi)$ **85.** $y = \cos(2x + 2\pi) - \dfrac{3}{2}$

87. (a) 4 sec   (b) 15 cycles/min

(c)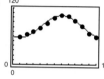

89. (a) $I(t) = 46.2 + 32.4\cos\left(\dfrac{\pi t}{6} - 3.67\right)$

(b) 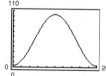   (c)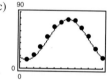

The model fits the data well.   The model fits the data well.

(d) Las Vegas: 80.6°; International Falls: 46.2°
The constant term gives the annual average temperature.

(e) 12; yes; One full period is one year.

(f) International Falls; amplitude; The greater the amplitude, the greater the variability in temperature.

91. (a) $\dfrac{1}{440}$ sec   (b) 440 cycles/sec

93. (a) 20 sec; It takes 20 seconds to complete one revolution on the Ferris wheel.

(b) 50 ft; The diameter of the Ferris wheel is 100 feet.

(c)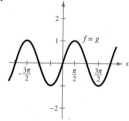

95. False. The graph of $f(x) = \sin(x + 2\pi)$ translates the graph of $f(x) = \sin x$ exactly one period to the left so that the two graphs look identical.

97.

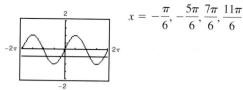

Conjecture: $\sin x = \cos\left(x - \dfrac{\pi}{2}\right)$

99.

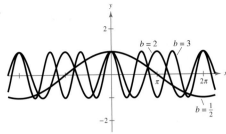

The value of b affects the period of the graph.
$b = \frac{1}{2} \rightarrow \frac{1}{2}$ cycle
$b = 2 \rightarrow 2$ cycles
$b = 3 \rightarrow 3$ cycles

101. (a) 0.4794, 0.4794   (b) 0.8417, 0.8415   (c) 0.5, 0.5
(d) 0.8776, 0.8776   (e) 0.5417, 0.5403
(f) 0.7074, 0.7071
The error increases as x moves farther away from 0.

Section 1.6   *(page 177)*

1. odd; origin   **3.** reciprocal   **5.** π
7. $(-\infty, -1] \cup [1, \infty)$ **9.** e, π **10.** c, 2π
11. a, 1   **12.** d, 2π **13.** f, 4   **14.** b, 4

15.

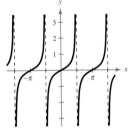

17.

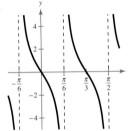

19.

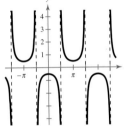

21.

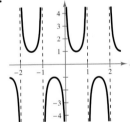

23.

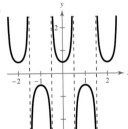

25.

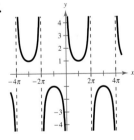

27.

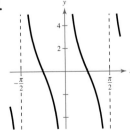

29.

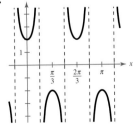

31.

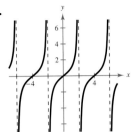

33.

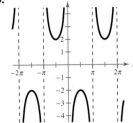

35.

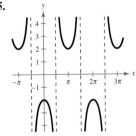

37.

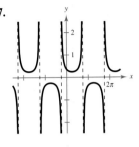

39.

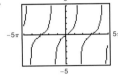

41.

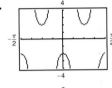

43.

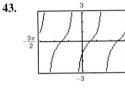

45.

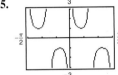

47.

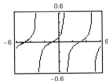

49. $-\dfrac{7\pi}{4}, -\dfrac{3\pi}{4}, \dfrac{\pi}{4}, \dfrac{5\pi}{4}$ **51.** $-\dfrac{4\pi}{3}, -\dfrac{\pi}{3}, \dfrac{2\pi}{3}, \dfrac{5\pi}{3}$

53. $-\dfrac{4\pi}{3}, -\dfrac{2\pi}{3}, \dfrac{2\pi}{3}, \dfrac{4\pi}{3}$ **55.** $-\dfrac{7\pi}{4}, -\dfrac{5\pi}{4}, \dfrac{\pi}{4}, \dfrac{3\pi}{4}$

57. Even   **59.** Odd   **61.** Odd   **63.** Even
65. d, $f \rightarrow 0$ as $x \rightarrow 0$.   **66.** a, $f \rightarrow 0$ as $x \rightarrow 0$.
67. b, $g \rightarrow 0$ as $x \rightarrow 0$.   **68.** c, $g \rightarrow 0$ as $x \rightarrow 0$.

69.

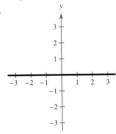

The functions are equal.

71.

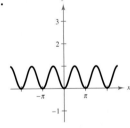

The functions are equal.

73.

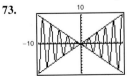

As $x \rightarrow \infty$, $g(x)$ oscillates.

75.

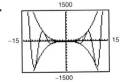

As $x \rightarrow \infty$, $f(x)$ oscillates.

77.

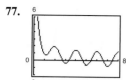

As $x \to 0$, $y \to \infty$.

79.

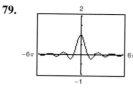

As $x \to 0$, $g(x) \to 1$.

81.

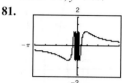

As $x \to 0$, $f(x)$ oscillates between 1 and -1.

83. (a) Period of $H(t)$: 12 mo   (b) Summer; winter
     Period of $L(t)$: 12 mo   (c) About 0.5 mo

85. $d = 7 \cot x$

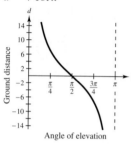

87. True. $y = \sec x$ is equal to $y = 1/\cos x$, and if the reciprocal
of $y = \sin x$ is translated $\pi/2$ units to the left, then

$$\frac{1}{\sin\left(x + \dfrac{\pi}{2}\right)} = \frac{1}{\cos x} = \sec x.$$

89. (a) As $x \to 0^+$, $f(x) \to \infty$.   (b) As $x \to 0^-$, $f(x) \to -\infty$.
    (c) As $x \to \pi^+$, $f(x) \to -\infty$.   (d) As $x \to \pi^-$, $f(x) \to \infty$.

91. (a) As $x \to \left(\dfrac{\pi}{2}\right)^+$, $f(x) \to -\infty$.

    (b) As $x \to \left(\dfrac{\pi}{2}\right)^-$, $f(x) \to \infty$.

    (c) As $x \to \left(-\dfrac{\pi}{2}\right)^+$, $f(x) \to \infty$.

    (d) As $x \to \left(-\dfrac{\pi}{2}\right)^-$, $f(x) \to -\infty$.

93. (a)

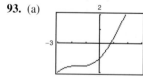

    0.7391

    (b) 1, 0.5403, 0.8576, 0.6543, 0.7935, 0.7014, 0.7640,
      0.7221, 0.7504, 0.7314, . . . ; 0.7391

Section 1.7   (page 186)

1. $y = \sin^{-1} x$; $-1 \le x \le 1$

3. $y = \tan^{-1} x$; $-\infty < x < \infty$; $-\dfrac{\pi}{2} < y < \dfrac{\pi}{2}$ **5.** $\dfrac{\pi}{6}$

7. $\dfrac{\pi}{3}$ **9.** $\dfrac{\pi}{6}$ **11.** $\dfrac{5\pi}{6}$ **13.** $-\dfrac{\pi}{3}$ **15.** $\dfrac{2\pi}{3}$ **17.** $-\dfrac{\pi}{3}$

19.

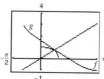

21. 1.19   **23.** -0.85 **25.** -1.25 **27.** 0.32
29. 1.99   **31.** 0.74   **33.** 1.07   **35.** 1.36   **37.** -1.52

39. $-\dfrac{\pi}{3}$, $-\dfrac{\sqrt{3}}{3}$, 1 **41.** $\theta = \arctan \dfrac{x}{4}$ **43.** $\theta = \arcsin \dfrac{x+2}{5}$

45. $\theta = \arccos \dfrac{x+3}{2x}$ **47.** 0.3   **49.** -0.1 **51.** 0

53. $\dfrac{3}{5}$ **55.** $\dfrac{\sqrt{5}}{5}$ **57.** $\dfrac{12}{13}$ **59.** $\dfrac{\sqrt{34}}{5}$ **61.** $\dfrac{\sqrt{5}}{3}$

63. 2   **65.** $\dfrac{1}{x}$ **67.** $\sqrt{1 - 4x^2}$ **69.** $\sqrt{1 - x^2}$

71. $\dfrac{\sqrt{9 - x^2}}{x}$ **73.** $\dfrac{\sqrt{x^2 + 2}}{x}$

75.

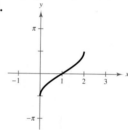

    Asymptotes: $y = \pm 1$

77. $\dfrac{9}{\sqrt{x^2 + 81}}$

79. $\dfrac{|x - 1|}{\sqrt{x^2 - 2x + 10}}$

81.

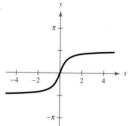

83.

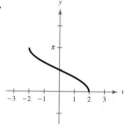

    The graph of g is a
    horizontal shift one unit
    to the right of f.

85.

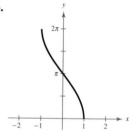

87.

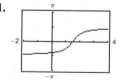

89.

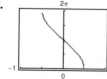

91.

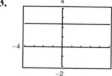

93.

95. $3\sqrt{2}\sin\left(2t+\dfrac{\pi}{4}\right)$

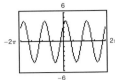

The graph implies that the identity is true.

97. $\dfrac{\pi}{2}$ **99.** $\dfrac{\pi}{2}$ **101.** π

103. (a) $\theta = \arcsin\dfrac{5}{s}$ (b) $0.13, 0.25$

105. (a)

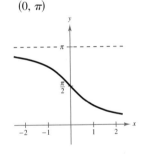

(b) 2 ft   (c) $\beta = 0$; As x increases, β approaches 0.

107. (a) $\theta \approx 26.0°$ (b) 24.4 ft

109. (a) $\theta = \arctan\dfrac{x}{20}$ (b) $14.0°, 31.0°$

111. False. $\dfrac{5\pi}{4}$ is not in the range of the arctangent.

113. False. $\sin^{-1}x$ is the inverse of $\sin x$, not the reciprocal.

115. Domain:
$(-\infty, \infty)$
Range:
$(0, \pi)$

117. Domain:
$(-\infty, -1]\cup[1, \infty)$
Range:
$[-\pi/2, 0)\cup(0, \pi/2]$

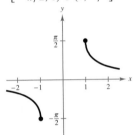

119. $\dfrac{\pi}{4}$ **121.** $\dfrac{3\pi}{4}$ **123.** $\dfrac{\pi}{6}$ **125.** $\dfrac{\pi}{3}$ **127.** 1.17

129. 0.19   **131.** 0.54   **133.** -0.12

135. (a) $\dfrac{\pi}{4}$ (b) $\dfrac{\pi}{2}$ (c) 1.25   (d) 2.03

137. (a) $f\circ f^{-1}$ $f^{-1}\circ f$

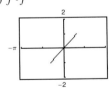   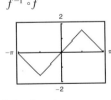

(b) The domains and ranges of the functions are restricted. The graphs of $f\circ f^{-1}$ and $f^{-1}\circ f$ differ because of the domains and ranges of f and f^{-1}.

Section 1.8   *(page 196)*

1. bearing   **3.** period

5. $a \approx 1.73$
$c \approx 3.46$
$B = 60°$

7. $a \approx 8.26$
$c \approx 25.38$
$A = 19°$

9. $c = 5$
$A \approx 36.87°$
$B \approx 53.13°$

11. $a \approx 49.48$
$A \approx 72.08°$
$B \approx 17.92°$

13. $a \approx 91.34$
$b \approx 420.70$
$B = 77°45'$

15. 3.00

17. 2.50   **19.** 214.45 ft   **21.** 19.7 ft

23. 19.9 ft   **25.** 11.8 km   **27.** 56.3°   **29.** 2.06°

31. (a) $\sqrt{h^2 + 34h + 10{,}289}$ (b) $\theta = \arccos\dfrac{100}{l}$

(c) 53.02 ft

33. (a) $l = 250$ ft, $A \approx 36.87°$, $B \approx 53.13°$ (b) 4.87 sec

35. 554 mi north; 709 mi east

37. (a) 104.95 nm south; 58.18 nm west

(b) S 36.7° W; distance $\approx$ 130.9 nm

39. N 56.31° W   **41.** (a) N 58° E   (b) 68.82 m

43. 35.3°   **45.** 29.4 in.   **47.** $d = 4\sin\pi t$

49. $d = 3\cos\dfrac{4\pi t}{3}$ **51.** $\omega = 528\pi$

53. (a) 9   (b) $\frac{3}{5}$ (c) 9   (d) $\frac{5}{12}$

55. (a) $\frac{1}{4}$ (b) 3   (c) 0   (d) $\frac{1}{6}$

57. (a)   (b) $\dfrac{\pi}{8}$ (c) $\dfrac{\pi}{32}$

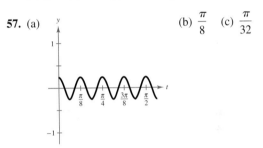

59. (a)   (b) 12; Yes, there are 12 months in a year.

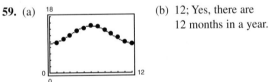

(c) 2.77; The maximum change in the number of hours of daylight

61. False. The scenario does not create a right triangle because the tower is not vertical.

Review Exercises   *(page 202)*

1. (a)   **3.** (a)

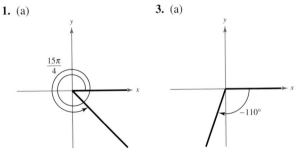

(b) Quadrant IV   (b) Quadrant III

(c) $\dfrac{23\pi}{4}, -\dfrac{\pi}{4}$ (c) 250°, $-470°$

5. 7.854   **7.** -0.589 **9.** 54.000°   **11.** $-200.535°$

13. 198° 24'   **15.** 48.17 in.   **17.** Area $\approx$ 339.29 in.^{2}

19. $\left(-\dfrac{1}{2}, \dfrac{\sqrt{3}}{2}\right)$ **21.** $\left(-\dfrac{\sqrt{3}}{2}, -\dfrac{1}{2}\right)$

CHAPTER 1

23. $\sin \dfrac{3\pi}{4} = \dfrac{\sqrt{2}}{2}$ $\csc \dfrac{3\pi}{4} = \sqrt{2}$

$\cos \dfrac{3\pi}{4} = -\dfrac{\sqrt{2}}{2}$ $\sec \dfrac{3\pi}{4} = -\sqrt{2}$

$\tan \dfrac{3\pi}{4} = -1$ $\cot \dfrac{3\pi}{4} = -1$

25. $\sin \dfrac{11\pi}{4} = \sin \dfrac{3\pi}{4} = \dfrac{\sqrt{2}}{2}$ **27.** $\sin\left(-\dfrac{17\pi}{6}\right) = \sin \dfrac{7\pi}{6} = -\dfrac{1}{2}$

29. -75.3130 **31.** 3.2361

33. $\sin \theta = \dfrac{4\sqrt{41}}{41}$

$\cos \theta = \dfrac{5\sqrt{41}}{41}$

$\tan \theta = \dfrac{4}{5}$

$\csc \theta = \dfrac{\sqrt{41}}{4}$

$\sec \theta = \dfrac{\sqrt{41}}{5}$

$\cot \theta = \dfrac{5}{4}$

35. 0.6494 **37.** 3.6722

39. (a) 3 (b) $\dfrac{2\sqrt{2}}{3}$ (c) $\dfrac{3\sqrt{2}}{4}$ (d) $\dfrac{\sqrt{2}}{4}$

41. 71.3 m

43. $\sin \theta = \dfrac{4}{5}$ $\csc \theta = \dfrac{5}{4}$

$\cos \theta = \dfrac{3}{5}$ $\sec \theta = \dfrac{5}{3}$

$\tan \theta = \dfrac{4}{3}$ $\cot \theta = \dfrac{3}{4}$

45. $\sin \theta = \dfrac{4}{5}$ $\csc \theta = \dfrac{5}{4}$

$\cos \theta = \dfrac{3}{5}$ $\sec \theta = \dfrac{5}{3}$

$\tan \theta = \dfrac{4}{3}$ $\cot \theta = \dfrac{3}{4}$

47. $\sin \theta = -\dfrac{\sqrt{11}}{6}$ **49.** $\sin \theta = \dfrac{\sqrt{21}}{5}$

$\cos \theta = \dfrac{5}{6}$ $\tan \theta = -\dfrac{\sqrt{21}}{2}$

$\tan \theta = -\dfrac{\sqrt{11}}{5}$ $\csc \theta = \dfrac{5\sqrt{21}}{21}$

$\csc \theta = -\dfrac{6\sqrt{11}}{11}$ $\sec \theta = -\dfrac{5}{2}$

$\cot \theta = -\dfrac{5\sqrt{11}}{11}$ $\cot \theta = -\dfrac{2\sqrt{21}}{21}$

51. $\theta' = 84°$ **53.** $\theta' = \dfrac{\pi}{5}$

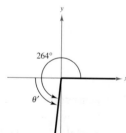

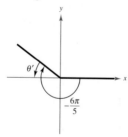

55. $\sin \dfrac{\pi}{3} = \dfrac{\sqrt{3}}{2}$; $\cos \dfrac{\pi}{3} = \dfrac{1}{2}$; $\tan \dfrac{\pi}{3} = \sqrt{3}$

57. $\sin(-150°) = -\dfrac{1}{2}$; $\cos(-150°) = -\dfrac{\sqrt{3}}{2}$;

$\tan(-150°) = \dfrac{\sqrt{3}}{3}$

59. -0.7568 **61.** 0.9511

63. 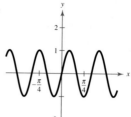   **65.**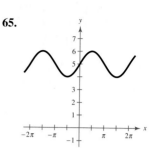

67.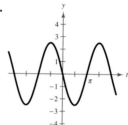

69. (a) $y = 2 \sin 528\pi x$ (b) 264 cycles/sec

71. 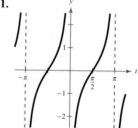   **73.**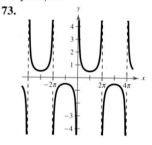

75.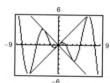

As $x \to +\infty$, $f(x)$ oscillates.

77. $-\dfrac{\pi}{2}$ **79.** $\dfrac{\pi}{6}$ **81.** -0.98 **83.** 0.09

85.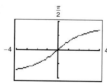

87. $\dfrac{4}{5}$ **89.** $\dfrac{4\sqrt{15}}{15}$ **91.** $\dfrac{\sqrt{4-x^2}}{x}$

93.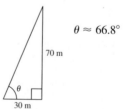

$\theta \approx 66.8°$

70 m

30 m

95. 1221 mi, $85.6°$

97. False. For each θ there corresponds exactly one value of y.

99. The function is undefined because $\sec \theta = 1/\cos \theta$.

101. The ranges of the other four trigonometric functions are $(-\infty, \infty)$ or $(-\infty, -1] \cup [1, \infty)$.

103. Answers will vary.

Chapter Test   *(page 205)*

1. (a)

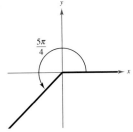

(b) $\dfrac{13\pi}{4}, -\dfrac{3\pi}{4}$ (c) $225°$

2. 3500 rad/min   **3.** About 709.04 ft²

4. $\sin \theta = \dfrac{3\sqrt{10}}{10}$ $\csc \theta = \dfrac{\sqrt{10}}{3}$

$\cos \theta = -\dfrac{\sqrt{10}}{10}$ $\sec \theta = -\sqrt{10}$

$\tan \theta = -3$ $\cot \theta = -\dfrac{1}{3}$

5. For $0 \le \theta < \dfrac{\pi}{2}$:   For $\pi \le \theta < \dfrac{3\pi}{2}$:

$\sin \theta = \dfrac{3\sqrt{13}}{13}$ $\sin \theta = -\dfrac{3\sqrt{13}}{13}$

$\cos \theta = \dfrac{2\sqrt{13}}{13}$ $\cos \theta = -\dfrac{2\sqrt{13}}{13}$

$\csc \theta = \dfrac{\sqrt{13}}{3}$ $\csc \theta = -\dfrac{\sqrt{13}}{3}$

$\sec \theta = \dfrac{\sqrt{13}}{2}$ $\sec \theta = -\dfrac{\sqrt{13}}{2}$

$\cot \theta = \dfrac{2}{3}$ $\cot \theta = \dfrac{2}{3}$

6. $\theta' = 25°$

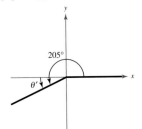

7. Quadrant III   **8.** $150°, 210°$ **9.** 1.33, 1.81

10. $\sin \theta = -\dfrac{4}{5}$ **11.** $\sin \theta = \dfrac{21}{29}$

$\tan \theta = -\dfrac{4}{3}$ $\cos \theta = -\dfrac{20}{29}$

$\csc \theta = -\dfrac{5}{4}$ $\tan \theta = -\dfrac{21}{20}$

$\sec \theta = \dfrac{5}{3}$ $\csc \theta = \dfrac{29}{21}$

$\cot \theta = -\dfrac{3}{4}$ $\cot \theta = -\dfrac{20}{21}$

12.   **13.**

   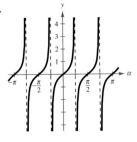

14.

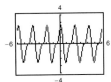

Period: 2

15.

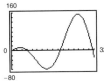

Not periodic

16. $a = -2, b = \dfrac{1}{2}, c = -\dfrac{\pi}{4}$ **17.** $\dfrac{\sqrt{55}}{3}$

18.

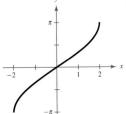

19. $309.3°$ **20.** $d = -6 \cos \pi t$

Problem Solving   *(page 207)*

1. (a) $\dfrac{11\pi}{2}$ rad or $990°$ (b) About 816.42 ft

3. (a) 4767 ft   (b) 3705 ft

(c) $w \approx 2183$ ft, $\tan 63° = \dfrac{w + 3705}{3000}$

5. (a)

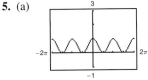

Even

(b)

Even

7. $h = 51 - 50 \sin\left(8\pi t + \dfrac{\pi}{2}\right)$

9. (a)

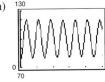

(b) Period $= \dfrac{3}{4}$ sec;

Answers will vary.

(c) 20 mm; Answers will vary.   (d) 80 beats/min

(e) Period $= \dfrac{15}{16}$ sec; $\dfrac{32\pi}{15}$

11. (a)

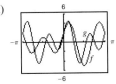

(b) Period of f: 2π;

Period of g: π

(c) Yes, because the sine and cosine functions are periodic.

13. (a) $40.5°$ (b) $x \approx 1.71$ ft; $y \approx 3.46$ ft   (c) About 1.75 ft

(d) As you move closer to the rock, d must get smaller and smaller. The angles θ_1 and θ_2 will decrease along with the distance y, so d will decrease.

CHAPTER 1

Chapter 2

Section 2.1   *(page 215)*

1. $\tan u$ **3.** $\cot u$ **5.** $\cot^2 u$

7. $\sin x = \dfrac{1}{2}$

$\cos x = \dfrac{\sqrt{3}}{2}$

$\tan x = \dfrac{\sqrt{3}}{3}$

$\csc x = 2$

$\sec x = \dfrac{2\sqrt{3}}{3}$

$\cot x = \sqrt{3}$

9. $\sin x = \dfrac{3}{5}$

$\cos x = \dfrac{4}{5}$

$\tan x = \dfrac{3}{4}$

$\csc x = \dfrac{5}{3}$

$\sec x = \dfrac{5}{4}$

$\cot x = \dfrac{4}{3}$

11. $\sin x = \dfrac{\sqrt{15}}{4}$

$\cos x = \dfrac{1}{4}$

$\tan x = \sqrt{15}$

$\csc x = \dfrac{4\sqrt{15}}{15}$

$\sec x = 4$

$\cot x = \dfrac{\sqrt{15}}{15}$

13. $\sin \theta = -1$

$\cos \theta = 0$

$\tan \theta$ is undefined.

$\csc \theta = -1$

$\sec \theta$ is undefined.

$\cot \theta = 0$

15. c   **16.** b   **17.** f   **18.** a   **19.** e   **20.** d

21. $\sin^2 x$ **23.** $\sec x + 1$ **25.** $\sin^4 x$

27. $\csc^2 x(\cot x + 1)$ **29.** $(3 \sin x + 1)(\sin x - 2)$

31. $(\csc x - 1)(\csc x + 2)$ **33.** $1 + 2 \sin x \cos x$

35. $\csc \theta$ **37.** $\cos^2 \phi$ **39.** $\sin^2 x$ **41.** $\tan x$

43. $\sec \beta$ **45.** $2 \csc^2 x$ **47.** $-\cot x$ **49.** $1 + \cos y$

51. $\csc x$ **53.** $3 \sin \theta$ **55.** $2 \tan \theta$

57. $3 \cos \theta = 3;\ \sin \theta = 0;\ \cos \theta = 1$ **59.** $0 \le \theta \le \pi$

61. $\mu = \tan \theta$ **63.** True. For example, $\sin(-x) = -\sin x$.

65. $\infty, 0$ **67.** Not an identity because $\dfrac{\sin k\theta}{\cos k\theta} = \tan k\theta$

69. $a \sec \theta$

71. $\cos \theta = \pm\sqrt{1 - \sin^2 \theta}$

$\tan \theta = \pm\dfrac{\sin \theta}{\sqrt{1 - \sin^2 \theta}}$

$\cot \theta = \pm\dfrac{\sqrt{1 - \sin^2 \theta}}{\sin \theta}$

$\sec \theta = \pm\dfrac{1}{\sqrt{1 - \sin^2 \theta}}$

$\csc \theta = \dfrac{1}{\sin \theta}$

Section 2.2   *(page 222)*

1. identity   **3.** $\tan u$ **5.** $\cos^2 u$ **7.** $-\csc u$

9–49. Answers will vary.

51. In the first line, $\cot(x)$ is substituted for $\cot(-x)$, which is incorrect; $\cot(-x) = -\cot(x)$.

53. (a)

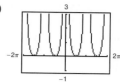

(b)

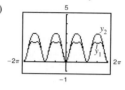

Identity

(c) Answers will vary.

55. (a)

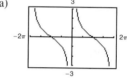

(b)

Not an identity

(c) Answers will vary.

57. (a)

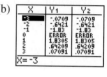

(b)

Identity

(c) Answers will vary.

59–61. Answers will vary.   **63.** 1

65. Answers will vary.

67. True. Many different techniques can be used to verify identities.

69. False. $\sin x^2 = \sin(x \cdot x) \ne \sin^2 x = (\sin x)(\sin x)$

71. The equation is not an identity because $\sin \theta = \pm\sqrt{1 - \cos^2 \theta}$.

Possible answer: $\dfrac{7\pi}{4}$

73. The equation is not an identity because $1 - \cos^2 \theta = \sin^2 \theta$.

Possible answer: $-\dfrac{\pi}{2}$

Section 2.3   *(page 231)*

1. isolate   **3.** quadratic   **5–9.** Answers will vary.

11. $\dfrac{\pi}{3} + 2n\pi, \dfrac{2\pi}{3} + 2n\pi$ **13.** $\dfrac{2\pi}{3} + 2n\pi, \dfrac{4\pi}{3} + 2n\pi$

15. $\dfrac{\pi}{6} + n\pi, \dfrac{5\pi}{6} + n\pi$ **17.** $\dfrac{\pi}{3} + n\pi, \dfrac{2\pi}{3} + n\pi$

19. $\dfrac{\pi}{8} + \dfrac{n\pi}{2}, \dfrac{3\pi}{8} + \dfrac{n\pi}{2}$ **21.** $\dfrac{n\pi}{3}, \dfrac{\pi}{4} + n\pi$

23. $n\pi, \dfrac{3\pi}{2} + 2n\pi$ **25.** $0, \dfrac{\pi}{2}, \pi, \dfrac{3\pi}{2}$

27. $0, \pi, \dfrac{\pi}{6}, \dfrac{5\pi}{6}, \dfrac{7\pi}{6}, \dfrac{11\pi}{6}$ **29.** $\dfrac{\pi}{3}, \dfrac{5\pi}{3}, \pi$ **31.** No solution

33. $\pi, \dfrac{\pi}{3}, \dfrac{5\pi}{3}$ **35.** $\dfrac{\pi}{6}, \dfrac{5\pi}{6}, \dfrac{7\pi}{6}, \dfrac{11\pi}{6}$ **37.** $\dfrac{\pi}{2}$

39. $\dfrac{\pi}{6} + n\pi, \dfrac{5\pi}{6} + n\pi$ **41.** $\dfrac{\pi}{12} + \dfrac{n\pi}{3}$

43. $\dfrac{\pi}{2} + 4n\pi, \dfrac{7\pi}{2} + 4n\pi$ **45.** $3 + 4n$

47. $-2 + 6n, 2 + 6n$ **49.** $2.678, 5.820$ **51.** $1.047, 5.236$

53. $0.860, 3.426$ **55.** $0, 2.678, 3.142, 5.820$

57. $0.983, 1.768, 4.124, 4.910$

59. $0.3398, 0.8481, 2.2935, 2.8018$

61. $1.9357, 2.7767, 5.0773, 5.9183$

63. $\arctan(-4) + \pi, \arctan(-4) + 2\pi, \arctan 3, \arctan 3 + \pi$

65. $\dfrac{\pi}{4}, \dfrac{5\pi}{4}$, arctan 5, arctan 5 + π **67.** $\dfrac{\pi}{3}, \dfrac{5\pi}{3}$

69. arctan$\left(\frac{1}{3}\right)$, arctan$\left(\frac{1}{3}\right)$ + π, arctan$\left(-\frac{1}{3}\right)$ + π, arctan$\left(-\frac{1}{3}\right)$ + 2π

71. arccos$\left(\frac{1}{4}\right)$, 2π − arccos$\left(\frac{1}{4}\right)$

73. $\dfrac{\pi}{2}$, arcsin$\left(-\dfrac{1}{4}\right)$ + 2π, arcsin$\left(\dfrac{1}{4}\right)$ + π

75. −1.154, 0.534   **77.** 1.110

79. (a)

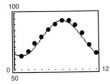

(b) $\dfrac{\pi}{3} \approx 1.0472$

$\dfrac{5\pi}{3} \approx 5.2360$

0

$\pi \approx 3.1416$

Maximum: (1.0472, 1.25)
Maximum: (5.2360, 1.25)
Minimum: (0, 1)
Minimum: (3.1416, −1)

81. (a)

(b) $\dfrac{\pi}{4} \approx 0.7854$

$\dfrac{5\pi}{4} \approx 3.9270$

Maximum: (0.7854, 1.4142)
Minimum: (3.9270, −1.4142)

83. (a)

(b) $\dfrac{\pi}{4} \approx 0.7854$

$\dfrac{5\pi}{4} \approx 3.9270$

$\dfrac{3\pi}{4} \approx 2.3562$

$\dfrac{7\pi}{4} \approx 5.4978$

Maximum: (0.7854, 0.5)
Maximum: (3.9270, 0.5)
Minimum: (2.3562, −0.5)
Minimum: (5.4978, −0.5)

85. 1

87. (a) All real numbers x except $x = 0$ (b) y-axis symmetry
   (c) y approaches 1.   (d) Four solutions: $\pm\pi, \pm 2\pi$

89. 0.04 sec, 0.43 sec, 0.83 sec

91. January, November, December

93. (a) and (c)

(b) $H = 15.65 \cos\left(\dfrac{\pi}{6}t - \dfrac{7\pi}{6}\right) + 77.95$ (d) 77.95°F

The model fits the data well.

(e) Above 86°: June through September
   Below 86°: October through May

95. (a)

(b) $0.6 < x < 1.1$

$A \approx 1.12$

97. 1

99. True. The first equation has a smaller period than the second
equation, so it will have more solutions in the interval $[0, 2\pi)$.

101. The equation would become $\cos^2 x = 2$; this is not the correct
method to use when solving equations.

103. (a)

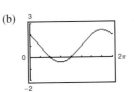

Graphs intersect when $x = \dfrac{\pi}{2}$
and $x = \pi$.

(b)

x-intercepts: $\left(\dfrac{\pi}{2}, 0\right), (\pi, 0)$

Both methods produce the same x-values.

Section 2.4   *(page 239)*

1. $\sin u \cos v - \cos u \sin v$ **3.** $\dfrac{\tan u + \tan v}{1 - \tan u \tan v}$

5. $\cos u \cos v + \sin u \sin v$

7. (a) $\dfrac{\sqrt{2} - \sqrt{6}}{4}$ (b) $\dfrac{\sqrt{2} + 1}{2}$

9. (a) $\dfrac{\sqrt{6} + \sqrt{2}}{4}$ (b) $\dfrac{\sqrt{2} - \sqrt{3}}{2}$

11. $\sin\dfrac{11\pi}{12} = \dfrac{\sqrt{2}}{4}\left(\sqrt{3} - 1\right)$ **13.** $\sin\dfrac{17\pi}{12} = -\dfrac{\sqrt{2}}{4}\left(\sqrt{3} + 1\right)$

$\cos\dfrac{11\pi}{12} = -\dfrac{\sqrt{2}}{4}\left(\sqrt{3} + 1\right)$ $\cos\dfrac{17\pi}{12} = \dfrac{\sqrt{2}}{4}\left(1 - \sqrt{3}\right)$

$\tan\dfrac{11\pi}{12} = -2 + \sqrt{3}$ $\tan\dfrac{17\pi}{12} = 2 + \sqrt{3}$

15. $\sin 105° = \dfrac{\sqrt{2}}{4}\left(\sqrt{3} + 1\right)$ **17.** $\sin 195° = \dfrac{\sqrt{2}}{4}\left(1 - \sqrt{3}\right)$

$\cos 105° = \dfrac{\sqrt{2}}{4}\left(1 - \sqrt{3}\right)$ $\cos 195° = -\dfrac{\sqrt{2}}{4}\left(\sqrt{3} + 1\right)$

$\tan 105° = -2 - \sqrt{3}$ $\tan 195° = 2 - \sqrt{3}$

19. $\sin\dfrac{13\pi}{12} = \dfrac{\sqrt{2}}{4}\left(1 - \sqrt{3}\right)$

$\cos\dfrac{13\pi}{12} = -\dfrac{\sqrt{2}}{4}\left(1 + \sqrt{3}\right)$

$\tan\dfrac{13\pi}{12} = 2 - \sqrt{3}$

21. $\sin\left(-\dfrac{13\pi}{12}\right) = \dfrac{\sqrt{2}}{4}\left(\sqrt{3} - 1\right)$

$\cos\left(-\dfrac{13\pi}{12}\right) = -\dfrac{\sqrt{2}}{4}\left(\sqrt{3} + 1\right)$

$\tan\left(-\dfrac{13\pi}{12}\right) = -2 + \sqrt{3}$

23. $\sin 285° = -\dfrac{\sqrt{2}}{4}\left(\sqrt{3} + 1\right)$

$\cos 285° = \dfrac{\sqrt{2}}{4}\left(\sqrt{3} - 1\right)$

$\tan 285° = -\left(2 + \sqrt{3}\right)$

25. $\sin(-165°) = -\dfrac{\sqrt{2}}{4}(\sqrt{3} - 1)$

$\cos(-165°) = -\dfrac{\sqrt{2}}{4}(1 + \sqrt{3})$

$\tan(-165°) = 2 - \sqrt{3}$

27. $\sin 1.8$ **29.** $\sin 75°$ **31.** $\tan 15°$ **33.** $\cos(3x - 2y)$

35. $\dfrac{\sqrt{3}}{2}$ **37.** $\dfrac{\sqrt{3}}{2}$ **39.** $-\sqrt{3}$ **41.** $-\dfrac{63}{65}$ **43.** $-\dfrac{63}{16}$

45. $\frac{65}{56}$ **47.** $\frac{3}{5}$ **49.** $-\frac{44}{117}$ **51.** $-\frac{125}{44}$ **53.** 1   **55.** 0

57–63. Proofs   **65.** $-\sin x$ **67.** $-\cos \theta$ **69.** $\dfrac{\pi}{6}, \dfrac{5\pi}{6}$

71. $\dfrac{5\pi}{4}, \dfrac{7\pi}{4}$ **73.** $0, \dfrac{\pi}{3}, \pi, \dfrac{5\pi}{3}$ **75.** $\dfrac{\pi}{4}, \dfrac{7\pi}{4}$ **77.** $\dfrac{\pi}{2}, \pi, \dfrac{3\pi}{2}$

79. (a) $y = \dfrac{5}{12}\sin(2t + 0.6435)$ (b) $\dfrac{5}{12}$ ft (c) $\dfrac{1}{\pi}$ cycle/sec

81. True. $\sin(u \pm v) = \sin u \cos v \pm \cos u \sin v$

83. False. $\tan\left(x - \dfrac{\pi}{4}\right) = \dfrac{\tan x - 1}{1 + \tan x}$

85. (a) All real numbers h except $h = 0$

(b)

h	0.5	0.2	0.1
$f(h)$	0.267	0.410	0.456
$g(h)$	0.267	0.410	0.456

h	0.05	0.02	0.01
$f(h)$	0.478	0.491	0.496
$g(h)$	0.478	0.491	0.496

(c)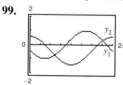

(d) As $h \to 0^+$, $f \to 0.5$ and $g \to 0.5$.

87–89. Answers will vary.

91. (a) $\sqrt{2}\sin\left(\theta + \dfrac{\pi}{4}\right)$ (b) $\sqrt{2}\cos\left(\theta - \dfrac{\pi}{4}\right)$

93. (a) $13\sin(3\theta + 0.3948)$ (b) $13\cos(3\theta - 1.1760)$

95. $\sqrt{2}\sin\theta + \sqrt{2}\cos\theta$ **97.** $15°$

99.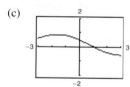

No, $y_1 \neq y_2$ because their graphs are different.

101. (a) and (b) Proofs

Section 2.5 (page 249)

1. $2\sin u \cos u$ **3.** $\tan^2 u$

5. $\dfrac{1}{2}[\sin(u + v) + \sin(u - v)]$ **7.** $0, \dfrac{\pi}{3}, \pi, \dfrac{5\pi}{3}$

9. $0, \dfrac{2\pi}{3}, \dfrac{4\pi}{3}$ **11.** $0, \dfrac{\pi}{2}, \pi, \dfrac{3\pi}{2}$

13. $\dfrac{\pi}{2}, \dfrac{\pi}{6}, \dfrac{5\pi}{6}, \dfrac{7\pi}{6}, \dfrac{3\pi}{2}, \dfrac{11\pi}{6}$ **15.** $3\sin 2x$ **17.** $3\cos 2x$

19. $4\cos 2x$ **21.** $\sin 2u = -\frac{24}{25}$, $\cos 2u = \frac{7}{25}$, $\tan 2u = -\frac{24}{7}$

23. $\sin 2u = \frac{15}{17}$, $\cos 2u = \frac{8}{17}$, $\tan 2u = \frac{15}{8}$

25. $8\cos^4 x - 8\cos^2 x + 1$ **27.** $\frac{1}{8}(3 + 4\cos 2x + \cos 4x)$

29. $\dfrac{(3 - 4\cos 4x + \cos 8x)}{(3 + 4\cos 4x + \cos 8x)}$ **31.** $\dfrac{1}{8}(1 - \cos 8x)$

33. $\sin 75° = \dfrac{1}{2}\sqrt{2 + \sqrt{3}}$ **35.** $\sin \dfrac{\pi}{8} = \dfrac{1}{2}\sqrt{2 - \sqrt{2}}$

$\cos 75° = \dfrac{1}{2}\sqrt{2 - \sqrt{3}}$ $\cos \dfrac{\pi}{8} = \dfrac{1}{2}\sqrt{2 + \sqrt{2}}$

$\tan 75° = 2 + \sqrt{3}$ $\tan \dfrac{\pi}{8} = \sqrt{2} - 1$

37. (a) Quadrant I

(b) $\sin\dfrac{u}{2} = \dfrac{3}{5}$, $\cos\dfrac{u}{2} = \dfrac{4}{5}$, $\tan\dfrac{u}{2} = \dfrac{3}{4}$

39. (a) Quadrant II

(b) $\sin\dfrac{u}{2} = \dfrac{\sqrt{26}}{26}$, $\cos\dfrac{u}{2} = -\dfrac{5\sqrt{26}}{26}$, $\tan\dfrac{u}{2} = -\dfrac{1}{5}$

41. $|\sin 3x|$ **43.** $-|\tan 4x|$

45. π **47.** $\dfrac{\pi}{3}, \pi, \dfrac{5\pi}{3}$

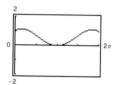   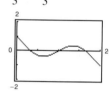

49. $\frac{1}{2}(\cos 2\theta - \cos 8\theta)$ **51.** $\frac{1}{2}(\cos(-2\theta) + \cos 6\theta)$

53. $2\cos 4\theta \sin \theta$ **55.** $2\cos 4x \cos 2x$ **57.** $\dfrac{\sqrt{6}}{2}$

59. $-\sqrt{2}$

61. $0, \dfrac{\pi}{4}, \dfrac{\pi}{2}, \dfrac{3\pi}{4}, \pi, \dfrac{5\pi}{4}, \dfrac{3\pi}{2}, \dfrac{7\pi}{4}$ **63.** $\dfrac{\pi}{6}, \dfrac{5\pi}{6}$

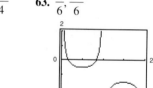

65–71. Answers will vary.

73. (a) $\cos\theta = \dfrac{M^2 - 2}{M^2}$ (b) π (c) 0.4482

(d) 760 mi/h; 3420 mi/h

75. $x = 2r(1 - \cos\theta)$

77. False. For $u < 0$,

$\sin 2u = -\sin(-2u)$

$= -2\sin(-u)\cos(-u)$

$= -2(-\sin u)\cos u$

$= 2\sin u \cos u$.

79. Answers will vary.

Review Exercises (page 253)

1. $\tan x$ **3.** $\cot x$

5. $\sin\theta = \dfrac{2\sqrt{13}}{13}$

$\cos\theta = \dfrac{3\sqrt{13}}{13}$

$\csc\theta = \dfrac{\sqrt{13}}{2}$

$\cot\theta = \dfrac{3}{2}$

7. $\sin^2 x$ **9.** 1 **11.** $\tan u \sec u$ **13.** $\cot^2 x$

15. $-2\tan^2\theta$ **17.** $5\cos\theta$ **19–25.** Answers will vary.

27. $\dfrac{\pi}{3} + 2n\pi, \dfrac{2\pi}{3} + 2n\pi$ **29.** $\dfrac{\pi}{6} + n\pi$

31. $\dfrac{\pi}{3} + n\pi, \dfrac{2\pi}{3} + n\pi$ **33.** $0, \dfrac{2\pi}{3}, \dfrac{4\pi}{3}$ **35.** $0, \dfrac{\pi}{2}, \pi$

37. $\dfrac{\pi}{8}, \dfrac{3\pi}{8}, \dfrac{9\pi}{8}, \dfrac{11\pi}{8}$ **39.** $\dfrac{\pi}{2}$

41. $0, \dfrac{\pi}{8}, \dfrac{3\pi}{8}, \dfrac{5\pi}{8}, \dfrac{7\pi}{8}, \dfrac{9\pi}{8}, \dfrac{11\pi}{8}, \dfrac{13\pi}{8}, \dfrac{15\pi}{8}$

43. $0, \pi, \arctan 2, \arctan 2 + \pi$

45. $\arctan(-3) + \pi, \arctan(-3) + 2\pi, \arctan 2, \arctan 2 + \pi$

47. $\sin 285° = -\dfrac{\sqrt{2}}{4}(\sqrt{3}+1)$ **49.** $\sin\dfrac{25\pi}{12} = \dfrac{\sqrt{2}}{4}(\sqrt{3}-1)$

$\cos 285° = \dfrac{\sqrt{2}}{4}(\sqrt{3}-1)$ $\cos\dfrac{25\pi}{12} = \dfrac{\sqrt{2}}{4}(\sqrt{3}+1)$

$\tan 285° = -2 - \sqrt{3}$ $\tan\dfrac{25\pi}{12} = 2 - \sqrt{3}$

51. $\sin 15°$ **53.** $-\frac{24}{25}$ **55.** -1

57–59. Answers will vary. **61.** $\dfrac{\pi}{4}, \dfrac{7\pi}{4}$

63. $\sin 2u = \frac{24}{25}$

$\cos 2u = -\frac{7}{25}$

$\tan 2u = -\frac{24}{7}$

65.

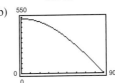

67. $\dfrac{1 - \cos 4x}{1 + \cos 4x}$

69. $\sin(-75°) = -\frac{1}{2}\sqrt{2 + \sqrt{3}}$

$\cos(-75°) = \frac{1}{2}\sqrt{2 - \sqrt{3}}$

$\tan(-75°) = -2 - \sqrt{3}$

71. (a) Quadrant II

(b) $\sin\dfrac{u}{2} = \dfrac{2\sqrt{5}}{5}, \cos\dfrac{u}{2} = -\dfrac{\sqrt{5}}{5}, \tan\dfrac{u}{2} = -2$

73. $-|\cos 5x|$ **75.** $\frac{1}{2}[\sin 10\theta - \sin(-2\theta)]$

77. $2\cos\dfrac{11\theta}{2}\cos\dfrac{\theta}{2}$ **79.** $\theta = 15°$ or $\dfrac{\pi}{12}$

81. False. If $(\pi/2) < \theta < \pi$, then $\cos(\theta/2) > 0$. The sign of $\cos(\theta/2)$ depends on the quadrant in which $\theta/2$ lies.

83. True. $4\sin(-x)\cos(-x) = 4(-\sin x)\cos x$

$= -4\sin x\cos x$

$= -2(2\sin x\cos x)$

$= -2\sin 2x$

85. No. For an equation to be an identity, the equation must be true for all real numbers x. $\sin\theta = \frac{1}{2}$ has an infinite number of solutions but is not an identity.

Chapter Test (page 255)

1. $\sin\theta = -\dfrac{6\sqrt{61}}{61}$ $\csc\theta = -\dfrac{\sqrt{61}}{6}$

$\cos\theta = -\dfrac{5\sqrt{61}}{61}$ $\sec\theta = -\dfrac{\sqrt{61}}{5}$

$\tan\theta = \dfrac{6}{5}$ $\cot\theta = \dfrac{5}{6}$

2. 1 **3.** 1 **4.** $\csc\theta\sec\theta$

5. $\theta = 0, \dfrac{\pi}{2} < \theta \le \pi, \dfrac{3\pi}{2} < \theta < 2\pi$

6.

7–12. Answers will vary.

$y_1 = y_2$

13. $\frac{1}{8}(3 - 4\cos x + \cos 2x)$ **14.** $\tan 2\theta$

15. $2(\sin 5\theta + \sin\theta)$ **16.** $-2\sin 2\theta\sin\theta$

17. $0, \dfrac{3\pi}{4}, \pi, \dfrac{7\pi}{4}$ **18.** $\dfrac{\pi}{6}, \dfrac{\pi}{2}, \dfrac{5\pi}{6}, \dfrac{3\pi}{2}$ **19.** $\dfrac{\pi}{6}, \dfrac{5\pi}{6}, \dfrac{7\pi}{6}, \dfrac{11\pi}{6}$

20. $\dfrac{\pi}{6}, \dfrac{5\pi}{6}, \dfrac{3\pi}{2}$ **21.** $0, 2.596$ **22.** $\dfrac{\sqrt{2} - \sqrt{6}}{4}$

23. $\sin 2u = -\frac{20}{29}, \cos 2u = -\frac{21}{29}, \tan 2u = \frac{20}{21}$

24. Day 123 to day 223 **25.** $t = 0.26$ min

0.58 min

0.89 min

1.20 min

1.52 min

1.83 min

Problem Solving (page 259)

1. $\sin\theta = \pm\sqrt{1 - \cos^2\theta}$ $\sec\theta = \dfrac{1}{\cos\theta}$

$\tan\theta = \pm\dfrac{\sqrt{1 - \cos^2\theta}}{\cos\theta}$ $\cot\theta = \pm\dfrac{\cos\theta}{\sqrt{1 - \cos^2\theta}}$

$\csc\theta = \pm\dfrac{1}{\sqrt{1 - \cos^2\theta}}$

3. Answers will vary. **5.** $u + v = w$

7. (a) $A = 100\sin\dfrac{\theta}{2}\cos\dfrac{\theta}{2}$ (b) $A = 50\sin\theta; \theta = \dfrac{\pi}{2}$

9. (a) $F = \dfrac{0.6W\cos\theta}{\sin 12°}$

(b)

(c) Maximum: $\theta = 0°$

Minimum: $\theta = 90°$

11. (a) High tides: 6:12 A.M., 6:36 P.M.

Low tides: 12:00 A.M., 12:24 P.M.

(b) The water depth never falls below 7 feet.

(c)
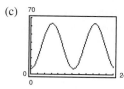

70

0

0   24

13. (a) $n = \frac{1}{2}\left(\cot\frac{\theta}{2} + \sqrt{3}\right)$ (b) $\theta \approx 76.5°$

15. (a) $\frac{\pi}{6} \le x \le \frac{5\pi}{6}$ (b) $\frac{2\pi}{3} \le x \le \frac{4\pi}{3}$

(c) $\frac{\pi}{2} < x < \pi, \frac{3\pi}{2} < x < 2\pi$

(d) $0 \le x \le \frac{\pi}{4}, \frac{5\pi}{4} \le x < 2\pi$

Chapter 3

Section 3.1   *(page 268)*

1. oblique   3. angles; side

5. $A = 30°, a \approx 14.14, c \approx 27.32$

7. $C = 120°, b \approx 4.75, c \approx 7.17$

9. $B = 60.9°, b \approx 19.32, c \approx 6.36$

11. $B = 42°4', a \approx 22.05, b \approx 14.88$

13. $C = 80°, a \approx 5.82, b \approx 9.20$

15. $C = 83°, a \approx 0.62, b \approx 0.51$

17. $B \approx 21.55°, C \approx 122.45°, c \approx 11.49$

19. $A \approx 10°11', C \approx 154°19', c \approx 11.03$

21. $B \approx 9.43°, C = 25.57°, c \approx 10.53$

23. $B \approx 18°13', C \approx 51°32', c \approx 40.06$

25. $B \approx 48.74°, C \approx 21.26°, c \approx 48.23$ 27. No solution

29. Two solutions:

$B \approx 72.21°, C \approx 49.79°, c \approx 10.27$

$B \approx 107.79°, C \approx 14.21°, c \approx 3.30$

31. No solution   33. $B = 45°, C = 90°, c \approx 1.41$

35. (a) $b \le 5, b = \frac{5}{\sin 36°}$ (b) $5 < b < \frac{5}{\sin 36°}$

(c) $b > \frac{5}{\sin 36°}$

37. (a) $b \le 10.8, b = \frac{10.8}{\sin 10°}$ (b) $10.8 < b < \frac{10.8}{\sin 10°}$

(c) $b > \frac{10.8}{\sin 10°}$

39. 10.4   41. 20   43. 1675.2   45. 3204.5

47. 24.1 m   49. 16.1°   51. 240°

53. 3.2 mi   55. $d = \frac{2\sin\theta}{\sin(\phi - \theta)}$

57. True. If an angle of a triangle is obtuse (greater than 90°), then the other two angles must be acute and therefore less than 90°. The triangle is oblique.

59. False. When just three angles are known, the triangle cannot be solved.

61. (a) $A = 20\left(15\sin\frac{3\theta}{2} - 4\sin\frac{\theta}{2} - 6\sin\theta\right)$

(b)
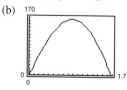

170

0

0   1.7

(c) Domain: $0 \le \theta \le 1.6690$

The domain would increase in length and the area would have a greater maximum value.

Section 3.2   *(page 275)*

1. Cosines   3. $b^2 = a^2 + c^2 - 2ac\cos B$

5. $A \approx 38.62°, B \approx 48.51°, C \approx 92.87°$

7. $A \approx 26.38°, B \approx 36.34°, C \approx 117.28°$

9. $B \approx 23.79°, C \approx 126.21°, a \approx 18.59$

11. $B \approx 29.44°, C \approx 100.56°, a \approx 23.38$

13. $A \approx 30.11°, B \approx 43.16°, C \approx 106.73°$

15. $A \approx 92.94°, B \approx 43.53°, C \approx 43.53°$

17. $B \approx 27.46°, C \approx 32.54°, a \approx 11.27$

19. $A \approx 141°45', C \approx 27°40', b \approx 11.87$

21. $A = 27°10', C = 27°10', b \approx 65.84$

23. $A \approx 33.80°, B \approx 103.20°, c \approx 0.54$

	a	b	c	d	θ	φ
25.	5	8	12.07	5.69	45°	135°
27.	10	14	20	13.86	68.2°	111.8°
29.	15	16.96	25	20	77.2°	102.8°

31. Law of Cosines; $A \approx 102.44°, C \approx 37.56°, b \approx 5.26$

33. Law of Sines; No solution

35. Law of Sines; $C = 103°, a \approx 0.82, b \approx 0.71$

37. 43.52   39. 10.4   41. 52.11   43. 0.18

45.

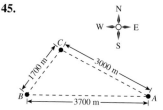

N 37.1° E, S 63.1° E

47. 373.3 m   49. 43.3 mi

51. (a) N 59.7° E   (b) N 72.8° E   53. 63.7 ft   55. 72.3°

57. 46,837.5 ft²   59. $83,336.37

61. False. For s to be the average of the lengths of the three sides of the triangle, s would be equal to $(a + b + c)/3$.

63. $c^2 = a^2 + b^2$; The Pythagorean Theorem is a special case of the Law of Cosines.

65. The Law of Cosines can be used to solve the single-solution case of SSA. There is no method that can solve the no-solution case of SSA.

67. Proof

Section 3.3   *(page 287)*

1. directed line segment   3. magnitude

5. magnitude; direction   7. unit vector   9. resultant

11. $\|\mathbf{u}\| = \|\mathbf{v}\| = \sqrt{17}$, slope$_\mathbf{u}$ = slope$_\mathbf{v}$ = $\frac{1}{4}$

$\mathbf{u}$ and $\mathbf{v}$ have the same magnitude and direction, so they are equal.

13. $\mathbf{v} = \langle 1, 3\rangle, \|\mathbf{v}\| = \sqrt{10}$ 15. $\mathbf{v} = \langle 4, 6\rangle; \|\mathbf{v}\| = 2\sqrt{13}$

17. $\mathbf{v} = \langle 0, 5\rangle; \|\mathbf{v}\| = 5$ 19. $\mathbf{v} = \langle 8, 6\rangle; \|\mathbf{v}\| = 10$

21. $\mathbf{v} = \langle -9, -12\rangle; \|\mathbf{v}\| = 15$ 23. $\mathbf{v} = \langle 16, 7\rangle; \|\mathbf{v}\| = \sqrt{305}$

25.
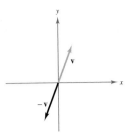

27.
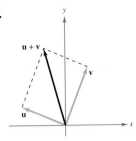

29.
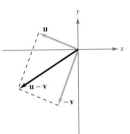

31. (a) $\langle 3, 4 \rangle$ (b) $\langle 1, -2 \rangle$
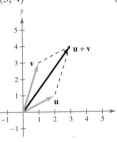
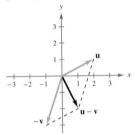

(c) $\langle 1, -7 \rangle$
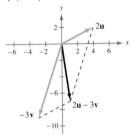

33. (a) $\langle -5, 3 \rangle$ (b) $\langle -5, 3 \rangle$
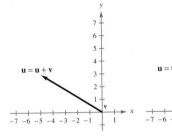
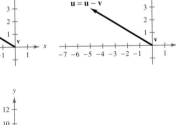

(c) $\langle -10, 6 \rangle$
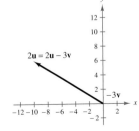

35. (a) $3\mathbf{i} - 2\mathbf{j}$ (b) $-\mathbf{i} + 4\mathbf{j}$
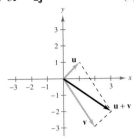
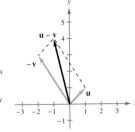

(c) $-4\mathbf{i} + 11\mathbf{j}$
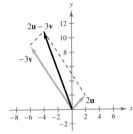

37. (a) $2\mathbf{i} + \mathbf{j}$ (b) $2\mathbf{i} - \mathbf{j}$
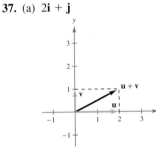

(c) $4\mathbf{i} - 3\mathbf{j}$
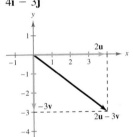

39. $\langle 1, 0 \rangle$ **41.** $\left\langle -\dfrac{\sqrt{2}}{2}, \dfrac{\sqrt{2}}{2} \right\rangle$ **43.** $\dfrac{\sqrt{2}}{2}\mathbf{i} + \dfrac{\sqrt{2}}{2}\mathbf{j}$

45. $\mathbf{j}$ **47.** $\dfrac{\sqrt{5}}{5}\mathbf{i} - \dfrac{2\sqrt{5}}{5}\mathbf{j}$ **49.** $\mathbf{v} = \langle -6, 8 \rangle$

51. $\mathbf{v} = \left\langle \dfrac{18\sqrt{29}}{29}, \dfrac{45\sqrt{29}}{29} \right\rangle$ **53.** $5\mathbf{i} - 3\mathbf{j}$ **55.** $6\mathbf{i} - 3\mathbf{j}$

57. $\mathbf{v} = \left\langle 3, -\dfrac{3}{2} \right\rangle$ **59.** $\mathbf{v} = \langle 4, 3 \rangle$
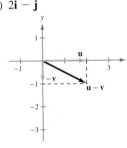
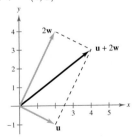

61. $\mathbf{v} = \left\langle \frac{7}{2}, -\frac{1}{2} \right\rangle$

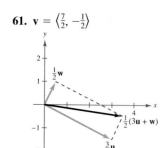

63. $\|\mathbf{v}\| = 6\sqrt{2}$; $\theta = 315°$ **65.** $\|\mathbf{v}\| = 3$; $\theta = 60°$

67. $\mathbf{v} = \langle 3, 0 \rangle$ **69.** $\mathbf{v} = \left\langle -\frac{7\sqrt{3}}{4}, \frac{7}{4} \right\rangle$

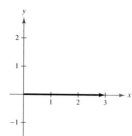

71. $\mathbf{v} = \left\langle \sqrt{6}, \sqrt{6} \right\rangle$ **73.** $\mathbf{v} = \left\langle \frac{9}{5}, \frac{12}{5} \right\rangle$

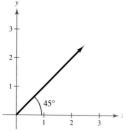

75. $\langle 5, 5 \rangle$ **77.** $\left\langle 10\sqrt{2} - 50, 10\sqrt{2} \right\rangle$
79. $90°$ **81.** $62.7°$
83. Vertical ≈ 125.4 ft/sec, horizontal ≈ 1193.4 ft/sec
85. $12.8°$; 398.32 N   **87.** $71.3°$; 228.5 lb
89. $T_L \approx 15{,}484$ lb   **91.** $T_{AC} \approx 1758.8$ lb
      $T_R \approx 19{,}786$ lb $T_{BC} \approx 1305.4$ lb
93. 3154.4 lb   **95.** 20.8 lb   **97.** $19.5°$
99. 1928.4 ft-lb   **101.** N $21.4°$ E; 138.7 km/h
103. True. See Example 1.   **105.** True. $a = b = 0$
107. Proof   **109.** $\langle 1, 3 \rangle$ or $\langle -1, -3 \rangle$
111. (a) $5\sqrt{5 + 4 \cos \theta}$
      (b)

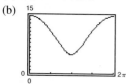

      (c) Range: $[5, 15]$
          Maximum is 15 when $\theta = 0$.
          Minimum is 5 when $\theta = \pi$.
      (d) The magnitudes of $\mathbf{F}_1$ and $\mathbf{F}_2$ are not the same.
113. Answers will vary.

Section 3.4   *(page 297)*

1. dot product   **3.** $\dfrac{\mathbf{u} \cdot \mathbf{v}}{\|\mathbf{u}\| \|\mathbf{v}\|}$ **5.** $\left(\dfrac{\mathbf{u} \cdot \mathbf{v}}{\|\mathbf{v}\|^2} \right)\mathbf{v}$ **7.** -19
9. -11 **11.** 6 **13.** -12 **15.** 18; scalar
17. $\langle 24, -12 \rangle$; vector   **19.** $\langle -126, -126 \rangle$; vector
21. $\sqrt{10} - 1$; scalar   **23.** -12; scalar   **25.** 17
27. $5\sqrt{41}$ **29.** 6 **31.** $90°$ **33.** $143.13°$
35. $60.26°$ **37.** $90°$ **39.** $\dfrac{5\pi}{12}$
41.

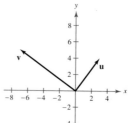

43.

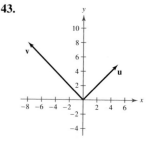

      About $91.33°$ $90°$
45. $26.57°, 63.43°, 90°$ **47.** $41.63°, 53.13°, 85.24°$
49. -20 **51.** -229.1 **53.** Not orthogonal
55. Not orthogonal   **57.** Orthogonal
59. $\frac{1}{37}\langle 84, 14 \rangle, \frac{1}{37}\langle -10, 60 \rangle$ **61.** $\frac{45}{229}\langle 2, 15 \rangle, \frac{6}{229}\langle -15, 2 \rangle$
63. $\langle 3, 2 \rangle$ **65.** $\langle 0, 0 \rangle$ **67.** $\langle -5, 3 \rangle, \langle 5, -3 \rangle$
69. $\frac{2}{3}\mathbf{i} + \frac{1}{2}\mathbf{j}, -\frac{2}{3}\mathbf{i} - \frac{1}{2}\mathbf{j}$ **71.** 32
73. (a) $\$35{,}727.50$
          This value gives the total amount paid to the employees.
      (b) Multiply $\mathbf{v}$ by 1.02.
75. (a) Force $= 30{,}000 \sin d$
      (b)

d	$0°$	$1°$	$2°$	$3°$	$4°$	$5°$
Force	0	523.6	1047.0	1570.1	2092.7	2614.7

d	$6°$	$7°$	$8°$	$9°$	$10°$
Force	3135.9	3656.1	4175.2	4693.0	5209.4

      (c) $29{,}885.8$ lb
77. 735 N-m   **79.** 779.4 ft-lb   **81.** $10{,}282{,}651.78$ N-m
83. 1174.62 ft-lb   **85–87.** Answers will vary.
89. False. Work is represented by a scalar.   **91.** Proof
93. (a) $\mathbf{u}$ and $\mathbf{v}$ are parallel.   (b) $\mathbf{u}$ and $\mathbf{v}$ are orthogonal.

Review Exercises   *(page 302)*

1. $C = 72°, b \approx 12.21, c \approx 12.36$
3. $A = 26°, a \approx 24.89, c \approx 56.23$
5. $C = 66°, a \approx 2.53, b \approx 9.11$
7. $B = 108°, a \approx 11.76, c \approx 21.49$
9. $A \approx 20.41°, C \approx 9.59°, a \approx 20.92$
11. $B \approx 39.48°, C \approx 65.52°, c \approx 48.24$
13. 19.06 **15.** 47.23 **17.** 31.1 m   **19.** 31.01 ft
21. $A \approx 27.81°, B \approx 54.75°, C \approx 97.44°$
23. $A \approx 16.99°, B \approx 26.00°, C \approx 137.01°$
25. $A \approx 29.92°, B \approx 86.18°, C \approx 63.90°$
27. $A = 36°, C = 36°, b \approx 17.80$

29. $A \approx 45.76°, B \approx 91.24°, c \approx 21.42$
31. Law of Sines; $A \approx 77.52°, B \approx 38.48°, a \approx 14.12$
33. Law of Cosines; $A \approx 28.62°, B \approx 33.56°, C \approx 117.82°$
35. About 4.3 ft, about 12.6 ft   **37.** 615.1 m   **39.** 7.64
41. 8.36   **43.** $\|u\| = \|v\| = \sqrt{61}$, slope$_u$ = slope$_v$ = $\frac{5}{6}$
45. $\langle 7, -5 \rangle$ **47.** $\langle 7, -7 \rangle$ **49.** $\langle -4, 4\sqrt{3} \rangle$
51. (a) $\langle -4, 3 \rangle$ (b) $\langle 2, -9 \rangle$

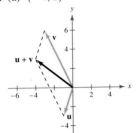
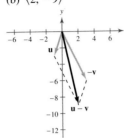

(c) $\langle -4, -12 \rangle$ (d) $\langle -14, 3 \rangle$

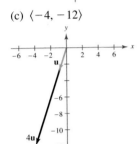
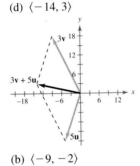

53. (a) $\langle -1, 6 \rangle$ (b) $\langle -9, -2 \rangle$

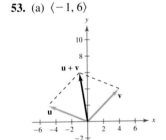
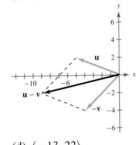

(c) $\langle -20, 8 \rangle$ (d) $\langle -13, 22 \rangle$

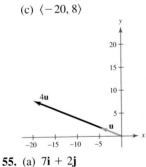
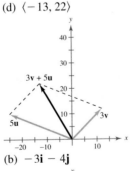

55. (a) $7i + 2j$ (b) $-3i - 4j$

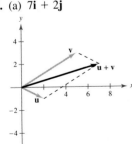
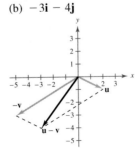

(c) $8i - 4j$ (d) $25i + 4j$

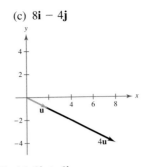
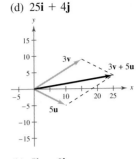

57. (a) $3i + 6j$ (b) $5i - 6j$

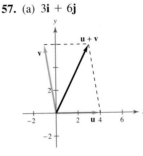
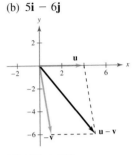

(c) $16i$ (d) $17i + 18j$

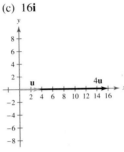
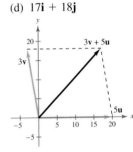

59. $\langle 30, 9 \rangle$ **61.** $\langle 22, -7 \rangle$

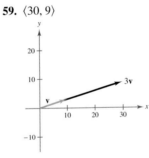
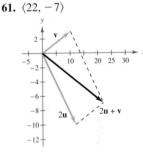

63. $\langle -10, -37 \rangle$

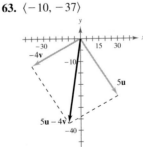

65. $-i + 5j$ **67.** $6i + 4j$ **69.** $\|v\| = 7; \theta = 60°$
71. $\|v\| = \sqrt{41}; \theta = 38.7°$ **73.** $\|v\| = 3\sqrt{2}; \theta = 225°$
75. 422.30 mi/h; 130.4°   **77.** 115.5 lb each   **79.** 45
81. -2 **83.** 40; scalar   **85.** $4 - 2\sqrt{5}$; scalar
87. $\langle 72, -36 \rangle$; vector   **89.** 38; scalar   **91.** $\dfrac{11\pi}{12}$
93. 160.5°   **95.** Orthogonal   **97.** Not orthogonal
99. $-\frac{13}{17}\langle 4, 1 \rangle, \frac{16}{17}\langle -1, 4 \rangle$ **101.** $\frac{5}{2}\langle -1, 1 \rangle, \frac{9}{2}\langle 1, 1 \rangle$

CHAPTER 3

103. 48   **105.** 72,000 ft-lb

107. True. Sin 90° is defined in the Law of Sines.

109. $\dfrac{a}{\sin A} = \dfrac{b}{\sin B} = \dfrac{c}{\sin C}$ **111.** Direction and magnitude

113. a; The angle between the vectors is acute.

115. The diagonal of the parallelogram with **u** and **v** as its adjacent sides

Chapter Test   *(page 306)*

1. Law of Sines; $C = 88°$, $b \approx 27.81$, $c \approx 29.98$

2. Law of Sines; $A = 42°$, $b \approx 21.91$, $c \approx 10.95$

3. Law of Sines

Two solutions: $B \approx 29.12°$, $C \approx 126.88°$, $c \approx 22.03$
$B \approx 150.88°$, $C \approx 5.12°$, $c \approx 2.46$

4. Law of Cosines; No solution

5. Law of Sines; $A \approx 39.96°$, $C \approx 40.04°$, $c \approx 15.02$

6. Law of Cosines; $A \approx 21.90°$, $B \approx 37.10°$, $c \approx 78.15$

7. 2052.5 m^2 **8.** 606.3 mi; 29.1°   **9.** $\langle 14, -23 \rangle$

10. $\left\langle \dfrac{18\sqrt{34}}{17}, -\dfrac{30\sqrt{34}}{17} \right\rangle$

11. $\langle -4, 12 \rangle$ **12.** $\langle 8, 2 \rangle$

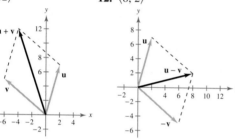

13. $\langle 28, 20 \rangle$ **14.** $\langle -4, 38 \rangle$

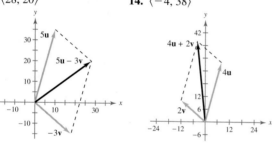

15. $\left\langle \dfrac{24}{25}, -\dfrac{7}{25} \right\rangle$ **16.** 14.9°; 250.15 lb   **17.** 135°

18. Yes   **19.** $\dfrac{37}{26}\langle 5, 1 \rangle$; $\dfrac{29}{26}\langle -1, 5 \rangle$ **20.** About 104 lb

Cumulative Test for Chapters 1–3   *(page 307)*

1. (a)

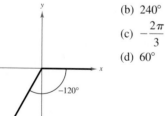

(b) 240°

(c) $-\dfrac{2\pi}{3}$

(d) 60°

(e) $\sin(-120°) = -\dfrac{\sqrt{3}}{2}$ $\csc(-120°) = -\dfrac{2\sqrt{3}}{3}$

$\cos(-120°) = -\dfrac{1}{2}$ $\sec(-120°) = -2$

$\tan(-120°) = \sqrt{3}$ $\cot(-120°) = \dfrac{\sqrt{3}}{3}$

2. $-83.1°$ **3.** $\dfrac{20}{29}$

4.

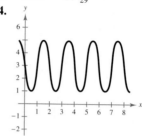

5.

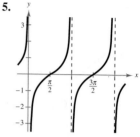

6.

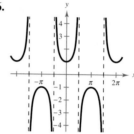

7. $a = -3$, $b = \pi$, $c = 0$

8.

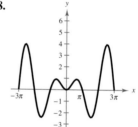

9. 4.9

10. $\dfrac{3}{4}$

11. $\sqrt{1 - 4x^2}$

12. 1

13. $2 \tan \theta$

14–16. Answers will vary.

17. $\dfrac{\pi}{3}, \dfrac{\pi}{2}, \dfrac{3\pi}{2}, \dfrac{5\pi}{3}$

18. $\dfrac{\pi}{6}, \dfrac{5\pi}{6}, \dfrac{7\pi}{6}, \dfrac{11\pi}{6}$ **19.** $\dfrac{3\pi}{2}$ **20.** $\dfrac{16}{63}$ **21.** $\dfrac{4}{3}$

22. $\dfrac{\sqrt{5}}{5}, \dfrac{2\sqrt{5}}{5}$ **23.** $\dfrac{5}{2}\left(\sin\dfrac{5\pi}{2} - \sin\pi\right)$ **24.** $-2\sin 8x \sin x$

25. Law of Sines; $B \approx 26.39°$, $C \approx 123.61°$, $c \approx 14.99$

26. Law of Cosines; $B \approx 52.48°$, $C \approx 97.52°$, $a \approx 5.04$

27. Law of Sines; $B = 60°$, $a \approx 5.77$, $c \approx 11.55$

28. Law of Cosines; $A \approx 26.28°$, $B \approx 49.74°$, $C \approx 103.98°$

29. Law of Sines; $C = 109°$, $a \approx 14.96$, $b \approx 9.27$

30. Law of Cosines; $A \approx 6.75°$, $B \approx 93.25°$, $c \approx 9.86$

31. 41.48 in.^{2} **32.** 599.09 m^2 **33.** $7\mathbf{i} + 8\mathbf{j}$

34. $\left\langle \dfrac{\sqrt{2}}{2}, \dfrac{\sqrt{2}}{2} \right\rangle$ **35.** -5 **36.** $-\dfrac{1}{13}\langle 1, 5 \rangle$; $\dfrac{21}{13}\langle 5, -1 \rangle$

37. About 395.8 rad/min; about 8312.7 in./min

38. 42π yd$^2 \approx 131.95$ yd^2 **39.** 5 ft   **40.** 22.6°

41. $d = 4\cos\dfrac{\pi}{4}t$ **42.** 32.6°; 543.9 km/h   **43.** 425 ft-lb

Problem Solving   *(page 313)*

1. 2.01 ft

3. (a)

A   75 mi   B
30°   135°   15°
x   75°
y
60°   Lost party

(b) Station A: 27.45 mi;
Station B: 53.03 mi

(c) 11.03 mi; S 21.7° E

5. (a) (i) $\sqrt{2}$ (ii) $\sqrt{5}$ (iii) 1
(iv) 1   (v) 1   (vi) 1
(b) (i) 1   (ii) $3\sqrt{2}$ (iii) $\sqrt{13}$
(iv) 1   (v) 1   (vi) 1
(c) (i) $\dfrac{\sqrt{5}}{2}$ (ii) $\sqrt{13}$ (iii) $\dfrac{\sqrt{85}}{2}$
(iv) 1   (v) 1   (vi) 1
(d) (i) $2\sqrt{5}$ (ii) $5\sqrt{2}$ (iii) $5\sqrt{2}$
(iv) 1   (v) 1   (vi) 1

7. $\mathbf{u} \cdot \mathbf{v} = 0$ and $\mathbf{u} \cdot \mathbf{w} = 0$
$$\mathbf{u} \cdot (c\mathbf{v} + d\mathbf{w}) = \mathbf{u} \cdot c\mathbf{v} + \mathbf{u} \cdot d\mathbf{w}$$
$$= c(\mathbf{u} \cdot \mathbf{v}) + d(\mathbf{u} \cdot \mathbf{w})$$
$$= 0$$

9. (a) $\mathbf{u} = \langle 0, -120 \rangle$, $\mathbf{v} = \langle 40, 0 \rangle$
(b)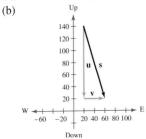
(c) 126.5 miles per hour; The magnitude gives the actual rate of the skydiver's fall.
(d) $71.57°$

(e)
Up
140
120
100
80
60 **u**
40 **s**
20 **v**
W ←——————→ E
−60 −20 20 40 60 80 100
Down

123.7 mi/h

Chapter 4

Section 4.1   *(page 321)*

1. real   **3.** pure imaginary   **5.** principal square
7. $a = -12, b = 7$ **9.** $a = 6, b = 5$ **11.** $8 + 5i$
13. $2 - 3\sqrt{3}i$ **15.** $4\sqrt{5}i$ **17.** 14   **19.** $-1 - 10i$
21. $0.3i$ **23.** $10 - 3i$ **25.** 1   **27.** $3 - 3\sqrt{2}i$
29. $-14 + 20i$ **31.** $\frac{1}{6} + \frac{7}{6}i$ **33.** $5 + i$ **35.** $108 + 12i$
37. 24   **39.** $-13 + 84i$ **41.** -10 **43.** $9 - 2i, 85$
45. $-1 + \sqrt{5}i, 6$ **47.** $-2\sqrt{5}i, 20$ **49.** $\sqrt{6}, 6$
51. $-3i$ **53.** $\frac{8}{41} + \frac{10}{41}i$ **55.** $\frac{12}{13} + \frac{5}{13}i$ **57.** $-4 - 9i$
59. $-\frac{120}{1681} - \frac{27}{1681}i$ **61.** $-\frac{1}{2} - \frac{5}{2}i$ **63.** $\frac{62}{949} + \frac{297}{949}i$
65. $-2\sqrt{3}$ **67.** -15 **69.** $\left(21 + 5\sqrt{2}\right) + \left(7\sqrt{5} - 3\sqrt{10}\right)i$
71. $1 \pm i$ **73.** $-2 \pm \frac{1}{2}i$ **75.** $-\frac{5}{2}, -\frac{3}{2}$ **77.** $2 \pm \sqrt{2}i$
79. $\dfrac{5}{7} \pm \dfrac{5\sqrt{15}}{7}$ **81.** $-1 + 6i$ **83.** $-14i$
85. $-432\sqrt{2}i$ **87.** i **89.** 81
91. (a) $z_1 = 9 + 16i, z_2 = 20 - 10i$
(b) $z = \dfrac{11{,}240}{877} + \dfrac{4630}{877}i$
93. False. When the complex number is real, the number equals its conjugate.

95. False.
$$i^{44} + i^{150} - i^{74} - i^{109} + i^{61} = 1 - 1 + 1 - i + i = 1$$
97. $i, -1, -i, 1, i, -1, -i, 1$; The pattern repeats the first four results. Divide the exponent by 4.
When the remainder is 1, the result is i.
When the remainder is 2, the result is -1.
When the remainder is 3, the result is $-i$.
When the remainder is 0, the result is 1.
99. $\sqrt{-6}\sqrt{-6} = \sqrt{6}i\sqrt{6}i = 6i^2 = -6$ **101.** Proof

Section 4.2   *(page 328)*

1. Fundamental Theorem; Algebra   **3.** conjugates
5. Three solutions   **7.** Four solutions
9. No real solutions   **11.** Two real solutions
13. Two real solutions   **15.** No real solutions   **17.** $\pm\sqrt{5}$
19. $-5 \pm \sqrt{6}$ **21.** 4   **23.** $-1 \pm 2i$ **25.** $\frac{1}{2} \pm i$
27. $\pm\sqrt{7}, \pm i$ **29.** $\pm\sqrt{6}, \pm i$
31. (a)
(b) $4, \pm 4i$
33. (a)
(b) $\pm\sqrt{2}i$
(c) The number of real zeros and the number of x-intercepts are the same.

35. $(x + 6i)(x - 6i)$; $\pm 6i$
37. $(x - 1 - 4i)(x - 1 + 4i)$; $1 \pm 4i$
39. $(x + 3)(x - 3)(x + 3i)(x - 3i)$; $\pm 3, \pm 3i$
41. $(z - 1 + i)(z - 1 - i)$; $1 \pm i$
43. $(x + 3)\left(x + \sqrt{3}\right)\left(x - \sqrt{3}\right)$; $-3, \pm\sqrt{3}$
45. $(x - 4)(x + 4i)(x - 4i)$; $4, \pm 4i$
47. $(2x - 1)\left(x + 3\sqrt{2}i\right)\left(x - 3\sqrt{2}i\right)$; $\frac{1}{2}, \pm 3\sqrt{2}i$
49. $x(x - 6)(x + 4i)(x - 4i)$; $0, 6, \pm 4i$
51. $(x + i)(x - i)(x + 3i)(x - 3i)$; $\pm i, \pm 3i$ **53.** $-\frac{3}{2}, \pm 5i$
55. $\pm 2i, 1, -\frac{1}{2}$ **57.** $-3 \pm i, \frac{1}{4}$ **59.** $-4, 3 \pm i$
61. $2, -3 \pm \sqrt{2}i, 1$ **63.** $f(x) = x^3 - x^2 + 25x - 25$
65. $f(x) = x^3 - 12x^2 + 46x - 52$
67. $f(x) = 3x^4 - 17x^3 + 25x^2 + 23x - 22$
69. $f(x) = -x^3 + x^2 - 4x + 4$
71. $f(x) = -3x^3 + 9x^2 - 3x - 15$
73. $f(x) = -2x^3 + 5x^2 - 10x + 4$
75. $f(x) = x^3 - 6x^2 + 4x + 40$
77. $f(x) = x^3 - 3x^2 + 6x + 10$
79. $f(x) = x^3 - 2x^2 - 4x - 16$
81. $f(x) = \frac{1}{2}x^4 + \frac{1}{2}x^3 - 2x^2 + x - 6$

CHAPTER 4

83. (a)

t	0	0.5	1	1.5	2	2.5	3
h	0	20	32	36	32	20	0

; No

(b) When you set $h = 64$, the resulting equation yields imaginary roots. So, the projectile will not reach a height of 64 feet.

(c)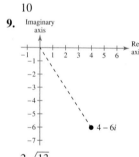

The graphs do not intersect, so the projectile does not reach 64 feet.

(d) The results all show that it is not possible for the projectile to reach a height of 64 feet.

85. (a) $P = -0.0001x^2 + 60x - 150,000$

(b) \$8,600,000   (c) \$115

(d) It is not possible to have a profit of 10 million dollars.

87. False. The most complex zeros it can have is two, and the Linear Factorization Theorem guarantees that there are three linear factors, so one zero must be real.

89. Answers will vary   **91.** $x^2 - 2ax + a^2 + b^2$

93. r_1, r_2, r_3 **95.** $5 + r_1, 5 + r_2, 5 + r_3$

97. The zeros cannot be determined.

Section 4.3   (page 336)

1. real; imaginary   **3.** trigonometric form; modulus; argument

5.

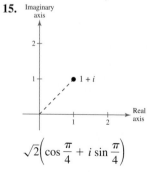

$-6 + 8i$

10

7.

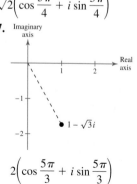

$-7i$

7

9.

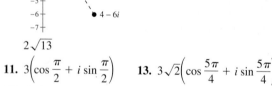

$4 - 6i$

$2\sqrt{13}$

11. $3\left(\cos\dfrac{\pi}{2} + i\sin\dfrac{\pi}{2}\right)$ **13.** $3\sqrt{2}\left(\cos\dfrac{5\pi}{4} + i\sin\dfrac{5\pi}{4}\right)$

15.

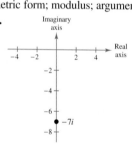

$1 + i$

$\sqrt{2}\left(\cos\dfrac{\pi}{4} + i\sin\dfrac{\pi}{4}\right)$

17.

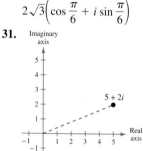

$1 - \sqrt{3}i$

$2\left(\cos\dfrac{5\pi}{3} + i\sin\dfrac{5\pi}{3}\right)$

19.

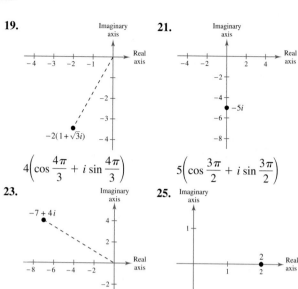

$-2(1 + \sqrt{3}i)$

$4\left(\cos\dfrac{4\pi}{3} + i\sin\dfrac{4\pi}{3}\right)$

21.

$-5i$

$5\left(\cos\dfrac{3\pi}{2} + i\sin\dfrac{3\pi}{2}\right)$

23.

$-7 + 4i$

$\sqrt{65}\,(\cos 2.62 + i\sin 2.62)$

25.

$2(\cos 0 + i\sin 0)$

27.

$3 + \sqrt{3}i$

$2\sqrt{3}\left(\cos\dfrac{\pi}{6} + i\sin\dfrac{\pi}{6}\right)$

29.

$-3 - i$

$\sqrt{10}\,(\cos 3.46 + i\sin 3.46)$

31.

$5 + 2i$

$\sqrt{29}\,(\cos 0.38 + i\sin 0.38)$

33.

$-8 - 5\sqrt{3}i$

$\sqrt{139}\,(\cos 3.97 + i\sin 3.97)$

35. $1 + \sqrt{3}i$

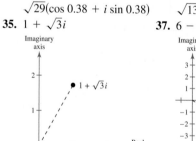

$1 + \sqrt{3}i$

37. $6 - 2\sqrt{3}i$

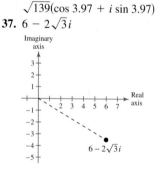

$6 - 2\sqrt{3}i$

39. $-\dfrac{9\sqrt{2}}{8} + \dfrac{9\sqrt{2}}{8}i$ **41.** 7

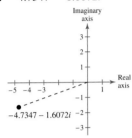

43. $-4.7347 - 1.6072i$

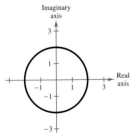

45. $4.6985 + 1.7101i$ **47.** $-1.8126 + 0.8452i$

49. $12\left(\cos\dfrac{\pi}{3} + i\sin\dfrac{\pi}{3}\right)$ **51.** $\dfrac{10}{9}(\cos 150° + i\sin 150°)$

53. $\cos 50° + i\sin 50°$ **55.** $\frac{1}{3}(\cos 30° + i\sin 30°)$

57. $\cos\dfrac{2\pi}{3} + i\sin\dfrac{2\pi}{3}$ **59.** $6(\cos 330° + i\sin 330°)$

61. (a) $\left[2\sqrt{2}\left(\cos\dfrac{\pi}{4} + i\sin\dfrac{\pi}{4}\right)\right]\left[\sqrt{2}\left(\cos\dfrac{7\pi}{4} + i\sin\dfrac{7\pi}{4}\right)\right]$

   (b) $4(\cos 0 + i\sin 0) = 4$ (c) 4

63. (a) $\left[2\left(\cos\dfrac{3\pi}{2} + i\sin\dfrac{3\pi}{2}\right)\right]\left[\sqrt{2}\left(\cos\dfrac{\pi}{4} + i\sin\dfrac{\pi}{4}\right)\right]$

   (b) $2\sqrt{2}\left(\cos\dfrac{7\pi}{4} + i\sin\dfrac{7\pi}{4}\right) = 2 - 2i$

   (c) $-2i - 2i^2 = -2i + 2 = 2 - 2i$

65. (a) $[5(\cos 0.93 + i\sin 0.93)] \div \left[2\left(\cos\dfrac{5\pi}{3} + i\sin\dfrac{5\pi}{3}\right)\right]$

   (b) $\dfrac{5}{2}(\cos 1.97 + i\sin 1.97) \approx -0.982 + 2.299i$

   (c) About $-0.982 + 2.299i$

67. (a) $[5(\cos 0 + i\sin 0)] \div \left[\sqrt{13}(\cos 0.98 + i\sin 0.98)\right]$

   (b) $\dfrac{5}{\sqrt{13}}(\cos 5.30 + i\sin 5.30) \approx 0.769 - 1.154i$

   (c) $\dfrac{10}{13} - \dfrac{15}{13}i \approx 0.769 - 1.154i$

69.   **71.**

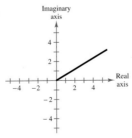

73. (a) $E = 24(\cos 30° + i\sin 30°)$ volts
   (b) $E = 12\sqrt{3} + 12i$ volts   (c) $|E| = 24$ volts
75–77. Answers will vary.

Section 4.4   *(page 342)*

1. DeMoivre's   **3.** $\sqrt[n]{r}\left(\cos\dfrac{\theta + 2\pi k}{n} + i\sin\dfrac{\theta + 2\pi k}{n}\right)$

5. $-4 - 4i$ **7.** $8i$ **9.** $1024 - 1024\sqrt{3}i$

11. $\dfrac{125}{2} + \dfrac{125\sqrt{3}}{2}i$ **13.** -1 **15.** $608.0204 + 144.6936i$

17. $-597 - 122i$ **19.** $-43\sqrt{5} + 4i$ **21.** $\dfrac{81}{2} + \dfrac{81\sqrt{3}}{2}i$

23. $32.3524 - 120.7407i$ **25.** $32i$ **27.** 27

29. $1 + i, -1 - i$ **31.** $-\dfrac{\sqrt{6}}{2} + \dfrac{\sqrt{6}}{2}i, \dfrac{\sqrt{6}}{2} - \dfrac{\sqrt{6}}{2}i$

33. $-1.5538 + 0.6436i, 1.5538 - 0.6436i$

35. $\dfrac{\sqrt{6}}{2} + \dfrac{\sqrt{2}}{2}i, -\dfrac{\sqrt{6}}{2} - \dfrac{\sqrt{2}}{2}i$

37. (a) $\sqrt{5}(\cos 60° + i\sin 60°)$ (b)
   $\sqrt{5}(\cos 240° + i\sin 240°)$
   (c) $\dfrac{\sqrt{5}}{2} + \dfrac{\sqrt{15}}{2}i, -\dfrac{\sqrt{5}}{2} - \dfrac{\sqrt{15}}{2}i$

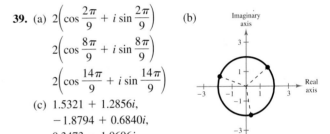

39. (a) $2\left(\cos\dfrac{2\pi}{9} + i\sin\dfrac{2\pi}{9}\right)$ (b)
   $2\left(\cos\dfrac{8\pi}{9} + i\sin\dfrac{8\pi}{9}\right)$
   $2\left(\cos\dfrac{14\pi}{9} + i\sin\dfrac{14\pi}{9}\right)$
   (c) $1.5321 + 1.2856i,$
   $-1.8794 + 0.6840i,$
   $0.3473 - 1.9696i$

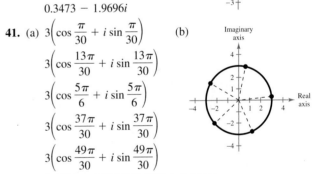

41. (a) $3\left(\cos\dfrac{\pi}{30} + i\sin\dfrac{\pi}{30}\right)$ (b)
   $3\left(\cos\dfrac{13\pi}{30} + i\sin\dfrac{13\pi}{30}\right)$
   $3\left(\cos\dfrac{5\pi}{6} + i\sin\dfrac{5\pi}{6}\right)$
   $3\left(\cos\dfrac{37\pi}{30} + i\sin\dfrac{37\pi}{30}\right)$
   $3\left(\cos\dfrac{49\pi}{30} + i\sin\dfrac{49\pi}{30}\right)$
   (c) $2.9836 + 0.3136i, 0.6237 + 2.9344i,$
   $-2.5981 + 1.5i, -2.2294 - 2.0074i, 1.2202 - 2.7406i$

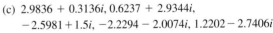

43. (a) $3\left(\cos\dfrac{\pi}{8} + i\sin\dfrac{\pi}{8}\right)$ (b)
   $3\left(\cos\dfrac{5\pi}{8} + i\sin\dfrac{5\pi}{8}\right)$
   $3\left(\cos\dfrac{9\pi}{8} + i\sin\dfrac{9\pi}{8}\right)$
   $3\left(\cos\dfrac{13\pi}{8} + i\sin\dfrac{13\pi}{8}\right)$
   (c) $2.7716 + 1.1481i,$
   $-1.1481 + 2.7716i,$
   $-2.7716 - 1.1481i, 1.1481 - 2.7716i$

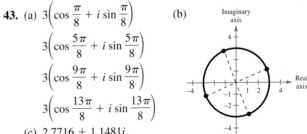

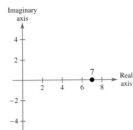

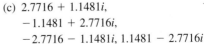

CHAPTER 4

45. (a) $5\left(\cos\dfrac{4\pi}{9} + i\sin\dfrac{4\pi}{9}\right)$ (b)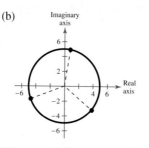

$5\left(\cos\dfrac{10\pi}{9} + i\sin\dfrac{10\pi}{9}\right)$

$5\left(\cos\dfrac{16\pi}{9} + i\sin\dfrac{16\pi}{9}\right)$

(c) $0.8682 + 4.9240i$,

$-4.6985 - 1.7101i$,

$3.8302 - 3.2139i$

47. (a) $2(\cos 0 + i\sin 0)$ (b)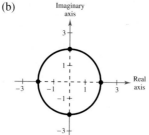

$2\left(\cos\dfrac{\pi}{2} + i\sin\dfrac{\pi}{2}\right)$

$2(\cos \pi + i\sin \pi)$

$2\left(\cos\dfrac{3\pi}{2} + i\sin\dfrac{3\pi}{2}\right)$

(c) $2, 2i, -2, -2i$

49. (a) $\cos 0 + i\sin 0$ (b)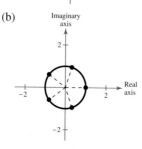

$\cos\dfrac{2\pi}{5} + i\sin\dfrac{2\pi}{5}$

$\cos\dfrac{4\pi}{5} + i\sin\dfrac{4\pi}{5}$

$\cos\dfrac{6\pi}{5} + i\sin\dfrac{6\pi}{5}$

$\cos\dfrac{8\pi}{5} + i\sin\dfrac{8\pi}{5}$

(c) $1, 0.3090 + 0.9511i, -0.8090 + 0.5878i,$

$-0.8090 - 0.5878i, 0.3090 - 0.9511i$

51. (a) $5\left(\cos\dfrac{\pi}{3} + i\sin\dfrac{\pi}{3}\right)$ (b)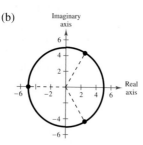

$5(\cos \pi + i\sin \pi)$

$5\left(\cos\dfrac{5\pi}{3} + i\sin\dfrac{5\pi}{3}\right)$

(c) $\dfrac{5}{2} + \dfrac{5\sqrt{3}}{2}i, -5,$

$\dfrac{5}{2} - \dfrac{5\sqrt{3}}{2}i$

53. (a) $\sqrt{2}\left(\cos\dfrac{7\pi}{20} + i\sin\dfrac{7\pi}{20}\right)$ (b)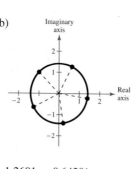

$\sqrt{2}\left(\cos\dfrac{3\pi}{4} + i\sin\dfrac{3\pi}{4}\right)$

$\sqrt{2}\left(\cos\dfrac{23\pi}{20} + i\sin\dfrac{23\pi}{20}\right)$

$\sqrt{2}\left(\cos\dfrac{31\pi}{20} + i\sin\dfrac{31\pi}{20}\right)$

$\sqrt{2}\left(\cos\dfrac{39\pi}{20} + i\sin\dfrac{39\pi}{20}\right)$

(c) $0.6420 + 1.2601i, -1 + i, -1.2601 - 0.6420i,$

$0.2212 - 1.3968i, 1.3968 - 0.2212i$

55. $\cos\dfrac{3\pi}{8} + i\sin\dfrac{3\pi}{8}$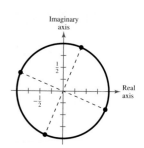

$\cos\dfrac{7\pi}{8} + i\sin\dfrac{7\pi}{8}$

$\cos\dfrac{11\pi}{8} + i\sin\dfrac{11\pi}{8}$

$\cos\dfrac{15\pi}{8} + i\sin\dfrac{15\pi}{8}$

57. $\cos\dfrac{\pi}{6} + i\sin\dfrac{\pi}{6}$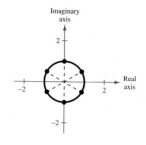

$\cos\dfrac{\pi}{2} + i\sin\dfrac{\pi}{2}$

$\cos\dfrac{5\pi}{6} + i\sin\dfrac{5\pi}{6}$

$\cos\dfrac{7\pi}{6} + i\sin\dfrac{7\pi}{6}$

$\cos\dfrac{3\pi}{2} + i\sin\dfrac{3\pi}{2}$

$\cos\dfrac{11\pi}{6} + i\sin\dfrac{11\pi}{6}$

59. $3\left(\cos\dfrac{\pi}{5} + i\sin\dfrac{\pi}{5}\right)$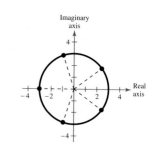

$3\left(\cos\dfrac{3\pi}{5} + i\sin\dfrac{3\pi}{5}\right)$

$3(\cos \pi + i\sin \pi)$

$3\left(\cos\dfrac{7\pi}{5} + i\sin\dfrac{7\pi}{5}\right)$

$3\left(\cos\dfrac{9\pi}{5} + i\sin\dfrac{9\pi}{5}\right)$

61. $4(\cos 0 + i\sin 0)$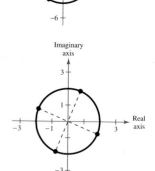

$4\left(\cos\dfrac{2\pi}{3} + i\sin\dfrac{2\pi}{3}\right)$

$4\left(\cos\dfrac{4\pi}{3} + i\sin\dfrac{4\pi}{3}\right)$

63. $2\left(\cos\dfrac{3\pi}{8} + i\sin\dfrac{3\pi}{8}\right)$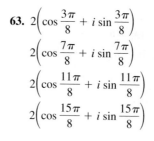

$2\left(\cos\dfrac{7\pi}{8} + i\sin\dfrac{7\pi}{8}\right)$

$2\left(\cos\dfrac{11\pi}{8} + i\sin\dfrac{11\pi}{8}\right)$

$2\left(\cos\dfrac{15\pi}{8} + i\sin\dfrac{15\pi}{8}\right)$

65. $2\left(\cos\dfrac{\pi}{8} + i\sin\dfrac{\pi}{8}\right)$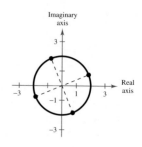

$2\left(\cos\dfrac{5\pi}{8} + i\sin\dfrac{5\pi}{8}\right)$

$2\left(\cos\dfrac{9\pi}{8} + i\sin\dfrac{9\pi}{8}\right)$

$2\left(\cos\dfrac{13\pi}{8} + i\sin\dfrac{13\pi}{8}\right)$

67. $\sqrt[6]{2}\left(\cos\dfrac{7\pi}{12} + i\sin\dfrac{7\pi}{12}\right)$

$\sqrt[6]{2}\left(\cos\dfrac{5\pi}{4} + i\sin\dfrac{5\pi}{4}\right)$

$\sqrt[6]{2}\left(\cos\dfrac{23\pi}{12} + i\sin\dfrac{23\pi}{12}\right)$

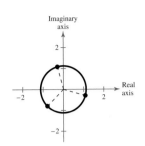

69. $\sqrt[12]{2}\left(\cos\dfrac{5\pi}{24} + i\sin\dfrac{5\pi}{24}\right)$

$\sqrt[12]{2}\left(\cos\dfrac{13\pi}{24} + i\sin\dfrac{13\pi}{24}\right)$

$\sqrt[12]{2}\left(\cos\dfrac{7\pi}{8} + i\sin\dfrac{7\pi}{8}\right)$

$\sqrt[12]{2}\left(\cos\dfrac{29\pi}{24} + i\sin\dfrac{29\pi}{24}\right)$

$\sqrt[12]{2}\left(\cos\dfrac{37\pi}{24} + i\sin\dfrac{37\pi}{24}\right)$

$\sqrt[12]{2}\left(\cos\dfrac{15\pi}{8} + i\sin\dfrac{15\pi}{8}\right)$

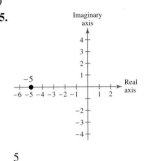

71. (a) prisoner set   (b) escape set
(c) escape set   (d) prisoner set

73. False. The complex number needs to be converted to trigonometric form before using DeMoivre's Theorem.
$\left(4 + \sqrt{6}i\right)^8 \approx \left[\sqrt{22}(\cos 0.55 + i\sin 0.55)\right]^8$

75. The given equation can be written as
$x^4 = -16 = 16(\cos\pi + i\sin\pi)$
which means that you solve the equation by finding the four fourth roots of -16. Each of these roots has the form
$\sqrt[4]{16}\left(\cos\dfrac{\pi + 2\pi k}{4} + i\sin\dfrac{\pi + 2\pi k}{4}\right).$

77. Answers will vary.

79. (a) $i, -2i$ (b) $\left(-1 \pm \sqrt{2}\right)i$
(c) $\dfrac{\sqrt{2}}{2} - \left(1 + \dfrac{\sqrt{6}}{2}\right)i, -\dfrac{\sqrt{2}}{2} + \left(-1 + \dfrac{\sqrt{6}}{2}\right)i$

81–83. Answers will vary.

Review Exercises *(page 346)*

1. $6 + 2i$ **3.** $4\sqrt{3}i$ **5.** $-1 + 3i$ **7.** $3 + 7i$
9. $11 + 44i$ **11.** $40 + 65i$ **13.** $-4 - 46i$
15. $-45 + 28i$ **17.** $\dfrac{10}{3}i$ **19.** $\dfrac{23}{17} + \dfrac{10}{17}i$ **21.** $\dfrac{21}{13} - \dfrac{1}{13}i$
23. $\pm\dfrac{\sqrt{3}}{3}i$ **25.** $1 \pm 3i$ **27.** $-10 + i$ **29.** i
31. Five solutions   **33.** Four solutions
35. Two real solutions   **37.** No real solutions
39. $0, 2$ **41.** $\dfrac{3}{2} \pm \dfrac{\sqrt{11}i}{2}$ **43.** $-4 \pm \sqrt{6}$
45. $-\dfrac{3}{4} \pm \dfrac{\sqrt{39}i}{4}$ **47.** $16.8°C$
49. $\dfrac{1}{2}\left(2x + 1 - \sqrt{5}i\right)\left(2x + 1 + \sqrt{5}i\right); -\dfrac{1}{2} \pm \dfrac{\sqrt{5}}{2}i$
51. $(2x - 3)(x + 5i)(x - 5i); \dfrac{3}{2}, \pm 5i$
53. $\left(2x + \sqrt{5}\right)\left(2x - \sqrt{5}\right)\left(x + \sqrt{2}i\right)\left(x - \sqrt{2}i\right); \pm\dfrac{\sqrt{5}}{2}, \pm\sqrt{2}i$
55. $-7, 2$ **57.** $-5, 1 \pm 2i$ **59.** $-\dfrac{1}{2}, 5 \pm 3i$
61. $-2, 3, -3 \pm \sqrt{5}i$ **63.** $f(x) = 12x^4 - 19x^3 + 9x - 2$

65. $f(x) = x^3 - 7x^2 + 13x - 3$
67. $f(x) = 3x^4 - 14x^3 + 17x^2 - 42x + 24$
69. $f(x) = x^4 + 27x^2 + 50$
71. $f(x) = 2x^3 - 14x^2 + 24x - 20$

73.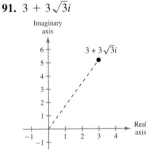

8

75.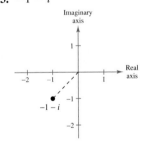

5

77.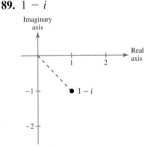

$\sqrt{34}$

79. $8(\cos 0 + i\sin 0)$ **81.** $3\left(\cos\dfrac{3\pi}{2} + i\sin\dfrac{3\pi}{2}\right)$

83. $5\sqrt{2}\left(\cos\dfrac{7\pi}{4} + i\sin\dfrac{7\pi}{4}\right)$ **85.** $6\left(\cos\dfrac{5\pi}{6} + i\sin\dfrac{5\pi}{6}\right)$

87. $\sqrt{3} + i$ **89.** $1 - i$

91. $3 + 3\sqrt{3}i$ **93.** $-1 - i$

95. $28\left(\cos\dfrac{7\pi}{12} + i\sin\dfrac{7\pi}{12}\right)$ **97.** $\dfrac{1}{2}\left(\cos\dfrac{\pi}{2} + i\sin\dfrac{\pi}{2}\right)$

99. (a) $z_1 = \sqrt{2}\left(\cos\dfrac{\pi}{4} + i\sin\dfrac{\pi}{4}\right)$

$z_2 = \sqrt{2}\left(\cos\dfrac{7\pi}{4} + i\sin\dfrac{7\pi}{4}\right)$

(b) $z_1 z_2 = 2(\cos 0 + i\sin 0)$

$\dfrac{z_1}{z_2} = \left(\cos\dfrac{\pi}{2} + i\sin\dfrac{\pi}{2}\right)$

101. (a) $z_1 = 4\left(\cos\dfrac{11\pi}{6} + i \sin\dfrac{11\pi}{6}\right)$

$z_2 = 10\left(\cos\dfrac{3\pi}{2} + i \sin\dfrac{3\pi}{2}\right)$

(b) $z_1 z_2 = 40\left(\cos\dfrac{4\pi}{3} + i \sin\dfrac{4\pi}{3}\right)$

$\dfrac{z_1}{z_2} = \dfrac{2}{5}\left(\cos\dfrac{\pi}{3} + i \sin\dfrac{\pi}{3}\right)$

103. $\dfrac{625}{2} + \dfrac{625\sqrt{3}}{2}i$ **105.** $2035 - 828i$ **107.** $-8 - 8i$

109. (a) $3\left(\cos\dfrac{\pi}{4} + i \sin\dfrac{\pi}{4}\right)$ (b)

$3\left(\cos\dfrac{7\pi}{12} + i \sin\dfrac{7\pi}{12}\right)$

$3\left(\cos\dfrac{11\pi}{12} + i \sin\dfrac{11\pi}{12}\right)$

$3\left(\cos\dfrac{5\pi}{4} + i \sin\dfrac{5\pi}{4}\right)$

$3\left(\cos\dfrac{19\pi}{12} + i \sin\dfrac{19\pi}{12}\right)$

$3\left(\cos\dfrac{23\pi}{12} + i \sin\dfrac{23\pi}{12}\right)$

(c) $2.1213 + 2.1213i, -0.7765 + 2.8978i,$
$-2.8978 + 0.7765i, -2.1213 - 2.1213i,$
$0.7765 - 2.8978i, 2.8978 - 0.7765i$

111. (a) $2\left(\cos\dfrac{\pi}{4} + i \sin\dfrac{\pi}{4}\right)$ (b)

$2\left(\cos\dfrac{3\pi}{4} + i \sin\dfrac{3\pi}{4}\right)$

$2\left(\cos\dfrac{5\pi}{4} + i \sin\dfrac{5\pi}{4}\right)$

$2\left(\cos\dfrac{7\pi}{4} + i \sin\dfrac{7\pi}{4}\right)$

(c) $\sqrt{2} + \sqrt{2}i, -\sqrt{2} + \sqrt{2}i,$
$-\sqrt{2} - \sqrt{2}i, \sqrt{2} - \sqrt{2}i$

113. $3\left(\cos\dfrac{\pi}{4} + i \sin\dfrac{\pi}{4}\right) = \dfrac{3\sqrt{2}}{2} + \dfrac{3\sqrt{2}}{2}i$

$3\left(\cos\dfrac{3\pi}{4} + i \sin\dfrac{3\pi}{4}\right) = -\dfrac{3\sqrt{2}}{2} + \dfrac{3\sqrt{2}}{2}i$

$3\left(\cos\dfrac{5\pi}{4} + i \sin\dfrac{5\pi}{4}\right) = -\dfrac{3\sqrt{2}}{2} - \dfrac{3\sqrt{2}}{2}i$

$3\left(\cos\dfrac{7\pi}{4} + i \sin\dfrac{7\pi}{4}\right) = \dfrac{3\sqrt{2}}{2} - \dfrac{3\sqrt{2}}{2}i$

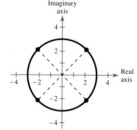

115. $2\left(\cos\dfrac{\pi}{2} + i \sin\dfrac{\pi}{2}\right) = 2i$

$2\left(\cos\dfrac{7\pi}{6} + i \sin\dfrac{7\pi}{6}\right) = -\sqrt{3} - i$

$2\left(\cos\dfrac{11\pi}{6} + i \sin\dfrac{11\pi}{6}\right) = \sqrt{3} - i$

117. False.
$\sqrt{-18}\sqrt{-2} = 3\sqrt{2}i\,\sqrt{2}i$ and $\sqrt{(-18)(-2)} = \sqrt{36}$
$= 3\sqrt{4}i^2$ $= 6$
$= 6i^2$
$= -6$

119. False. A fourth-degree polynomial with real coefficients has four zeros, and complex zeros occur in conjugate pairs.

121. (a) $4(\cos 60° + i \sin 60°)$ (b) -64
$4(\cos 180° + i \sin 180°)$
$4(\cos 300° + i \sin 300°)$

123. $z_1 z_2 = -4, \dfrac{z_1}{z_2} = -\cos 2\theta - i \sin 2\theta$

Chapter Test   *(page 349)*

1. $-5 + 10i$ **2.** $-3 + 5i$ **3.** $-65 + 72i$ **4.** 43

5. $\dfrac{32}{73} + \dfrac{12}{73}i$ **6.** $\dfrac{1}{2} \pm \dfrac{\sqrt{5}}{2}i$ **7.** Five solutions

8. Four solutions

9. $(x - 6)\left(x + \sqrt{5}i\right)\left(x - \sqrt{5}i\right); 6, \pm\sqrt{5}i$

10. $\left(x + \sqrt{6}\right)\left(x - \sqrt{6}\right)(x + 2i)(x - 2i); \pm\sqrt{6}, \pm 2i$

11. $\pm 2, \pm\sqrt{2}i$ **12.** $\dfrac{3}{2}, 2 \pm i$

13. $x^4 - 15x^3 + 73x^2 - 119x$

14. $x^4 - 8x^3 + 28x^2 - 60x + 63$

15. No. If $a + bi, b \neq 0$, is a zero, its conjugate $a - bi$ is also a zero.

16. $4\sqrt{2}\left(\cos\dfrac{7\pi}{4} + i \sin\dfrac{7\pi}{4}\right)$ **17.** $-3 + 3\sqrt{3}i$

18. $-\dfrac{6561}{2} - \dfrac{6561\sqrt{3}}{2}i$ **19.** $5832i$

20. $4\sqrt[4]{2}\left(\cos\dfrac{\pi}{12} + i \sin\dfrac{\pi}{12}\right)$

$4\sqrt[4]{2}\left(\cos\dfrac{7\pi}{12} + i \sin\dfrac{7\pi}{12}\right)$

$4\sqrt[4]{2}\left(\cos\dfrac{13\pi}{12} + i \sin\dfrac{13\pi}{12}\right)$

$4\sqrt[4]{2}\left(\cos\dfrac{19\pi}{12} + i \sin\dfrac{19\pi}{12}\right)$

21. $3\left(\cos\dfrac{\pi}{6} + i\sin\dfrac{\pi}{6}\right)$

$3\left(\cos\dfrac{5\pi}{6} + i\sin\dfrac{5\pi}{6}\right)$

$3\left(\cos\dfrac{3\pi}{2} + i\sin\dfrac{3\pi}{2}\right)$

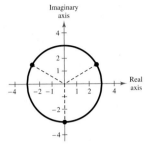

22. No. When you set $h = 125$, the resulting equation yields imaginary roots. So, the projectile will not reach a height of 125 feet.

Problem Solving *(page 351)*

1. (a) $z^3 = 8$ for all three complex numbers.

(b) $z^3 = 27$ for all three complex numbers.

(c) The cube roots of a positive real number "a" are:

$\sqrt[3]{a}, \dfrac{-\sqrt[3]{a} + \sqrt[3]{a}\sqrt{3}i}{2}$, and $\dfrac{-\sqrt[3]{a} - \sqrt[3]{a}\sqrt{3}i}{2}$.

3. (a) $f(x) = -2x^3 + 3x^2 + 11x - 6$

(b)

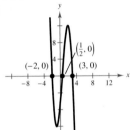

(c) (Equations and graphs will vary.) There are infinitely many possible functions for f.

5. Answers will vary.

7. (a) $0 < k < 4$ (b) $k < 0$ (c) $k > 4$

9. (a) Not correct because f has $(0, 0)$ as an intercept.

(b) Not correct because the function must be at least a fourth-degree polynomial.

(c) Correct function

(d) Not correct because k has $(-1, 0)$ as an intercept.

11. (a) Yes (b) No (c) Yes

13. (a) $1 + i, 3 + i$ (b) $1 - i, 2 + 3i$

(c) $1 + i, -\frac{7}{2} + 3i$ (d) $4 + 5i, -\frac{1}{3} - \frac{1}{3}i$

15. Answers will vary.

Chapter 5

Section 5.1 *(page 362)*

1. algebraic **3.** One-to-One **5.** $A = P\left(1 + \dfrac{r}{n}\right)^{nt}$

7. 0.863 **9.** 0.006 **11.** 1767.767

13. d **14.** c **15.** a **16.** b

17.

x	-2	-1	0	1	2
$f(x)$	4	2	1	0.5	0.25

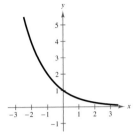

19.

x	-2	-1	0	1	2
$f(x)$	36	6	1	0.167	0.028

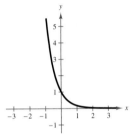

21.

x	-2	-1	0	1	2
$f(x)$	0.125	0.25	0.5	1	2

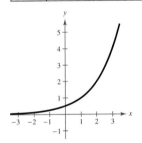

23. $x = 2$ **25.** $x = -5$

27. Shift the graph of f one unit up.

29. Reflect the graph of f in the origin.

31.

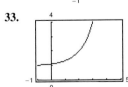

33.

35. 24.533 **37.** 7166.647

39.

x	-2	-1	0	1	2
$f(x)$	0.135	0.368	1	2.718	7.389

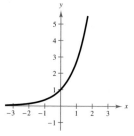

41.

x	-8	-7	-6	-5	-4
$f(x)$	0.055	0.149	0.406	1.104	3

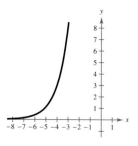

43.

x	-2	-1	0	1	2
$f(x)$	4.037	4.100	4.271	4.736	6

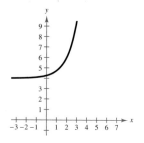

45. 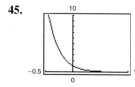   **47.**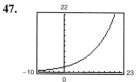

49.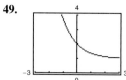

51. $x = \frac{1}{3}$ **53.** $x = 3, -1$

55.

n	1	2	4	12
A	\$1828.49	\$1830.29	\$1831.19	\$1831.80

n	365	Continuous
A	\$1832.09	\$1832.10

57.

n	1	2	4	12
A	\$5477.81	\$5520.10	\$5541.79	\$5556.46

n	365	Continuous
A	\$5563.61	\$5563.85

59.

t	10	20	30
A	\$17,901.90	\$26,706.49	\$39,841.40

t	40	50
A	\$59,436.39	\$88,668.67

61.

t	10	20	30
A	\$22,986.49	\$44,031.56	\$84,344.25

t	40	50
A	\$161,564.86	\$309,484.08

63. \$104,710.29   **65.** \$35.45

67. (a)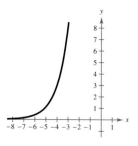

(b)

t	20	21	22	23
P (in millions)	342.748	345.604	348.485	351.389

t	24	25	26	27
P (in millions)	354.318	357.271	360.249	363.251

t	28	29	30	31
P (in millions)	366.279	369.331	372.410	375.513

t	32	33	34	35
P (in millions)	378.643	381.799	384.981	388.190

t	36	37	38	39
P (in millions)	391.425	394.687	397.977	401.294

t	40	41	42	43
P (in millions)	404.639	408.011	411.412	414.840

t	44	45	46	47
P (in millions)	418.298	421.784	425.300	428.844

t	48	49	50
P (in millions)	432.419	436.023	439.657

(c) 2038

69. (a) 16 g   (b) 1.85 g

(c)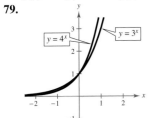

71. (a) $V(t) = 49,810\left(\frac{7}{8}\right)^t$ (b) $29,197.71

73. True. As $x \to -\infty$, $f(x) \to -2$ but never reaches -2.

75. $f(x) = h(x)$ **77.** $f(x) = g(x) = h(x)$

79.   (a) $x < 0$ (b) $x > 0$

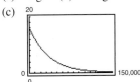

81.

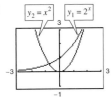

As the x-value increases, y_1 approaches the value of e.

83. (a)   (b)

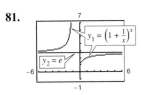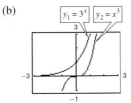

In both viewing windows, the constant raised to a variable power increases more rapidly than the variable raised to a constant power.

85. c, d

Section 5.2   *(page 372)*

1. logarithmic   **3.** natural; e **5.** $x = y$ **7.** $4^2 = 16$

9. $32^{2/5} = 4$ **11.** $\log_5 125 = 3$ **13.** $\log_4 \frac{1}{64} = -3$

15. 6   **17.** 0   **19.** 2   **21.** -0.058 **23.** 1.097

25. 7   **27.** 1   **29.** $x = 5$ **31.** $x = 7$

33.   **35.**

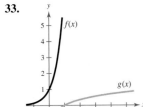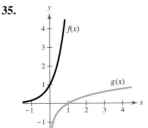

37. c   **38.** d   **39.** b   **40.** a

41.

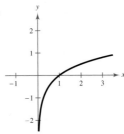

Domain: $(0, \infty)$
x-intercept: $(1, 0)$
Vertical asymptote: $x = 0$

43.

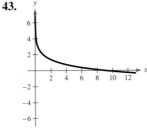

Domain: $(0, \infty)$
x-intercept: $(9, 0)$
Vertical asymptote: $x = 0$

45.

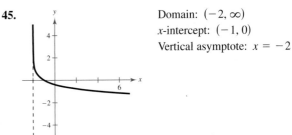

Domain: $(-2, \infty)$
x-intercept: $(-1, 0)$
Vertical asymptote: $x = -2$

47.

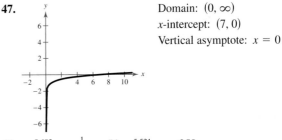

Domain: $(0, \infty)$
x-intercept: $(7, 0)$
Vertical asymptote: $x = 0$

49. $e^{-0.693\ldots} = \frac{1}{2}$ **51.** $e^{5.521\ldots} = 250$

53. $\ln 7.3890\ldots = 2$ **55.** $\ln 0.406\ldots = -0.9$

57. 2.913   **59.** -23.966 **61.** 5   **63.** $-\frac{5}{6}$

65.

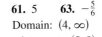
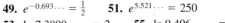

Domain: $(4, \infty)$
x-intercept: $(5, 0)$
Vertical asymptote: $x = 4$

67.

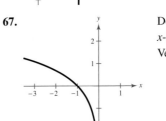

Domain: $(-\infty, 0)$
x-intercept: $(-1, 0)$
Vertical asymptote: $x = 0$

69. 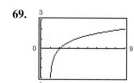   **71.**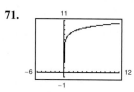

73. $x = 8$ **75.** $x = -5, 5$

77. (a) 30 yr; 10 yr   (b) \$323,179; \$199,109
   (c) \$173,179; \$49,109
   (d) $x = 750$; The monthly payment must be greater than \$750.

79. (a)

r	0.005	0.010	0.015	0.020	0.025	0.030
t	138.6	69.3	46.2	34.7	27.7	23.1

As the rate of increase r increases, the time t in years for the population to double decreases.

(b)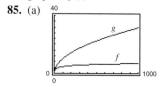

81. (a) 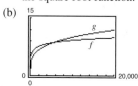   (b) 80   (c) 68.1   (d) 62.3

83. False. Reflecting $g(x)$ in the line $y = x$ will determine the graph of $f(x)$.

85. (a)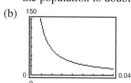

$g(x)$; The natural log function grows at a slower rate than the square root function.

(b)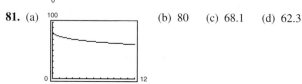

$g(x)$; The natural log function grows at a slower rate than the fourth root function.

87. (a) False   (b) True   (c) True   (d) False
89. Answers will vary.

Section 5.3   *(page 379)*

1. change-of-base   **3.** $\dfrac{1}{\log_b a}$ **4.** c   **5.** a   **6.** b

7. (a) $\dfrac{\log 16}{\log 5}$ (b) $\dfrac{\ln 16}{\ln 5}$ **9.** (a) $\dfrac{\log \frac{3}{10}}{\log x}$ (b) $\dfrac{\ln \frac{3}{10}}{\ln x}$
11. 1.771   **13.** -1.048 **15.** $\frac{3}{2}$ **17.** $-3 - \log_5 2$
19. $6 + \ln 5$ **21.** 2   **23.** $\frac{3}{4}$ **25.** 4
27. -2 is not in the domain of $\log_2 x$.   **29.** 4.5   **31.** $-\frac{1}{2}$
33. 7   **35.** 2   **37.** $\ln 4 + \ln x$ **39.** $4 \log_8 x$
41. $1 - \log_5 x$ **43.** $\frac{1}{2} \ln z$ **45.** $\ln x + \ln y + 2 \ln z$
47. $\ln z + 2 \ln(z - 1)$ **49.** $\frac{1}{2} \log_2(a - 1) - 2 \log_2 3$

51. $\frac{1}{3} \ln x - \frac{1}{3} \ln y$ **53.** $2 \ln x + \frac{1}{2} \ln y - \frac{1}{2} \ln z$
55. $2 \log_5 x - 2 \log_5 y - 3 \log_5 z$ **57.** $\frac{3}{4} \ln x + \frac{1}{4} \ln(x^2 + 3)$
59. 1.1833   **61.** 1.0686   **63.** 1.9563   **65.** 2.5646
67. $\ln 2x$ **69.** $\log_2 x^2 y^4$ **71.** $\log_3 \sqrt[4]{5x}$
73. $\log \dfrac{x}{(x + 1)^2}$ **75.** $\log \dfrac{xz^3}{y^2}$ **77.** $\ln \dfrac{x}{(x + 1)(x - 1)}$
79. $\ln \sqrt[3]{\dfrac{x(x + 3)^2}{x^2 - 1}}$ **81.** $\log_8 \dfrac{\sqrt[3]{y(y + 4)^2}}{y - 1}$
83. $\log_2 \frac{32}{4} = \log_2 32 - \log_2 4$; Property 2
85. $\beta = 10(\log I + 12)$; 60 dB   **87.** 70 dB
89. $\ln y = \frac{1}{4} \ln x$ **91.** $\ln y = -\frac{1}{4} \ln x + \ln \frac{5}{2}$
93. $y = 256.24 - 20.8 \ln x$
95. (a) and (b)   (c)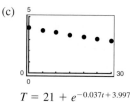

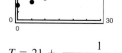

$T = 21 + e^{-0.037t + 3.997}$
The results are similar.

(d) 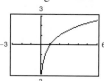   (e) Answers will vary.

$$T = 21 + \dfrac{1}{0.001t + 0.016}$$

97. False; $\ln 1 = 0$ **99.** False; $\ln(x - 2) \neq \ln x - \ln 2$
101. False; $u = v^2$

103. $f(x) = \dfrac{\log x}{\log 2} = \dfrac{\ln x}{\ln 2}$ **105.** $f(x) = \dfrac{\log x}{\log \frac{1}{4}} = \dfrac{\ln x}{\ln \frac{1}{4}}$

107. *Sample answers:*

(a) $\ln(1 + 3) \overset{?}{=} \ln(1) + \ln(3)$
   $1.39 \neq 0 + 1.10$
   $\ln u + \ln v = \ln(uv)$, but $\ln(u + v) \neq \ln u + \ln v$

(b) $\ln(3 - 1) \overset{?}{=} \ln(3) - \ln(1)$
   $0.69 \neq 1.10 - 0$
   $\ln u - \ln v = \ln \dfrac{u}{v}$, but $\ln(u - v) \neq \ln u - \ln v$

(c) $(\ln 2)^3 \overset{?}{=} 3 \ln 2$
   $0.33 \neq 2.08$
   $n(\ln u) = \ln u^n$, but $(\ln u)^n \neq n(\ln u)$

109. $\ln 1 = 0$
$\ln 2 \approx 0.6931$
$\ln 3 \approx 1.0986$
$\ln 4 \approx 1.3862$
$\ln 5 \approx 1.6094$
$\ln 6 \approx 1.7917$
$\ln 8 \approx 2.0793$
$\ln 9 \approx 2.1972$
$\ln 10 \approx 2.3025$
$\ln 12 \approx 2.4848$
$\ln 15 \approx 2.7080$
$\ln 16 \approx 2.7724$
$\ln 18 \approx 2.8903$
$\ln 20 \approx 2.9956$

Section 5.4 *(page 389)*

1. (a) $x = y$ (b) $x = y$ (c) x (d) x
3. (a) Yes   (b) No   5. (a) Yes   (b) No   (c) No
7. 2   9. 2   11. $e^{-1} \approx 0.368$ 13. 64
15. $(3, 8)$ 17. $2, -1$ 19. $\dfrac{\ln 5}{\ln 3} \approx 1.465$
21. $\ln 28 \approx 3.332$ 23. $\dfrac{\ln 80}{2\ln 3} \approx 1.994$
25. $3 - \dfrac{\ln 565}{\ln 2} \approx -6.142$ 27. $\dfrac{1}{3}\log\left(\dfrac{3}{2}\right) \approx 0.059$
29. $\dfrac{\ln 12}{3} \approx 0.828$ 31. 0   33. $\dfrac{\ln \frac{8}{3}}{3\ln 2} + \dfrac{1}{3} \approx 0.805$
35. $\dfrac{\ln 3}{\ln 2 - \ln 3} \approx -2.710$ 37. $0, \dfrac{\ln 4}{\ln 5} \approx 0.861$
39. $\ln 5 \approx 1.609$ 41. $2\ln 75 \approx 8.635$
43. $\dfrac{\ln 4}{365 \ln\left(1 + \dfrac{0.065}{365}\right)} \approx 21.330$ 45. $e^{-3} \approx 0.050$
47. $\dfrac{e^{2.1}}{6} \approx 1.361$ 49. $\dfrac{e^{10/3}}{5} \approx 5.606$ 51. $e^{-4/3} \approx 0.264$
53. $2(3^{11/6}) \approx 14.988$ 55. No solution   57. No solution
59. No solution   61. 2
63.

3.328

65.

-0.478

67.

20.086

69.

1.482

71. (a) 27.73 yr   (b) 43.94 yr   73. $-1, 0$
75. 1   77. $e^{-1/2} \approx 0.607$ 79. $e^{-1} \approx 0.368$
81. (a) $y = 100$ and $y = 0$; The range falls between 0% and 100%.
(b) Males: 69.51 in.   Females: 64.49 in.
83. 12.76 in.
85. (a)

x	0.2	0.4	0.6	0.8	1.0
y	162.6	78.5	52.5	40.5	33.9

(b)

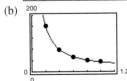

The model appears to fit the data well.
(c) 1.2 m
(d) No. According to the model, when the number of g's is less than 23, x is between 2.276 meters and 4.404 meters, which isn't realistic in most vehicles.

87. $\log_b uv = \log_b u + \log_b v$
True by Property 1 in Section 5.3.
89. $\log_b(u - v) = \log_b u - \log_b v$
False.
$1.95 \approx \log(100 - 10) \neq \log 100 - \log 10 = 1$
91. Yes. See Exercise 57.
93. For $rt < \ln 2$ years, double the amount you invest. For $rt > \ln 2$ years, double your interest rate or double the number of years, because either of these will double the exponent in the exponential function.
95. (a) 7%   (b) 7.25%   (c) 7.19%   (d) 7.45%
The investment plan with the greatest effective yield and the highest balance after 5 years is plan (d).

Section 5.5 *(page 399)*

1. $y = ae^{bx}; y = ae^{-bx}$ 3. normally distributed
5. (a) $P = \dfrac{A}{e^{rt}}$ (b) $t = \dfrac{\ln\left(\dfrac{A}{P}\right)}{r}$

Initial Investment	Annual % Rate	Time to Double	Amount After 10 years
7. $1000	3.5%	19.8 yr	$1419.07
9. $750	8.9438%	7.75 yr	$1834.37
11. $6376.28	4.5%	15.4 yr	$10,000.00

13. $303,580.52
15. (a) 7.27 yr   (b) 6.96 yr   (c) 6.93 yr   (d) 6.93 yr
17. (a)

r	2%	4%	6%	8%	10%	12%
t	54.93	27.47	18.31	13.73	10.99	9.16

(b)

r	2%	4%	6%	8%	10%	12%
t	55.48	28.01	18.85	14.27	11.53	9.69

19.

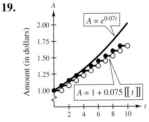

Continuous compounding

Half-life (years)	Initial Quantity	Amount After 1000 Years
21. 1599	10 g	6.48 g
23. 5715	2.26 g	2 g

25. $y = e^{0.7675x}$ 27. $y = 5e^{-0.4024x}$
29. (a)

Year	1980	1990	2000	2010
Population	106.1	143.15	196.25	272.37

(b) 2017
(c) No; The population will not continue to grow at such a quick rate.
31. $k = 0.2988$; About 5,309,734 hits
33. About 800 bacteria

35. (a) $V = -300t + 1150$ (b) $V = 1150e^{-0.368799t}$

(c)

The exponential model depreciates faster.

(d)

t	1 yr	3 yr
$V = -300t + 1150$	850	250
$V = 1150e^{-0.368799t}$	795	380

(e) Answers will vary.

37. (a) About 12,180 yr old   (b) About 4797 yr old

39. (a)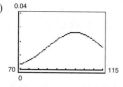

(b) 100

41. (a) 1998: 55,557 sites
2003: 147,644 sites
2006: 203,023 sites

(b)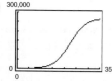

(c) and (d) 2010

43. (a) 203 animals   (b) 13 mo

(c)

Horizontal asymptotes: $p = 0, p = 1000$. The population size will approach 1000 as time increases.

45. (a) $10^{6.6} \approx 3,981,072$ (b) $10^{5.6} \approx 398,107$
(c) $10^{7.1} \approx 12,589,254$

47. (a) 20 dB   (b) 70 dB   (c) 40 dB   (d) 120 dB

49. 95%   **51.** 4.64   **53.** 1.58×10^{-6} moles/L

55. $10^{5.1}$ **57.** 3:00 A.M.

59. (a) 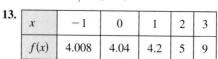   (b) $t \approx 21$ yr; Yes

61. False. The domain can be the set of real numbers for a logistic growth function.

63. False. The graph of $f(x)$ is the graph of $g(x)$ shifted five units up.

65. Answers will vary.

Review Exercises   *(page 406)*

1. 0.164   **3.** 0.337   **5.** 1456.529
7. Shift the graph of f one unit up.

9. Reflect f in the x-axis and shift one unit up.

11.

x	-1	0	1	2	3
$f(x)$	8	5	4.25	4.063	4.016

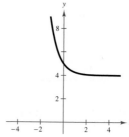

13.

x	-1	0	1	2	3
$f(x)$	4.008	4.04	4.2	5	9

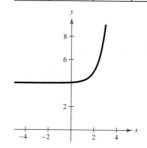

15.

x	-2	-1	0	1	2
$f(x)$	3.25	3.5	4	5	7

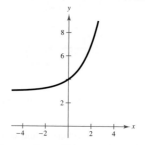

17. $x = 1$ **19.** $x = 4$ **21.** 2980.958   **23.** 0.183

25.

x	-2	-1	0	1	2
$h(x)$	2.72	1.65	1	0.61	0.37

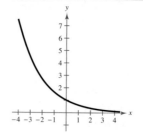

27.

x	-3	-2	-1	0	1
$f(x)$	0.37	1	2.72	7.39	20.09

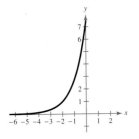

29. (a) 0.154   (b) 0.487   (c) 0.811

31.

n	1	2	4	12
A	\$6719.58	\$6734.28	\$6741.74	\$6746.77

n	365	Continuous
A	\$6749.21	\$6749.29

33. $\log_3 27 = 3$ **35.** $\ln 2.2255\ldots = 0.8$ **37.** 3
39. -2 **41.** $x = 7$ **43.** $x = -5$
45. Domain: $(0, \infty)$ **47.** Domain: $(-5, \infty)$
   x-intercept: $(1, 0)$ x-intercept: $(9995, 0)$
   Vertical asymptote: $x = 0$ Vertical asymptote: $x = -5$

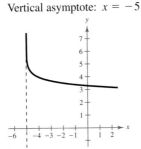

49. 3.118   **51.** 0.25
53. Domain: $(0, \infty)$ **55.** Domain: $(-\infty, 0), (0, \infty)$
   x-intercept: $(e^{-3}, 0)$ x-intercepts: $(\pm 1, 0)$
   Vertical asymptote: $x = 0$ Vertical asymptote: $x = 0$

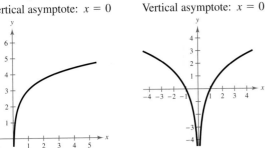

57. 53.4 in.   **59.** (a) and (b) 2.585
61. (a) and (b) -2.322 **63.** $\log 2 + 2 \log 3 \approx 1.255$
65. $2 \ln 2 + \ln 5 \approx 2.996$ **67.** $1 + 2 \log_5 x$
69. $2 - \frac{1}{2} \log_3 x$ **71.** $2 \ln x + 2 \ln y + \ln z$
73. $\log_2 5x$ **75.** $\ln \dfrac{x}{\sqrt[4]{y}}$ **77.** $\log_3 \dfrac{\sqrt{x}}{(y+8)^2}$

79. (a) $0 \le h < 18{,}000$
    (b)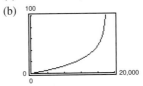
    Vertical asymptote: $h = 18{,}000$
    (c) The plane is climbing at a slower rate, so the time required increases.
    (d) 5.46 min
81. 3   **83.** $\ln 3 \approx 1.099$ **85.** $e^4 \approx 54.598$ **87.** 1, 3
89. $\dfrac{\ln 32}{\ln 2} = 5$
91.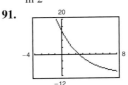
    2.447
93. $\frac{1}{3}e^{8.2} \approx 1213.650$ **95.** $3e^2 \approx 22.167$
97. No solution   **99.** 0.900
101.   **103.**

1.482   0, 0.416, 13.627
105. 73.2 yr   **107.** e   **108.** b   **109.** f   **110.** d
111. a   **112.** c   **113.** $y = 2e^{0.1014x}$
115.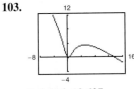

71
117. (a) 10^{-6} W/m²   (b) $10\sqrt{10}$ W/m²
    (c) 1.259×10^{-12} W/m²
119. True by the inverse properties.

Chapter Test   (page 409)

1. 2.366   **2.** 687.291   **3.** 0.497   **4.** 22.198
5.

x	-1	$-\frac{1}{2}$	0	$\frac{1}{2}$	1
$f(x)$	10	3.162	1	0.316	0.1

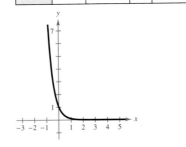

6.

x	-1	0	1	2	3
$f(x)$	-0.005	-0.028	-0.167	-1	-6

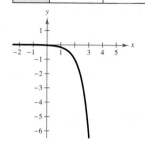

7.

x	-1	$-\frac{1}{2}$	0	$\frac{1}{2}$	1
$f(x)$	0.865	0.632	0	-1.718	-6.389

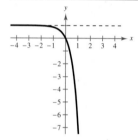

8. (a) -0.89 (b) 9.2
9. Domain: $(0, \infty)$
x-intercept: $(10^{-6}, 0)$
Vertical asymptote: $x = 0$

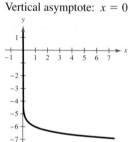

10. Domain: $(4, \infty)$
x-intercept: $(5, 0)$
Vertical asymptote: $x = 4$

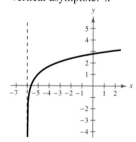

11. Domain: $(-6, \infty)$
x-intercept: $(e^{-1} - 6, 0)$
Vertical asymptote: $x = -6$

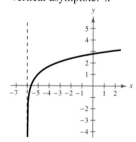

12. 1.945 **13.** -0.167 **14.** -11.047
15. $\log_2 3 + 4\log_2 a$ **16.** $\ln 5 + \frac{1}{2}\ln x - \ln 6$
17. $3\log(x - 1) - 2\log y - \log z$ **18.** $\log_3 13y$
19. $\ln \dfrac{x^4}{y^4}$ **20.** $\ln \dfrac{x^3 y^2}{x + 3}$ **21.** -2
22. $\dfrac{\ln 44}{-5} \approx -0.757$ **23.** $\dfrac{\ln 197}{4} \approx 1.321$

24. $e^{1/2} \approx 1.649$ **25.** $e^{-11/4} \approx 0.0639$ **26.** 20
27. $y = 2745e^{0.1570t}$ **28.** 55%
29. (a)

x	$\frac{1}{4}$	1	2	4	5	6
H	58.720	75.332	86.828	103.43	110.59	117.38

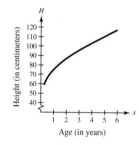

(b) 103 cm; 103.43 cm

Problem Solving   (page 411)

1.

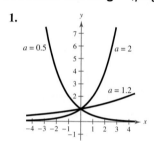

$y = 0.5^x$ and $y = 1.2^x$
$0 < a \leq e^{1/e}$

3. As $x \to \infty$, the graph of e^x increases at a greater rate than the graph of x^n.
5. Answers will vary.
7. (a)

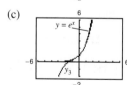

(b)

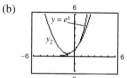

(c)

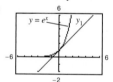

9.

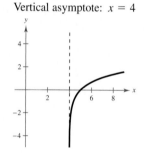

$f^{-1}(x) = \ln\left(\dfrac{x + \sqrt{x^2 + 4}}{2}\right)$

11. c   **13.** $t = \dfrac{\ln c_1 - \ln c_2}{\left(\dfrac{1}{k_2} - \dfrac{1}{k_1}\right)\ln \dfrac{1}{2}}$

15. (a) $y_1 = 252,606(1.0310)^t$

(b) $y_2 = 400.88t^2 - 1464.6t + 291,782$

(c)
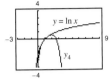

(d) The exponential model is a better fit. No, because the model is rapidly approaching infinity.

17. $1, e^2$

19. $y_4 = (x - 1) - \frac{1}{2}(x - 1)^2 + \frac{1}{3}(x - 1)^3 - \frac{1}{4}(x - 1)^4$

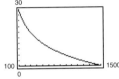

The pattern implies that
$$\ln x = (x - 1) - \frac{1}{2}(x - 1)^2 + \frac{1}{3}(x - 1)^3 - \cdots.$$

21.

$17.7 \text{ ft}^3/\text{min}$

23. (a)
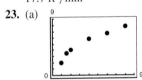

25. (a)
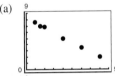

(b)–(e) Answers will vary.   (b)–(e) Answers will vary.

Chapter 6

Section 6.1   *(page 418)*

1. inclination   **3.** $\left| \dfrac{m_2 - m_1}{1 + m_1 m_2} \right|$ **5.** $\dfrac{\sqrt{3}}{3}$ **7.** -1

9. $\sqrt{3}$ **11.** 0.2660   **13.** 3.2236   **15.** -4.1005

17. $\dfrac{3\pi}{4}$ rad, $135°$ **19.** $\dfrac{\pi}{4}$ rad, $45°$ **21.** 0.6435 rad, $36.9°$

23. 1.9513 rad, $111.8°$ **25.** $\dfrac{\pi}{6}$ rad, $30°$ **27.** $\dfrac{5\pi}{6}$ rad, $150°$

29. 1.0517 rad, $60.3°$ **31.** 2.1112 rad, $121.0°$

33. 1.6539 rad, $94.8°$ **35.** $\dfrac{3\pi}{4}$ rad, $135°$ **37.** $\dfrac{\pi}{4}$ rad, $45°$

39. $\dfrac{5\pi}{6}$ rad, $150°$ **41.** 1.2490 rad, $71.6°$

43. 2.4669 rad, $141.3°$ **45.** 1.1071 rad, $63.4°$

47. 0.1974 rad, $11.3°$ **49.** 1.4289 rad, $81.9°$

51. 0.9273 rad, $53.1°$ **53.** 0.8187 rad, $46.9°$

55. $(1, 5) \leftrightarrow (4, 5)$: slope $= 0$
$(4, 5) \leftrightarrow (3, 8)$: slope $= -3$
$(3, 8) \leftrightarrow (1, 5)$: slope $= \frac{3}{2}$
$(1, 5)$: $56.3°$; $(4, 5)$: $71.6°$; $(3, 8)$: $52.1°$

57. $(-4, -1) \leftrightarrow (3, 2)$: slope $= \frac{3}{7}$
$(3, 2) \leftrightarrow (1, 0)$: slope $= 1$
$(1, 0) \leftrightarrow (-4, -1)$: slope $= \frac{1}{5}$
$(-4, -1)$: $11.9°$; $(3, 2)$: $21.8°$; $(1, 0)$: $146.3°$

59. $\dfrac{\sqrt{2}}{2} \approx 0.7071$ **61.** $\dfrac{3\sqrt{5}}{5} \approx 1.3416$ **63.** $\dfrac{\sqrt{2}}{2} \approx 0.7071$

65. 0   **67.** $\dfrac{4\sqrt{10}}{5} \approx 2.5298$ **69.** 1

71. $\dfrac{5\sqrt{34}}{34} \approx 0.8575$ **73.** $\dfrac{8\sqrt{5}}{5} \approx 3.5777$

75. (a)   **77.** (a)

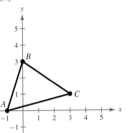   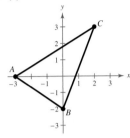

(b) $\dfrac{11\sqrt{17}}{17}$ (c) $\dfrac{11}{2}$ (b) $\dfrac{19\sqrt{34}}{34}$ (c) $\dfrac{19}{2}$

79. (a)   (b) $\dfrac{\sqrt{5}}{5}$ (c) 1

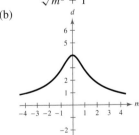

81. $2\sqrt{2}$ **83.** 0.1003, 1054 ft   **85.** $31.0°$

87. $\alpha \approx 33.69°$; $\beta \approx 56.31°$

89. True. The inclination of a line is related to its slope by $m = \tan \theta$. If the angle is greater than $\pi/2$ but less than π, then the angle is in the second quadrant, where the tangent function is negative.

91. False. The inclination is the positive angle measured counterclockwise from the x-axis.

93. (a) $d = \dfrac{4}{\sqrt{m^2 + 1}}$

(b)
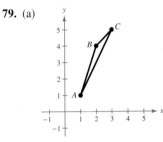

(c) $m = 0$

(d) The graph has a horizontal asymptote of $d = 0$. As the slope becomes larger, the distance between the origin and the line, $y = mx + 4$, becomes smaller and approaches 0.

95. The inclination of a line measures the angle of intersection (measured counterclockwise) of a line and the x-axis. The angle between two lines is the acute angle of their intersection, which must be less than $\pi/2$.

Section 6.2   *(page 426)*

1. conic   **3.** locus   **5.** axis   **7.** focal chord
9. e   **10.** b   **11.** d   **12.** f   **13.** a   **14.** c
15. $x^2 = \frac{3}{2}y$ **17.** $x^2 = 2y$ **19.** $y^2 = -8x$
21. $x^2 = -4y$ **23.** $y^2 = 4x$ **25.** $x^2 = \frac{8}{3}y$
27. $y^2 = -\frac{25}{2}x$

29. Vertex: $(0, 0)$
    Focus: $\left(0, \frac{1}{2}\right)$
    Directrix: $y = -\frac{1}{2}$

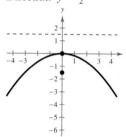

31. Vertex: $(0, 0)$
    Focus: $\left(-\frac{3}{2}, 0\right)$
    Directrix: $x = \frac{3}{2}$

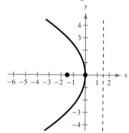

33. Vertex: $(0, 0)$
    Focus: $\left(0, -\frac{3}{2}\right)$
    Directrix: $y = \frac{3}{2}$

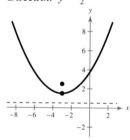

35. Vertex: $(1, -2)$
    Focus: $(1, -4)$
    Directrix: $y = 0$

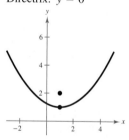

37. Vertex: $\left(-3, \frac{3}{2}\right)$
    Focus: $\left(-3, \frac{5}{2}\right)$
    Directrix: $y = \frac{1}{2}$

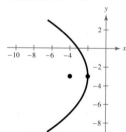

39. Vertex: $(1, 1)$
    Focus: $(1, 2)$
    Directrix: $y = 0$

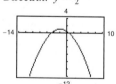

41. Vertex: $(-2, -3)$
    Focus: $(-4, -3)$
    Directrix: $x = 0$

43. Vertex: $(-2, 1)$
    Focus: $\left(-2, -\frac{1}{2}\right)$
    Directrix: $y = \frac{5}{2}$

45. Vertex: $\left(\frac{1}{4}, -\frac{1}{2}\right)$
    Focus: $\left(0, -\frac{1}{2}\right)$
    Directrix: $x = \frac{1}{2}$

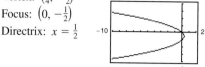

47. $(x - 3)^2 = -(y - 1)$ **49.** $y^2 = 4(x + 4)$
51. $(y - 3)^2 = 8(x - 4)$ **53.** $x^2 = -8(y - 2)$
55. $(y - 2)^2 = 8x$ **57.** $4x - y - 8 = 0$
59. $4x - y + 2 = 0$ **61.** $y^2 = 640x$ **63.** $y^2 = 6x$
65. (a) $x^2 = 12{,}288y$ (in feet)   (b) 22.6 ft
67. $x^2 = -\frac{25}{4}(y - 48)$ **69.** About 19.6 m
71. (a) $(0, 45)$ (b) $y = \frac{1}{180}x^2$
73. (a) $17{,}500\sqrt{2}$ mi/h $\approx 24{,}750$ mi/h
    (b) $x^2 = -16{,}400(y - 4100)$
75. (a) $x^2 = -49(y - 100)$ (b) 70 ft
77. False. If the graph crossed the directrix, then there would exist points closer to the directrix than the focus.
79. $m = \dfrac{x_1}{2p}$
81.

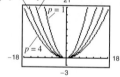

Single point $(0, 0)$; A single point is formed when a plane intersects only the vertex of the cone.

83. (a) $p = 3$ $p = 2$ 21

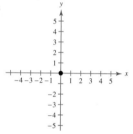

    As p increases, the graph becomes wider.
    (b) $(0, 1), (0, 2), (0, 3), (0, 4)$ (c) 4, 8, 12, 16; $4|p|$
    (d) It is an easy way to determine two additional points on the graph.

Section 6.3   *(page 436)*

1. ellipse; foci   **3.** minor axis   **5.** b   **6.** c   **7.** a
8. d   **9.** $\dfrac{x^2}{4} + \dfrac{y^2}{16} = 1$ **11.** $\dfrac{x^2}{49} + \dfrac{y^2}{45} = 1$
13. $\dfrac{x^2}{49} + \dfrac{y^2}{24} = 1$ **15.** $\dfrac{x^2}{9} + \dfrac{y^2}{36} = 1$ **17.** $\dfrac{x^2}{36} + \dfrac{5y^2}{9} = 1$
19. $\dfrac{(x - 2)^2}{1} + \dfrac{(y - 3)^2}{9} = 1$ **21.** $\dfrac{(x - 4)^2}{16} + \dfrac{(y - 2)^2}{1} = 1$
23. $\dfrac{(x - 2)^2}{9} + \dfrac{y^2}{5} = 1$ **25.** $\dfrac{(x - 1)^2}{9} + \dfrac{(y - 3)^2}{4} = 1$
27. $\dfrac{x^2}{60} + \dfrac{(y - 2)^2}{64} = 1$ **29.** $\dfrac{(x - 3)^2}{36} + \dfrac{(y - 2)^2}{32} = 1$
31. $\dfrac{(x - 2)^2}{4} + \dfrac{(y - 2)^2}{1} = 1$

33. Center: $(0, 0)$
Vertices: $(\pm 5, 0)$
Foci: $(\pm 3, 0)$
Eccentricity: $\frac{3}{5}$

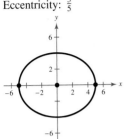

35. Center: $(0, 0)$
Vertices: $(0, \pm 3)$
Foci: $(0, \pm 2)$
Eccentricity: $\frac{2}{3}$

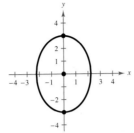

37. Center: $(4, -1)$
Vertices: $(4, -6), (4, 4)$
Foci: $(4, 2), (4, -4)$
Eccentricity: $\frac{3}{5}$

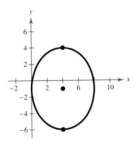

39. Center: $(-5, 1)$
Vertices: $\left(-\frac{7}{2}, 1\right), \left(-\frac{13}{2}, 1\right)$
Foci: $\left(-5 \pm \frac{\sqrt{5}}{2}, 1\right)$
Eccentricity: $\frac{\sqrt{5}}{3}$

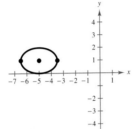

41. Center: $(-2, 3)$
Vertices: $(-2, 6), (-2, 0)$
Foci: $(-2, 3 \pm \sqrt{5})$
Eccentricity: $\frac{\sqrt{5}}{3}$

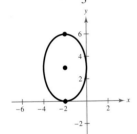

43. Center: $(4, 3)$
Vertices: $(14, 3), (-6, 3)$
Foci: $(4 \pm 4\sqrt{5}, 3)$
Eccentricity: $\frac{2\sqrt{5}}{5}$

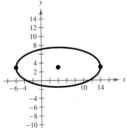

45. Center: $\left(-\frac{3}{2}, \frac{5}{2}\right)$
Vertices: $\left(-\frac{3}{2}, \frac{5}{2} \pm 2\sqrt{3}\right)$
Foci: $\left(-\frac{3}{2}, \frac{5}{2} \pm 2\sqrt{2}\right)$
Eccentricity: $\frac{\sqrt{6}}{3}$

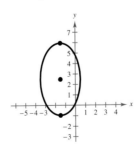

47. Center: $(1, -1)$
Vertices: $\left(\frac{9}{4}, -1\right), \left(-\frac{1}{4}, -1\right)$
Foci: $\left(\frac{7}{4}, -1\right), \left(\frac{1}{4}, -1\right)$
Eccentricity: $\frac{3}{5}$

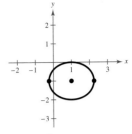

49.

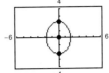

Center: $(0, 0)$
Vertices: $(0, \pm\sqrt{5})$
Foci: $(0, \pm\sqrt{2})$

51.

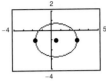

Center: $\left(\frac{1}{2}, -1\right)$
Vertices: $\left(\frac{1}{2} \pm \sqrt{5}, -1\right)$
Foci: $\left(\frac{1}{2} \pm \sqrt{2}, -1\right)$

53. $\dfrac{x^2}{25} + \dfrac{y^2}{16} = 1$

55. (a) $\dfrac{x^2}{2352.25} + \dfrac{y^2}{529} = 1$ (b) About 85.4 ft

57. About 229.8 mm   **59.** $e \approx 0.052$

61.

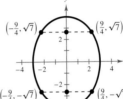

63.

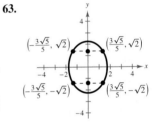

65. False. The graph of $(x^2/4) + y^4 = 1$ is not an ellipse. The degree of y is 4, not 2.

67. $\dfrac{(x - 6)^2}{324} + \dfrac{(y - 2)^2}{308} = 1$

69. (a) $A = \pi a(20 - a)$ (b) $\dfrac{x^2}{196} + \dfrac{y^2}{36} = 1$

(c)

a	8	9	10	11	12	13
A	301.6	311.0	314.2	311.0	301.6	285.9

$a = 10$, circle

(d)

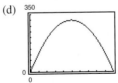

The shape of an ellipse with a maximum area is a circle. The maximum area is found when $a = 10$ (verified in part c) and therefore $b = 10$, so the equation produces a circle.

71. Proof

Section 6.4   (page 446)

1. hyperbola; foci   **3.** transverse axis; center   **5.** b

6. c   **7.** a   **8.** d   **9.** $\dfrac{y^2}{4} - \dfrac{x^2}{12} = 1$

11. $\dfrac{(x - 4)^2}{4} - \dfrac{y^2}{12} = 1$ **13.** $\dfrac{(y - 5)^2}{16} - \dfrac{(x - 4)^2}{9} = 1$

15. $\dfrac{y^2}{9} - \dfrac{4(x-2)^2}{9} = 1$ **17.** $\dfrac{(y-2)^2}{4} - \dfrac{x^2}{4} = 1$

19. Center: $(0,0)$

Vertices: $(\pm 1, 0)$

Foci: $(\pm\sqrt{2}, 0)$

Asymptotes: $y = \pm x$

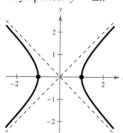

21. Center: $(0,0)$

Vertices: $(0, \pm 5)$

Foci: $\left(0, \pm\sqrt{106}\right)$

Asymptotes: $y = \pm\frac{5}{9}x$

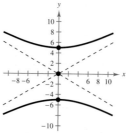

23. Center: $(0,0)$

Vertices: $(0, \pm 1)$

Foci: $\left(0, \pm\sqrt{5}\right)$

Asymptotes: $y = \pm\frac{1}{2}x$

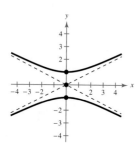

25. Center: $(1, -2)$

Vertices: $(3, -2), (-1, -2)$

Foci: $\left(1 \pm \sqrt{5}, -2\right)$

Asymptotes:

$y = -2 \pm \frac{1}{2}(x - 1)$

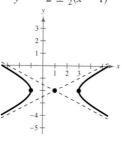

27. Center: $(2, -6)$

Vertices: $\left(2, -\dfrac{17}{3}\right), \left(2, -\dfrac{19}{3}\right)$

Foci: $\left(2, -6 \pm \dfrac{\sqrt{13}}{6}\right)$

Asymptotes: $y = -6 \pm \frac{2}{3}(x - 2)$

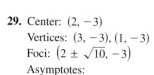

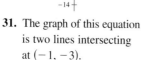

29. Center: $(2, -3)$

Vertices: $(3, -3), (1, -3)$

Foci: $\left(2 \pm \sqrt{10}, -3\right)$

Asymptotes:

$y = -3 \pm 3(x - 2)$

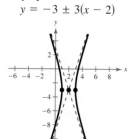

31. The graph of this equation is two lines intersecting at $(-1, -3)$.

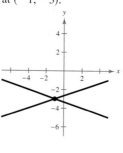

33. Center: $(0,0)$

Vertices: $\left(\pm\sqrt{3}, 0\right)$

Foci: $\left(\pm\sqrt{5}, 0\right)$

Asymptotes: $y = \pm\dfrac{\sqrt{6}}{3}x$

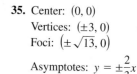

35. Center: $(0,0)$

Vertices: $(\pm 3, 0)$

Foci: $\left(\pm\sqrt{13}, 0\right)$

Asymptotes: $y = \pm\dfrac{2}{3}x$

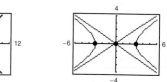

37. Center: $(1, -3)$

Vertices: $\left(1, -3 \pm \sqrt{2}\right)$

Foci: $\left(1, -3 \pm 2\sqrt{5}\right)$

Asymptotes: $y = -3 \pm \frac{1}{3}(x - 1)$

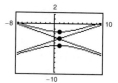

39. $\dfrac{x^2}{1} - \dfrac{y^2}{25} = 1$ **41.** $\dfrac{17y^2}{1024} - \dfrac{17x^2}{64} = 1$

43. $\dfrac{(x-2)^2}{1} - \dfrac{(y-2)^2}{1} = 1$ **45.** $\dfrac{(x-3)^2}{9} - \dfrac{(y-2)^2}{4} = 1$

47. (a) $\dfrac{x^2}{1} - \dfrac{y^2}{169/3} = 1$ (b) About 2.403 ft

49. $\dfrac{x^2}{98,010,000} - \dfrac{y^2}{13,503,600} = 1$

51. (a) $x \approx 110.3$ mi   (b) 57.0 mi   (c) 0.00129 sec

(d) The ship is at the position $(144.2, 60)$.

53. Ellipse   **55.** Hyperbola   **57.** Hyperbola

59. Parabola   **61.** Ellipse   **63.** Parabola

65. Parabola   **67.** Circle

69. True. For a hyperbola, $c^2 = a^2 + b^2$. The larger the ratio of b to a, the larger the eccentricity of the hyperbola, $e = c/a$.

71. False. When $D = -E$, the graph is two intersecting lines.

73. Answers will vary.   **75.** $y = 1 - 3\sqrt{\dfrac{(x-3)^2}{4} - 1}$

77.

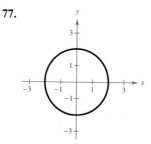

The equation $y = x^2 + C$ is a parabola that could intersect the circle in zero, one, two, three, or four places depending on its location on the y-axis.

(a) $C > 2$ and $C < -\frac{17}{4}$ (b) $C = 2$

(c) $-2 < C < 2$, $C = -\frac{17}{4}$ (d) $C = -2$

(e) $-\frac{17}{4} < C < -2$

Section 6.5   *(page 455)*

1. rotation; axes   **3.** invariant under rotation   **5.** $(3, 0)$

7. $\left(\dfrac{3 + \sqrt{3}}{2}, \dfrac{3\sqrt{3} - 1}{2}\right)$ **9.** $\left(\dfrac{3\sqrt{2}}{2}, -\dfrac{\sqrt{2}}{2}\right)$

11. $\left(\dfrac{2\sqrt{3} + 1}{2}, \dfrac{2 - \sqrt{3}}{2}\right)$

13. $\dfrac{(y')^2}{2} - \dfrac{(x')^2}{2} = 1$

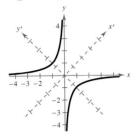

15. $\dfrac{\left(y' + \dfrac{3\sqrt{2}}{2}\right)^2}{12} - \dfrac{\left(x' + \dfrac{\sqrt{2}}{2}\right)^2}{12} = 1$

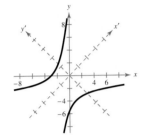

17. $\dfrac{(x')^2}{6} + \dfrac{(y')^2}{3/2} = 1$ **19.** $\dfrac{(x')^2}{1} + \dfrac{(y')^2}{4} = 1$

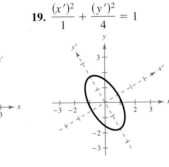

21. $(x')^2 = y'$ **23.** $(x' - 1)^2 = 6\left(y' + \dfrac{1}{6}\right)$

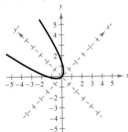

25.

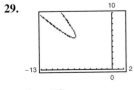

$\theta \approx 37.98°$

27.

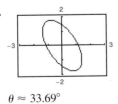

$\theta \approx 33.69°$

29.

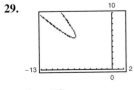

$\theta = 45°$

31. e   **32.** f   **33.** b   **34.** a   **35.** d   **36.** c

37. (a) Parabola

(b) $y = \dfrac{(8x - 5) \pm \sqrt{(8x - 5)^2 - 4(16x^2 - 10x)}}{2}$

(c)

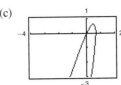

39. (a) Ellipse

(b) $y = \dfrac{6x \pm \sqrt{36x^2 - 28(12x^2 - 45)}}{14}$

(c)

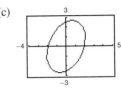

41. (a) Hyperbola

(b) $y = \dfrac{6x \pm \sqrt{36x^2 + 20(x^2 + 4x - 22)}}{-10}$

(c)

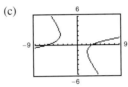

43. (a) Parabola

(b) $y = \dfrac{-(4x - 1) \pm \sqrt{(4x - 1)^2 - 16(x^2 - 5x - 3)}}{8}$

(c)

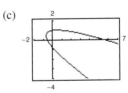

45.

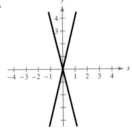

47.

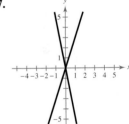

49.

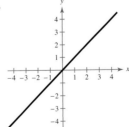

51.

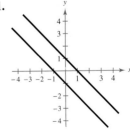

CHAPTER 6

53.

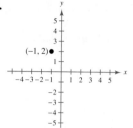

(−1, 2)●

55. $(14, -8), (6, -8)$

57. $(1, 0)$ **59.** $(1, \sqrt{3}), (1, -\sqrt{3})$ **61.** $(2, 2), (2, 4)$

63. (a) $(y' + 9)^2 = 9(x' - 12)$ (b) 2.25 ft

65. True. The graph of the equation can be classified by finding the discriminant. For a graph to be a hyperbola, the discriminant must be greater than zero. If $k \geq \frac{1}{4}$, then the discriminant would be less than or equal to zero.

67. Major axis: 4; Minor axis: 2

Section 6.6   *(page 463)*

1. plane curve   **3.** eliminating; parameter

5. (a)

t	0	1	2	3	4
x	0	1	$\sqrt{2}$	$\sqrt{3}$	2
y	3	2	1	0	−1

(b)

(c) $y = 3 - x^2$

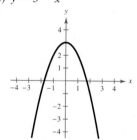

The graph of the rectangular equation shows the entire parabola rather than just the right half.

7. (a)

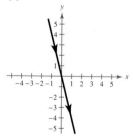

(b) $y = -4x$

9. (a)

(b) $y = 3x + 4$

11. (a)

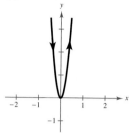

(b) $y = 16x^2$

13. (a)

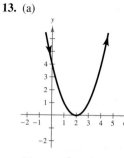

(b) $y = x^2 - 4x + 4$

15. (a)

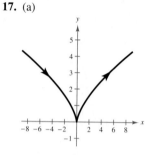

(b) $y = 1 - x^2, \quad x \geq 0$

17. (a)

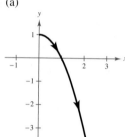

(b) $y = x^{2/3}$

19. (a)

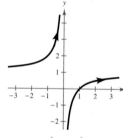

(b) $y = \dfrac{(x - 1)}{x}$

21. (a)

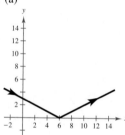

(b) $y = \left| \dfrac{x}{2} - 3 \right|$

23. (a)

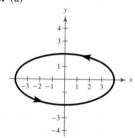

(b) $\dfrac{x^2}{16} + \dfrac{y^2}{4} = 1$

25. (a)

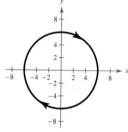

(b) $\dfrac{x^2}{36} + \dfrac{y^2}{36} = 1$

27. (a)

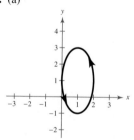

(b) $\dfrac{(x - 1)^2}{1} + \dfrac{(y - 1)^2}{4} = 1$

29. (a)

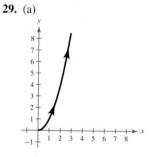

(b) $y = x^2, \quad x > 0$

31. (a)

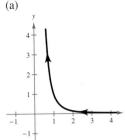

(b) $y = \dfrac{1}{x^3}$, $x > 0$

33. (a)

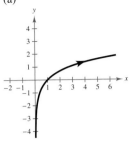

(b) $y = \ln x$

35.

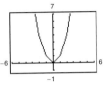

37.

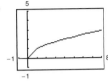

39.

41.

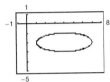

43.

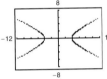

45.

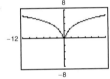

47. Each curve represents a portion of the line $y = 2x + 1$.

Domain	Orientation
(a) $(-\infty, \infty)$	Left to right
(b) $[-1, 1]$	Depends on θ
(c) $(0, \infty)$	Right to left
(d) $(0, \infty)$	Left to right

49. $y - y_1 = m(x - x_1)$

51. $\dfrac{(x - h)^2}{a^2} + \dfrac{(y - k)^2}{b^2} = 1$

53. $x = 3t$
      $y = 6t$

55. $x = 3 + 4\cos\theta$
      $y = 2 + 4\sin\theta$

57. $x = 5\cos\theta$
      $y = 3\sin\theta$

59. $x = 4\sec\theta$
      $y = 3\tan\theta$

61. (a) $x = t,\ y = 3t - 2$ (b) $x = -t + 2,\ y = -3t + 4$

63. (a) $x = t,\ y = 2 - t$ (b) $x = -t + 2,\ y = t$

65. (a) $x = t,\ y = \frac{1}{2}(t - 1)$ (b) $x = -t + 2,\ y = -\frac{1}{2}(t - 1)$

67. (a) $x = t,\ y = t^2 + 1$ (b) $x = -t + 2,\ y = t^2 - 4t + 5$

69. (a) $x = t,\ y = 3t^2 + 1$
      (b) $x = -t + 2,\ y = 3t^2 - 12t + 13$

71. (a) $x = t,\ y = 1 - 2t^2$
      (b) $x = 2 - t,\ y = -2t^2 + 8t - 7$

73. (a) $x = t,\ y = \dfrac{1}{t}$ (b) $x = -t + 2,\ y = -\dfrac{1}{t - 2}$

75. (a) $x = t,\ y = e^t$ (b) $x = -t + 2,\ y = e^{-t+2}$

77.

79.

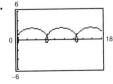

81.

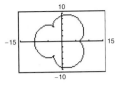

83.

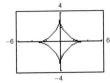

85.

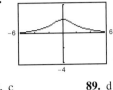

87. b
      Domain: $[-2, 2]$
      Range: $[-1, 1]$

88. c
Domain:
$[-4, 4]$
Range:
$[-6, 6]$

89. d
Domain:
$(-\infty, \infty)$
Range:
$(-\infty, \infty)$

90. a
Domain:
$(-\infty, \infty)$
Range:
$[-2, 2]$

91. f
Domain: $[-3, 6]$
Range: $\left[-3\sqrt{3}, 3\sqrt{3}\right]$

92. e
Domain: $(-\infty, \infty)$
Range: $(-2, 2)$

93. (a)

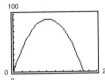

(b)

Maximum height: 90.7 ft   Maximum height: 204.2 ft
Range: 209.6 ft   Range: 471.6 ft

(c)

(d)

Maximum height: 60.5 ft   Maximum height: 136.1 ft
Range: 242.0 ft   Range: 544.5 ft

95. (a) $x = (146.67\cos\theta)t$
      $y = 3 + (146.67\sin\theta)t - 16t^2$

(b)   No

(c)   Yes

(d) $19.3°$

97. (a) $x = (\cos 35°)v_0 t$
      $y = 7 + (\sin 35°)v_0 t - 16t^2$
(b) About 54.09 ft/sec
(c)   22.04 ft

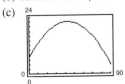

(d) About 2.03 sec

CHAPTER 6

99. (a) $h = 7$, $v_0 = 40$, $\theta = 45°$
$x = (40 \cos 45°)t$
$y = 7 + (40 \sin 45°)t - 16t^2$

(b)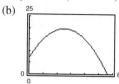

(c) Maximum height: 19.5 ft
   Range: 56.2 ft

101. $x = a\theta - b \sin \theta$
$y = a - b \cos \theta$

103. True
$x = t$
$y = t^2 + 1 \Longrightarrow y = x^2 + 1$
$x = 3t$
$y = 9t^2 + 1 \Longrightarrow y = x^2 + 1$

105. False. The parametric equations $x = t^2$ and $y = t$ give the rectangular equation $x = y^2$, so y is not a function of x.

107. Parametric equations are useful when graphing two functions simultaneously on the same coordinate system. For example, they are useful when tracking the path of an object so that the position and the time associated with that position can be determined.

109. Yes. The orientation would change.

111. $-1 < t < \infty$

Section 6.7   (page 471)

1. pole   **3.** polar

5. 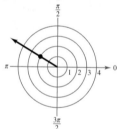   **7.**

$\left(2, -\dfrac{7\pi}{6}\right), \left(-2, -\dfrac{\pi}{6}\right)$ $\left(4, \dfrac{5\pi}{3}\right), \left(-4, -\dfrac{4\pi}{3}\right)$

9. 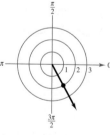   **11.**

$(2, \pi), (-2, 0)$ $\left(-2, -\dfrac{4\pi}{3}\right), \left(2, \dfrac{5\pi}{3}\right)$

13. 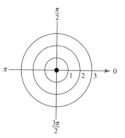   **15.**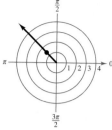

$\left(0, \dfrac{5\pi}{6}\right), \left(0, -\dfrac{\pi}{6}\right)$ $(\sqrt{2}, -3.92), (-\sqrt{2}, -0.78)$

17. 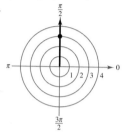   **19.** $(0, 0)$ **21.** $(0, 3)$

$(-3, 4.71), (3, 1.57)$

23. $(-\sqrt{2}, \sqrt{2})$ **25.** $\left(\dfrac{\sqrt{2}}{2}, \dfrac{\sqrt{2}}{2}\right)$ **27.** $(\sqrt{3}, 1)$

29. $\left(-\dfrac{3}{2}, \dfrac{3\sqrt{3}}{2}\right)$ **31.** $(-1.84, 0.78)$ **33.** $(-1.1, -2.2)$

35. $(1.53, 1.29)$ **37.** $(-1.20, -4.34)$ **39.** $(-0.02, 2.50)$

41. $(-3.60, 1.97)$ **43.** $\left(\sqrt{2}, \dfrac{\pi}{4}\right)$ **45.** $\left(3\sqrt{2}, \dfrac{5\pi}{4}\right)$

47. $(6, \pi)$ **49.** $\left(5, \dfrac{3\pi}{2}\right)$ **51.** $(5, 2.21)$ **53.** $\left(\sqrt{6}, \dfrac{5\pi}{4}\right)$

55. $\left(2, \dfrac{11\pi}{6}\right)$ **57.** $(3\sqrt{13}, 0.98)$ **59.** $(13, 1.18)$

61. $(\sqrt{13}, 5.70)$ **63.** $(\sqrt{29}, 2.76)$ **65.** $(\sqrt{7}, 0.86)$

67. $\left(\dfrac{17}{6}, 0.49\right)$ **69.** $\left(\dfrac{\sqrt{85}}{4}, 0.71\right)$ **71.** $r = 3$

73. $\theta = \dfrac{\pi}{4}$ **75.** $r = 10 \sec \theta$ **77.** $r = \csc \theta$

79. $r = \dfrac{-2}{3 \cos \theta - \sin \theta}$ **81.** $r^2 = 16 \sec \theta \csc \theta = 32 \csc 2\theta$

83. $r = a$ **85.** $r = 2a \cos \theta$ **87.** $r^2 = \cos 2\theta$

89. $r = \cot^2 \theta \csc \theta$ **91.** $x^2 + y^2 - 4y = 0$

93. $x^2 + y^2 + 2x = 0$ **95.** $\sqrt{3}x + y = 0$

97. $\dfrac{\sqrt{3}}{3}x + y = 0$ **99.** $x^2 + y^2 = 16$ **101.** $y = 4$

103. $x = -3$ **105.** $x^2 + y^2 - x^{2/3} = 0$

107. $(x^2 + y^2)^2 = 2xy$ **109.** $(x^2 + y^2)^2 = 6x^2y - 2y^3$

111. $x^2 + 4y - 4 = 0$ **113.** $4x^2 - 5y^2 - 36y - 36 = 0$

115. $2x - 3y = 6$

117. $x^2 + y^2 = 36$ **119.** $-\sqrt{3}x + 3y = 0$

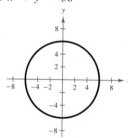   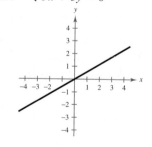

121. $x^2 + (y - 1)^2 = 1$ **123.** $(x + 3)^2 + y^2 = 9$

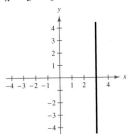

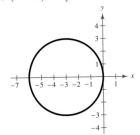

23.

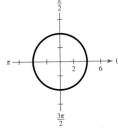

25.

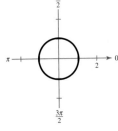

125. $x - 3 = 0$

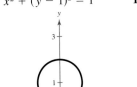

27.

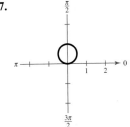

29.

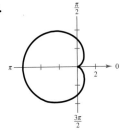

127. (a) $r = 30$

(b) $(30, 5\pi/6)$; 30 represents the distance of the passenger car from the center, and $5\pi/6 = 150°$ represents the angle to which the car has rotated.

(c) $(-25.98, 15)$; The car is about 25.98 feet to the left of the center and 15 feet above the center.

129. True. Because r is a directed distance, the point (r, θ) can be represented as $(r, \theta \pm 2\pi n)$.

131. False. If $r_1 = -r_2$, then (r_1, θ) and (r_2, θ) are different points.

133. $(x - h)^2 + (y - k)^2 = h^2 + k^2$

Radius: $\sqrt{h^2 + k^2}$

Center: (h, k)

135. (a) Answers will vary.

(b) $(r_1, \theta_1), (r_2, \theta_2)$ and the pole are collinear.

$d = \sqrt{r_1^2 + r_2^2 - 2r_1 r_2} = |r_1 - r_2|$

This represents the distance between two points on the line $\theta = \theta_1 = \theta_2$.

(c) $d = \sqrt{r_1^2 + r_2^2}$

This is the result of the Pythagorean Theorem.

Section 6.8   (page 479)

1. $\theta = \dfrac{\pi}{2}$ **3.** convex limaçon   **5.** lemniscate

7. Rose curve with 4 petals   **9.** Limaçon with inner loop

11. Rose curve with 3 petals   **13.** Polar axis

15. $\theta = \dfrac{\pi}{2}$ **17.** $\theta = \dfrac{\pi}{2}$, polar axis, pole

19. Maximum: $|r| = 20$ when $\theta = \dfrac{3\pi}{2}$

Zero: $r = 0$ when $\theta = \dfrac{\pi}{2}$

21. Maximum: $|r| = 4$ when $\theta = 0, \dfrac{\pi}{3}, \dfrac{2\pi}{3}$

Zeros: $r = 0$ when $\theta = \dfrac{\pi}{6}, \dfrac{\pi}{2}, \dfrac{5\pi}{6}$

31.

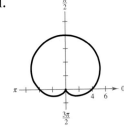

33.

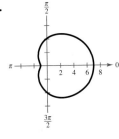

35.

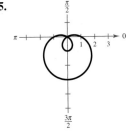

37.

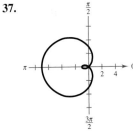

39.

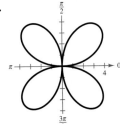

41.

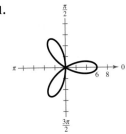

43.

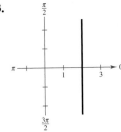

45.

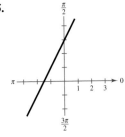

CHAPTER 6

47.

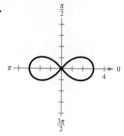

49.

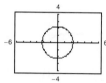

51.

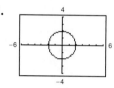

53.

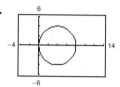

55.

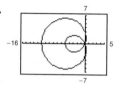

57.

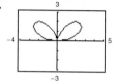

59.

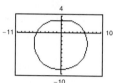

$0 \leq \theta < 2\pi$

61.

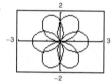

63.

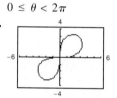

$0 \leq \theta < 4\pi$ $0 \leq \theta < \pi/2$

65.

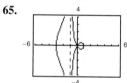

67.

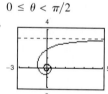

69. (a)

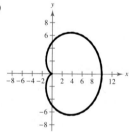

Cardioid

(b) 0 radians

71. True. The equation is of the form $r = a \sin n\theta$, where n is odd.

73. (a)

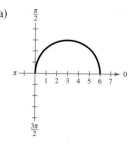

(b)

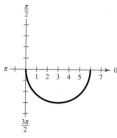

Upper half of circle   Lower half of circle

(c)

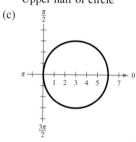

(d)

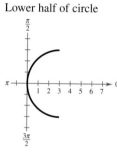

Full circle   Left half of circle

75. Answers will vary.

77. (a) $r = 2 - \dfrac{\sqrt{2}}{2}(\sin \theta - \cos \theta)$ (b) $r = 2 + \cos \theta$

(c) $r = 2 + \sin \theta$ (d) $r = 2 - \cos \theta$

79. (a)

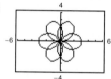

(b)

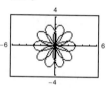

$0 \leq \theta < 4\pi$ $0 \leq \theta < 4\pi$

(c) Yes. Explanations will vary.

Section 6.9 *(page 485)*

1. conic   **3.** vertical; right

5. $e = 1$: $r = \dfrac{2}{1 + \cos \theta}$, parabola

$e = 0.5$: $r = \dfrac{1}{1 + 0.5 \cos \theta}$, ellipse

$e = 1.5$: $r = \dfrac{3}{1 + 1.5 \cos \theta}$, hyperbola

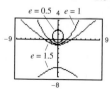

7. $e = 1$: $r = \dfrac{2}{1 - \sin \theta}$, parabola

$e = 0.5$: $r = \dfrac{1}{1 - 0.5 \sin \theta}$, ellipse

$e = 1.5$: $r = \dfrac{3}{1 - 1.5 \sin \theta}$, hyperbola

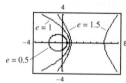

9. e   **10.** c   **11.** d   **12.** f   **13.** a   **14.** b

15. Parabola

17. Parabola

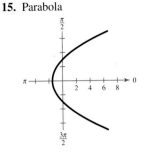

19. Ellipse

21. Ellipse

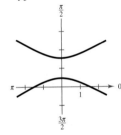

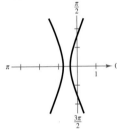

23. Hyperbola

25. Hyperbola

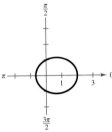

27.

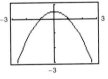

Parabola

29.

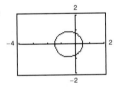

Ellipse

31.

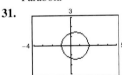

Ellipse

33.

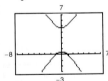

Hyperbola

35.

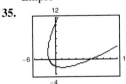

37.

39. $r = \dfrac{1}{1 - \cos \theta}$ **41.** $r = \dfrac{1}{2 + \sin \theta}$

43. $r = \dfrac{2}{1 + 2 \cos \theta}$ **45.** $r = \dfrac{2}{1 - \sin \theta}$

47. $r = \dfrac{10}{1 - \cos \theta}$ **49.** $r = \dfrac{10}{3 + 2 \cos \theta}$

51. $r = \dfrac{20}{3 - 2 \cos \theta}$ **53.** $r = \dfrac{9}{4 - 5 \sin \theta}$

55. Answers will vary.

57. $r = \dfrac{9.2930 \times 10^7}{1 - 0.0167 \cos \theta}$

Perihelion: 9.1404×10^7 mi
Aphelion: 9.4508×10^7 mi

59. $r = \dfrac{1.0820 \times 10^8}{1 - 0.0068 \cos \theta}$

Perihelion: 1.0747×10^8 km
Aphelion: 1.0895×10^8 km

61. $r = \dfrac{1.4039 \times 10^8}{1 - 0.0934 \cos \theta}$

Perihelion: 1.2840×10^8 mi
Aphelion: 1.5486×10^8 mi

63. $r = \dfrac{2.494}{1 + 0.995 \sin \theta}$; $r \approx 1.25$ astronomical units

65. True. The graphs represent the same hyperbola.

67. True. The conic is an ellipse because the eccentricity is less than 1.

69. The original equation graphs as a parabola that opens downward.
  (a) The parabola opens to the right.
  (b) The parabola opens up.
  (c) The parabola opens to the left.
  (d) The parabola has been rotated.

71. Answers will vary.

73. $r^2 = \dfrac{24{,}336}{169 - 25 \cos^2 \theta}$ **75.** $r^2 = \dfrac{144}{25 \cos^2 \theta - 9}$

77. $r^2 = \dfrac{144}{25 \cos^2 \theta - 16}$

79. (a) Ellipse
  (b) The given polar equation, r, has a vertical directrix to the left of the pole. The equation r_1 has a vertical directrix to the right of the pole, and the equation r_2 has a horizontal directrix below the pole.
  (c)

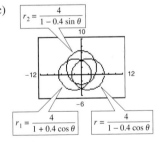

$r_2 = \dfrac{4}{1 - 0.4 \sin \theta}$

$r_1 = \dfrac{4}{1 + 0.4 \cos \theta}$ $r = \dfrac{4}{1 - 0.4 \cos \theta}$

Review Exercises   *(page 490)*

1. $\dfrac{\pi}{4}$ rad, 45°   **3.** 1.1071 rad, 63.43°   **5.** 0.4424 rad, 25.35°

7. 0.6588 rad, 37.75°   **9.** $4\sqrt{2}$ **11.** Hyperbola

13. $y^2 = 16x$ **15.** $(y - 2)^2 = 12x$

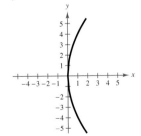

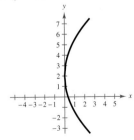

17. $y = -4x - 2$ **19.** $6\sqrt{5}$ m

21. $\dfrac{(x-3)^2}{25} + \dfrac{y^2}{16} = 1$ **23.** $\dfrac{(x-2)^2}{4} + \dfrac{(y-1)^2}{1} = 1$

25. The foci occur 3 feet from the center of the arch.

27. Center: $(-1, 2)$

Vertices:

$(-1, 9), (-1, -5)$

Foci: $\left(-1, 2 \pm 2\sqrt{6}\right)$

Eccentricity: $\dfrac{2\sqrt{6}}{7}$

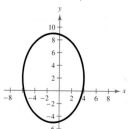

29. Center: $(1, -4)$

Vertices:

$(1, 0), (1, -8)$

Foci: $\left(1, -4 \pm \sqrt{7}\right)$

Eccentricity: $\dfrac{\sqrt{7}}{4}$

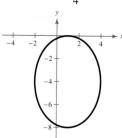

31. $\dfrac{y^2}{1} - \dfrac{x^2}{3} = 1$ **33.** $\dfrac{x^2}{16} - \dfrac{y^2}{9} = 1$

35. Center: $(5, -3)$

Vertices:

$(11, -3), (-1, -3)$

Foci: $\left(5 \pm 2\sqrt{13}, -3\right)$

Asymptotes:

$y = -3 \pm \frac{2}{3}(x - 5)$

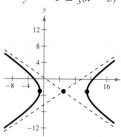

37. Center: $(1, -1)$

Vertices:

$(5, -1), (-3, -1)$

Foci: $(6, -1), (-4, -1)$

Asymptotes:

$y = -1 \pm \frac{3}{4}(x - 1)$

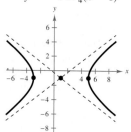

39. 72 mi   **41.** Hyperbola   **43.** Ellipse

45. $\dfrac{(y')^2}{6} - \dfrac{(x')^2}{6} = 1$ **47.** $\dfrac{(x')^2}{3} + \dfrac{(y')^2}{2} = 1$

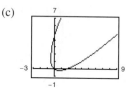

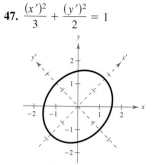

49. (a) Parabola

(b) $y = \dfrac{24x + 40 \pm \sqrt{(24x+40)^2 - 36(16x^2 - 30x)}}{18}$

(c)

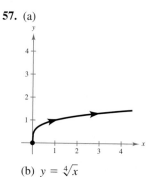

51. (a) Parabola

(b) $y = \dfrac{-(2x - 2\sqrt{2}) \pm \sqrt{(2x - 2\sqrt{2})^2 - 4(x^2 + 2\sqrt{2}x + 2)}}{2}$

(c)

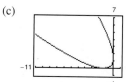

53. (a)

t	-2	-1	0	1	2
x	-8	-5	-2	1	4
y	15	11	7	3	-1

(b)

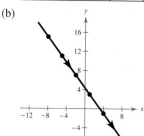

55. (a)

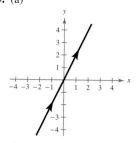

(b) $y = 2x$

57. (a)

(b) $y = \sqrt[4]{x}$

59. (a)

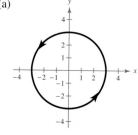

(b) $x^2 + y^2 = 9$

61. (a) $x = t, y = 2t + 3$

(b) $x = t - 1, y = 2t + 1$

(c) $x = 3 - t, y = 9 - 2t$

63. (a) $x = t, y = t^2 + 3$

(b) $x = t - 1, y = t^2 - 2t + 4$

(c) $x = 3 - t, y = t^2 - 6t + 12$

65. (a) $x = t, y = 2t^2 + 2$

(b) $x = t - 1, y = 2t^2 - 4t + 4$

(c) $x = 3 - t, y = 2t^2 - 12t + 20$

67.

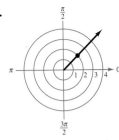

69.

$\left(2, -\dfrac{7\pi}{4}\right), \left(-2, \dfrac{5\pi}{4}\right)$ (7, 1.05), (−7, −2.09)

71. $\left(-\dfrac{1}{2}, -\dfrac{\sqrt{3}}{2}\right)$ **73.** $\left(-\dfrac{3\sqrt{2}}{2}, \dfrac{3\sqrt{2}}{2}\right)$ **75.** $\left(1, \dfrac{\pi}{2}\right)$

77. $\left(2\sqrt{13}, 0.9828\right)$ **79.** $r = 9$ **81.** $r = 6\sin\theta$

83. $r^2 = 10\csc 2\theta$ **85.** $x^2 + y^2 = 25$

87. $x^2 + y^2 = 3x$ **89.** $x^2 + y^2 = y^{2/3}$

91. Symmetry: $\theta = \dfrac{\pi}{2}$, polar axis, pole

Maximum value of $|r|$: $|r| = 6$ for all values of θ

No zeros of r

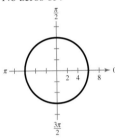

93. Symmetry: $\theta = \dfrac{\pi}{2}$, polar axis, pole

Maximum value of $|r|$: $|r| = 4$ when $\theta = \dfrac{\pi}{4}, \dfrac{3\pi}{4}, \dfrac{5\pi}{4}, \dfrac{7\pi}{4}$

Zeros of r: $r = 0$ when $\theta = 0, \dfrac{\pi}{2}, \pi, \dfrac{3\pi}{2}$

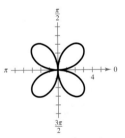

95. Symmetry: polar axis

Maximum value of $|r|$: $|r| = 4$ when $\theta = 0$

Zeros of r: $r = 0$ when $\theta = \pi$

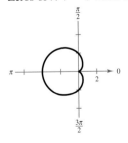

97. Symmetry: $\theta = \dfrac{\pi}{2}$

Maximum value of $|r|$: $|r| = 8$ when $\theta = \dfrac{\pi}{2}$

Zeros of r: $r = 0$ when $\theta = 3.4814, 5.9433$

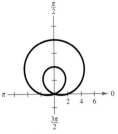

99. Symmetry: $\theta = \dfrac{\pi}{2}$, polar axis, pole

Maximum value of $|r|$: $|r| = 3$ when $\theta = 0, \dfrac{\pi}{2}, \pi, \dfrac{3\pi}{2}$

Zeros of r: $r = 0$ when $\theta = \dfrac{\pi}{4}, \dfrac{3\pi}{4}, \dfrac{5\pi}{4}, \dfrac{7\pi}{4}$

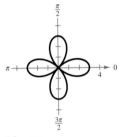

101. Limaçon

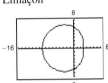

103. Rose curve

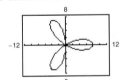

105. Hyperbola

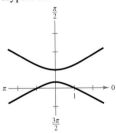

107. Ellipse

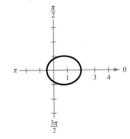

109. $r = \dfrac{4}{1 - \cos\theta}$ **111.** $r = \dfrac{5}{3 - 2\cos\theta}$

113. $r = \dfrac{7961.93}{1 - 0.937\cos\theta}$; 10,980.11 mi

115. False. The equation of a hyperbola is a second-degree equation.

117. False. $(2, \pi/4)$, $(-2, 5\pi/4)$, and $(2, 9\pi/4)$ all represent the same point.

119. (a) The graphs are the same.   (b) The graphs are the same.

Chapter Test   *(page 493)*

1. 0.3805 rad, 21.8°   **2.** 0.8330 rad, 47.7°   **3.** $\dfrac{7\sqrt{2}}{2}$

4. Parabola: $y^2 = 2(x - 1)$
Vertex: $(1, 0)$
Focus: $\left(\frac{3}{2}, 0\right)$

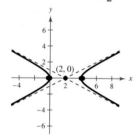

5. Hyperbola: $\dfrac{(x - 2)^2}{4} - y^2 = 1$

Center: $(2, 0)$
Vertices: $(0, 0), (4, 0)$
Foci: $\left(2 \pm \sqrt{5}, 0\right)$

Asymptotes: $y = \pm\dfrac{1}{2}(x - 2)$

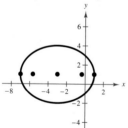

6. Ellipse: $\dfrac{(x + 3)^2}{16} + \dfrac{(y - 1)^2}{9} = 1$

Center: $(-3, 1)$
Vertices: $(1, 1), (-7, 1)$
Foci: $\left(-3 \pm \sqrt{7}, 1\right)$

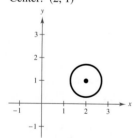

7. Circle: $(x - 2)^2 + (y - 1)^2 = \frac{1}{2}$
Center: $(2, 1)$

8. $(x - 2)^2 = \dfrac{4}{3}(y + 3)$ **9.** $\dfrac{y^2}{2/5} - \dfrac{x^2}{18/5} = 1$

10. (a) $45°$
(b)

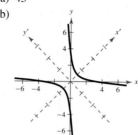

11.

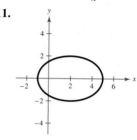

$\dfrac{(x - 2)^2}{9} + \dfrac{y^2}{4} = 1$

12. (a) $x = t, y = 3 - t^2$
(b) $x = t - 2, y = -t^2 + 4t - 1$

13. $\left(\sqrt{3}, -1\right)$

14. $\left(2\sqrt{2}, \dfrac{7\pi}{4}\right), \left(-2\sqrt{2}, \dfrac{3\pi}{4}\right), \left(2\sqrt{2}, -\dfrac{\pi}{4}\right)$

15. $r = 3 \cos \theta$

16.

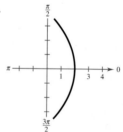

Parabola

17.

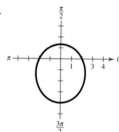

Ellipse

18.

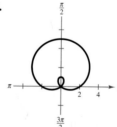

Limaçon with inner loop

19.

Rose curve

20. Answers will vary. For example: $r = \dfrac{1}{1 + 0.25 \sin \theta}$

21. Slope: 0.1511; Change in elevation: 789 ft   **22.** No; Yes

Cumulative Test for Chapters 4–6   *(page 494)*

1. $6 - 7i$ **2.** $-2 - 3i$ **3.** $-21 - 20i$ **4.** 4

5. $\frac{2}{13} + \frac{10}{13}i$ **6.** $-2, \pm 2i$ **7.** $-7, 0, 3$

8. $x^4 + x^3 - 33x^2 + 45x + 378$

9. $2\sqrt{2}\left(\cos \dfrac{3\pi}{4} + i \sin \dfrac{3\pi}{4}\right)$ **10.** $-12\sqrt{3} + 12i$

11. $-8 + 8\sqrt{3}\,i$ **12.** -64

13. $\cos 0 + i \sin 0$

$\cos \dfrac{2\pi}{3} + i \sin \dfrac{2\pi}{3}$

$\cos \dfrac{4\pi}{3} + i \sin \dfrac{4\pi}{3}$

14. $3\left(\cos \dfrac{\pi}{8} + i \sin \dfrac{\pi}{8}\right)$

$3\left(\cos \dfrac{5\pi}{8} + i \sin \dfrac{5\pi}{8}\right)$

$3\left(\cos \dfrac{9\pi}{8} + i \sin \dfrac{9\pi}{8}\right)$

$3\left(\cos \dfrac{13\pi}{8} + i \sin \dfrac{13\pi}{8}\right)$

15. Reflect f in the x-axis and y-axis, and shift three units to the right.

16. Reflect f in the x-axis, and shift four units up.

17. 1.991   **18.** -0.067 **19.** 1.717   **20.** 0.390

21. 0.906   **22.** -1.733 **23.** -4.087

24. $\ln(x + 4) + \ln(x - 4) - 4 \ln x,\quad x > 4$

25. $\ln \dfrac{x^2}{\sqrt{x + 5}},\quad x > 0$ **26.** $\dfrac{\ln 12}{2} \approx 1.242$

27. $\dfrac{\ln 9}{\ln 4} + 5 \approx 6.585$ **28.** $\dfrac{64}{5} = 12.8$

29. $\frac{1}{2}e^8 \approx 1490.479$

30.

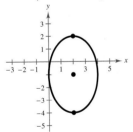

Horizontal asymptotes: $y = 0,\ y = 1000$

31. 6.3 h   **32.** 2019   **33.** 81.87°   **34.** $\dfrac{11\sqrt{5}}{5}$

35. Ellipse; $\dfrac{(x - 2)^2}{4} + \dfrac{(y + 1)^2}{9} = 1$

Center: $(2, -1)$

Vertices: $(2, 2), (2, -4)$

Foci: $\left(2, -1 \pm \sqrt{5}\right)$

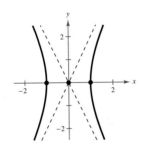

36. Hyperbola; $x^2 - \dfrac{y^2}{4} = 1$

Center: $(0, 0)$

Vertices: $(1, 0), (-1, 0)$

Foci: $\left(\sqrt{5}, 0\right), \left(-\sqrt{5}, 0\right)$

Asymptotes: $y = \pm 2x$

37. Circle;

$(x + 1)^2 + (y - 3)^2 = 22$

Center: $(-1, 3)$

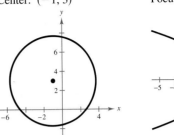

38. Parabola; $y^2 = -2(x + 1)$

Vertex: $(-1, 0)$

Focus: $\left(-\frac{3}{2}, 0\right)$

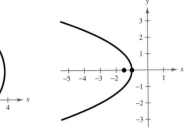

39. $x^2 - 4x + y^2 + 8y - 48 = 0$ **40.** $\dfrac{y^2}{16} - \dfrac{x^2}{9} = 1$

41. (a) 45°

(b)

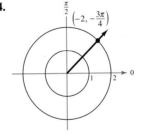

42.

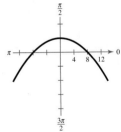

$\dfrac{(x - 3)^2}{16} + y^2 = 1$

43. (a) $x = t,\ y = 1 - t$ (b) $x = 2 - t,\ y = t - 1$

44.

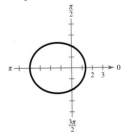

$\left(-2, \dfrac{5\pi}{4}\right), \left(2, -\dfrac{7\pi}{4}\right),$ and $\left(2, \dfrac{\pi}{4}\right)$

45. $r = 16 \sin \theta$ **46.** $9x^2 - 16y^2 + 20x + 4 = 0$

47. Ellipse

48. Parabola

49. (a) iii   (b) i   (c) ii

Problem Solving *(page 499)*

1. (a) 1.2016 rad   (b) 2420 ft, 5971 ft   **3.** $A = \dfrac{4a^2 b^2}{a^2 + b^2}$

5. (a) Because $d_1 + d_2 \le 20$, by definition, the outer bound that the boat can travel is an ellipse. The islands are the foci.

(b) Island 1: $(-6, 0)$;
Island 2: $(6, 0)$

(c) 20 mi; Vertex: $(10, 0)$

(d) $\dfrac{x^2}{100} + \dfrac{y^2}{64} = 1$

7. Proof

9. (a) The first set of parametric equations models projectile motion along a straight line. The second set of parametric equations models projectile motion of an object launched at a height of h units above the ground that will eventually fall back to the ground.

(b) $y = (\tan\theta)x$; $\quad y = h + x\tan\theta - \dfrac{16x^2\sec^2\theta}{v_0^2}$

(c) In the first case, the path of the moving object is not affected by a change in the velocity because eliminating the parameter removes v_0.

11.
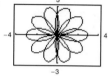
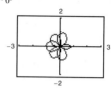

$r = 3\sin\left(\dfrac{5\theta}{2}\right)$

$r = -\cos\left(\sqrt{2}\,\theta\right),$
$-2\pi \le \theta \le 2\pi$

Sample answer: If n is a rational number, then the curve has a finite number of petals. If n is an irrational number, then the curve has an infinite number of petals.

13. (a)
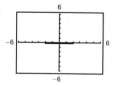

The graph is a line between -2 and 2 on the x-axis.

(b)
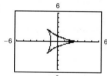

The graph is a three-sided figure with counterclock-wise orientation.

(c)
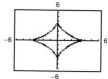

The graph is a four-sided figure with counterclock-wise orientation.

(d)
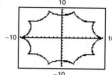

The graph is a 10-sided figure with counterclock-wise orientation.

(e)
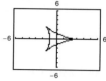

The graph is a three-sided figure with clockwise orientation.

(f)
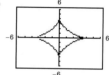

The graph is a four-sided figure with clockwise orientation.

15.
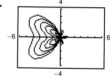
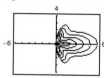

For $n \ge 1$, a bell is produced.
For $n \le -1$, a heart is produced.
For $n = 0$, a rose curve is produced.

Index

APPENDIX:

MICHIGAN STATE UNIVERSITY MATH 114 TRIGONOMETRY SUPPLEMENT WITH PRACTICE PROBLEMS

By Sharon V. Griffin
Academic Specialist
Michigan State University

1.   **The location of the terminal side of any angle in standard position**
     Note: Four of the five quadrant angles in radian measure are irrational numbers, therefore they do not terminate or have a repeating pattern. Approximations are provided on the figure below.

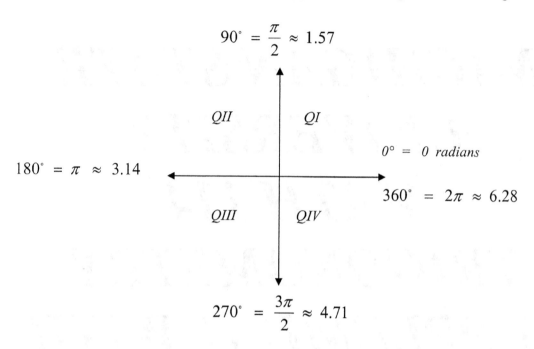

$$90° = \frac{\pi}{2} \approx 1.57$$

QII   *QI*

$$0° = 0 \; radians$$

$$180° = \pi \approx 3.14$$

$$360° = 2\pi \approx 6.28$$

QIII   *QIV*

$$270° = \frac{3\pi}{2} \approx 4.71$$

Example problem
In what quadrant is the terminal side of angle $\theta = 5$ in standard position?

Example explanation
- Since there is no symbol, the angle given is in radian measure.
- An angle in standard position has its initial side on the positive x-axis and its vertex at the origin.
- A positive angle travels counterclockwise. A negative angle travels clockwise.
- Also note an angle in radian measure does not need to be a multiple of π. [$5 \neq 5\pi$]

Example solution
Using the figure above $4.71 < 5 < 6.28$.
The terminal side of angle θ is in quadrant *QIV* .

2.   **Conversion between radian measure, degree measure and revolutions (rotations)**
     1 *revolution* or 1 *rotation* $= 2\pi \; radians = 360°$.

     There is no symbol for radian measure. An angle in degree measure has a degree sign °.

3.   **Conversion of angles from degree decimal notation to degree-minute-second (D°M'S")
     notation and D°M'S" notation to degree decimal notation**
     60 seconds = 1 minute. 60 minutes = 1 degree.

Example problem

Convert 126.3475° to DMS notation.

Example explanation

The integer degrees will remain the same:	126°
Use the conversion factor on the decimal part:	$\left(.3475°\right)\left(\dfrac{60'}{1°}\right) = 20.85'$
The integer minutes will remain the same:	20'
Use the conversion factor on the decimal part:	$\left(.85'\right)\left(\dfrac{60''}{1'}\right) = 51''$

Example solution   126.3475° = 126° 20' 51"

4.   **Finding the arc length, radius/diameter or central angle**

- s is the arc length.
- r is the radius.
- θ is the central angle.
- The units on the arc length and the radius must be the same in the formula.
- The central angle must be in **RADIAN** measure in the formula.

Example problems (special cases)

(A)   Cleveland, OH has latitude 41°30' North and Tampa, FL has latitude 28° North. Find the distance between the two cities. (Let the radius of the earth be 4000 miles. Assume cities are on the same longitude.)   Provide at least ±0.01 decimal place accuracy.

Example (A) explanation

- Find the distance between the two latitudes by subtraction:

  41.5° − 28° = 13.5°

- Use conversion factor to change the difference to radian measure:

  $$\left(13.5°\right)\left(\frac{\pi}{180°}\right) = .075\pi$$

- The distance between the cities is the arc length:

  $$s = \left(4000\right)\left(.075\pi\right)$$

Example (A) solution

$$s = 300\pi \ miles \approx 942.48 \ miles$$

(B)   The minute hand of a clock is 5 inches long. How far does the tip of the minute hand move between 9:40 AM and 10:13 AM? Provide at least ±0.01 decimal place accuracy.

Example (B) explanation

- The minute hand of the clock is the radius: $r = 5\ inches$
- Find how much time passed: *Time passed was* 33 min.
- This time is part of a revolution.
  On a clock one revolution is 60 minutes.
- Convert time to revolutions:

$$(33\,min)\left(\frac{1\ rev}{60\ min}\right) = \frac{33}{60} = \frac{11}{20}\ rev$$

- Convert revolutions to radian measure:

$$\left(\frac{11}{20}\ rev\right)\left(\frac{2\pi}{1\ rev}\right) = \frac{22\pi}{20} = \frac{11\pi}{10}\ radians$$

- The distance the minute hand moved is the arc length:

$$s = (5)\left(\frac{11\pi}{10}\right)$$

Example (B) solution

$$s = \frac{55\pi}{10} = \frac{11\pi}{2}\,inches \approx 17.28\ inches$$

5.   **Positive and negative coterminal angles**
      Coterminal angles have the same initial and terminal sides. To find coterminal angles, add or subtract "rotations" using 2π in radian measure or $360°$ in degree measure.

Example problem

Find **one positive** and **one negative** coterminal angle for $\theta = -\dfrac{25\pi}{14}$ in standard position.

Example explanation/solution

- Add 2π (will need a common denominator) to the angle until a positive angle is reached for the positive coterminal angle.

$$-\frac{25\pi}{14} + 2\pi = -\frac{25\pi}{14} + \frac{28\pi}{14} = \frac{3\pi}{14}$$

- Subtract 2π (will need a common denominator) to the angle until a negative angle is reached for the negative coterminal angle.

$$-\frac{25\pi}{14} - 2\pi = -\frac{25\pi}{14} - \frac{28\pi}{14} = -\frac{53\pi}{14}$$

4

6.   The Unit Circle

Know the entire UNIT CIRCLE (including quadrant angles) and be able to indicate angles (in EXACT RADIAN measure in terms of π) with their corresponding EXACT (x, y) coordinates.

THE UNIT CIRCLE

The equation of the unit circle is $x^2 + y^2 = 1$ with the center of the circle at the origin (0, 0) and the radius is 1. On the unit circle (only), for a given angle θ, the x coordinate is the numerical value of cos θ and the y coordinate is the numerical value of sin θ.

Here is the unit circle with its common angles (in radians and degrees) with the associated EXACT ordered pairs (x, y).

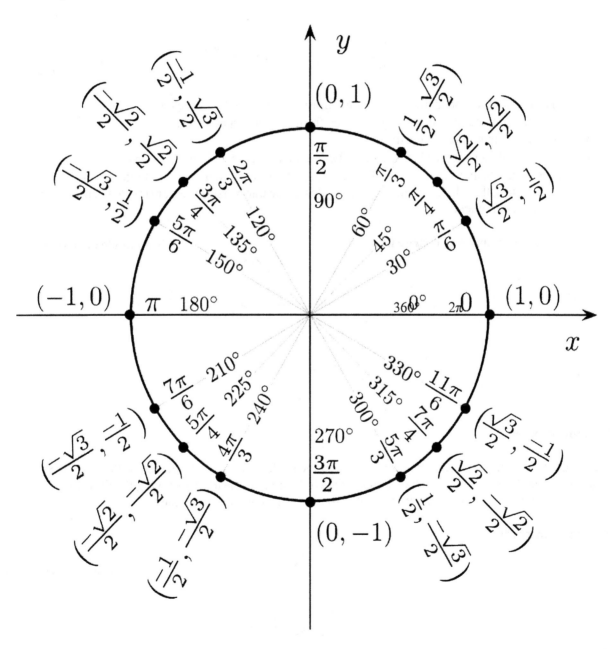

7. Right triangles and the six basic trig functions

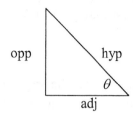

$$(\text{opp})^2 + (\text{adj})^2 = (\text{hyp})^2$$

adj means adjacent to the specified angle θ
opp means opposite of the specified angle θ
hyp is the hypotenuse. The hypotenuse is across from the 90° angle and is the longest side of a right triangle.

Trig Function **Definitions**:

$$\sin \theta = \frac{opp}{hyp} \qquad \cos \theta = \frac{adj}{hyp} \qquad \tan \theta = \frac{opp}{adj}$$

$$\csc \theta = \frac{hyp}{opp} \qquad \sec \theta = \frac{hyp}{adj} \qquad \cot \theta = \frac{adj}{opp}$$

- An angle of elevation is the angle formed from the horizontal looking up.
- An angle of depression is the angle formed from the horizontal looking down.
- The line of sight refers to the diagonal.
- The angle of elevation and the angle of depression are alternate interior angles.
- When alternate interior angles are formed from a transversal (diagonal line) intersecting two parallel lines, then the alternate (opposite sides of the transversal) interior (inside the parallel lines) angles are congruent.

∠ α is an angle of depression. **∠ θ is an angle of elevation.**

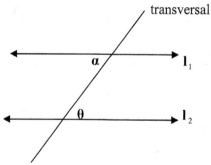

Example problem

From a point 75 feet away from the base of a tower, the angle of elevation to the top of the tower is 59°. What is the height of the tower? Draw a fully labeled illustration, clearly identifying the angle of elevation, as part of your solution. Provide at least ± 0.01 decimal place accuracy.

Example illustration

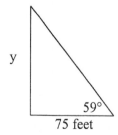

y is height of tower.

Example explanation

- The relationship between the angle, given side and unknown side is tangent

$$\tan 59° = \frac{opp}{adj} = \frac{y}{75}$$

$$75\tan 59° = y$$

Example solution   The height of the control tower is ≈ 124.82 *feet*

8. **Know all the basic trigonometric identities--Pythagorean, quotient, reciprocal, even/odd, and cofunctions of complements and how to use them.**

- Complementary angles are positive angles that sum to $90°$ or $\dfrac{\pi}{2}$ radians.

- Supplementary angles are positive angles that sum to $180°$ or π radians.

Example problem

Use trigonometric identities and the Complementary Angle Theorem to find the **EXACT VALUE** of: [Show all intermediate steps. No decimal approximations allowed in work or solution.]

$$\tan 75° \;+\; \frac{\cos(15°)}{\sin(-15°)} \;-\; 4\!\left(\sin^2(70°)+\sin^2(20°)\right)$$

Example explanation/solution

- Use odd property for sine $\tan\quad 75°\qquad\qquad +\qquad\qquad \dfrac{\cos(15°)}{-\sin(15°)}$

    $-\;4\!\left(\sin^2(70°)+\sin^2(20°)\right)$

-

- Use quotient identity for cotangent $\tan 75° \;-\; \cot(15°) \;-\; 4\!\left(\sin^2(70°)+\sin^2(20°)\right)$

- Use cofunctions of complements are equal $\tan 75° \;-\; \tan(75°) \;-\; 4\!\left(\sin^2(70°)+\cos^2(70°)\right)$

- Use Pythagorean identity $\tan 75° \;-\; \tan(75°) \;-\; 4(1)$

- Simplify $0 - 4 = -4$

DEVELOPING THE PYTHAGOREAN TRIGONOMETRIC IDENTITIES

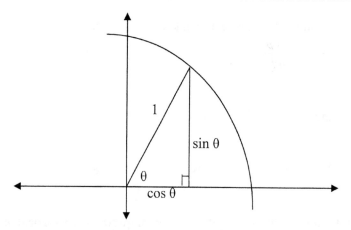

Pythagorean Identities:

From the above illustration of a right triangle inscribed in the first quadrant of the **unit circle**, the Pythagorean Theorem is used yielding the following basic trigonometric identity:

(1) $\sin^2 \theta + \cos^2 \theta = 1$

Using this first Pythagorean identity, divide both sides by $\cos^2 \theta$ yielding

$$\frac{\sin^2 \theta + \cos^2 \theta}{\cos^2 \theta} = \frac{1}{\cos^2 \theta}$$

Now split the fraction on the left side yielding

$$\frac{\sin^2 \theta}{\cos^2 \theta} + \frac{\cos^2 \theta}{\cos^2 \theta} = \frac{1}{\cos^2 \theta}$$

Using the quotient and reciprocal identities, the second Pythagorean identity simplifies to

(2) $\tan^2 \theta + 1 = \sec^2 \theta$

Using this first Pythagorean identity again, divide both sides by $\sin^2 \theta$ yielding

$$\frac{\sin^2 \theta + \cos^2 \theta}{\sin^2 \theta} = \frac{1}{\sin^2 \theta}$$

Now split the fraction on the left side yielding

$$\frac{\sin^2 \theta}{\sin^2 \theta} + \frac{\cos^2 \theta}{\sin^2 \theta} = \frac{1}{\sin^2 \theta}$$

Using the quotient and reciprocal identities, the third Pythagorean identity simplifies to

(3) $1 + \cot^2 \theta = \csc^2 \theta$

8

8. **The quadrant location of the terminal side of an angle in standard position given conditions/constraints.**

Example problem

In what quadrant is the terminal side of an angle given: $\cos\theta < 0 \quad and \quad \csc\theta < 0$

Example explanation

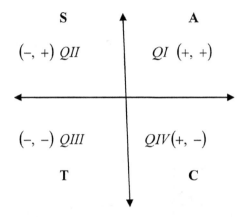

- **The six basic trig functions have positive values according to the diagram above and explanation below:**

  A stands for all six basic trig functions
  C stands for cosine and its reciprocal secant
  T stands for tangent and its reciprocal cotangent
  S stands for sine and its reciprocal cosecant

- Cosine is negative in *QII and QIII*.
- Cosecant, the reciprocal of sine, is negative in *QIII and QIV*.

Example solution   Therefore, for both conditions to be true, the terminal side of angle θ is in *QIII*.

9. **Finding reference angles**
   The reference angle is the acute angle formed by the terminal side of an angle and the closest x-axis quadrant angle. The given angle has the same trig values as the reference angle except the signs $+/-$ depend on the quadrant location of the terminal side.

Example problem   Find the reference angle for $\theta = \dfrac{5\pi}{8}$.

Example explanation/solution

- Locate the terminal side of the given angle: $\theta = \dfrac{5\pi}{8}$ is in *QII*

- Locate the closest x-axis angle   Closest quadrant angle to *QII* is π

- The reference angle = $\lvert the\ \ difference\ \ between\ \ the\ \ angles\rvert$ $\left\lvert \pi - \dfrac{5\pi}{8} \right\rvert = \dfrac{3\pi}{8}$

9

10. Reference triangles when given specific trigonometric values and conditions

A REFERENCE TRIANGLE MUST BE IN THE CORRECT QUADRANT.
You must draw quadrant lines.
The hypotenuse comes diagonal from the origin.
The triangle connects to the closest part of the x-axis.
The side values, whether they are positive or negative, depend on the signs for each quadrant.
The hypotenuse value is ALWAYS positive.

Example problem

Find the EXACT value of $\cos\theta$ given $\sin\theta = \dfrac{5}{6}$ *and* $\sec\theta < 0$. Use a reference triangle.

Example explanation

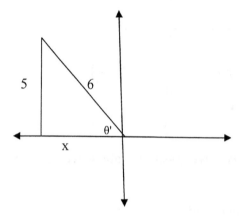

- Sine is positive and secant is negative in *QII*. Therefore, the reference triangle is drawn in *QII*.

- $\sin\theta = \dfrac{opp}{hyp}$

- θ' is the reference angle.

- Use the Pythagorean Theorem to find the missing side. Use the quadrant location to place the sign on the numerical value.

$$5^2 + x^2 = 6^2$$

$$x^2 = 36 - 25 = 11$$

$$x = -\sqrt{11} \qquad\qquad (x \text{ is negative in } QII.)$$

Example solution

$$\cos\theta = \dfrac{adj}{hyp} = -\dfrac{\sqrt{11}}{6}$$

Additional problems

1.   For $\theta = -860°$,

      a.   Determine the quadrant of the terminal side of the angle in standard position.

      b.   Convert the angle to revolutions and exact radian measure (in terms of π.)

2.   For $\beta = \dfrac{29\pi}{20}$,

      a.   Determine the quadrant of the terminal side of the angle in standard position.

      b.   Convert the angle to degree measure.

3.   For $\alpha = -7.9$,

      a.   Determine the quadrant of the terminal side of the angle in standard position.

      b.   Convert the angle to degree decimal and DMS notation.

4.   Through how many radians does the minute hand of a clock rotate from 12:41 PM to 1:23 PM? How far does it move in that time period if the minute hand of the clock is 3 inches?

5.   Given the radius of the earth is approximately 4000 miles, find the distance between City A with latitude 47.6° North and City B with latitude 37.8° North.

6.   Find the reference angle for $\beta = -\dfrac{13\pi}{9}$ in standard position

7.   Find the quadrant location of the terminal side of angle θ in standard position given $\cot\theta < 0 \ and \ \sec\theta > 0$.

8.   Determine the **EXACT trigonometric function VALUES**. Show all work including identities, coterminal and /or reference angles.

    a) $\tan\dfrac{5\pi}{2}$ b) $\csc\left(-\dfrac{3\pi}{4}\right)$ c) $\cot\left(\dfrac{5\pi}{6}\right)$

    d) $\cos(3\pi)$ e) $\sec\left(-\dfrac{5\pi}{6}\right)$ f) $\sin\left(\dfrac{13\pi}{3}\right)$

9.   The point $(-1, -5)$ is on the terminal side of an angle α in standard position. Find the **EXACT VALUE** of sin α, sec α and tan α. Draw the reference triangle as part of work.

10.   Given $\csc\theta = -\dfrac{65}{16}$ and $\cos\theta < 0$, find the **EXACT VALUES** of the following trigonometric functions of θ. **Show all work** including reference triangle and/or basic trig identities and definitions.

    a) $\sin\theta$ b) $\sin(-\theta)$ c) $\cos\theta$

    d) $\csc\left(\dfrac{\pi}{2}-\theta\right)$ e) $\cot\left(\dfrac{\pi}{2}-\theta\right)$ f) $\cot(-\theta)$

11. Use trigonometric identities and the Complementary Angle Theorem to find the **EXACT VALUE**
     of: [Show all intermediate steps. No decimal approximations allowed in work or solution.]

$$\left(\sin^2\left(7°\right) + \sin^2\left(83°\right)\right) - \sin\left(-7°\right)\sec\left(-83°\right)$$

12. State the EXACT coordinates of the point on the unit circle corresponding to the given angle:

a. $\dfrac{5\pi}{6}$ (,   )   b. $\dfrac{3\pi}{2}$ (,   )   c. $-\dfrac{\pi}{6}$ (,   )

d. $\dfrac{13\pi}{4}$ (,   )   e. $\dfrac{5\pi}{2}$ (,   )   f. $-\dfrac{5\pi}{3}$ (,   )

13. State the angle between $[0, 2\pi)$ on the unit circle in EXACT radian measure corresponding to the
     given point:

a. $\left(\dfrac{\sqrt{2}}{2}, -\dfrac{\sqrt{2}}{2}\right)$ _____   b. $\left(-\dfrac{\sqrt{3}}{2}, \dfrac{1}{2}\right)$ _____

c. $(1, 0)$ _____   d. $\left(\dfrac{1}{2}, \dfrac{\sqrt{3}}{2}\right)$ _____

14. Use the unit circle and trigonometric identities to find the EXACT VALUE of each trigonometric
     function:

a. $\cos\left(\dfrac{5\pi}{3}\right)$ b. $\sin\left(-\dfrac{7\pi}{6}\right)$ c. $\tan\left(\dfrac{\pi}{3}\right)$ d. $\sec\left(\dfrac{5\pi}{6}\right)$

e. $\csc\left(\dfrac{13\pi}{4}\right)$ f. $\cos\left(-\dfrac{7\pi}{2}\right)$ g. $\sin\left(\dfrac{11\pi}{6}\right)$ h. $\cot\left(\dfrac{2\pi}{3}\right)$

15. Any point (x, y) such that $x^2 + y^2 = 1$ is on the **unit** circle.
     Are the following points on the unit circle?

a. $(-3, 1)$ b. $\left(\dfrac{1}{5}, \dfrac{4}{5}\right)$ c. $\left(-\dfrac{5}{13}, -\dfrac{12}{13}\right)$ d. $\left(\dfrac{1}{4}, -\dfrac{\sqrt{15}}{4}\right)$

16. Given a point (x, y) in **quadrant IV** on the **unit** circle, if the x-coordinate is $\dfrac{\sqrt{7}}{4}$ then the

     y-coordinate is _____.(Simplify radical.)

17. How long should an escalator be if it is to make an angle of 32° with the floor and carry the
     people a vertical distance of 25 feet between floors?

SINE AND COSINE GRAPH PROPERTIES

Function	Domain	Range	Symmetry	x-intercepts
Sine	$(-\infty, \infty)$	$[-1, 1]$	odd	$x = n\pi$ (n is an integer)
Cosine	$(-\infty, \infty)$	$[-1, 1]$	even	$x = \dfrac{\pi}{2} + n\pi$ (n is an integer)

GIVEN THE GRAPH, FINDING A SINUSOIDAL FUNCTION OF THE FORM
$$y = a\sin(bx - c) + d \ \ or \ \ y = a\cos(bx - c) + d$$

1) **Find Range:**   [ymin, ymax]

2) **Find Amplitude:** $|a| = \dfrac{(y\max - y\min)}{2}$

Amplitude is half the distance between the maximum and minimum values of the function. Since amplitude is a distance, it is always positive. Only **sine and cosine** functions have amplitude.

3) **Is graph reflected?**   Use answers to (2) and (3) to write "a."
    If graph is not reflected "a" is positive.
    If graph is reflected, "a" is negative.

4) **Is range [-a, a]?**   Yes, then d = 0. No, then d ≠ 0.

5) **Vertical Shift:** $d = \dfrac{(y\max + y\min)}{2}$

6) **Dash in y = "d"**   **(Help—not part of graph)**

7) **Read Period (using x values)** [Note: The period has a positive numerical value.]

    Max to max: Find two consecutive maximums. What are the locations (x-values) for these maximums? Subtract the smallest x value from the largest x value to find the period.

Min to min: Find two consecutive minimums. What are the locations (x-values) for these minimums? Subtract the smallest x value from the largest x value to find the period.

Exact intercept to exact intercept: Use every other intercept to find the period.

(Case I.) If d = 0, read every other x-intercept. Subtract smallest x value from largest x value to find period.

(Case II.) If d ≠ 0, from the dashed line y = d, read every other intercept (do not use the x-axis). Subtract smallest x value from largest x value to find period.

8)   **Set Period** $= \dfrac{2\pi}{|b|}$. **Solve for** b.   **Note: In this course, $b > 0$**

New Period $= \dfrac{original\ or\ fundamental\ period}{|b|}$. For sine and cosine, the fundamental period is 2π.

---

9)   **Read Phase Shift from graph. (From x-axis or y = "d" if shifted up/down)**

10)   **Set Phase Shift** $= \dfrac{c}{b}$. **Solve for** c.

The phase shift is the horizontal shift (right or left).

---

11)   **Write the specified function:** $y = a\sin(bx - c) + d$ *or* $y = a\cos(bx - c) + d$

PRACTICE:

1.   For each graph, find the range, amplitude, period, vertical shift and phase shift. Find **a, b, c and d** then write the function as $y = a\cos(bx - c) + d$. Let the y-scale and x-scale = 1.

a)

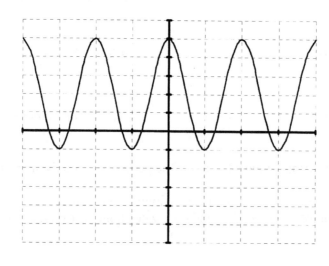

b)

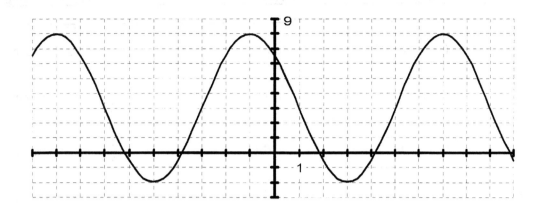

2. For the graph, find the range, amplitude, period, vertical shift and phase shift. Find **a, b, c and d** then write the function as $y = a\cos(bx - c) + d$.

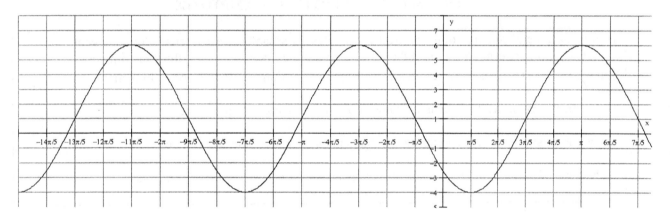

3. For each of the graphs in (1), now change the x-scale as specified and do the problems again. What remains the same? What changes?

a. The x-scale is π.   b. The x-scale is $\dfrac{\pi}{3}$.   c. The x-scale is $\dfrac{3\pi}{2}$.

d. The x-scale is $\dfrac{1}{4}$.   e. The x-scale is 2.   f. The x-scale is $\dfrac{2\pi}{5}$.

4. For graphs (1b) and (2), write each function as $y = a\sin(bx - c) + d$. [Let the y-scale and x-scale = 1 for graph (1b).] What changes? What remains the same?

15

Function	Range	Symmetry	vertical asymptotes
cosecant	$(-\infty, -1] \cup [1, \infty)$	odd	$x = n\pi$ (n is an integer)
secant	$(-\infty, -1] \cup [1, \infty)$	even	$x = \dfrac{\pi}{2} + n\pi$ (n is an integer)
tangent	$(-\infty, \infty)$	odd	$x = \dfrac{\pi}{2} + n\pi$ (n is an integer)
cotangent	$(-\infty, \infty)$	odd	$x = n\pi$ (n is an integer)

TRANSFORMED VERTICAL ASYMPTOTES

To find the general equation of the vertical asymptotes for secant, cosecant, tangent or cotangent—
Set the **original general equation** of the vertical asymptotes $= (bx - c)$ in the transformation.
Solve for x. This provides the new general asymptote equation for the transformed function.

The vertical asymptotes for secant and tangent occur where cosine equals zero.

The vertical asymptotes for cosecant and cotangent occur where sine equals zero.

To find specific consecutive vertical asymptotes, choose consecutive values of integer n.

PRACTICE:
1. For each function, determine the period, phase shift, range (in interval notation) and
    general equation of the vertical asymptotes. Graph at least one full cycle, Label five EXACT
    consecutive key points (x, y) and/or vertical asymptote equations in ONE cycle. Place a
    scale on the x and y axes. Dash in asymptotes.

(a) $f(x) = 3\sec(\pi x) - 1$

(b) $g(x) = 7\csc\left(4x - \dfrac{3\pi}{2}\right)$

(c) $h(x) = 5\tan\left(2x + \dfrac{\pi}{4}\right)$

(d) $p(x) = 3 - 5\sec\left(\dfrac{x}{4} + \dfrac{3\pi}{2}\right)$

(e) $m(x) = 10\cot\left(x - \dfrac{\pi}{6}\right)$

(f) $r(x) = 4\tan\left(\dfrac{\pi x}{3}\right)$

(g) $s(x) = 2 - 6\csc\left(\dfrac{\pi x}{2} + \pi\right)$

(h) $y = -8\cot(\pi x + \pi)$

Note: Dividing the new period by 4 gives the correct spacing for the key elements (points and/or vertical asymptotes). Use the LCD (least common denominator) of the phase shift and this spacing to aid in finding an appropriate x axis scale.

INVERSE TRIG FUNCTIONS
The solution to an inverse trig problem is an **ANGLE** within the specified range.

INVERSE TRIG FUNCTION RANGE REVIEW:

The range for $Sin^{-1}(x)$ is $\left[-\dfrac{\pi}{2}, \dfrac{\pi}{2}\right]$.

The range for $Cos^{-1}(x)$ is $\left[0, \pi\right]$.

The range for $Tan^{-1}(x)$ is $\left(-\dfrac{\pi}{2}, \dfrac{\pi}{2}\right)$.

PRACTICE:

Find the **EXACT VALUE** of each (if possible)

1. $\arccos\left(\dfrac{\sqrt{2}}{2}\right)$

2. $Sin^{-1}\left(-\dfrac{\sqrt{3}}{2}\right)$

3. $\arctan\left(\sqrt{3}\right)$

4. $Sin^{-1}\left(\dfrac{3}{2}\right)$

5. $Tan^{-1}\left(-\dfrac{\sqrt{3}}{3}\right)$

6. $Sin^{-1}\left(-\dfrac{\sqrt{3}}{2}\right)$

Find the **EXACT VALUE** of each. Show all work including reference triangles when needed.

7. $Sin^{-1}\left(\cos\dfrac{5\pi}{6}\right)$

8. $\csc\left(\arccos -\dfrac{\sqrt{3}}{2}\right)$

9. $Cos^{-1}\left(\cos\dfrac{4\pi}{3}\right)$

10. $Sin^{-1}\left(\sin\dfrac{7\pi}{4}\right)$

11. $Sin^{-1}\left(\sin\dfrac{2\pi}{3}\right)$

12. $Tan^{-1}\left(\tan\dfrac{7\pi}{10}\right)$

13. $\cot\left(\arcsin\left(-\dfrac{4}{7}\right)\right)$

14. $\cos\left(Sin^{-1}-\dfrac{12}{13}\right)$

17

SIMPLE HARMONIC MOTION (SHM) REVIEW:
$$d = a\sin(\omega t) \quad or \quad d = a\cos(\omega t)$$

If a SHM model is a **sine** function, then at time $t = 0$, the object is starting from rest. Then moving up (if "a" is positive) or down (if "a" is negative).

If the SHM model is a **cosine** function then at time $t = 0$, the object is either pulled up (if "a" is positive) or pulled down (if "a" is negative) and then released.

The object is at equilibrium when d = 0.

From the graph of the function, if the displacement **d** at time **t** is between equilibrium and the minimum or equilibrium and the maximum, then the object is **moving away** from equilibrium. If the displacement **d** at time **t** is between the minimum and equilibrium or the maximum and equilibrium, then the object is **moving toward** equilibrium.

The maximum displacement from rest is the amplitude = $|a|$.

The units are those of length (feet, inches, yards, cm, etc.)

The time required for one oscillation is the period (denoted P in textbook) = $\dfrac{2\pi}{|\omega|}$.

The units are those of time (seconds, minutes, hours).

The number of oscillations per unit time is the frequency.

The frequency f is the reciprocal of the period therefore frequency $f = \dfrac{1}{period} = \dfrac{|\omega|}{2\pi}$.

The units are oscillations per unit time (seconds, minutes, hours).

Additional Application Problems:
[For SHM, Assume that the positive direction of motion is up and there is no friction.]

1.   The simple harmonic motion of an object is given by $d = 7\cos(\pi t)$ where d is in **centimeters** and t is in **seconds**.

   a)   Find the maximum displacement, the frequency and the time required for one oscillation.

   b)   Graph at least one cycle of the function.

   c)   What is the displacement when $t = \dfrac{3}{4}$ sec ?

2.   An object attached to a coiled spring is pulled down a distance of 8 feet from its resting position and then released. Assuming that the motion is simple harmonic with $\dfrac{2}{7\pi}$ oscillations per second, write an equation that relates the displacement d of the object from its rest position after time t .

3. You are standing **midway** between buildings A and B. The angle of elevation to the top of building A is 24°. The angle of elevation to the top of building B is 50°. If building A is 36 feet tall, how tall is building B?

4. From an office building window, the angle of depression to the roof of the neighboring building is 28° and the angle of depression to the base of the neighboring building is 65°. If the neighboring building is 33 feet tall, how far apart are the buildings?

5. An object, attached to a coiled spring, in SHM, is pulled up a distance of 7 millimeters from its rest position and released. If the time required for one oscillation is $\frac{1}{3}$ hour, write an equation that relates the displacement d of the object from its rest position after time t.

6. The simple harmonic motion of an object is given by $d = -12\sin\left(\frac{t}{5}\right)$ where d is in **centimeters** and t is in minutes.

   a) Find the maximum displacement, the frequency and the time required for one oscillation.

   b) Graph at least one cycle of the function.

   c) What is the displacement when $t = \frac{5\pi}{3}$ sec ?

7. A restaurant is situated due west of a library. A school is 12 km south of the restaurant. From the school, the bearing to the library is N63°E. How far is the restaurant from the library? [Draw illustration, showing bearing]

8. An object, attached to a coiled spring, in SHM, from its rest is moving down. If the maximum displacement is 9 feet and the time required for one oscillation is 2 minutes, write an equation that relates the displacement d of the object from its rest position after time t.

9. A bicyclist leaves the road and travels 5 miles due east along a trail and then turns and travels 3 miles due south. What is the bearing **from** the bicyclist to the initial location on the road?

10. Given that triangles ABC and DEF are similar, **find the height of Δ ABC:**

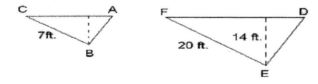

11. Two flagpoles are 50 feet apart. One pole is 60 feet tall and the other is 40 feet tall. Find the angle of depression of the top of the shorter pole from the top of the taller pole.

VERIFYING TRIG IDENTITIES
(process for this course)

To verify a trig identity, choose the most complicated side.
Treat the complicated side as an EXPRESSION (not an equation) and simplify using--
    (1) Basic trig identities (Pythagorean, quotient, reciprocal, cofunctions of complements and even/odd properties) and
    (2) Algebraic techniques (LCD, FOIL, factoring, a "1" of the conjugate, simplifying complex fractions, combining like terms, etc.)

PRACTICE:

VERIFY the trig identities algebraically. Show EVERY trig and algebraic STEP.
Work going DOWN. Work on ONE SIDE (EXCLUSIVELY). WRITE in all ANGLES.

1. $\dfrac{\tan^2 x}{\sec x + 1} = \dfrac{1 - \cos x}{\cos x}$

2. $\dfrac{\cos\left(\dfrac{\pi}{2} - x\right)}{1 + \cos(-x)} - \dfrac{1}{\tan(-x)} = \csc x$

3. $\dfrac{\cot\left(\dfrac{\pi}{2} - x\right)}{\sec^2 x - \tan^2 x - \tan(-x)} = \dfrac{\sin x}{\sin x + \cos x}$

4. $\sin x\left(\tan\left(\dfrac{\pi}{2} - x\right) - \csc(-x)\right) - \left(\csc^2 x - \cot^2 x\right) = \cos x$

5. $1 + \dfrac{\sin(-x)}{\sec\left(\dfrac{\pi}{2} - x\right) + \cot(-x)} = -\cos x$

20

TRIGONOMETRIC EQUATIONS

Equation Review:

1) To solve a trinomial equation write in the form $ax^2 + bx + c = 0$.

   By the zero property, if n and m are real numbers, variables or algebraic expressions, if $nm = 0$ then $n = 0$ or $m = 0$ (or both).

   The trinomial is now prepared to be factored or use the quadratic formula: $x = \dfrac{-b \pm \sqrt{b^2 - 4ac}}{2a}$.

2) If $a^2 = b$ then $a = \pm\sqrt{b}$

3) $a + \dfrac{b}{c} = \dfrac{ac + b}{c}$ $\qquad\qquad\qquad c \neq 0$

4) $\dfrac{a}{b} + \dfrac{c}{d} = \dfrac{ad + bc}{bd}$ $\qquad\qquad b \neq 0, \quad d \neq 0$

5) $ab + ac = a(b + c)$

6) $\dfrac{a + b}{c} = \dfrac{a}{c} + \dfrac{b}{c}$ $\qquad\qquad\qquad c \neq 0$

Solving Trigonometric Equations

1) Solve by simplifying, extracting roots, factoring, substitution, and using basic trigonometric identities and formulas.
2) If substitution is involved, such as $\sin(w\theta)$ where $m \neq 1$, use substitution (let $u = w\theta$).
   Find the general solution in terms of u.
   Re-substitute $w\theta$.

   Divide entire general solution by w $\left(or \ multiply \ by \ \dfrac{1}{w} \right)$. This is the θ general solution.

   Plug in values for "n" (starting at $n = 0$) until you have all answers in $[0, 2\pi)$.
   If w is an integer, you will yield "w times the regular # of answers on $[0, 2\pi)$."
3) If EXACT VALUES required, use UNIT CIRCLE values.
4) If the solutions are NOT EXACT values, use the inverse feature on your calculator in combination with positive coterminal angles within $[0, 2\pi)$, the reference angle and the quadrant location. The reference angle is always the QUADRANT ONE (positive) angle.
5) If given secant, cosecant, or cotangent, use reciprocal trigonometric function (except when 0 will be in the denominator) then solve.
6) For the general solution, use the fundamental period times integer n (for most problems).
   If it is a multiple angle $w\theta$ problem, the fundamental period times n will also be divided by w.

Trig Equation Problem Type	Solution Quadrant Location	Work for $[0, 2\pi)$ Answers
$\sin\theta$ = positive #	QI, QII	QI angle QII: π - Ref. angle
$\cos\theta$ = positive #	QI, QIV	QI angle QIV: 2π - Ref. angle
$\tan\theta$ = positive #	QI, QIII	QI angle QIII: π + Ref. angle
$\sin\theta$ = negative #	QIII, QIV	QIII: π + Ref. angle QIV: 2π - Ref. angle
$\cos\theta$ = negative #	QII, QIII	QII: π - Ref. angle QIII: π + Ref. angle
$\tan\theta$ = negative #	QII, QIV	QII: π - Ref. angle QIV: 2π - Ref. angle

COMBINING CONCEPTS:

For each simple harmonic motion (SHM) model:
a)   Find the time for one oscillation.

b)   Find the frequency.

c)   Find the maximum displacement from rest.

d)   Find the general solution for the object at rest (d = 0).

e)   MULTIPLE CHOICE: The object at time t = 0 seconds is—

i)   from rest moving down   ii)   from rest moving up

iii)   pulled down and then released   iv)   pulled up and then released

1. $d = -9\cos(2t)$ where d is in **centimeters** and t is in **seconds**.

2. $d = 6\sin\left(\dfrac{\pi t}{3}\right)$ where d is in **inches** and t is in **minutes**.

3. $d = -5\sin(4\pi t)$ where d is in **millimeters** and t is in **minutes**.

4. $d = 10\cos\left(\dfrac{t}{7}\right)$ where d is in **feet** and t is in **hours**.

PRACTICE:

Solve the equation algebraically on [0, 2π). All solutions should be EXACT values where possible; otherwise, provide ±0.0001 decimal place accuracy.

1. $3\csc\theta - 4\sin\theta = 0$

2. $\sqrt{3}\tan(3\theta) = 1$

3. $(5\sin\theta + 1)(\sin 4\theta + 1) = 0$

4. $2\sec^2(2\theta) = 8$

5. $\sec^2\theta - 2\tan\theta - 4 = 0$

6. $12\sin^2\theta - 9 = 0$

7. $\cos\theta + \sin\theta\tan\theta = 2$

8. $15\cos^2\theta + \cos\theta = 6$

9. $\tan^3\theta + \tan^2\theta - 3\tan\theta - 3 = 0$

10. $\cos^2(2\theta) + \cos(2\theta) = 0$

For the following equations, solve for θ in radian measure.
Give EXACT values where possible; otherwise provide ±0.0001 decimal place accuracy.

11. $-6\cos^2\theta + 8 = 7\sin\theta$

12. $\sqrt{3}\csc(9\theta) + 2 = 0$

13. $13 + 7\sin\theta = 15\cos^2\theta$

14. $\cot^2\theta - 7\csc\theta = -11$

15. $\tan^2\theta + \sec\theta - 1 = 0$

16. $\csc^2\theta - 7\cot\theta - 9 = 0$

17. $\cot^2\theta - 7\csc\theta + 11 = 0$

18. $3\tan^2(4\theta) - 1 = 0$

19. $2\tan\theta\sec\theta + 3\tan\theta = 0$

23

SUM and DIFFERENCE FORMULAS
And
DOUBLE and HALF ANGLE FORMULAS

FORMULA REVIEW:

$\sin(u + v) = \sin u \cos v + \cos u \sin v$

$\cos(u + v) = \cos u \cos v - \sin u \sin v$

$\sin(u - v) = \sin u \cos v - \cos u \sin v$

$\cos(u - v) = \cos u \cos v + \sin u \sin v$

$\tan(u + v) = \dfrac{\tan u + \tan v}{1 - \tan u \tan v}$

$\tan(u - v) = \dfrac{\tan u - \tan v}{1 + \tan u \tan v}$

$\sin 2u = 2 \sin u \cos u$

$\tan(2u) = \dfrac{2 \tan u}{1 - \tan^2 u}$

$\cos(2u) = \cos^2 u - \sin^2 u = 2 \cos^2 u - 1 = 1 - 2 \sin^2 u$

$\sin \dfrac{u}{2} = \pm \sqrt{\dfrac{1 - \cos u}{2}}$

$\cos \dfrac{u}{2} = \pm \sqrt{\dfrac{1 + \cos u}{2}}$

$\tan \dfrac{u}{2} = \dfrac{1 - \cos u}{\sin u} = \dfrac{\sin u}{1 + \cos u} = \pm \sqrt{\dfrac{1 - \cos u}{1 + \cos u}}$

IN COMBINATION WITH TRIG EQUATIONS

PRACTICE:
Solve the equation algebraically on $[0, 2\pi)$. All solutions should be **EXACT** values where possible; otherwise, provide ± 0.0001 decimal place accuracy.

1. $\cos(2\theta) + 3 \sin \theta = 2$

2. $\cos(\pi - \theta) + \sin\left(\theta - \dfrac{\pi}{2}\right) = 1$

For the following equations, solve for θ in radian measure.
Give **EXACT** values where possible; otherwise provide ± 0.0001 decimal place accuracy.

3. $\cos(2\theta) = -5 \sin \theta - 2$

4. $\sin\left(\dfrac{\pi}{4} + \theta\right) + \sin\left(\theta - \dfrac{\pi}{4}\right) = -1$

24

IN COMBINATION WITH INVERSE TRIG FUNCTIONS/REFERENCE TRIANGLES

Find the **EXACT VALUE** of each of the following.
Use the formulas, unit circle and reference triangles where needed.

1. $\cos\left(Tan^{-1} \sqrt{3} \;\; + \;\;\; Sin^{-1}\left(-\dfrac{2}{5}\right)\right)$

2. $\tan\left(Sin^{-1}\left(\dfrac{\sqrt{51}}{10}\right) \;-\; \dfrac{5\pi}{6} \right)$

3. $\sin\left(Cos^{-1}\dfrac{2}{3} \;-\; Tan^{-1}(-7)\right)$

4. $\csc\left(Tan^{-1}\;(-9) \;+\; Cos^{-1}\left(-\dfrac{1}{4}\right)\right)$

5. $\tan\left[Sin^{-1}\left(-\dfrac{1}{9}\right) \;+\; Cos^{-1}\left(-\dfrac{3}{8}\right)\right].$

6. $\cos\left[2Tan^{-1}\left(-\dfrac{5\sqrt{2}}{11}\right)\right]$

7. $\tan\left[2Sin^{-1}\left(-\dfrac{2}{11}\right)\right]$

8. $\sin\left[\dfrac{1}{2}Cos^{-1}\left(-\dfrac{1}{6}\right)\right]$

25

RIGHT TRIANGLES REVISITED and OBLIQUE TRIANGLES

1. Angles are often given Greek names such as alpha α, beta β, and gamma γ. These names correspond to the angle names A, B and C.

2. The sum of the interior angles of any triangle is 180°.

3. The sum of the two smallest sides of a triangle must be larger than the largest side; otherwise, it is not a triangle.

4. Two acute angles are complementary if their measures sum to 90°.
   Two positive angles are supplementary if their measures sum to 180°.

5. There can only be one right or obtuse angle in a triangle.

6. The largest angle is across from the longest side. The smallest angle is across from the smallest side. The middle angle is across from the middle length side.

7. When parallel lines are cut by a transversal (diagonal), the alternate interior angles are congruent.

8. The Pythagorean Theorem and the six trigonometric definitions are based on right triangles. Only a right triangle has a hypotenuse.

9. An angle of elevation is the angle formed from the horizontal looking up. The diagonal is called the line of sight. The angle of depression is the angle formed from the horizontal looking down. Angles of elevation and angles of depression are alternate interior angles.

10. A navigational land bearing (as presented in this textbook) is the angle and direction from the North-South line (y-axis). The angle is between 0° and 90°. The bearing starts with the direction North or South, followed by the angle, then by the direction West or East. The word "due" means straight. The other key word in an application involving finding a bearing is the word "from." In this course, air navigational bearings will also be explained this way instead of the alternate explanation in this textbook.

11. The sides of a triangle are lengths or distances. If given information with time and speeds or rates, use the formula distance = rate × time to find the lengths of the triangle sides.

Solve the triangle means find all missing angles and side lengths.

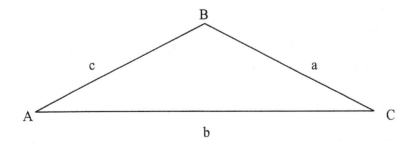

An oblique triangle does not have a right (90°) angle.

For these triangles, the Law of Sines or the Law of Cosines (depending on given information) is used to solve the triangle.

Use the Law of Sines when given:

1) Two angles and any side (AAS or ASA)
2) Two sides and the angle opposite one of them (SSA) [Ambiguous case; 0, 1, 2 triangles]

The Law of Sines is $\dfrac{\sin A}{a} = \dfrac{\sin B}{b} = \dfrac{\sin C}{c}$ or $\dfrac{a}{\sin A} = \dfrac{b}{\sin B} = \dfrac{c}{\sin C}$

There must be **ONE complete** "fraction" to use the Law of Sines. If the case is AAS or ASA, finding the third angle will produce a complete fraction. The complete fraction is used in conjunction with one of the other fractions to find a missing angle or side.

Use the Law of Cosines when given:

1) SSS (side-side-side)
2) SAS (side-included angle-side).

The Law of Cosines is

$$a^2 = b^2 + c^2 - 2bc \cos A \qquad \textbf{OR}$$

$$b^2 = a^2 + c^2 - 2ac \cos B \qquad \textbf{OR}$$

$$c^2 = a^2 + b^2 - 2ab \cos C$$

AAS or ASA:

1) Find missing angle using 180° - sum of other two angles.
2) Find missing side using Law of Sines.
3) Find remaining missing side using Law of Sines.

SSA (Ambiguous case): 0, 1, or 2 triangles

1) Find first missing angle using Law of Sines.
   Note: If the sine of the angle is less than –1 or greater than 1, there is NO triangle.
   Remember the range of the sine function is $[-1, 1]$.

2) Find remaining missing angle using 180° - sum of other two angles.

3) Find remaining missing side using Law of Sines.

4) CHECK to see if there are 2 triangles:
   a) Treat FIRST angle FOUND in step (1) as the reference angle.
   b) Subtract FIRST angle FOUND from 180°. This is the NEW first angle found for the (possible) 2nd triangle. (The original FIRST angle FOUND and this NEW angle are supplements.)
   c) Now find NEW third angle. [Note: Given information is the same for both triangles.]
      If third angle is positive, there are two triangles. Go to step (d).
      If third angle is negative, there is only one triangle. STOP.
   d) Find remaining missing side for 2nd triangle.

SSS

1) Find the first angle using the Law of Cosines.
2) Find next angle using Law of Cosines.
3) Find remaining angle using 180° - sum of other two angles.

IMPORTANT NOTE!

The Law of Sines can be used to find the second angle in a SSS triangle provided the first angle found is the LARGEST angle.

If the first angle found is NOT the largest angle, using the Law of Sines to find the LARGEST angle will yield the reference angle if the LARGEST angle is obtuse. $\text{Sin}^{-1}(positive\ \#)$ is restricted so that the answer will always be a quadrant I angle. If the actual angle is obtuse, then the value of $\text{Sin}^{-1}(positive\ \#)$ is the reference angle. The actual obtuse angle measure would be 180° - the reference angle.

SAS

1) Find missing side using Law of Cosines.
2) Find missing angle using Law of Cosines.
   Note: Using Law of Sines may give you reference angle instead of actual angle.
3) Find remaining angle using 180° - sum of other two angles.

Area Formulas:	Area = ½ absinC or ½ bcsinA or ½ acsinB

(Heron's Formula) Area = $\sqrt{s(s-a)(s-b)(s-c)}$ where s is the semiperimeter.

$$s = \frac{(a+b+c)}{2}$$ **Area has squared units.**

PRACTICE:

1. Draw the following **bearings**:

   S25°E   N10°W   N81° E   S64°W

2. A bicyclist leaves the road and travels 3 miles due east along a trail and then turns and travels 5 miles due south. What is the bearing **from** the bicyclist to the initial location on the road?

3. A triangular plot of land has sides 25 feet, 30 feet and 35 feet. Approximate the largest angle between the sides. What is the area of the triangular plot?

4.   **Solve** the following triangle(s) ABC. Solutions should be accurate to ±0.1 decimal place accuracy:

i) $b = 15$ $c = 12$ $C = 18°$
ii) $a = 10.1$ $b = 22.3$ $C = 94.2°$
iii) $a = 12$ $c = 7$ $C = 23°$
iv) $a = 6$ $b = 7$ $c = 8$
v) $\alpha = 25.2°$ $\beta = 93.1°$ $c = 11$

5.   Find the **area** of the triangle ABC. Solutions should be accurate to ±0.1 decimal place accuracy:

$a = 4$ cm $A = 35.2°$ $B = 74.1°$

A fully labeled illustration is needed for all word problems. This includes the sides given, the angles/bearings given and a variable name for the unknown(s) you are seeking.

---

6.   A surveyor walks 5 mi due east along the border of a triangular property, turns and heads N18°E for 3 miles to the upper corner of the property. What is the area of this property?

7.   The course for a charity walk through a college campus starts at administrative building *A* and proceeds in the direction N50°W to biology building *B*, then in the direction N19°E to chemistry lab *C*, and then back to building *A*. *C* is 6 miles directly north of *A*. What is the total distance of the walking course?

8.   To measure the height of a landmark, two sightings are taken. From "x" feet away from the landmark, the first angle of elevation is 56°. When the person backs up 20 feet and measures the angle again, the angle of elevation is 43°. Find the height of the landmark.

9.   A ship travels 62 nautical miles due east, then adjusts its course to a heading of S73°E. After traveling 20 nautical miles in this direction, how far is the ship from its point of departure? What is the bearing from the ship to its point of departure?

10.   The remaining gift wrapping paper is a triangle with sides 11, 10 and 8 inches. If the surface area of the boxed gift is 38 square inches, is there enough wrapping paper? Show all mathematical work.

11.   An ATV (all-terrain vehicle) travels N23°E for 2 km and then turns and travels N72°W for 6 km.
      a)   How far is the ATV from its starting point?
      b)   What is the bearing of the ATV from its initial location?

12.   From an office building window, the angle of depression to the roof of the neighboring building is 28° and the angle of depression to the base of the neighboring building is 65°. If the neighboring building is 33 feet tall, how far apart are the buildings?

This is the MSU Math 114 Final Exam from Fall 2012.

Exactly 2 hours time was allowed to take this final exam.
Problem #14 was matching (20 choices) with no work required.
Problems 15 – 20 were multiple choice with no partial credit.

1a) Solve the triangle ABC. Provide ±0.1 decimal place accuracy on solution.

   a = 12   b = 15   c = 7.5

1b) Find the area of triangle ABC. Provide ±0.1 decimal place accuracy on solution.

2. Point A is 2 miles from point B at the base of the mountain. A cable car carries passengers from point A to point C at the top of the mountain. The angles of elevation to C from A and B are 24° and 72°, respectively.

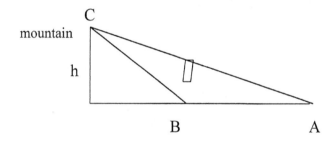

2a) Find the distance between A and C.
    Provide ±0.1 decimal place accuracy on solution.

2b) Find the height **h** of the mountain.
    Provide ±0.1 decimal place accuracy on solution.

3. Ferry boat *Amelia* leaves a lighthouse in the direction of N63°E and travels for 4.5 miles.   Barge *Bo* leaves the same lighthouse traveling in the direction S12°W. *Amelia* and *Bo* are 10 miles apart.
    [Solution should have ± 0.1 decimal place accuracy.]
    [Draw a fully labeled illustration, including both bearings, as part of your solution.]

3a) How far is barge *Bo* from the lighthouse?

3b) What is the bearing from ferry boat *Amelia* to barge *Bo*?

4.   Given: $\tan \alpha = 2\sqrt{2}$ and $\cos \alpha < 0$,

$\sec \beta = \dfrac{13}{5}$ and $\sin \beta < 0$.

Find the EXACT value of the following.
Show all work, including reference triangles.

4a) $\sin(\alpha + \beta)$

4b) $\cos\left(\dfrac{\beta}{2}\right)$

5.   Find the EXACT value of: $\tan\left(2Cos^{-1}\left(-\dfrac{20}{29}\right)\right)$

Show all work. Draw a reference triangle. Simplify completely.

6.   CONVERT the point from rectangular to EXACT polar coordinates then
PLOT and label the polar coordinates. Show all work.

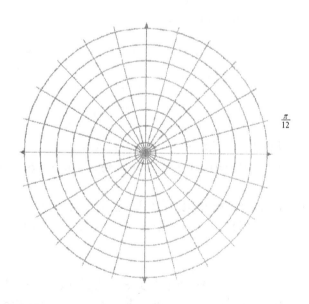

$\frac{\pi}{12}$

RECTANGULAR   **TO**   **POLAR**

$\left(2, 2\sqrt{3}\right)$ $\left(\quad,\quad\right)$

31

7. For the given graph, write the function $f(x) = a \cos (bx - c) + d$
   and find the following. Show all work for each EXACT value including formulas.

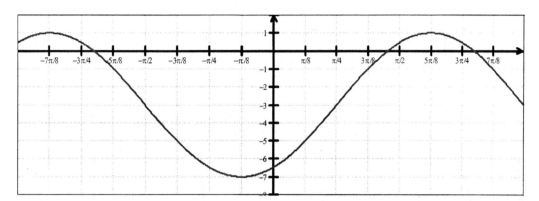

Amplitude _____   Period_____

Phase Shift _____   Range _____ (in interval notation)

a = _____

b = _____

c = _____

d = _____

$f(x) =$ _____

8. Given the function $f(x) = -7 \sec\left(2x - \dfrac{\pi}{9}\right) - 4$, find the EXACT
   values of the following. Show all work.

8a) Period _____

8b) Phase Shift _____

8c) Range (in interval notation) _____

8d) General formula for vertical asymptotes _____
    (in terms of integer n)

8e)   Multiple Choice. CIRCLE the LETTER of the correct graph of the function:

A   B   C

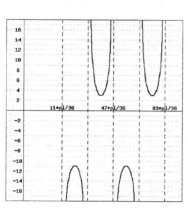

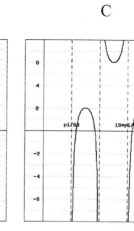

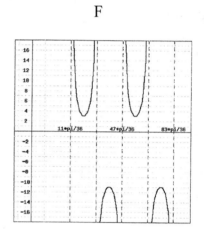

D   E   F

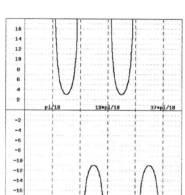

9.   VERIFY the trigonometric identity. Work going down.
     Work on ONE side only. Write in all angles. Show every step.

$$\frac{\sin\left(\dfrac{\pi}{2} - \theta\right)}{\sec(-\theta) + \tan(-\theta)} = 1 + \sin\theta$$

10. Find the EXACT values of θ in $[0,\ 2\pi)$. Show all work.

$$\sin\theta + \cos^2\theta = 1$$

11. Find ALL solutions for θ in radian measure. Find EXACT solutions
     where possible, otherwise, provide ± 0.0001 accuracy. State reference angle.

$$6\cos^2\theta + 7\cos\theta + 2 = 0$$

12. A racing car travels in a **circular** motion around a judge's stand. When the
     car travels 6 miles, it has made a (central) angle of 120° at the judge's stand. What
     is the **diameter** of the track?   Solution should have ± 0.1 decimal place accuracy.

13. An object, attached to a coiled spring, in simple harmonic motion, from its rest is
     moving down. If the maximum displacement is 9 feet and the time required
     for one oscillation is 8π minutes, find the equation that relates the displacement d of
     the object from its rest position after time t. [Show all work.]

14. Determine the **EXACT VALUE(S)** of each of the following.

i.   _____ $Sin^{-1}\left(-\dfrac{\sqrt{3}}{2}\right)$

ii. _____Angle from unit circle on $[0, 2\pi)$ that corresponds to coordinates $\left(\dfrac{\sqrt{3}}{2}, -\dfrac{1}{2}\right)$.

iii. _____All values for θ on $[0, 2\pi)$ such that $\sin\theta = -\dfrac{\sqrt{3}}{2}$.

iv. _____$Tan^{-1}\left(\tan\dfrac{7\pi}{6}\right)$.

v.   _____The complement of $\dfrac{\pi}{6}$.

vi. _____A coterminal angle of $\dfrac{19\pi}{4}$.

vii. _____$Cos^{-1}\left(\tan\dfrac{3\pi}{4}\right)$.

15.	The exact value of csc θ given $(-3, -4)$ is on the terminal side of an angle θ in standard position is _____.

16.	Given that triangles ABC and ADE are similar, **the length of side ED** (*height h of balloon*) is _____ .

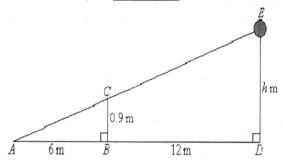

17.	The rectangular representation of the polar equation $r = 2\cos\theta$ is _____ .

18.	An airplane pilot measures the angle of depression to a ship to be 52°. The pilot is flying at an altitude of 35,000 feet. The horizontal distance between the plane and the ship to the nearest mile is _____ .

19.	All solutions for $\cos(5\theta) = -1$ (where n is an integer) are _____ .

20.	The polar coordinates $\left(-6, -\dfrac{\pi}{6}\right)$ location is quadrant _____ and has rectangular coordinates _____ .

Math 114 Sample Final (Fall 2012) answers:

1a.	$A = 52.4°$	$B = 97.9°$	$C = 29.7°$

1b.	*Area* ≈ 44.6 *square units*

2a.	2.6 *miles*	2b.	1.0 *mile*

3a.   6.5 *miles*   3b. $S\,32.5°W$

#3

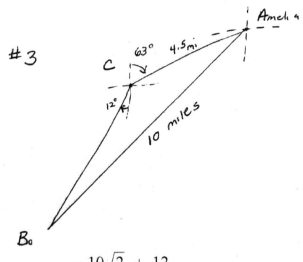

#4

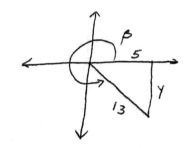

4a. $\dfrac{-10\sqrt{2}\ +\ 12}{39}$ 4b $-\dfrac{3\sqrt{13}}{13}$

Note: Reference triangle for α is in quadrant *QIII*.
        Reference triangle for β is in quadrant *QIV*.

5. $\tan(2\theta)\ =\ \dfrac{840}{41}$

#5

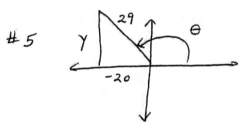

Note: Reference triangle is in quadrant *QII*.

6.   Rectangular coordinates are in quadrant *QI*, therefore, polar coordinates are in the same quadrant

      on 4ᵗʰ circle from the center and on terminal side of $\dfrac{\pi}{3}$. Two possible polar representations are

      $\left(4,\ \dfrac{\pi}{3}\right)\ or\ \left(-4,\ \dfrac{4\pi}{3}\right)$.

7.   Amplitude = 4   Period = $\dfrac{3\pi}{2}$

      Phase Shift = $\dfrac{5\pi}{8}$ *or* $-\dfrac{7\pi}{8}$ if **a** is positive

      Phase Shift = $-\dfrac{\pi}{8}$ if **a** is negative   Range = $\left[-7,\ 1\right]$

36

Math 114 Sample Final (Fall 2012) answers: (continued)

For   a = 4 $b = \dfrac{4}{3}$ $c = \dfrac{5\pi}{6}$ or $-\dfrac{7\pi}{6}$ d = −3

$$f(x) = 4\cos\left(\frac{4}{3}x - \frac{5\pi}{6}\right) - 3 \quad \text{or} \quad f(x) = 4\cos\left(\frac{4}{3}x + \frac{7\pi}{6}\right) - 3$$

For   a = −4 $b = \dfrac{4}{3}$ $c = -\dfrac{\pi}{6}$ d = −3

$$f(x) = -4\cos\left(\frac{4}{3}x + \frac{\pi}{6}\right) - 3$$

8a.   Period = π 8b.   Phase shift $= \dfrac{\pi}{18}$ 8c.   Range: $(-\infty, -11] \cup [3, \infty)$

8d. $x = \dfrac{11\pi}{36} + \dfrac{n\pi}{2}$ n is an integer   8e.   Graph F

9. $\dfrac{\cos\theta}{\sec\theta - \tan\theta}$

$\dfrac{\cos\theta}{\dfrac{1}{\cos\theta} - \dfrac{\sin\theta}{\cos\theta}}$

$\dfrac{\dfrac{\cos\theta}{1 - \sin\theta}}{\cos\theta}$

$\dfrac{\cos^2\theta}{1 - \sin\theta}$

$\dfrac{1 - \sin^2\theta}{1 - \sin\theta}$

$\dfrac{(1 - \sin\theta)(1 + \sin\theta)}{1 - \sin\theta}$

$1 + \sin\theta \quad = \quad 1 + \sin\theta$

10. Use Pythagorean Identity substitution $\cos^2\theta = 1 - \sin^2\theta$. Factor and solve: $\left\{0, \pi, \dfrac{\pi}{2}\right\}$

11. Factor or use quadratic formula.

Reference angle for non-EXACT answers is 0.8411

Solutions are: $\dfrac{2\pi}{3} + 2\pi n,$ $\dfrac{4\pi}{3} + 2\pi n,$

$2.3005 + 2\pi n,$ $3.9827 + 2\pi n$

12. $\dfrac{18}{\pi}$ *or approximately 5.7 miles*

13. $y = -9\sin\left(\dfrac{t}{4}\right)$

14i. $-\dfrac{\pi}{3}$ 14ii. $\dfrac{11\pi}{6}$

14iii. $\left\{\dfrac{4\pi}{3}, \dfrac{5\pi}{3}\right\}$ 14iv. $\dfrac{\pi}{6}$

14v. $\dfrac{\pi}{3}$ 14vi. $\dfrac{3\pi}{4}$

14vii. π

15. $-\dfrac{5}{4}$ 16. $2.7\,m$

17. $x^2 + y^2 = 2x$ 18. $27{,}345\ miles$

19. $\theta = \dfrac{\pi}{5} + \dfrac{2\pi n}{5}$ 20. $QII; \left(-3\sqrt{3}, 3\right)$

38

Definition of the Six Trigonometric Functions

Right triangle definitions, where $0 < \theta < \pi/2$

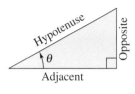

$$\sin \theta = \frac{\text{opp}}{\text{hyp}} \qquad \csc \theta = \frac{\text{hyp}}{\text{opp}}$$

$$\cos \theta = \frac{\text{adj}}{\text{hyp}} \qquad \sec \theta = \frac{\text{hyp}}{\text{adj}}$$

$$\tan \theta = \frac{\text{opp}}{\text{adj}} \qquad \cot \theta = \frac{\text{adj}}{\text{opp}}$$

Circular function definitions, where θ is any angle

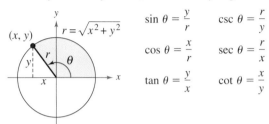

$$\sin \theta = \frac{y}{r} \qquad \csc \theta = \frac{r}{y}$$

$$\cos \theta = \frac{x}{r} \qquad \sec \theta = \frac{r}{x}$$

$$\tan \theta = \frac{y}{x} \qquad \cot \theta = \frac{x}{y}$$

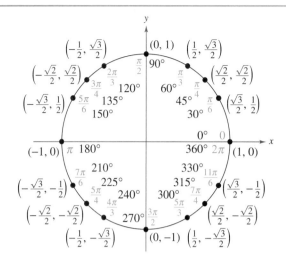

Reciprocal Identities

$$\sin u = \frac{1}{\csc u} \qquad \cos u = \frac{1}{\sec u} \qquad \tan u = \frac{1}{\cot u}$$

$$\csc u = \frac{1}{\sin u} \qquad \sec u = \frac{1}{\cos u} \qquad \cot u = \frac{1}{\tan u}$$

Quotient Identities

$$\tan u = \frac{\sin u}{\cos u} \qquad \cot u = \frac{\cos u}{\sin u}$$

Pythagorean Identities

$$\sin^2 u + \cos^2 u = 1$$
$$1 + \tan^2 u = \sec^2 u \qquad 1 + \cot^2 u = \csc^2 u$$

Cofunction Identities

$$\sin\left(\frac{\pi}{2} - u\right) = \cos u \qquad \cot\left(\frac{\pi}{2} - u\right) = \tan u$$

$$\cos\left(\frac{\pi}{2} - u\right) = \sin u \qquad \sec\left(\frac{\pi}{2} - u\right) = \csc u$$

$$\tan\left(\frac{\pi}{2} - u\right) = \cot u \qquad \csc\left(\frac{\pi}{2} - u\right) = \sec u$$

Even/Odd Identities

$$\sin(-u) = -\sin u \qquad \cot(-u) = -\cot u$$
$$\cos(-u) = \cos u \qquad \sec(-u) = \sec u$$
$$\tan(-u) = -\tan u \qquad \csc(-u) = -\csc u$$

Sum and Difference Formulas

$$\sin(u \pm v) = \sin u \cos v \pm \cos u \sin v$$
$$\cos(u \pm v) = \cos u \cos v \mp \sin u \sin v$$
$$\tan(u \pm v) = \frac{\tan u \pm \tan v}{1 \mp \tan u \tan v}$$

Double-Angle Formulas

$$\sin 2u = 2 \sin u \cos u$$
$$\cos 2u = \cos^2 u - \sin^2 u = 2 \cos^2 u - 1 = 1 - 2 \sin^2 u$$
$$\tan 2u = \frac{2 \tan u}{1 - \tan^2 u}$$

Power-Reducing Formulas

$$\sin^2 u = \frac{1 - \cos 2u}{2}$$

$$\cos^2 u = \frac{1 + \cos 2u}{2}$$

$$\tan^2 u = \frac{1 - \cos 2u}{1 + \cos 2u}$$

Sum-to-Product Formulas

$$\sin u + \sin v = 2 \sin\left(\frac{u + v}{2}\right) \cos\left(\frac{u - v}{2}\right)$$

$$\sin u - \sin v = 2 \cos\left(\frac{u + v}{2}\right) \sin\left(\frac{u - v}{2}\right)$$

$$\cos u + \cos v = 2 \cos\left(\frac{u + v}{2}\right) \cos\left(\frac{u - v}{2}\right)$$

$$\cos u - \cos v = -2 \sin\left(\frac{u + v}{2}\right) \sin\left(\frac{u - v}{2}\right)$$

Product-to-Sum Formulas

$$\sin u \sin v = \frac{1}{2}[\cos(u - v) - \cos(u + v)]$$

$$\cos u \cos v = \frac{1}{2}[\cos(u - v) + \cos(u + v)]$$

$$\sin u \cos v = \frac{1}{2}[\sin(u + v) + \sin(u - v)]$$

$$\cos u \sin v = \frac{1}{2}[\sin(u + v) - \sin(u - v)]$$

FORMULAS FROM GEOMETRY

Triangle:

$h = a \sin \theta$

$\text{Area} = \dfrac{1}{2}bh$

$c^2 = a^2 + b^2 - 2ab \cos \theta$ (Law of Cosines)

Right Triangle:

Pythagorean Theorem
$c^2 = a^2 + b^2$

Equilateral Triangle:

$h = \dfrac{\sqrt{3}s}{2}$

$\text{Area} = \dfrac{\sqrt{3}s^2}{4}$

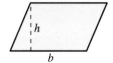

Parallelogram:

$\text{Area} = bh$

Trapezoid:

$\text{Area} = \dfrac{h}{2}(a + b)$

Circle:

$\text{Area} = \pi r^2$

$\text{Circumference} = 2\pi r$

Sector of Circle:

$\text{Area} = \dfrac{\theta r^2}{2}$

$s = r\theta$

θ in radians

Circular Ring:

$\text{Area} = \pi(R^2 - r^2)$

$\qquad = 2\pi pw$

$p = $ average radius,

$w = $ width of ring

Sector of Circular Ring:

$\text{Area} = \theta pw$

$p = $ average radius,

$w = $ width of ring,

θ in radians

Ellipse:

$\text{Area} = \pi ab$

$\text{Circumference} \approx 2\pi \sqrt{\dfrac{a^2 + b^2}{2}}$

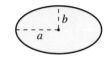

Cone:

$\text{Volume} = \dfrac{Ah}{3}$

$A = $ area of base

Right Circular Cone:

$\text{Volume} = \dfrac{\pi r^2 h}{3}$

$\text{Lateral Surface Area} = \pi r \sqrt{r^2 + h^2}$

Frustum of Right Circular Cone:

$\text{Volume} = \dfrac{\pi(r^2 + rR + R^2)h}{3}$

$\text{Lateral Surface Area} = \pi s(R + r)$

Right Circular Cylinder:

$\text{Volume} = \pi r^2 h$

$\text{Lateral Surface Area} = 2\pi rh$

Sphere:

$\text{Volume} = \dfrac{4}{3}\pi r^3$

$\text{Surface Area} = 4\pi r^2$

Wedge:

$A = B \sec \theta$

$A = $ area of upper face,

$B = $ area of base

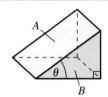